CYTOKINES AND INFLAMMATION

Editor

Edward S. Kimball
Oncology and Endocrinology Research
Janssen Research Foundation
Spring House, Pennsylvania

CRC Press
Boca Raton Ann Arbor Boston London

Library of Congress Cataloging-in-Publication Data

Cytokines and inflammation / editor, Edward S. Kimball.
  p.   cm.
  Includes bibliographical references and index.
  ISBN 0-8493-8806-6
  1. Cytokines. 2. Inflammation — Pathophysiology. I. Kimball,
Edward S.
  [DNLM: 1. Cytokines — physiology. 2. Inflammation —
physiopathology. 3. Inflammation — therapy.   QW 700 C997]
QR185.8.C95C983   1991
616'.0473—dc20
DNLM/DLC
for Library of Congress   91-191
                                                                    CIP

Developed by Telford Press

This book represents information obtained from authentic and highly regarded sources. Reprinted material is quoted with permission, and sources are indicated. A wide variety of references are listed. Every reasonable effort has been made to give reliable data and information, but the author and the publisher cannot assume responsibility for the validity of all materials or for the consequences of their use.

All rights reserved. This book, or any parts thereof, may not be reproduced in any form without written consent from the publisher.

Direct all inquiries to CRC Press, Inc., 2000 Corporate Blvd., N.W., Boca Raton, Florida, 33431.

© 1991 by CRC Press, Inc.

International Standard Book Number 0-8493-8806-6

Library of Congress Card Number 91-191
Printed in the United States

Preface

Cytokines are intercellular regulatory proteins that mediate a multiplicity of immunologic as well as nonimmunologic biological functions. Insufficient production of cytokines, such as interleukin-1 (IL-1) or interleukin-2 (IL-2), or the inability to adequately respond to these as well as to other immunomodulatory cytokines, oftentimes results in a state of immunosuppression. Conversely, exuberant cytokine production may cause severe shock, autoimmune disease, or immunopathology associated with a given disease. The latter may be the case in diseases such as rheumatoid arthritis or in syndromes such as ARDS and granuloma formation.

There is a voluminous and ever-expanding body of literature that supports a role for immunomodulatory cytokines in inflammatory diseases and the pathology that accompanies them. IL-1 and tumor necrosis factor (TNF) are potent inducers of prostaglandins, thromboxanes, leukotrienes, and collagenolytic enzymes. They are pyrogenic. They induce the production of other cytokines, act in synergy with them and with one another, and cause the production and release of acute phase proteins. TNF and IL-1 are the two most frequently reported inflammatory cytokines, although IL-6 is rapidly closing the literature citation gap, and IL-8 is also gaining. Progress is being made at such a startling rate that when this volume was conceived 18 months ago, the various anagrams for neutrophil activating factor/macrophage-derived chemotactic factor/ neutrophil activating peptide/etc. had not even been formally designated as IL-8, and no detailed discussion of this peptide is included here. Perhaps that will be rectified in a future edition. Nevertheless, this volume does include a discussion of the more recently described leukemia inhibitory factor (LIF), which is related to IL-6, and also discusses the recently described IL-1 receptor antagonist, which should be undergoing clinical trials for rheumatoid arthritis.

Because of the diversity of biological functions mediated by cytokines, no single edition can adequately review this literature unless one limits its scope. Therefore, instead of addressing every aspect of cytokine biology, this volume examines the role of cytokines in inflammation only.

The involvement of cytokines in specific areas of inflammatory disease is presented: granulomatous responses, lung disease, hepatic dysfunction and the acute phase, arthritis and accompanying bone remodeling, neurogenic inflammation, and shock. By concentrating on individual aspects, it was also possible to provide more extensive information concerning cytokines other than IL-1 and TNF. Thus, there are relevant discussions of the roles of granulocyte-monocyte colony-stimulating factor (GM-CSF), IL-6, IL-2, transforming growth factor beta (TGFβ), epidermal growth factor (EGF) and LIF. Where possible, the medical treatments that affect cytokine activity were discussed.

The control of cytokine production and biological activity as a therapeutic modality is an endeavor occupying the resources of most major pharmaceutical companies and is a principal raison d'etre for the biotech industry. Two approaches have been examined. One is to seek to discover endogenous biological response modifiers, usually proteins, which control cytokine function or production; the other is to synthesize heterocyclic compounds in the organic chemistry laboratory. The results of both approaches are discussed in separate chapters, along with a third chapter that deals with second messenger pathways involved in IL-1 and TNF production.

I hope that by organizing the material in this way, researchers in industry and academia can gain a better appreciation of the complexity of the mechanisms that operate in inflammatory disease, as well as an appreciation of the strengths and shortcomings of current concepts for therapy.

Edward S. Kimball
Spring House, Pennsylvania

About the Editor

Dr. Edward S. Kimball is a Research Fellow in Oncology and Endocrinology Research in the Janssen Research Foundation where his research is concerned with the immunobiology of cytokines in inflammatory disease and cancer, as well as mechanisms of controlling cytokine production and physiological responses to them.

Dr. Kimball holds degrees from CCNY (B.S. Chemistry), Northeastern University (M.S., Chemistry) and the University of Pennsylvania (Ph.D., Immunology). He worked seven years at the National Institutes of Health, the last four of which were with the Biological Response Modifiers Program of the National Cancer Institute at the Frederick Cancer Research and Development Facility. His interest in cytokine research began there when he studied physiological fluids from normal individuals and cancer patients for cytokines and growth factors; in the process he demonstrated the presence of biologically active IL-1 and peptide fragments in body fluids which appeared to have been derived from IL-1. He has also published a number of studies examining the involvement of the cytokine system with the nervous system in inflammatory and autoimmune diseases. The editor is a member of the American Association of Immunologists, the Leukocyte Culture Society, and the New York Academy of Sciences.

Acknowledgments

The authors deserve special acknowledgment for the level of excellence they brought to this project, for their willingness to devote their valuable time and expertise, and for striving to make their chapters as current and as timely as possible, given the constraints of producing a book.

Chapter Authors

HEINZ BAUMANN, PhD
Department of Cell and Molecular Biology, Roswell Park Memorial Institute, Buffalo, New York

CONSTANCE BRINCKERHOFF, PhD
Departments of Medicine and Biochemistry, Dartmouth Medical School, Hanover, New Hampshire

JOSEPH G. CANNON, PhD
Department of Medicine, Tufts University Medical Center, Boston, Massachusetts

ANNE M. DELANEY, PhD
Department of Medicine, Dartmouth Medical School, Hanover, New Hampshire

COLIN DUNN, PhD
Department of Hypersensitivity Diseases Research, Pharmaceutical Research & Development, The Upjohn Company, Kalamazoo, Michigan

JACK GAULDIE, PhD
Department of Pathology, McMaster University, Hamilton, Ontario, Canada

SIMEON GOLDBLUM, MD
Department of Infectious Diseases, University of Maryland School of Medicine, Baltimore, Maryland

EDWARD S. KIMBALL, PhD
Oncology and Endocrinology Research, Janssen Research Foundation, Spring House, Pennsylvania

ELIZABETH KOVACS, PhD
Loyola University, Stritch School of Medicine, Department of Cell Biology, Neurobiology, and Anatomy, Maywood, Illinois

JOSEPH A. LORENZO, PhD
Department of Research, VA Medical Center, Newington, Connecticut

CRAIG J. MCCLAIN, MD
Division of Digestive Diseases and Nutrition, University of Kentucky College of Medicine, Lexington, Kentucky

FRANCIS J. PERSICO, PHD
  Department of Experimental Therapeutics, R. W. Johnston Pharmaceutical Research Institute, Raritan, New Jersey

ALAN SHAW, PHD
  Merck Sharp & Dohme, West Point, Pennsylvania

STEVEN SHEDLOFSKY
  Department of Medicine, University of Kentucky College of Medicine, VA Hospital, Lexington Kentucky

*To my lovely wife Ann
and our sons, Aaron and Matthew*

Contents

Preface iii

About the Editor v

Chapter Authors vii

1. Cytokines as Mediators of Chronic Inflammatory Disease
   by Colin J. Dunn 1

2. Naturally Occurring Inhibitors of Cytokines
   by Alan Shaw 35

3. Low Molecular Weight Inhibitors of Interleukin-1
   by Francis J. Persico 59

4. Control of IL-1 and TNFα Production at the Level of Second Messenger Pathways
   by Elizabeth J. Kovacs 89

5. Cytokines and Growth Factors in Arthritic Diseases: Mechanisms of Cell Proliferation and Matrix Degradation in Rheumatoid Arthritis
   by Constance E. Brinckerhoff and Anne M. Delany 109

6. Cytokines and Bone Metabolism: Resorption and Formation
   by Joseph A. Lorenzo 145

7. Involvement of Cytokines in Neurogenic Inflammation
   by Edward S. Kimball 169

8. The Role of Cytokines in Acute Pulmonary Vascular Endothelial Injury
   by Simeon E. Goldblum 191

9. Hepatic Dysfunction due to Cytokines
   by Steven I. Shedlofsky and Craig J. McClain 235

10. Cytokines and Acute Phase Protein Expression
    by Jack Gauldie and Heinz Baumann 275

11. Cytokines and Shock
    by Joseph G. Cannon 307

Index 331

1. Cytokines as Mediators of Chronic Inflammatory Disease

COLIN J. DUNN

INTRODUCTION

Historical Overview of Inflammation

Before embarking on a review of "cytokines" in disease processes, it would be remiss not to give a brief overview of the inflammatory response, since it accounts for a major portion of known pathologies.

The original signs indicative of inflammation were well recognized as early as the year 30 BC by Celsus (redness, swelling, heat, and pain); Galen (AD 130—200) and later Hunter added a most important fifth sign, "loss of function", an observation often forgotten in the "mediator/factor-riddled" age of modern studies. Many hypotheses and mediators have since been put forward to explain the mechanism(s) by which these phenomena arise following tissue injury. Major contributions were made by Addison and Cohnheim, who were the first to observe adherence of leukocytes to, and migration from, the microvasculature during early stages of tissue trauma.[1] Cohnheim's proposal that "without blood vessels there is no injury" was certainly no understatement, although in lower animal forms there are exceptions to this generally accepted rule. It also evokes close scrutiny of many techniques currently used to study inflammation *in vitro*, which by their very nature exclude the contribution of a circulatory system (blood vessels and lymphatics) and associated hemodynamics.

Such "old" observations have stood the test of time and have played a substantial part in the development of modern day concepts on the mechanisms of inflammatory disease. However, the major breakthrough occurred around the turn of the century when Metchnikoff, and Wright and Douglas, generated the controversial cellular vs. humoral theories of inflammation;[1] unfortunately the divided schools of thought actually hindered further progress in this field to the extent that the real significance of immunology was not fully appreciated until the late 1950s.[2]

With the passage of time, it became clear that a basic sequence of events was key to virtually all types of inflammatory response regardless of the provocative stimulus, which range from foreign substances that may be immunogenic (e.g., bacterial cell components, immune complexes, "foreign" proteins, autoimmunity) to those that are nonimmunogenic (e.g., colloidal carbon, insoluble crystalline materials, other indigestible particulates). Although the ensuing response is a result of immune or nonimmune mechanisms, the final outcome is ostensibly the same — namely gross inflammation. In keeping with historical observations, the initial reaction comprises immediate vasodilatation followed by transient vasoconstriction. Microvascular endothelium becomes more permeable to plasma proteins that leak out into the extravascular compartment, establishing an inflammatory exudate (swelling, edema) that in itself may disrupt local tissue function.[3] This is inevitably accompanied by margination and firm adhesion of leukocytes to the vascular endothelium, culminating in accumulation of large numbers of migrating leukocytes in the edematous extravascular tissues. The presence of these cells may add to the tissue damage either by releasing degradative enzymes and by a variety of other mediators that have been assigned "proinflammatory" functions. The generation of such mediators is facilitated by leukocytes interacting with exudate proteins, other leukocytes, and tissue cells[1] and will be the subject of further discussion later in this review.

In acute inflammation, leukocyte infiltration consists largely of polymorphonuclear cells. The reaction may subside and heal or proceed into a persistent "chronic" phase characterized by continuous infiltration and division of mononuclear phagocytes (monocytes) that undergo activation with subsequent transformation into "aggressive" macrophages, "secretory" epithelioid cells, and multinucleated giant cells, all of which play important roles in both host defense and development of disease. The nature of the inciting agent determines the eventual outcome; thus, digestible materials may be rapidly eliminated by phagocytes, leading to resolution of the acute inflammatory response. Alternatively, stimuli that are either indigestible or provoke an immune response may evoke a persistent, chronic inflammatory reaction such as that seen in granulomatous inflammation. The consequences of this "uncontrolled" reaction may be devastating, forming the basis for many pathological conditions. So far I have discussed events occurring at the local inflamed tissue site. Irrespective of the acute or chronic nature of the tissue injury, a distinct sequence of events is triggered systemically; hepatocytes are stimulated to secrete "acute phase reactants" into the bloodstream (see below for details); bone marrow progenitor cells are activated in order to increase output of appropriate leukocytes to meet the demand of excessive recruitment of these cells to the inflamed tissue;[4] endocrinological changes also occur in an attempt to maintain homeostasis and may be more pronounced in the most severe reactions.

From the above it is evident that phagocytic leukocytes are central to the inflammatory process. Their interactions with other leukocytes and connective tissue cells in the generation of, and response to, inflammatory "cytokines" will be the main subject of this review.

Cytokine Biology

The term "cytokine" is vague and needs clarification. Many mediators of inflammatory/pathological processes have been elucidated over the past century; some have been well defined (histamine, serotonin, kinins, prostaglandins, leukotrienes, etc.) and as a result have been subjected to rigorous testing to assess their true relevance in a given pathological condition. Other so-called mediators of inflammation remained less well characterized partly due to the complexity of their structure and the lack of appropriate technology to provide the essential information for identification. As a result we saw the evolution of a plethora of factors responsible for a wide range of biological responses; witness the entry of lymphokines, a group of chemically ill-defined molecules secreted from sensitized lymphocytes following antigen stimulation. For many years remarkable progress was made in this popular area of immunology, to the extent that "lymphokines" and the various functions ascribed to them began to form the very building blocks of "immunoinflammatory" disease. Of these, the most important were macrophage activation factors (MAF), T and B lymphocyte activation/growth factors, specific chemotactic factors, etc. Many of these have subsequently been characterized and are commonly known to us now as interleukins (see below) and interferons, although MAF and migration inhibitory and chemotactic lymphokines still survive as distinct entities.[5-7] With the evolution of biotechnology, identification of the specific molecular structures of these and other pathological "mediators" necessitated some kind of systematic nomenclature in this explosive area of biology. Initially the term "interleukin" (IL-) was deemed most appropriate, since it described the communication between leukocytes; numerical subdivision (IL-1, IL-2, etc.) delineated the specific physiochemical characterization associated with a variety of functions or phenomena for each individual interleukin. What had not been perceived at this time, however, was the fact that these molecules also interacted with and, in certain cases, were derived from nonhematogenous cells. This should not have come as a surprise, since Yoshida et al.[8] and Pick[9] had made the observation many years ago that lymphokines were *not* exclusive to the tissues of the immune system. For these reasons, the broader term "cytokine" was adopted for these mediators; in this review I shall focus mainly on interleukin-1α and β (IL-1), interleukin-2 (IL-2), interferon (IFN α,β,γ), and tumor necrosis factor α (TNFα, for which there is no IL- designation).

Interleukin-1 and TNFα

Interleukin-1 was first described under the guise of several acronyms (leukocyte endogenous mediator-LEM; endogenous pyrogen-EP) as a substance derived from polymorphonuclear leukocytes, which was responsible for fever induction during inflammation and sepsis.[10] This 17 Kd peptide (α and β forms) is produced and secreted by macrophages, as well as by many other connective tissue cells, and has numerous functions (see 11 for review); IL-1 was later found to be identical to lymphocyte activating factor (LAF) through the serendipitous findings of Gery et al.[12] The evidence for IL-1 as a pivotal mediator of immune as well as inflammatory reactions is compelling, as shown by its ability to induce acute phase reactants, fever, hyperadhesiveness of endothelium for leukocytes, leukocytosis, and lymphocyte proliferation.[11] Local injection of recombinant IL-1α or β provokes acute inflammation characterized by edema, neutrophil infiltration, and rapid resolution of the lesion.[13,14] IL-1 has also been shown to induce a Schwartzman-like response (thrombohemorrhagic lesion) when administered 24 hr after a "priming" injection of IL-1, although lipopolysaccharide will substitute as the primer.[15] The *in vivo* effects of importance to its general pathological role are the ability of IL-1 to stimulate the thrombogenic potential of vascular endothelium, via induction of endothelial procoagulant "tissue factor" synthesis, and production of tissue plasma activator inhibitor.[16,17] An autocrine secretory mechanism has been suggested for IL-1 in monocytes and endothelial cells[18,19] thus serving to amplify the endothelial cell-associated changes described above. Leukocyte-endothelial cell interaction may also propagate local inflammation via induction of endothelial cell pathological changes through IL-1β and TNFα released from the phagocytes.[20] Wound healing is critical in the resolution of all kinds of tissue trauma. Although there is good evidence for the involvement of IL-1 in wound healing,[21] several other tissue growth factors may be equally important in this respect, as shown by *in vivo* studies,[22] where PDGF may actually mediate the effect of IL-1.[23]

Not all biologic responses induced by IL-1 are desirable; for the most part, those described above could be considered essential to survival of the host, i.e., maintenance of a fully functional inflammatory and immune system. However, excessive or perpetual triggering of the inflammatory response leads to the development of undesirable effects, culminating in disease. It is therefore obvious from the above that uncontrolled production of IL-1 may be pathological, both systemically and locally, through intravascular coagulation/thrombosis, vasculitis, and excessive tissue trauma leading to organ dysfunction.[24] IL-1 and TNFα may be major factors directly responsible for the destruction of bone and cartilage in arthritic diseases.[25-28]

Tumor necrosis factor α (TNFα, cachectin) has a remarkably similar profile of biological activity to that of IL-1.[29] As its name suggests, TNFα was associated with necrosis of certain tumors and has been implicated as the primary agent responsible for inducing shock and related syndromes.[30] Profound metabolic disturbances are also attributable to TNFα, probably accounting for the "wasting" (cachexia) seen in many chronic diseases.[31]

Interleukin-2

IL-2 is quite distinct from IL-1 and TNFα in that its major function is to stimulate T lymphocyte proliferation and maturation; activated T-cells express high affinity IL-2 receptors that mediate the IL-2-driven process.[32] Recent studies have shown that B lymphocytes and macrophages express receptors for IL-2, rendering them susceptible to stimulation by this cytokine.[33,34] Systemic administration of high-dose IL-2 in tumor therapy provided evidence of the multiple pathological consequences of this cytokine if present in excess, and will be discussed in more detail below.

Interferons

It is impossible to give a brief review of the many biological actions of the interferons, a class of molecules that exerts potent antiviral effects. For a detailed review, the reader is referred to Bocci[35] and Browning.[36] However, some important general comments should be made here. Interferons α and β possess antitumor and antiviral activity[37] at least in part via enhancing natural killer cell activity; they also induce class I antigen expression and (like IL-1, IL-2, and TNFα) cause fever.[38] According to the time of administration, or endogenous release, they are potent anti-inflammatory agents.[39] Interferon gamma is antiproliferative, activates macrophages and endothelial cells, and induces the expression of class I and II antigens;[36] identity with the earlier-described lymphokine, macrophage activating factor (MAF), has been suggested.[40] Perhaps the most intriguing property of this cytokine is the synergistic effects on other cytokines, suggesting its potential significance in the pathogenesis of a wide variety of diseases.[38, 40-42]

PATHOLOGICAL EVENTS RELATED TO CYTOKINES

From the information above it is impossible to classify cytokines as having purely physiological or pathological roles in host defense. Each cytokine has to be evaluated individually based on evidence obtained primarily from *in vitro* cell/organ culture systems. Further clues to the "pathogenicity" of these substances are being gathered from (1) detection at the site of pathology, (2) their ability to mimic specific pathological responses following local or systemic administration, and (3) suppression of the pathological response by antagonists. Of these approaches, both (1) and (2) are well underway; information on the development and effectiveness of antagonists, however, is sparse and restricted mainly to the experimental stage at present.

Evidence for the role of a specific group of cytokines (IL-1α and β, IL-2, TNFα, and IFNs) in pathological processes will be reviewed; this selection was determined partly because we have sufficient experience with these cytokines to draw more concrete and less speculative conclusions as to their significance in pathological responses. Accordingly, it would be pretentious to attempt to put the remaining cytokines into some meaningful perspective, although they will be discussed in the context of their interactions with the cytokines above, as appropriate.

Localization of Cytokines in Pathological Tissues

Having established a precedent for the potentially-harmful effects of cytokines, it was imperative to determine the relationship, if any, between these substances and various experimental and clinical diseases. In this respect many studies have clearly demonstrated the actual presence of the cytokine in question at the site of injury using immunocytochemistry or cytokine assays (bioassay, radioimmunoassay, ELISA), or the potential for cytokine production using *in situ* hybridization techniques. Evidence rapidly accumulated indicating that measurable levels of IL-1 and TNFα could be detected in a range of experimental and clinical diseases both in affected tissues and inflammatory fluids (see "Future Directions" below for references). Other studies showed enhanced production of cytokines (IL-1/IL-2/IFNs) in peripheral blood or from "primed" circulating mononuclear cells of diseased patients. This circumstantial association of certain cytokines with tissue injury/disease was reinforced by the more compelling studies demonstrating the specific cytological localization of these mediators within the lesion itself.

Of the studies to date, the following exemplify the current status of the temporal "cytokine-disease" relationship.

Although several cell types are able to produce IL-1 and TNFα, phagocytic leukocytes are by far the most important sources in inflammatory lesions. Polymorphonuclear leukocytes have been clearly shown to release significant quantities of IL-1 transiently in experimental acute inflammation.[43] During the chronic phase of inflammation, when monocyte infiltration and macrophage activation/proliferation occurs, it is this mononuclear cell that is the most abundant source of IL-1 and TNFα. Both IL-1α and TNFα are thought to be "presented" as effector molecules on the macrophage membrane,[44,45] whereas IL-1β is secreted into the extracellular milieu in which it exerts its pathological effects. The concept that membrane bound IL-1α may be the biologically active form of this particular cytokine has recently been challenged by Minnich-Carruth et al.,[46] who claim that IL-1 is slowly secreted off the macrophage membrane. Even though this will be a contentious issue in the future, these findings may change little in the concept that high concentrations of IL-1α achieved in the local microenvironment would be of considerable importance in the initiation or perpetuation of inflammatory disease. It is important to recognize the potential significance of appropriate negative-feedback control mechanisms via secretion of IL-1 inhibitors by infiltrating polymorphonuclear and mononuclear phagocytes.[47-49]

Chensue et al.[50] demonstrated that granulomatous macrophages initially synthesize IL-1, but switch to preferential production of TNFα, which correlates with the appearance of Ia + ve macrophages in the later stages of granuloma formation. This applied only to the severe chronic granulomatous response (induced by Schistosome hypersensitivity), since macrophages from granulation tissue provoked by relatively inert sephadex beads failed to produce significant amounts of IL-1 or TNFα at any time. Resolution of granulomatous responses was associated with failure of macrophages to produce either cytokine; these data therefore implied at least that both IL-1 (early granuloma) and TNFα (later development of lesion) were involved in some causal way to the evolution of a pathological response and that TNFα was more important for chronicity. It was also concluded that the severity of the lesion was reflected by the presence of IL-1 and TNFα. Duff et al.[51] provided evidence supporting this view, where messenger RNA synthesis for IL-1β and TNFα was evaluated in chronic rheumatoid synovitis by *in situ* hybridization. A similar scenario of early and late production of IL-1 and TNFα, respectively, by macrophages was proposed by Vissers et al.;[52] the same pattern of cytokine response occurred in human macrophages cultured with glomerular basement membrane and immune complexes, which suggested the participation of these cytokines in glomerulonephritis. That these cytokines actually play a causal role in pathological responses remains to be proven; however, the plausibility of the concept is strongly supported by experimental evidence presented later in this review (see section on "the pathological effects of cytokines," below).

A slightly different view of IL-1 production by granulomatous cells was put forward by Montreewasuwat et al.,[53] who showed that IL-1 secretion was more likely to occur in epithelioid granulomata, since lesions without these cells, but containing large numbers of macrophages, did not produce IL-1; this observation, made in the chronic phase of granuloma formation, is of interest because epithelioid cells represent a specialized macrophage whose prime function is that of secretion rather than phagocytosis.[54] Therefore, the actual pattern of cytokine production and its significance in the pathological response to injurious stimuli may well vary according to the specific type of reaction, of which there are several variants for chronic inflammation (e.g., epithelioid, multinucleated giant cell, and macrophagic).

More clues to this question may be gleaned from recent attempts to define the association between cytokines and the cellular infiltrates of pulmonary and lymphatic sarcoid granulomata. Sarcoidosis is a poorly understood disease of unknown etiology, which is best described as an immunoinflammatory response characterized by uncontrolled proliferation of T lymphocytes, marked macrophage infiltration, and epithelioid cell/giant cell formation.[55] Using anti-IL-1 antibodies, Chilosi et al.[56] provided evidence supporting close association between proliferating macrophages and IL-1 in sarcoid granuloma; endothelial cells also stained positive for IL-1. Surprisingly, multinucleated giant cells, which are derived from fused macrophages,[57] were negative for IL-1. In accordance with the experimental granulomatous data from Motreewasuwat et al.,[53] Tsuda et al.[58] provided evidence that epithelioid cells from sarcoid lesions were distinctly positive for IL-1 by monoclonal antibody staining. Clearly these data suggest that IL-1 might be of significance in the pathogenesis of chronic inflammatory disease.

Follow-up studies of sarcoidosis revealed a strong positive correlation between IL-2, IL-2 receptor (IL-2R), IFN gamma, and epithelioid, as well as multinucleated giant cells,[58,59] using immunocytochemical localization. Thus, it appears from these studies that the macrophage, and related cells, may significantly influence at least this type of chronic immunoinflammatory response by providing or responding to cytokines capable of perpetuating the macrophage and T lymphocyte proliferation so characteristic of this disease. Surprisingly, the T lymphocyte component showed very weak staining for IL-2, IL-2R, and IFN γ, which was limited to a small subpopulation of T "helper" cells, indicating an apparent refractoriness of the lymphoid cells to these cytokines.

Immunocytochemical staining for cytokines in synovium from rheumatoid arthritis (RA) patients revealed an apparent association of IL-2 with T helper/suppressor/cytotoxic lymphocytes, whereas IFN γ was localized in T-cells, B-cells, and macrophages.[60] In spite of these findings, the intensity of staining indicated that only very low levels of cytokine were present, which was contrary to that which would be predicted. Firestein et al.[61] provided further data supporting these unusual observations, at least for IL-2; in fact, they concluded that IL-2 was distinctly absent from RA lesions. As discussed above, similar paradoxical findings were reported for sarcoidosis, which represents a challenge to *in vitro* dogma that would lead us to predict the opposite (i.e., strong association between T-cells and IL-2, etc.). However, Firestein et al.[61] did report significant production of CSF-1 (stimulating factor for monocyte generation in hematopoietic tissues) and a mast-cell stimulating factor from RA tissues; it would be premature to speculate on the meaning of these observations, and we are left with the more perplexing question concerning the paucity of good evidence for IL-2 and IFN γ. In this respect, the work of Symons et al.[62] may shed some light on the problem. These investigators reported substantially elevated IL-2R levels in the plasma and synovial fluid of RA patients, which showed a positive correlation with intensity of disease. Logically they conclude that such a high concentration of soluble IL-2R is a response to increased IL-2 production; interaction of IL-2R with pathological levels of IL-2 produced at the inflammatory site could serve as an important control mechanism, preventing subsequent induction of the well-documented detrimental systemic effects of this cytokine.[63] Both RA and sarcoidosis patients experience variable deficiencies in cell-mediated immune responses,[55,64] which may be linked to an overexuberant IL-2R response in these diseases, culminating in systemic immunosuppression. Further supportive evidence for a pathogenic role of IL-2 receptors is derived from experiments demonstrating abrogation of mouse delayed-type hypersensitivity responses by prior administration of anti-IL-2R antibody.[65]

It is worth mentioning recent intriguing studies carried out to determine the cell association and kinetics of "migration inhibition factor" (MIF) production in hypersensitivity. MIF probably represents a class of lymphokines that is generated from antigen-sensitized lymphocytes exposed to antigen challenge *in vitro* and *in vivo*.[40] Interferon gamma may represent at least part of the MIF family by functional and physicochemical criteria.[40] Malorny et al.[5] traced the appearance of MIF in a murine delayed-type hypersensitivity response by immunostaining pathological tissues with anti-MIF antibody (7D 10). As in the IL-2/IL-2R/IFN localization studies above, these investigators were surprised to find that the infiltrating lymphocytes (helper, suppressor, cytotoxic) were negative for anti-MIF antibody staining; in contrast, strong positive MIF staining was associated with endothelial cells during the early response, followed by increasing numbers of MIF-positive infiltrating macrophages in the chronic phase. This dual association of endothelial cell and macrophage MIF was not dependent on the immune system, as shown by its presence in the "nonspecific" inflammatory response to croton oil, and is reminiscent of the distinct classes of "antigen-dependent" and "antigen-independent" MIF identified in Arthus and tuberculin hypersensitivity responses by Yamamoto et al.[66] The view of Malorny et al.[5] that the presence of MIF-positive macrophage infiltrates is essential for an effective inflammatory response with resolution of the lesion is supported by the absence of these cells in persistent diseases, such as sarcoidosis, lepromatous leprosy, and RA,[67] and lends support to the therapeutic potential of MIF, or similar substances, in chronic intractable inflammatory disease; the viability of this approach has already been tested experimentally for MIF by Mizushima et al.;[68] similar work is underway using IFN γ and will be discussed in more detail later in this review, together with other interferons.

It may be pertinent at this point to add that MIF has often been suspected of at least overlapping with, if not being identical to, the macrophage activating factor (MAF) lymphokine. In this respect IL-4, which stimulates B-cell growth and immunoglobulin production,[69] is thought to possess MAF activity[70] and appears to be associated with fusion of macrophages into multinucleate giant cells in chronic inflammation.[71] We may well see other lymphokines reemerge as more well-defined molecular entities in the evolving "directory" of interleukins, giving a clearer perspective of the rationale for their existence and function in the pathophysiology of inflammatory disease.

The Pathological Effects of Cytokine Administration In Vivo

The ability of IL-1, TNFα, and TNFβ (lymphotoxin-LT) to provoke acute inflammatory responses following local subcutaneous or intradermal injection has been explored by many groups. The consensus is that IL-1α/β[15,72,73] and TNFβ, but not TNFα,[74] are potent proinflammatory agents, causing influx of polymorphonuclear leukocytes and increased vascular permeability that rapidly resolves (24 to 48 hr after injection). The true inflammatory potential of TNFα is, however, revealed when this cytokine is administered locally with synergizing suboptimal doses of IL-1α.[15] Repeated injection of TNFβ into rabbit skin results in abscess formation associated with local destruction of surrounding dermal tissue.[74] These effects, particularly those of TNFβ, are thought to play a significant part in the destruction of tumors via necrosis, in addition to transient nonspecific tissue injury.

IL-1 has been reported to play a key role in the Schwartzman reaction, which is characterized by intense polymorphonuclear leukocyte infiltration together with a thrombohemmorhagic response.[73] Movat et al.[15] subsequently established a requirement for both IL-1 and TNFα for the full expression of this response, reinforcing the aforementioned synergistic effects of these two cytokines; the relevance of these observations remains to be determined, but probably underscores the importance of concommitant IL-1 and TNFα release in initiating a "preparative" reaction for the induction of pathological responses associated with the Schwartzman phenomenon, such as disseminated intravascular coagulation following sepsis. The rationale behind the acute inflammatory properties of IL-1 and TNFα is easily justified in that these cytokines have profound stimulatory effects on vascular endothelium. Both induce the synthesis of procoagulant activity[75] and leukocyte adhesion molecules by vascular endothelium in vitro,[76,77] events pertinent to the cytokine-driven pathological responses described above. Numerous studies have since unequivocally confirmed the procoagulant effects of IL-1, TNFα, and endotoxin on vascular endothelial cells (ECs) in vivo.[16,17] Endothelial cell-leukocyte "adhesion" molecules have been demonstrated by immunocytochemical means in pathological lesions.[78,79] Dunn et al.[80] were also able to show extensive leukocyte adhesion, subendothelial leukocyte migration, and subsequent deendothelialization accompanied by formation of platelet microthrombi within the lumena of rabbit jugular veins exposed *in situ* to purified IL-1. Such information provided valuable back-up confirmatory data for the promising parallel *in vitro* studies.

As is usual in science, these studies provided more questions than answers; were the pathological effects of cytokines merely transient, acute phenomena? If so, what was the value of such observations and how did they fit into the context of chronic, destructive unremitting disease? It was precisely this challenge that prompted us to test the hypothesis that cytokines may actually initiate and propagate distinct chronic inflammatory tissue reactions, an area of experimental and clinical pathology that had so far received little attention in terms of specific "mechanistic/mediator" elucidation. The problem of designing an appropriate system to test the hypothesis was aided by the availabilty of "slow-release" technology, for which we are deeply indebted to the research endeavours of Dr. R. Langer and co-workers at the Massachusetts Institute of Technology. They developed and employed ethylene vinyl acetate copolymer (EVA) formulations for continual release of molecules of varied size and chemical composition.[81] We logically reasoned that subcutaneous implantation of cytokines incorporated into EVA disks might provide a model that would mimic the effects of continued production of these substances by resident tissue cells (e.g., histiocytes) or inflammatory leukocytes, including lymphocytes and macrophages, which are widely held to be actively engaged in the propagation of chronic inflammatory diseases (reviewed above in "INTRODUCTION"). Since this is the central theme of the present review, I shall spend some time on details of the methodology employed, the results obtained, and their significance to disease in general, as described by Dunn et al.[24]

Ethylene vinyl acetate copolymer disks were prepared as described by Rhine et al.[81] Solubilized EVA in methylene chloride was admixed with cytokine, or bovine serum albumin, in 10 mM Tris-glycerol (10% w/v) buffer solution (maximum aqueous content not exceeding 50%). The EVA-cytokine solution was cast in small glass petridishes in a total volume of 2 ml, freeze dried to remove all trace of methylene chloride, and stored at -20°C. Each disk was divided into quarters, each of which was implanted subcutaneously into the dorso-lateral flank of CF-1 mice (where necessary, LPS-resistant C3H/HeJ mice were used). We elected to test the following recombinant cytokines in this system: (1) human IL-1α (obtained from Dainippon, Japan); (2) human IL-1β prepared at the Upjohn Company according to the method described by Paslay et al.;[82] (3) human IL-2 and rat IFNγ (Amgen, CA.); (4) murine TNFα (Genzyme, CA). Specific activities of respective cytokines were rhIL-1α and β, 2×10^7 U/mg (mouse thymocyte proliferation); rhIL-2, 0.3×10^7 U/mg (HT-2 cell proliferation); rIFN γ, 1×10^7 U/mg (L-929 cytotoxicity); rTNFα 4×10^7 U/mg (L-929 cytotoxicity). *Escherichia coli* lipopolysaccharide (0.55.B5 phenol extract, Sigma, MO) was incorporated into EVA as an endotoxin control.

Implants were removed together with intact surrounding skin and abdominal muscle tissue, fixed in buffered formalin, and prepared for histological examination. Control implants (EVA-buffer-BSA) induced a mild fibrotic response between days 4 to 21 postimplantation, which resembled a typical nonspecific granulation reaction to the foreign implant; few inflammatory phagocytes were observed (Figure 1). In contrast, EVA-IL-1β implants (10^4 U per mouse) provoked a striking inflammatory response characterized by polymorphonuclear and mononuclear phagocyte infiltration as early as 4 d postimplantation; by d 7 the lesion took on the appearance of chronic inflammatory tissue, consisting of large activated macrophages, extensive angiogenesis, peripheral fibrosis, and decreased polymorphonuclear leukocyte content (Figure 2), which persisted up to 15 to 21 d. At no time were lymphocytes observed. Implants containing IL-1α gave weak inflammatory responses, but this was subsequently found to be due to failure of the EVA polymer to release this particular cytokine *in vivo*. That IL-1α is as potent as IL-1β in the genesis of chronic granulomatous responses has since been determined using subcutaneously implanted Alzet minipumps as a slow-release system. The possible contribution of low-level endotoxin to these lesions was ruled out by several factors, including the failure of large concentrations of LPS (150 to 1500 pg per mouse) to induce similar responses and the ability of endotoxin-resistant C3H/HeJ mice to support granulomatous reactions to EVA-IL-1 implants equivalent to those observed for normal mice. These findings indicate that that IL-1 alone (α or β) is able not only to initiate, but also to perpetuate, macrophage granulomatous lesions, supporting circumstantial data suggesting that this cytokine plays a significant role in chronic disease. The precise mechanisms by which IL-1 exerts this effect are unknown; however, of the multiple activities attributed to IL-1, the following are probably central to its chronic granulomatous potential. First, the enhancement of vascular endothelial cell (EC) adhesiveness for leukocytes by IL-1[42,76,83] could trigger the initial infiltration of these cells into normal tissue: our recent data using rabbit jugular veins clearly demonstrate that these phenomena do occur *in vivo* following local application of IL-1;[80] although IL-1 is not chemotactic, it may in some way cause the local release of leukocyte chemoattractants, as suggested by Matsushima et al.[84] Second, the assembly of granulomatous tissue is dependent in part on formation of new blood vessels (angiogenesis), which was a prominent feature of the EVA-IL-1-induced lesions. Prendergast et al.[85] provided good evidence, using the cornea of irradiated rabbits, that IL-1 itself causes a profound angiogenic reaction when released continually from locally implanted EVA polymers. Why slow-release IL-1 induces a chronic macrophage-like granuloma, rather than a repetitive acute polymorphonuclear infiltration that would result in necrosis, needs to be addressed. The recent demonstration that IL-1 stimulates production of the hematopoietic factor, granulocyte-macrophage colony-stimulating factor (GM-CSF), from ECs,[86] as well as other connective tissue cells,[87,88] suggests that this may represent an important link between systemic and local events in the recruitment of monocytes (as well as neutrophils) in tissue injury and may account for the abundant macrophage response elicited by slow-release IL-1. Current research by Bevilacqua et al.[83] suggests that the induction and persistence of the leukocyte adhesion molecule,

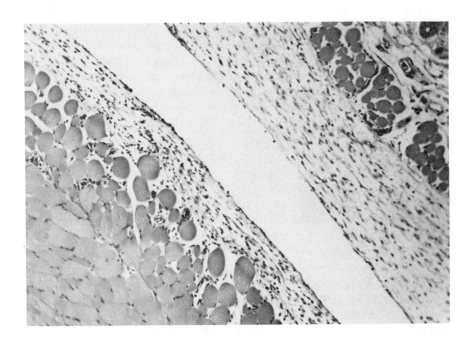

Figure 1A. Subcutaneous 7 d "control" lesion induced by EVA implants containing BSA (12.5 mg per mouse). Mild fibrotic reaction borders EVA implant (diagonal space) leaving juxtaposed abdominal muscle (left) and subcutaneous muscle (right) intact. (Hematoxylin-eosin; magnification × 100.)

ICAM-1, may be more important for adherence of leukocytes in chronic inflammation, compared with the more transiently expressed ELAM-1 adhesion molecule.

As observed with acute local administration of TNFα, slow release of this cytokine, using either EVA[89] or Alzet minipump subcutaneous implants, did not result in an impressive inflammatory response; the most significant lesions were seen only by Alzet pump delivery and comprised a sparse mononuclear leukocyte infiltrate. In view of the synergism between IL-1 and TNFα in acute inflammation, the slow-release chronic effects should be tested using combinations of both cytokines. Although TNFα has been shown to be a potent angiogenic cytokine,[90] we failed to provide any such supportive evidence using these slow-release systems.

In contrast, local slow release of IL-2 (10^2 to 10^4 U per mouse) from subcutaneously implanted EVA disks caused an early accumulation of perivascular lymphocytes 4 d postimplantation (Figure 3), which rapidly evolved into a lesion consisting primarily of proliferating lymphoid and macrophage foci between 7 to 15 d (Figure 4). Angiogenesis, though less intense than that seen in EVA-IL-1 lesions, was also evident. Our latest studies, using subcutaneously implanted Alzet minipumps to deliver IL-2, show even more impressive effects presumably due to the increased efficiency and improved release kinetics of IL-2 at the tissue site. This enabled us to extend our previous observations to include other pathological features, such as

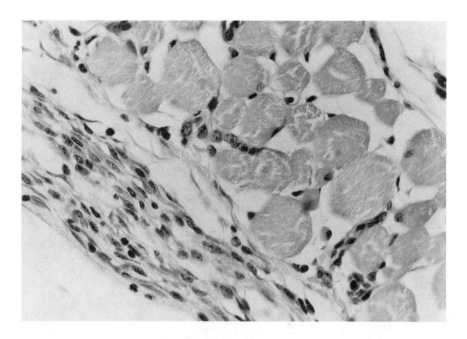

Figure 1B. Closer view of reaction surrounding implant indicates modest macrophage infiltration and fibroblast response to EVA implant. (Hematoxylin-eosin; magnification × 400.) (From *Monokines and Other Non-Lymphocytic Cytokines*, M. C. Powanda, et al., Eds., A. R. Liss, New York, 1988, 331. With permission.)

edema and distinct foci of eosinophil leukocytes interspersed in the mononuclear infiltrates. Numerous macrophages contained eosinophilic cytoplasmic inclusions, which stained negative for hemosiderin, an observation in common with the systemic effects of intermittent high-dose IL-2 (6×10^5 U per mouse b.i.d. for 7 d) shown by Anderson et al.,[63] for which there is at present no explanation. The appearance of a lesion with specific, well-defined components resembling those normally associated with hypersensitivity reactions was unexpected, since Kasahara et al.[91] were unable to induce a response to sephadex beads coated with IL-2 in mouse lung; they did detect a short-lived (maximal around 3 d) granulomatous response to IL-1-coated beads. There is, however, reference to the chemotactic[92,93] and proadhesive[94] effects of IL-2 for lymphoid cells *in vitro*, which might explain initiation of the early lymphocyte infiltration by IL-2 *in vivo*. The subsequent lymphocyte proliferation observed in these lesions accords well with the T-cell growth-inducing properties of IL-2;[32] release of IFNγ and other lymphokines from the IL-2-stimulated lymphocytes provides a plausible mechanism for the recruitment[95] and activation of monocytes,[96-98] which resulted in the impressive macrophage response to local slow release of IL-2. It is of interest that macrophages express receptors for IL-2,[59] which may serve as an additional mechanism perpetuating the granulomatous macrophage response. Whether angiogenesis was related to

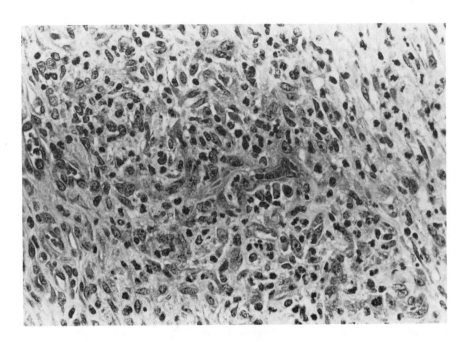

Figure 2. Subcutaneous 7 d lesion induced by EVA-rhIL-1β (10^4 U per mouse) implants; note intense monocyte macrophage infiltration, angiogenesis, and scattered neutrophilic accumulations proximal to EVA implant (left side, not shown). Peripheral fibroblast response is evident distal to implant (right side). (Hematoxylin-eosin; magnification × 400.) (From Dunn, C. J., et al., *Agents Actions*, 27:290, 1989. With permission.)

IL-2-induced T-cell factors, like those reported in specific delayed-type hypersensitivity reactions by Watt and Auerbach,[99] remains to be determined.

Circulating eosinophilia has been well documented following systemic treatment with excessive high doses of IL-2 and is believed to be related to specific mobilization of these cells from the bone marrow;[63] precisely how eosinophils are recruited into the IL-2-induced granulomatous lesion is open to conjecture. The obvious parallelism with known eosinophil T-cell granulomatous reactions tempts the speculation that IL-2, in some way, causes local production of eosinophil chemoattractants and growth factors, among which the recently described IL-3 and IL-5 are likely candidates.[100,101] In summary, it is clear that relatively low-dose IL-2 (10^2 to 10^4 U per mouse) is able to induce a chronic lymphoid-macrophage-eosinophil granulomatous response when continuously released in normal tissue; this opens up exciting new avenues of research to explore the true pathogenic potential of IL-2 in immunoinflammatory diseases, particulary chronic asthmatic and parasitic reactions.

Further slow-release studies with EVA-IFN γ subcutaneous implants[89] confirmed the remarkably sparse fibrotic response observed by Granstein et al.,[13] who

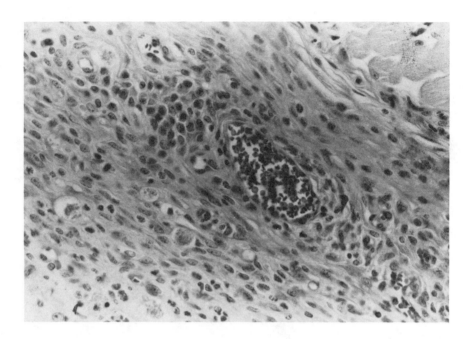

Figure 3. Subcutaneous lesion induced by EVA-rhIL-2 (10^3 U per mouse) implant showing early (4 d) perivascular lymphocyte accumulation (vasculitis). (Hematoxylin-eosin; magnification × 400.) (From Dunn, C. J., et al., *Agents Actions,* 27:290, 1989. With permission.)

showed that this cytokine exerted antiproliferative effects on the "normal" granulation reaction to the implant, accompanied by a significant reduction of collagen synthesis. The only remarkable feature we observed was the increased formation of foreign body giant cells in IFNγ lesions compared with controls. The failure of IFNγ to induce a local pathological response was surprising for several reasons. First, intralesional injection of IFNγ into leprosy patients with deficient cell-mediated reactivity provoked a striking mononuclear leukocyte infiltration and increased MHC II antigen expression resembling delayed-type hypersensitivity.[96] Similar effects were observed by Issekutz et al.[95] using a rat dermal DTH model; IL-1 augmented the IFNγ effect. The role of IFNγ in immunoinflammatory diseases may therefore be indirect, requiring the presence of other factors (cytokines, leukocytes) for full expression of its inflammatory potential. The recent work of Issekutz et al.[102] may give some insight to this concept, since intradermal lymphokine supernatants obtained from activated T-cells caused lymphocyte accumulation in rat skin. Fractionation studies led to the identification of IFNγ as the major active factor present, although an unidentified cofactor was also required for lymphocyte recruitment. Recombinant IFNγ alone induced lymphocyte infiltration when injected into rat skin;[95] the significance of this observation remains obscure in light of the slow-release IFNγ data presented above. These apparent discrepancies may reside in the

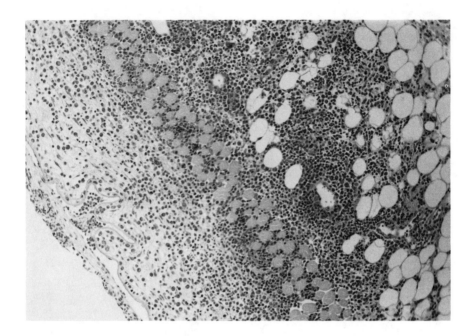

Figure 4A. Subcutaneous 7 d lesion induced by EVA-rhIL-2 (10^3 U per mouse) implants. Extensive mononuclear cell infiltration predominates and has traversed subcutaneous muscle bundles expanding in the adjacent loose connective tissue. EVA implant to the left is located by diagonal space in bottom left corner; epidermis located at top right side (not shown). (Hematoxylin-eosin; magnification × 100.)

fact that interferons α, β, and γ enhance or suppress pathological responses according to the timing of administration,[39] which is the major difference between the two experimental protocols. Thus, continuous local IFNγ release may favor antiproliferative activity, whereas acute administration causes preferential transient lymphocyte accumulation in normal tissues. A more plausible explanation is that Issekutz et al.[95] measured the migration of infused prelabeled peritoneal exudate lymphocytes that preferentially localize at inflammatory sites, thus biasing the apparent specific effects of IFN injection. Clearly, further work is required to define the most appropriate environment in which IFNγ most effectively unleashes its powerful pathogenic potential; the recent demonstration by Nickoloff et al.[103] and Makgoba et al.[104] of the association between IFNγ, expression of the lymphocyte adhesion molecule, ICAM-1, on epidermal cells, and lymphocyte infiltration in skin diseases suggests that we should have a good grasp of the molecular mechanisms involved in the near future.

A review of this nature would be incomplete without some discussion on the apparent "arthritogenic" potential of IL-1 and TNFα, as suggested by numerous *in vitro* studies; these cytokines clearly cause cartilage destruction by induction of chondrocyte collagenase synthesis and release and inhibition of proteoglycan syn-

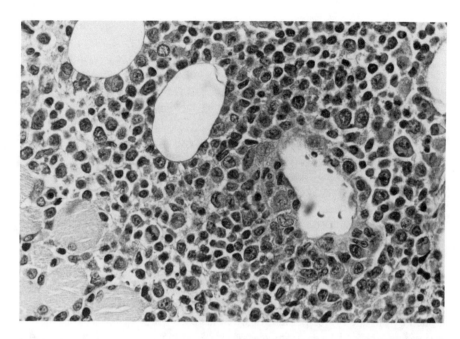

Figure 4B. Detailed view of response showing perivascular "palisading" of mononuclear cells and dense infiltration of predominantly lymphocytes and macrophages. Proliferation of these cells is evidenced by the presence of binucleate cells and mitosis within the loose subcutaneous connective tissue. (Hematoxylin-eosin; magnification × 400.) (From *Monokines and Other Non-Lymphocytic Cytokines*, M. C. Powanda, et al., Eds., A. R. Liss, New York, 1988, 332. With permission.)

thesis, resulting in the inevitable loss of matrix and cartilage integrity.[25,27,105] Other *in vitro* models have been used to present a strong case for the bone-resorbing effects of IL-1 and TNFα.[106,107] Unfortunately, there is little good evidence for either of these effects *in vivo*. Dingle et al.[108] have provided the most impressive case for IL-1α-mediated cartilage degradation following intraarticular injections of catabolin. Since catabolin is a heterogeneous mixture of substances, of which IL-1α is one, more studies are needed to elucidate the specificity of this response; there can be no doubt that these are exciting observations relevant to the arthritides, but we have to avoid the temptation to attribute these effects to specific cytokines simply because the *in vitro* data are so compelling. Other *in vivo* approaches to this dilemma have employed intra-articular injection of purified and recombinant IL-1 in a variety of doses and treatment regimens, none of which provide convincing evidence of specific cartilage or bone destruction;[109-111] synovitis is reported in every case, which is hardly surprising, in view of the inflammatory properties already described for IL-1 and TNFα in this review. Preliminary experiments in our laboratory indicate that continuous infusion of rIL-1 for 7 d in mice leaves articular cartilage and subchondral bone intact in the face of rampant synovitis; from these "negative" observations arises the provocative question: why and how does the articular cartilage and bone retain its integrity in such a hostile inflammatory environment? As in the case for IFNγ (above), the cytokines IL-1 and TNFα may be potent inducers of articular destruction only when presented to an appropriate environment conducive to their effects; these experiments need to be carefully designed, and it is my opinion that this approach will generate new insights into the pathogenesis of arthritic diseases in which cytokines will inevitably be involved. The recent experiments of Hom et al.[112] and Killar and Dunn,[113] where administration of IL-1 exacerbates murine autoimmune arthritis to type II collagen, reinforce this view, although the effector mechanism(s) are almost certainly systemic. The conversion of a "nonerosive" to "erosive" local reaction to intraarticular methylated bovine serum albumin in nonimmunized mice by subcutaneous injection of IL-1 also adds a new dimension to the systemic influence of this cytokine in the pathogenesis of arthritic disease.[114]

Systemic Manifestation of Disease by Cytokines

Of immense importance to disease in general are the unequivocal systemic, multiorgan effects of cytokines. Activation of the adrenal-pituitary axis is of paramount importance in stress and subsequent development of disease, as succinctly defined by the late Professor Hans Selye in his excellent series of lectures entitled *In Vivo: A Case for Supramolecular Biology*.[115] I am sure he would be gratified to learn that local injection of IL-1 results in an immediate and sustained increase in blood corticosterone levels[116] through stimulation of adrenocorticotropin hormone (ACTH) production from the pituitary via hypothalamic corticotropin releasing factor (CRF) release,[117,118] or direct induction of ACTH release through interaction of IL-1 with pituitary cells;[119a] inevitably these events are associated with concomitant thymic hypoplasia,[116] which we have confirmed in our IL-1 slow-release systems described above. Selye would also be intrigued to see the thrombohemorrhagic lesions resulting from IL-1 and TNFα administration, especially since he was the first to describe this phenomenon as one of the three major pathological tissue responses to the concerted actions of multiple, diverse groups of "stressors".[115] These observations need to be taken into account when assessing any systemic effects attributed to these cytokines; for example, the reported bone resorption and hypercalcemia induced by high-dose subcutaneous infusion of IL-1 by Sabatini et al.[120] may be a consequence of several interrelated events, such as endogenous corticosterone release and the demineralization effects of local inflammation (induced by the IL-1 infusion),[89] which have been described by Krempien et al.,[121] in addition to possible direct effects of IL-1 alone.

Both IL-1[122] and, to a greater extent, TNFα[30] reproduce many of the features of shock, when given systemically. The pathological sequelae of septic shock during lethal bacteremia can be successfully prevented by prior administration of anti-TNFα monoclonal antibody,[123] reinforcing the critical role of this cytokine particularly in endotoxic shock associated with sepsis. These areas are covered in greater detail by other authors in this book.

The diversity of systemic pathological effects resulting from intraperitoneal or intravenous administration of IL-2 in rodents and man has been extremely well documented partly due to the extensive use of this cytokine in tumor therapy.[26,124,125] High-dose IL-2 (6×10^5 U i.p. twice daily for 7 d) induces overt toxicity that is characterized by extensive pulmonary and hepatic lymphocyte infiltration and lymphoid pleural effusions.[63] Intravascular changes include eosinophilia, thrombocytopenia, and dramatic leakage of fluid from the capillary bed, commonly referred to as the capillary leak syndrome.[63,124] Whilst it is difficult to put these phenomena into some perspective due to the abnormally high doses of IL-2 used, certain features are reminiscent of our "local" slow-release IL-2 pathological findings (tissue lymphoid cell recruitment, edema, tissue eosinophilia),[89] where normal tissues were subjected to much lower doses of IL-2 (10^2 to 10^4 U per mouse). A dose-dependent increase in IL-2R-bearing lymphocytes, lymphocyte-activated killer cells, and natural killer cells appeared after continuous IL-2 infusion in cancer patients[26] and is further evidence of the potential disturbances that may occur in the immune function of patients with elevated blood IL-2 levels observed in various diseases (see below).

Of particular interest is the influence of systemically administered cytokines on the development of neurologic disease, reported by Bocci.[38] The prolonged administration of interferons, TNFα, and interleukins 1 and 2 results in central nervous system dysfunction, the features of which are similar for each cytokine (fever, decreased psychomotor activity, paresthesiae, and, in extreme cases, coma). Little is known about the mechanisms involved, but Bocci views the capillary endothelium as the central responsive cell from which cytokines stimulate a myriad of substances capable of adversely affecting neural function. Circumventricular organs (CVO) lining the cerebral ventricles are strategically placed to influence centers controlling thermoregulation, sleep, and hormonal secretion; Bocci proposes that the increased leakiness of the CVO capillary bed may provide a direct route of cytokine passage to these critical centers, dispelling some of the myths that the "blood brain barrier" (specialized "tight" capillary endothelium) would prevent cytokine access to the central nervous system. The ability of microglia to secrete IL-1[126] and the potent stimulatory effects of the neuropeptides, substance P and K, for IL-1, TNFα, and IL-6 synthesis and release from macrophages[127] suggests endogenous mechanisms for the local accumulation of cytokines within neural tissues. Since IL-1 enhances nerve growth factor synthesis during injury, it may be beneficial in nerve trauma.[128] It will be interesting to see how important cytokines are in the pathogenesis of CNS diseases, which is clearly an open area for future development.

CYTOKINES AND DISEASE THERAPY

Although the thrust of this review has been to investigate the association of cytokines with disease, two additional key points need to be made. First, cytokines are present in a variety of tissues under normal conditions[129-131] and may represent subtle local or systemic homeostatic mechanisms responsible for maintenance of many physiologic functions;[129] second, the so-called pathological potential exhibited by cytokines may be harnessed to therapeutic advantage in a variety of diseases (see Table 1 for summary). The most successful progress has been made in the field of tumor therapy, particularly using IFN and IL-2 treatment. The relatively low success rate should not deter future ventures; new information on the mode of action of cytokines would be lost if it were not combined with imaginative approaches designed to overcome requirements for high doses and toxic side effects, which are often major limiting factors with cytokine therapy. Interleukin-2 serves as one of the best examples; alone, systemic administration of IL-2 had little impact on tumor regression. However, infusion of *in vitro* expanded lymphocyte-activated killer (LAK) cells with IL-2 improved the efficacy of this cytokine;[132] recent advantages over this strategy have been attained by replacing the LAK cells with specific "tumor infiltrating lymphocytes" (TIL) during IL-2 infusion.[132]

Local administration of low-dose IL-2 may represent another alternative to avoid the toxicity problems observed with systemic IL-2 therapy,[133-135a] although this would clearly be restricted by accessibility to localized immunogenic tumors. We now know that IL-2 induces a specific pathological response in normal tissue following local slow release,[89] which in itself could combat tumor growth. In this respect, these observations place more emphasis on the pathogenic potential of IL-2, rather than on the molecule itself, as the key to successful therapy.

Induction of cytokine release represents an alternative means of specific treatment, as shown by the experimental success with interferon inducers as anti-inflammatory[136,137] and antifibrotic[138] agents. Interferon toxicity appears to vary according to the timing of administration,[35] which would be expected from the nyctohemeral rhythm of the endogenous IFN response cycle;[129] such an observation is not unique for IFN and clearly needs to be considered when designing disease therapy based on cytokine induction or administration.

Treatment of lepromatous leprosy[96] and chronic granulomatous disease[97,98] with IFNγ, or murine lupus nephritis with TNFα,[139] where these cytokines appear to be deficient, should encourage further research in the discovery of diseases that require cytokine replacement therapy.

The rapidly expanding literature on hemopoietic growth factors stresses their tremendous potential for restoring depleted progenitor cells following cytotoxic drug treatment and in diseases presenting with life-threatening leukopenia.[140] Interleukin-1, which probably has a "priming" effect, should also be considered among the variety of colony-stimulating growth factors (see Table 1). Information concerning the sustained effects of chronic administration of these cytokines supports the viability of this concept.[141,142]

One final note of caution: the dangers inherent in combination cytokine therapy cannot be overstated, as shown by the induction of lethal disseminated intravascular coagulation in mice administered TNFα and IFNγ;[143] unexpected toxicity can only be prevented by scientific vigilance and thorough preclinical testing.

Table 1. Cytokine and Cytokine-Related Therapeutic Approaches to Disease

Cytokine/antibody	Disease target	Species tested	Reference
IL-1	Radiation/cytotoxic injury	Rodent	160
	Bacterial infection	Rodent	161
TNFα	Autoimmune lupus nephritis	Rodent	139
TNFβ (LT)	Tumor destruction	Rodent	163
IFNs	Anti-inflammatory; immunoregulation	Rodent/human	39
IFNα, β, γ	Tumor destruction	Human	164
	Tumor and lymphocyte-induced angiogenesis	Rodent	165
IFNγ	Rheumatoid arthritis	Human	166, 167
	Lepromatous leprosy	Human	96
	Chronic granulomatous disease	Human	97, 98
	Experimental allergic encephalomyelitis	Rodent	168
IFN inducers			
Poly I:C	Fibrosis; transplantation	Rodent	138, 169
Tilorone	Adjuvant arthritis	Rodent	136
	DTH granuloma	Rodent	137
IL-2 + LAK cell or tumor infiltrating lymphocyte cotreatment	Tumor destruction	Rodent, human	132, 132a
Colony-stimulating factors: GM-CSF, G-CSF, M-CSF, Multi-CSF	Cytotoxic injury; bone marrow transplantation; myelodysplastic syndromes; AIDS neutropenia	Rodent and human	Reviewed in 140
CSF-1 (M-CSF)	Tumor destruction	Rodent	170
Basic FGF (bovine)	Cartilage repair	Rabbit	171
GM-CSF Ab + IL-3 Ab	Cerebral malaria	Rodent	172
IL-4 Ab	Allergy; parasitic infection	Rodent	162

FUTURE DIRECTIONS IN CYTOKINE RESEARCH

From the enormous amount of work reported on cytokine-related pathology, I have attempted to weave some semblance of a cohesive network linking specific cytokines to inflammatory diseases. The whole picture is far from complete; what are urgently needed are more definitive studies on the specific association between cytokines and disease mechanisms, a sentiment echoed by Nagy et al.[144] for cytokines and the "surfeit of factors" that await clarification as angiogenic agents *in vivo*. Whilst the evidence that these mediators are involved in pathological processes is beyond doubt, the degree to which they influence the course of events in the pathogenesis of disease is circumstantial and derived from empirical data.

Key advances in the future will rely heavily on quantitative and kinetic studies correlating the presence and localization of defined cytokines or cytokine patterns with specific diseases and disease syndromes. It is equally crucial to extend *in vivo* studies by continuing to define the disease-inducing capacity of cytokines; perhaps the most important approaches will reside in the development and use of specific antagonists to test the concepts already proposed for the central and critical role of cytokines in disease. In this respect, molecular biology is proving a tremendous asset in the elucidation and synthesis of natural cytokine antagonists, for which IL-1 is the best known example[47-49] (see Chapter 2 in this book). The development and therapeutic use of specific "functionally active" antibodies will be invaluable tools in defining the true significance of individual cytokines in pathological responses. Drugs such as Cyclosporin A, which prevents the synthesis of lymphokines by T-cells,[145-147] will continue to be extremely useful in this area of research. Cyclophilin, which binds Cyclosporin A, may be identical to a ubiquitous intracellular enzyme (PPIase) responsible for catalyzing protein refolding during protein synthesis; the inhibition of PPIase actvity by Cyclosporin A therefore represents a novel biochemical target for T-cell-lymphokine synthesis inhibition.[148,149]

Significant progress is being made in the identification of lymphocyte subsets and their apparent specificity for cytokine production; two major murine T-cell subsets, Th1 and Th2, selectively release either IL-2 and IFNγ or IL-4, respectively.[150,151] Similar distinctions are being drawn from human "naive" and "memory" T-cell subsets with respect to differential cytokine production,[152] although no evidence has yet been found to suggest a common identity between these and the murine subsets. Nevertheless, these observations represent important inroads for the elucidation of pathological responses driven by specific cellular elements of the immune response and will enable us to gain a better understanding of the cytokine "patterns" that are beginning to unfold in rheumatoid arthritis, sarcoidosis, and other diverse experimental and clinical immunopathological conditions described in this review.

As is usual for new molecular entities, it has become fashionable to measure cytokine levels in inflammatory fluids and the circulation during disease; altered circulating levels of cytokines, or changes in the potential of circulating leukocytes to produce cytokines, have been claimed to correlate with disease intensity, e.g., rheumatoid arthritis,[153] Kawasaki disease,[154] systemic lupus erythematosis,[155] multiple sclerosis,[156] severe burn injury,[157] and chronic fatigue immunodeficiency syndrome.[158,159] Such information yields, at best, only tentative conclusions, and a thorough review of the literature reveals occasional inconsistencies in the data for a given disease. However, different diseases often present with common clinical "syndromes", such as fever, fatigue, malaise, myalgia, arthralgia, sleep-pattern disturbances, edema, weight change, etc. These clinical signs correlate well with the reported systemic effects of the four cytokines discussed in the above section on "Pathological effects of cytokines" and may lend credence to the role of altered circulating cytokines in Selye's "general syndrome of just being sick" (observations recorded by Selye,1925; see Reference 115). We are often in danger of overlooking the obvious; I would suggest that these observations fall into this category and should be recognized as a significant advancement of our knowledge of the molecular biology of disease.

SUMMARY

An overview of the inflammatory process and its relationship to the development of disease is presented; the potential of cytokines as mediators in the pathogenesis of various nonimmune and immune-mediated diseases is discussed based on (1) their localization in pathological tissues and fluids, (2) the ability of cytokines to mimic local inflammatory disease when administered *in vivo*, and (3) cytokine-induced systemic disease. Major emphasis is placed on IL-1, TNFα, IL-2 and the interferons (IFN α, β, γ). The evidence that each of these cytokines plays a significant role in the evolution and perpetuation of immunoinflammatory disease is beyond doubt; typical features of acute pathologic responses are directly inducible by certain cytokines alone when given locally (IL-1, TNFα, IL-2); with the exception of TNFα, these cytokines provoke qualitatively distinct chronic granulomatous inflammatory reactions upon local slow release. The failure of TNFα or IFNγ to induce lesions under similar conditions suggests the requirement for a "primed" environment for full expression of their pathogenic potential; interaction with other cytokines may be essential, although this remains to be clarified, especially in chronic disease. These observations probably delineate the widely held view that chronic inflammatory disease is most likely due to multiple cytokine interactions. In contrast, each cytokine exhibits numerous pathologic effects following systemic administration; of these, central nervous system dysfunction is common, the mechanisms of which require further clarification. Aside from the obvious pathogenic effects, the therapeutic value of cytokines is discussed, emphasizing limitations of these studies and suggestions for future approaches.

ACKNOWLEDGMENTS

I would like to thank Mr. Robert Simmons for his expert help in the development of the micrographs; Ms. B. Mundo, who organized and typed the manuscript; Mrs. E. Dunn for her encouragement and assistance in assembling the references; and Dr. D. E. Tracey for his critical review of the scientific content.

REFERENCES

1. Dunn, C.J. and D.A. Willoughby, Leukocyte and macrophage migration inhibitory activities in inflammatory exudates — involvement of the coagulation system. *Lymphokines*, 4:231, 1981.
2. Silverstein, A.M., History of immunology. Cellular versus humoral immunity: determinants and consequences of an epic nineteenth century battle. *Cell. Immunol.*, 48:208, 1979.
3. Waksman, B.H., Pathogenetic mechanisms in multiple sclerosis: a summary. *Ann. N.Y. Acad. Sci.*, 436:125, 1984.

4. Williams, D.M. and N.W. Johnson, Alterations in peripheral blood leucocyte distribution in response to local inflammatory stimuli in the rat. *J. Pathol.*, 118:129, 1976.
5. Malorney, U., J. Knop, G. Burmeister, and C. Sorg, Immunohistochemical demonstration of migration inibitory factor (MIF) in experimental allergic contact dermatitis. *Clin. Exp. Immunol.*, 71:164, 1988.
6. Yoshida, T., *Basic and Clinical Aspects of Granulomatous Diseases*, D. Boros and T. Yoshida, Eds., Elsevier/North-Holland, Amsterdam, 1980, 81–96.
7. Honda, M. and H. Hayashi, Characterization of three macrophage chemotactic factors from PPD - induced delayed hypersensitivity reaction sites in guinea pigs, with special reference to a chemotactic lymphokine. *Am. J. Pathol.*, 108:171, 1982.
8. Yoshida, T., P.E. Bigazza, and S. Cohen, Biologic and antigenic similarity of virus-induced migration inhibition factor in coventional, lymphocyte-derived migration inhibition factor. *Proc. Natl. Acad. Sci. U.S.A.*, 72:1641, 1975.
9. Pick, E., The mechanism of action of macrophage migration inhibitory factor (MIF). A personal view. *Curr. Titles Immunol.*, 4:565, 1976.
10. Kampschmidt, R.F., The numerous postulated biological manifestations of interleukin-1. *J. Leukoc. Biol.*, 36:341, 1984.
11. Dinarello, C.A. and J.W. Mier, Lymphokines. *N. Engl. J. Med.*, 317:940, 1987.
12. Gery, I., R.K. Gershon, and B.H. Waksman, Potentiation of the T-lymphocyte response to mitogens. I. The reponding cell. *J. Exp. Med.*, 136:128, 1972.
13. Granstein, R.D., R. Margolis, S.B. Mizel, and D.N. Sauder, *The Physiologic Metabolic and Immunologic Actions of Interleukin-1*, M.J. Kluger, J.J. Oppenheim, and M.C. Powanda, Eds., A.R. Liss, New York, 1985, 25–29.
14. Beck, G., G.S. Habicht, J.L. Benach, and F. Mille, Interleukin-1 a common endogenous mediator of inflammation and the local Schwartzmann reaction. *J. Immunol.*, 136:3025, 1986.
15. Movat, H.Z., C.E. Burrowes, M.J. Cybulsky, and C.A. Dinarello, Acute inflammation and a Schwartzmann-like reaction induced by interleukin-1 and tumor necrosis factor. *Am. J. Pathol.*, 129:463, 1987.
16. Nawroth, P.P. and D.M. Stern, Implication of thrombin formation on the endothelial cell surface. *Semin. Thromb. Hemost.*, 12:197, 1986.
17. Emeis, J.J. and T. Kooistra, Interleukin-1 lipopolysaccharide induce an inhibitor of tissue-type plasminogen activator in vivo and in cultured endothelial cells. *J. Exp. Med.*, 163:1260, 1986.
18. Dinarello, C.A., T. Ikejima, S.J.C. Warner, S.F. Orencole, G. Lonnemann, J.G. Cannon, and P. Libby, Interleukin 1 induces interleukin 1. I. Induction of circulating interleukin 1 in rabbits in vivo and in human mononuclear cells in vitro. *J. Immunol.*, 139:1902, 1987.
19. Warner, S.J.C., K.R. Auger, and P. Libby, Interleukin 1 induces interleukin 1. II. Recombinant human interleukin 1 induces interleukin 1 production by adult human vascular endothelial cells. *J. Immunol.*, 139:1911, 1987.
20. Fleming, W.E., R.G. Schaub, M.R. Deibel, A.E. Berger, D.A. Wunderlich, M.M. Hardee, K.A. Richard, N.D. Staite, and C.J. Dunn, Inhibition of leukocyte modulation of endothelial cell functions by monoclonal antibodies to various cytokines. *FASEB J.*, 2:A1600, 1988.
21. Mizel, S.B., Regulation of immune and inflammatory responses by interleukin 1. *Clin. Immunol. Newsl.*, 3:123, 1982.

22. Sprugel. K.H., J.M. McPherson, A.W. Clowes, and R. Ross, Effects of growth factors in vivo. I. Cell ingrowth into porous subcutaneous chambers. *Am. J. Pathol.*, 129:601, 1987.
23. Raines, E.W., S.K. Dower, and R. Ross, Interleukin-1 mitofenic activity for fibroblasts and smooth muscle cells is due to PDGF-AA. *Science*, 243:393, 1989.
24. Dunn, C.J. and D.A. Willoughby, The inflammatory response — a review. *Handbk. Exp. Pharmacol.*, 87/I:465, 1989.
25. Krakauer, T., J.J. Oppenheim, and H.E. Jasin, Human interleukin-1 intermediates cartilage matrix degradation. *Cell. Immunol.*, 91:92, 1985.
26. Thompson, J.A., D.J. Lee, C.G. Lindgren, L.A. Benz, C. Collins, D. Levitt, and A. Fefer, Influence of dose and duration of infusion of interleukin-2 on toxicity and immunomodulation. *J. Clin. Oncol.*, 6:669, 1988.
27. Saklatvala, J., Tumor necrosis factor alpha stimulates resorption and inhibits synthesis of proteoglycan in cartilage. *Nature*, 32:547, 1986.
28. Bertolini, D.R., G.E. Nedwin, T.S. Bringman, D.D. Smith, and G.R. Mundy, Stimulation of bone resorption and inhibition of bone formation in vitro by human tumor necrosis factors. *Nature*, 319:516, 1986.
29. Le, J. and J. Vilcek, Biology of disease tumor necrosis and interleukin 1: cytokines with multiple overlapping biological activities. *Lab. Invest.*, 56:234, 1987.
30. Tracey, K.J., B. Beutler, S.F. Lowry, J. Merryweather, S. Wolpe, J.W. Milsark, R.J. Hariri, T.J. Fahey, III, A. Zentella, J.D. Albert, G.T. Shires, and A. Cerami, Shock and tissue injury induced by recombinant human cachectin. *Science*, 234:470, 1986.
31. Shalaby, M.R., D. Pennica, and M.A. Palladino, An overview of the history and biologic properties of tumor necrosis factors. *Springer Semin. Immmunopathol.*, 9:33, 1986.
32. Robb, R.J., Interleukin 2: the molecule and its function. *Immunol. Today*, 5:203, 1984.
33. Gallagher, G. and J. Willdridge, Evidence that interleukin 2 actively promotes both cell division and immunoglobulin secretion in human B cells. *Int. Arch. Allergy Appl. Immunol.*, 86:337, 1988.
34. Herrmann, F., S.A. Cannistra, A. Lindemann, D. Blohm, A. Rambaldi, R.H. Mertelsmann, and J.D. Griffin, Functional consequences of monocyte IL-2 reeptor expression. Induction of IL-1β secretion by IFNγ and IL-2. *J. Immunol.*, 142:139, 1989.
35. Bocci, V., The physiological interferon response. *Immunol. Today*, 6:7, 1985.
36. Browning, J., Interferons and rheumatoid arthritis: insight into interferon biology? *Immunol. Today*, 8:372,1987.
37. Herberman, R.B., Effects of biological response modifiers on effector cells with cytotoxic activity against tumors. *Semin. Oncol.*, 13:195, 1986.
38. Bocci, V., Central nervous system toxicity of interferons and other cytokines. *J. Biol. Regul. Homeost. Agents,* 2:107, 1988.
39. Billiau, A., Interferons and inflammation. *J. Interferon Res.*, 7:559, 1987.
40. Bonnem, E.M. and R.K. Oldham, Gamma-interferon: physiology and speculation on its role in medicine. *J. Biol. Response Mod.*, 6:275, 1987.
41. Ziff, M., Factors involved in cartilage injury. *J. Rheumatol.*, 11:13, 1983.
42. Pober, J.S., Cytokine-mediated activation of vascular endothelium. *Am. J. Pathol.*, 133:426, 1988.
43. Goto, K., S. Nakamura, F. Goto and M. Yoshinaga, Generation of an interleukin-1-like lymphocyte stimulating factor at inflammatory sites: correlation woth the infiltration of polymorphonuclear leucocytes. *Br. J. Exp. Pathol.*, 65:521, 1984.

44. Chensue, S.W., D.G. Remick, C. Shmyr-Forsch, T.F. Bells, and S.L. Kunkel, Immunohistochemical demonstration of cytoplasmic and membrane-associated tumor necrosis factor in murine macrophages. *Am. J. Pathol.*, 133:564, 1988
45. Wasik, M.A., R.P. Donnelly, and D.J. Beller, Lymphokine-independent induction of macrophage membrane IL-1 by autoreactive T cells recognizing either class I or class II MHC determinants. *J. Immunol.*, 141:3456, 1988.
46. Minnich-Carruth, L.L., J. Suttles, and S.B. Mizel, Evidence against the existence of a membrane form of murine IL-1α. *J. Immunol.*, 142:526, 1989.
47. Amento, E.P., J.T. Kurnick, and S.M. Krane, Interleukin-1 production by a human monocyte cell line is induced by a T lymphocyte product. *Immunobiology*, 163:276, 1982.
48. Fujiwara, H. and J.J. Ellner, Spontaneous production of a suppressor factor by the human macrophage-like cell line U937. *J. Immunol.*, 136:181, 1986.
49. Tiku, K., M.L. Tiku, S. Liu, and J.L. Skosey, Normal human neutrophils are a source of a specific interleukin 1 inhibitor. *J. Immunol.*, 136:3686, 1986.
50. Chensue,S.W., J.G. Otterness, G.I. Higashi, C. Shmyr-Forsch, and S.L. Kunkel, Monokine production by hypersensitivity (Shistosoma Mansoni egg) and foreign body (Sephadex Bead) - type granuloma macrophages. Evidence for sequential production of IL-1 and tumor necrosis factor. *J. Immunol.*, 142:1281, 1989.
51. Duff, G.W., E. Dickens, N. Wood, J. Manson, J. Symons, S. Poole, and F. diGiovani, *Monokines and Other Non-Lymphocytic Cytokines*, M.C. Powanda, J.J. Oppenheim, M.J. Kluger, and C.A. Dinarello, Eds, A.R. Liss, New York, 1988, 387–392.
52. Vissers, M.C.M., J.C. Fantone, R. Wiggins, and S.L. Kunkel, Glomerular basement membrane-containing immune complexes stimulate tumor necrosis factor and interleukin-1 production by human monocytes. *Am. J. Pathol.*, 134:1, 1989.
53. Montreewasuwat, J.S. and J.L. Turk, Interleukin-1 and prostaglandin production by cells of the mononuclear phagocyte system isolated from mycobacterial granulomas. *Cell. Immunol.*, 104:12, 1987.
54. Turk, J.L., P. Badenoch-Jones, and D. Parker, Ultrastructural observations on epethelioid cell granulomas induced by zirconium in the guineapig. *J. Pathol.*, 124:45, 1978.
55. Rohatgi, P.K. and R.A. Goldstein, Immunology of sarcoidosis. *Clin. Immunol. News.*, 3:115, 1982.
56. Chilosi, M., F. Menestrina, P. Capelli, L. Montagne, M. Lestani, G. Pazzola, A. Cipriani, C. Agostini, L. Trentin, R. Zambollo, and G., Semenzato, Immunohistochemical analysis of sarcoid granulomas. Evaluation of Ki67[+] and interleukin-1[+] cells. *Am. J. Pathol.*, 131:191, 1988.
57. Mariano, M. and W.G. Spector, The formation and properties of macrophage polykaryons (inflammatory giant cells). *J. Pathol.*, 113:1, 1973.
58. Tsuda, T., K. Sugisaki, Y. Abe, T. Yoshimatsu, T. Matsumoto, and E. Miyazaki, Immunohistochemical analysis of immunocytes, cytokines and cytokine receptor in sarcoid granulomas using monoclonal antibodies. *Nippon Kyobu Shikkan Gakkai Zasshi*, 26:851, 1988.
59. Hancock, W.W., L. Kobzik, A.J. Colby, C.J. O'Hara, A.G. Cooper, and J.J. Godleski, Detection of lymphokines and lymphokine receptors in pulmonary sarcoidosis. *Am. J. Pathol.*, 123:195, 1986.
60. Husby, G. and R.C. Williams, Jr., Immunohistochemical studies of interleukin-2 and γ-interferon in rheumatoid arthritis. *Arthritis Rheum.*, 28:174, 1985.
61. Firestein, G.S., W.D. Xu, K. Townsend, D. Broide, J. Alvaro-Gracia, A. Glasebrook, and N.J. Zvaifler, Cytokines in chronic inflammation and arthritis. *J. Exp. Med.*, 168:1573, 1988.

62. Symons, J.A., N.C. Wood, I.S. DiGiovine, and G.W. Duff, Soluble IL-2 receptor in rheumatorial arthritis. Correlation with disease activity, IL-1 and IL-2 inhibition. *J. Immunol.*, 141:2612, 1988.
63. Anderson, T.D., T.J. Hayes, M.K. Gately, J.M. Bontempo, L.L. Stern, and G.A. Truitt, Toxicology of human recombinant interleukin-2 in the mouse is mediated by Interleukin-activated lymphocytes. *Lab. Invest.*, 59:598, 1988.
64. Wilder, R.L., Immunopathogenesis of rheumatoid arthritis. *Immunol. Newsl.*, 6:1, 1985.
65. Kelley, V.E., G.N. Gaulton, and T.B. Strom, Inhibitory effects of anti-interleukin 2 receptor and anti-L3T4 antibodies on delayed type hypersensitivity: the role of complement and epitope. *J. Immunol.*, 138:2771, 1987.
66. Yamamoto, S., C.J. Dunn, and D.A. Willoughby, Studies on delayed hypersensitivity pleural exudates in guinea pigs. I. demonstration of substances in the cell-free exudate which cause inhibition of monuclear cell migration in vitro. *Immunology*, 30:505, 1976.
67. Sorg, C. and K. Odink, *Molecular Basis of Lymphokine Action — Exp. Biol. Med.*, D.R. Webb, C.W. Pierce, and S. Cohen, Eds., Humana Press, Clifton, NJ, 1987.
68. Mizushima. A., T. Baba, T. Ochiya, K. Yamaguchi, K. Onozaki, H. Yaoita, and K. Uyeno, Regulatory mechanisms of cutaneous delayed-type hypersensitivity 1. Suppression of cutaneous delayed-type hypersensitivity by migration inhibitory factor. *Cell. Immunol.*, 81:126, 1983.
69. Paul and O'Hara, B cell stimulating factor-1/Interleukin-4. *Ann. Rev. Immunol.*, 5:429, 1987.
70. Crawford, R., D.S. Finbloom, J. Ohara, W.E. Paul, and M. Melter, B-cell stimulatory factor-1 (interleukin 4) activates macrophages for increased tumoricidal activity and expression of Ia antigens. *J. Immunol.*, 139:135, 1987.
71. McInnes, A. and D.M. Rennick, Interleukin 4 induces cultured monocytes/macrophages to form giant multinucleated cells. *J. Exp. Med.*, 167:598, 1988.
72. Sauder et al., Epidermal cytokines: properties of epidermal cell thymocyte-acting factor (ETAF). *Lymphokine Res.*, 3:145, 1984.
73. Habicht, G.S. and G. Beck, IL-1 is an endogenous mediator of acute inflammation and the local Schwartzman reaction, in *The Physiologic, Metabolic, and Immunologic Actions of Interleukin-1*, M.J. Kluger, J.J. Oppenheim, and M.C. Powanda, Eds., A.R. Liss, New York, 1985, 12–23.
74. Averbook, B.J., R.S. Yamamoto, T.R. Ulich, E.W.B. Jeffes, J. Masunaka, and G.A. Granger, Purified native and recombinant human alpha lymphotoxin [tumor necrosis factor (TNF)-beta] induces inflammatory reactions in normal skin. *J. Clin. Immunol.*, 7:333, 1987.
75. Bevilacqua, M.P., J.S. Pober, G.R. Majeau, R.S. Cotran, and M.A. Gimbrone, Interleukin-1 (IL-1) induces biosynthesis and cell surface expression of procoagulant activity in human vascular endothelial cells. *J. Exp. Med.*, 160:618, 1984.
76. Dunn, C.J. and W.E. Fleming, Increased adhesion of polymorphonuclear leukocytes to vascular endothelium by specific interaction of endogenous (interleukin-1) and exogenous (lipopolysaccharide) substances with endothelial cells in vitro. *Eur. J. Rheumatol. Inflamm.*, 7:80, 1984.
77. Bevilacqua, M.P., J.S. Pober, M. Elyse Wheeler, R.S. Cotran, and M.A. Gimbrone, Interleukin-1 acts on cultured human vascular endothelium to increase the adhesion of polymorphonuclear leukocytes, monocytes, and related leukocyte cell lines. *J. Clin. Invest.*, 76:2003, 1985.

78. Cotran, R.S., M.A. Gimbrone, Jr., M.P. Bevilacqua, D.L. Hendrick, and J.S. Pober, Induction and detection of a human endothelial activation antigen in vivo. *J. Exp. Med.*, 164:661, 1986.
79. Cotran, R.S., J.S. Pober, M.A. Gimbrone, Jr., T.A. Springer, E.A. Wiebke, A.A. Gaspari, S.A. Rosenberg, and M.T. Lotze, Endothelial activation during interleukin-2 immunotherapy. A possible mechanism for the vascular leak syndrome. *J. Immunol.*, 140:1883, 1988.
80. Dunn, C.J., R.G. Schaub, W.E. Fleming, and A.J. Gibbons, *Leukocyte Emigration and Its Sequelae*, H. Mavat, Ed., S. Karger, Basel, Switzerland, 1987, 55–61.
81. Rhine, W.D., D.S.T. Hsieh, and R. Langer, Polymers for sustained macromolecule release: procedures to fabricate reproducible delivery systems and control release kinetics. *J. Pharm. Sci.*, 69:265, 1980.
82. Paslay, J.W., A.W. Yem, D.B. Carter, C.-C.S. Tomich, K.A. Curry, D.E. Tracey, and M.R. Deibel, Purification and preliminary characterization of recombinant human interleukin-1β. *Fed. Proc.*, 46:2282, 1987.
83. Bevilacqua, M.P., S. Stengelin, M.A. Gimbrone, Jr., and B. Seed, Endothelial leukocyte adhesion molecule 1: an inducible receptor for neutrophils related to complement regulatory proteins and lectins. *Science*, 243:1160, 1989.
84. Matsushima, K., K. Morishita, T. Yoshimura, S. Lavu, Y. Kobayashi, W. Lew, E. Appella, H.F. Kung, E.J. Leonard, and J.J. Oppenheim, Molecular cloning of a human monocyte-derived neutrophil chemotactic factor (MDNCF) and the induction of MDNCF mRNA by interleukin 1 and tumor necrosis factor. *J. Exp. Med.*, 167:1883, 1988.
85. Prendergast, R.A., G.A. Lutty, and C.A. Dinarello, Interleukin-1 induces corneal neovascularization. *Fed. Proc.*, 46:1200, 1987.
86. Sieff, C.A., S. Tsai, and D.V. Faller, Interleukin 1 induces cultured human endothelial cell production of granulocyte macrophage colony stimulating factor. *J. Clin. Invest.*, 79:48, 1987.
87. Zucali, J.R., C.A. Dinarello, D.J. Oblon, M.A. Gross, L. Anderson, and R.S. Weiner, Interleukin-1 stimulates fibroblasts to produce granulocyte macrophage colony stimulating activity and prostaglandin E_2. *J. Clin. Invest.*, 77:1857, 1986.
88. Kupper, T.S., F. Lee, N. Birchall, S. Clark, and S. Dower, Interleukin-1 binds to specific receptors on human keratinocytes and induces granulocyte macrophage colony-stimulating factor mRNA and protein. *J. Clin. Invest.*, 82:1787, 1988.
89. Dunn, C.J., M.M. Hardee, and N.D. Staite, Acute and chronic inflammatory responses to local administration of recombinant IL-1α, IL-1β, TNFα, IL-2 and "IFNγ in Mice". *Agents Actions*, 27:290, 1989.
90. Leibovich, S.J., P.J. Polverini, H.M. Shepard, D.M. Wiseman, V. Shively, and N. Nuseir, Macrophage-induced angiogenesis is mediated by tumor necrosis factor-α *Nature*, 329:630, 1987.
91. Kasahara, K., K. Kobayashi, Y. Shikama, I. Yoneya, K. Soezima, H. Ide, and T. Takahashi, Direct evidence for granuloma-inducing activity of interleukin-1. *Am. J. Pathol.*, 130:629, 1988.
92. Robbins, R.A., L. Klassenm J. Rasmussen, M.E.M. Clayton, and W.D. Russ, Interleukin-2-induced chemotaxis of human T-lymphocytes. *J. Lab. Clin. Med.*, 108:340, 1986.
93. Natuk, R.J. and R.M. Welsh, Chemotactic effect of human recombinant interleukin 2 on mouse activated large granular lyphocytes. *J. Immunol.*, 139:2737, 1987.
94. Aronson, F.R., P. Libby, E.P. Brandon, M.W. Janicka, and J.W. Mier, IL-2 rapidly induces natural killer cell adhesion to human endothelial cells. *J. Immunol.*, 141:158, 1988.

95. Issekutz, T.B., J.M. Stoltz, and P. Van der Meide, The recruitment of lymphocytes into the skin by T cell lymphokines: the role of γ-interferon. *Clin. Exp. Immunol.*, 73:70, 1988.
96. Kaplan, G., A. Nusrat, E.N. Sarno, C.K. Job, J. McElrath, J.A. Porto, and Z.A. Cohn, Cellular responses to the intradermal injection of recombinant human γ-interferon in lepromatous leprosy patients. *Am. J. Pathol.*, 128:345, 1987.
97. Sechler, J.M.G., H.L. Malech, C.J. White, and J.I. Gallin, Recombinant human interferon-α reconstitutes defective phagocyte function in patients with chronic granulomatous disease of childhood. *Proc. Natl. Acad. Sci. U.S.A.*, 85:4874, 1988.
98. Ezekowitz, A.R.B., M.C. Dinauer, H.S. Jaffe, S.H. Orkin, and P.E. Newburger, Partial correction of the phagocyte defect in patients with X-linked chronic granulomatous disease by subcutaneous interferon gamma. *N. Engl. J. Med.*, 319:146, 1988.
99. Watt, S.L. and R. Auerbach, A mitogenic factor for endothelial cells obtained from mouse secondary mixed leukocyte cultures. *J. Immunol.*, 136:197, 1986.
100. Lopez, A.F., L.B. To, Y.C. Yang, J.R. Gamble, M.F. Shannon, G.F. Burns, P.G. Dyson, C.A. Juttner, S. Clark, and M.A. Vadas, Stimulation of proliferation, differentiation and function of human cells by primate interleukin 3. *Proc. Natl. Acad. Sci. U.S.A.*, 84:2761, 1987.
101. Yamaguchi, Y., Y. Hayashi, Y. Sugama, Y. Miura, T. Kasahara, S. Kitamura, M. Torisu, S. Mifa, A. Tominag, K. Takatsu, and T. Suda, Highly purified murine interleukin 5 (IL-5) stimulates eosinophil function and prolongs in vitro survival. *J. Exp. Med.*, 167:1737, 1988.
102. Issekutz, T.B., J.M. Stoltz, and P.V.O. Meide, Lymphocyte recruitment in delayed-type hypersensitivity. The role of IFN-γ. *J. Immunol.*, 140:2989, 1988.
103. Nickoloff, B.J., The role of gamma interferon in epidermal trafficking of lymphocytes with emphasis on molecular and cellular adhesion events. *Arch. Dermatol.*, 124:1835, 1988.
104. Makgoba, M.W., M.E. Sanders, G.E. Ginther Luce, M.L. Dustin, T.A. Springer, E.A. Clark, P. Mannoni, and S. Shaw, ICAM-1 ligand for LFA-1-dependent adhesion of B, T and myeloid cells. *Nature*, 331:86, 1988.
105. Dodge, G.R. and A.R. Poole, Immunohistochemical detection and immunochemical analysis of type II collagen degradation in human normal rheumatoid and osteoarthritic articular cartilages and explants of bovine articular cartilage cultured with interleukin-1. *J. Clin. Invest.*, 83:647, 1987.
106. Thomson, B.M., G.R. Mundy, and T.J. Chambers, Tumor necrosis factors α and β induce osteoblastic cells to stimulate osteoclastic bone resorption. *J. Immunol.*, 138:775, 1987.
107. Gowen, M. and G.R. Mundy, Actions of recombinant interleukin-1, interleukin-2 and interferon γ on bone resorption in vitro. *J. Immunol.*, 136:2478, 1986.
108. Dingle, J.T., D.P. Page Thomas, B. King, and D.R. Bard, In vivo studies of articular tissue damage mediated by catabolin/interleukin 1. *Ann. Rheum. Dis.*, 46:527, 1987.
109. Pettipher, E.R., G.A. Higgs, and B. Henderson, Interleukin 1 induces leukocyte infiltration and cartilage proteoglycan degradation in the synovial joint. *Proc. Natl. Acad. Sci. U.S.A.*, 83:8749, 1986.
110. Meunier, P.C., L.W. Scwartz, L.D. Meunier, W.J. Johnson, and P.L. Simon, Interleukin-1 induced acute synovitis in the rabbit. *Fed. Proc.*, 46:736, 1987.
111. Caccese, R.G., J.F. Daniels, T. Hodge, J. Chang, and S.C. Gilman, Histomorphological changes in rat knee joints following intra-articular injection of interleukin-1. *Fed. Proc.*, 46:789, 1987.

112. Hom, J.T., A.M. Bendele, and D.G. Carlson, in vivo administration with IL-1 accelerates the development of collagen-induced arthritis in mice. *J. Immunol.*, 141:834, 1988.
113. Killar, L.M. and C.J. Dunn, Interleukin-1 potentiates the development of collagen-induced arthritic in mice. *Clin. Sci.*, 76:535, 1989.
114. Staite, N.D., K.A. Richard, D.G. Aspar, K.A. Franz, L.A. Galinet, and C.J. Dunn, Induction of an acute erosive monoarticular arthritis in mice with interleukin-1 and methylated bovine serum albumin. *Arthr. Rheum.*, 3:253–260, 1990.
115. Selye, H., Pathogen vs. soil in *In Vivo: A Case for Supramolecular Biology*, Liveright, New York, 1967, 98–129.
116. Morrissey, P.J., K. Charrier, A. Alpert, and L. Bressler, In vivo administration of IL-1 induces thymic hypoplasia and increased levels of serum corticosterone. *J. Immunol.*, 141:1456, 1988.
117. Berkenbosch, F., J. VanOers, A. delRey, F. Tilders, and H. Besedovsky, Corticotropin-releasing factor-producing neurons in the rat activated by interleukin-1. *Science*, 238:524, 1987.
118. Sapolsky, R.C. Rivier, G. Yamamoto, P. Plotsky, and W. Vale, Interleukin-1 stimulates the secretion of hypothalamic corticotropin releasing factor. *Science*, 238:522, 1987.
119. Bernton, E.W., J.W. Beach, J.W. Holaday, R.C. Smallridge, and H.G. Fein, Release of multiple hormones by a direct action of interleukin-1 on pituitary cells. *Science*, 238:519, 1987.
119a. Kehrer, P., D. Turnill, J.M. Dayer, A.F. Muller, and R.C. Gaillard, Human recombinant interleukin-1 beta and -alpha, but not recombinant tumor necrosis factor alpha stimulate ACTH release from rat anterior pituitary cells in vitro in a prostaglandin E2 and cAMP independent manner. *Neuroendocrinology*, 48:160, 1988.
120. Sabatini, M., B. Boyce, T. Aufdemorte, L. Bonewald, and G.R. Mundy, Infusions of recombinant human interleukins 1α and 1β cause hypercalcemia in normal mice. *Proc. Natl. Acad. Sci. U.S.A.*, 85:5235, 1988.
121. Krempien, B., S. Vukicevic, M. Vogel, A. Stavljenic, and R. Buchele, Cellular basis of inflammation-induced osteopenia in growing rats. *J. Bone Miner. Res.*, 3:573, 1988.
122. Okusawa, S., J.A. Gelfand, T. Ikejima, R.J. Conolly, and C.A. Dinarello, Interleukin 1 induces a shock-like state in rabbits. *J. Clin. Invest.*, 81:1162, 1988.
123. Tracey, K.J., Y. Fong, D.G. Hesse, K.R. Manogue, A.T. Lee, G.C. Kuo, S.F. Lowry, and A. Cerami, Anti-cachectin/TNF monoclonal antibodies prevent septic shock during lethal bacteremia. *Nature*, 330:662, 1987.
124. Rosenstein, M., S.E. Ettinghausen, and S.A. Rosenberg, Extravasation of intravascular fluid mediated by the systemic administration of recombinant interleukin 2. *J. Immunol.*, 137:1735, 1986.
125. Ettinghauser, S.E., J.G. Moore, D.E. White, L. Platanias, N.S. Young, and S.A. Rosenberg, Hematologic effects of immunotherapy with lymphokine-activated killer cells and recombinant interleukin-2 in cancer patients. *Blood*, 69:1654, 1987.
126. Giulian, D., T.J. Baker, L.N. Shih, and L.B. Lachman, Interleukin 1 of the central nervous system is produced by ameboid microglia. *J. Exp. Med.*, 164:594, 1986.
127. Lotz, M., J.H. Vaughan, and A. Carson, Effect of neuropeptides on production of inflammatory cytokines by human monocytes. *Science*, 241:1218, 1988.
128. Lindholm, D., R. Heumann, M. Meyer, and H. Thoenen, Interleukin-1 regulates synthesis of nerve growth factor in non-neuronal cells of rat sciatic nerve. *Nature*, 330:658, 1987.
129. Bocci, V., Roles of interferon produced in physiological conditions. A speculative review. *Immunology*, 64:1, 1988.

130. Takacs, L., E.J. Kovacs, M.R. Smith. H.A. Young, and S.K. Durum, Detection of IL-1α and IL-1β gene expression by in situ hybridization. Tissue localization of IL-1 mRNA in the normal C57BL/6 mouse. *J. Immunol.*, 141:3081, 1988.
131. Tovey, M.G., J. Content, J. Gresser, J. Gugenheim, B. Blanchard, J. Guymarho, P. Poupart, M. Gigou, A.Shaw, and W. Fiers, Genes for IFN-β-2 (IL-6), tumor necrosis factor, and IL-1 are expressed at high levels in the organs of normal individuals. *J. Immunol.*, 141:3106, 1988.
132. Rosenberg, S.A., B.S. Packard, P.M. Aebersold, D. Solomon, S.L. Topalian, S.T. Toy, P. Simon, M.T. Lotze, J.C. Chang, C.A. Seipp, C. Simpson, C. Carter, S. Bock, D. Schwartzentruber, J.P. Wei, and D.E. White, Use of tumor-infiltrating lymphocytes and interleukin-2 in the immunotherapy of patients with metastatic melanoma. A preliminary report. *N. Engl. J. Med.*, 319:1676, 1988.
132a. Editorial, Interleukin-2: sunrise for immunotherapy? *Lancet*, 2:308, 1989.
133. Dunn, C.J., N.D. Staite, and M.M. Hardee, Slow release of interleukin-2 from polymer implants. *Ann. Intern. Med.*, 109:761, 1988.
134. Bubenik, J. and M. Indrova, Cancer immunotherapy using local interleukin 2 administration. *Immunol. Lett.*, 16:305, 1987.
135. Nagatani, T., S. Kim, N. Baba, H. Miyamoto, H. Nakajima, and Y. Katoh, A case of cutaneous T cell lymphoma treated with recombinant interleukin 2 (rIL-2). *Acta Derm. Venereol.*, 68:504, 1988.
135a. Morikawa, K., F. Okada, M. Hosokawa, and H. Kobayashi, Enhancement of therapeutic effects of recombinant interleukin 2 on a transplantable rat fibrosarcoma by the use of a sustained release vehicle, pluronic gel. *Cancer Res.*, 47:37, 1987.
136. Chang, Y.-H., Adjuvant polyarthritis II. Suppression by tilorone. *J. Pharmacol. Exp. Ther.*, 203:156, 1977.
137. Dunn, C.J., L. Galinet, A. Gibbons, and S. Shields, Murine delayed-type hypersensitivity granuloma: an improved model for the identification and evaluation of different classes of anti-arthritic drugs, *Int. J. Immunopharamacol.*, 12:899, 1990.
138. Giri, S.N. and D.M. Hyde, Ameliorating effect of an interferon inducer polyinosinic-polycytidylic acid on bleomycin-induced lung fibrosis in hamsters. *Am. J. Pathol.*, 133:525, 1988.
139. Jacob, C.O. and O. McDevitt, Tumour necrosis factor-α in murine autoimmune "Lupus" nephritis. *Nature*, 331:356, 1988.
140. Metcalf, D., In vivo effects of recombinant colony stimulating factors. ISI atlas of science. *Immunology*, 1:238, 1988.
141. Cohen, A.M., K.M. Zsebo, H. Inoue, D. Hines, T.C. Boone, V.R. Chazin, L. Tsai, T. Ritch, and L.M. Souza, In vivo stimulation of granuloporesis by recombinant human granulocyte colony stimulation factor. *Proc. Natl. Acad. Sci. U.S.A.*, 84:2484, 1987.
142. Ulich, T.R., J. DelCastillo, K. Guo, and L. Souza, The hematologic effects of chronic administration of the monokines tumor necrosis factor, interleukin-1α and granulocyte-colony stimulating factor on bone marrow and circulation. *Am. J. Pathol.*, 134:149, 1989.
143. Tribble, H., M. Schneider, O. Bowersox, and J.E. Talmadge, Combination immunotherapy with recombinant human TNF and recombinant murine IFN gamma: increased therapy and toxicity. *Fed. Proc.*, 46:561, 1987.
144. Nagy, J.A., L.F. Brown, D.R. Seuger, N. Lanir, L. VandeWater, A.M. Dvorak, and H.F. Dvorak, Pathogenesis of tumor stoma generation: a critical role for leaky blood vessels and fibrin deposition. *Biochim. Biophys. Acta*, 948:305, 1989.

145. Elliot, J.F., Y. Lin, S.B. Mizel, R.C. Bleachley, D.G. Harnish, and V. Paetkau, Induction of interleukin-2 messenger RNA inhibited by cyclosporin A. *Science*, 226:1439, 1984.
146. Granelli-Piperino, A., K. Inaba, and R.M. Steinman, Stimulation of lymphokine release from T lymphoblasts. Requirements for mRNA synthesis and inhibition by cyclosporin. *Am. J. Med.*, 160:1792, 1984.
147. Kronke, M., W.J. Leonard, J.M. Depper, S.K. Arya, F. Wong-Staal, R.C. Gallo, T.A. Waldman, and W.C. Green, Cyclosporin A inhibits T-cell growth factor gene expression at the level of mRNA transcription. *Proc. Natl. Acad. Sci. U.S.A.*, 81:5214, 1984.
148. Fischer, G., B. Wittmann-Liebold, K. Lang, T. Kiefhaber, and F.X. Schmid, Cyclophilin and peptidyl-prolyl cis-trans isomerase are probably identical proteins. *Nature*, 337:476, 1989.
149. Takahashi, N., T. Hayaho, and M. Suzuki, Peptidyl-prolyl cis-trans isomerase is the cyclosporin A-binding protein cyclophilin. *Nature*, 337:473, 1989.
150. Mosmann, T.R., H. Cherwinski, M.W. Bond, M.A. Gierdlin, and R.L. Coffman, Two types of murine helper T cells clones. I. Definition according to profiles of lymphokine activities and secreted proteins. *J. Immunol.*, 136:2348, 1986.
151. Cherwinski, H.M., J.H. Shumacher, K.D. Brown, and T.R. Mosmann, Two types of mouse helper T cell clones. III. Further differences in lymphokine synthesis between T_{H1} and T_{H2} clones revealed by RNA hybridization, functionally monospecific bioassays, and monoclonal antibodies. *J. Exp. Med.*, 166:1229, 1987.
152. Sanders, M.E., M.W. Makgoba, and S. Shaw, Human naive and memory T cells: reinterpretation of helper-inducer and suppressor-inducer subsets. *Immunol. Today*, 9:195, 1988.
153. Eastgate, J.A., N.C. Wood, F.S. DiGiovine, J.A. Symons, F.M. Grinlinton, and G.W. Duff, Correlation of plasma interleukin 1 levels with disease activity in rheumatoid arthritis. *Lancet*, Sept. 24:706, 1988.
154. Maury, C.P.J., E. Salo, and P. Pelkonen, Circulating interleukin-1β in patients with Kawasaki disease. *N. Eng. J. Med.*, 319:1670, 1988.
155. McKenna, R.M., J.A. Wilkins, and R.J. Warnington, Lymphokine production in rheumatoid arthritis and systemic lupus erythematosus. *J. Rheumatol.*, 15:1639, 1988.
156. Beck, J., P. Rondot, L. Catinot, E. Falcoff, H. Kirchner, and J. Wietzerbin, Increased production of interferon gamma and tumor necrosis factor precedes clinical manifestation in multiple sclerosis: do cytokines trigger off exacerbations? *Acta Neurol. Scand.*, 78:318, 1988.
157. Rodrick, M.C., J.J. Wood, J.B. O'Mahoney, C.F. Davis, J.T. Grbic, R.H. Demling, N.M. Moss, J. Saporoschotz, A.Jordan, P. D'Eon, and J.A. Mannick, Mechanisms of immunosuppression associated with severe nonthermal traumatic injuries in man: production of interleukin 1 and 2. *J. Clin. Immunol.*, 6:310, 1986.
158. Behan, P.O. and W.M.H. Behan, Postviral fatigue syndrome. *Crit. Rev. Neurobiol.*, 4:157, 1988.
159. Dawson, J., Brainstorming the postviral fatigue syndrome. *Br. Med. J.*, 297:1151, 1988.
160. Castelli, M.P., P.L. Black, M. Schneider, R. Pennington, I. Abe, and J.E. Talmadge, Protective, restorative, and therapeutic properties of recombinant human IL-1 rodent models. *J. Immunol.*, 140:3830, 1988.
161. Ozaki, Y., T. Ohashi, A. Minami, and S. Nakamura, Enhanced resistance of mice to bacterial infection induced by recombinant human interleukin-1α. *Infect. Immun.*, 55:1436, 1987.

162. Finkelman, F.D., I.M. Katona, J.F. Urban, Jr., J. Holmes, J. O'Hara, A.S. Tung, J.V.G. Sample, and W.E. Paul, IL-4 is required to generate and sustain in vivo IgE responses. *J. Immunol.*, 141:2335, 1988.
163. Jeffes, E.W.B., B.J. Averbook, T.R. Ulrich, and G.A. Granger, Human alpha lymphotoxin (LT): studies examining the mechanism(s) of LT-induced inflammation and tumor destruction in vivo. *Lymphokine Res.*, 6:141, 1987.
164. Ratafia, M. and T. Purinton, Immunomodulators for cancer. *Am. Clin. Lab.*, 24, 1989.
165. Sidky, Y.A. and E.C. Borden, Inhibition of angiogenesis by interferons: effects on tumor- and lymphocyte-induced vascular responses. *Cancer Res.*, 47:5155, 1987.
166. Veys, E.M., H. Mielants, G. Verbruggen, J. Grosclaude, W. Meyer, A. Galazka, and J. Schindler, Interferon gamma in rheumatoid arthritis — a double blind study comparing human recombinant interferon gamma with placebo. *J. Rheumatol.*, 15:570, 1988.
167. Lemmel, E.M., H.J. Obert, and P.H. Hofschneider, Low-dose gamma interferon in treatment of rheumatoid arthritis. *Lancet*, March:598, 1988.
168. Billiau, A., H. Heremans, F. Vandekerckhove, R. Dijkmans, H. Sobis, E. Meulepas, and H. Carton, Enhancement of experimental allergic encephalomyelitis in mice by antibodies against IFNγ. *J. Immunol.*, 140:1506, 1988.
169. Peres, A., T.A. Seemayer, and W.S. Lapp, The effects of polyinosinic: polycytidylic acid (pl:C) on the GVH reaction: immunopathological observations. *Clin. Immunol. Immunopathol.*, 39:102, 1986.
170. Hume, D.A., R.E. Donahue, and I.J. Fidler, The therapeutic effect of human recombinant macrophage colony stimulating factor (CSF-1) in experimental murine metastatic melanoma. *Lymphokine Res.*, 8:69, 1989.
171. Cueves, P., J. Burgos, and A. Baird, Basic fibroblast growth factor (FGF)) promotes cartilage repair in vivo. *Biochem. Biophys. Res. Commun.*,156:611, 1988.
172. Grau, G.E., V. Kindler, P.F. Piguet, P.H. Lambert, and P. Vassalli, Prevention of experimental cerebral malaria by anticytokine antibodies. *J. Exp. Med.*, 168:1499, 1988.

2. Naturally Occurring Inhibitors of Cytokines

ALAN SHAW

INTRODUCTION

The cytokines as a family share the common characteristic of having been discovered as positive activities. Before the adoption of the rather anonymous "interleukin-n" nomenclature, we studied lymphocyte activating factor (IL-1), T-cell growth factor (IL-2), multiple colony stimulating factor (IL-3), B-cell growth factor (IL- 4), B-cell differentiation factor (IL-5), hybridoma growth factor (IL- 6), and so on. As the original descriptive names imply, these are factors that do something *positive* in *in vitro* assays. Another common characteristic of the cytokines is their extreme potency: specific activities are in the range of 10^7 to 10^8 units per mg in most cases. The functional potency of these molecules is in many cases even greater than these units per mg figures would imply, since in their natural context the producing cells and the target cells are in close proximity. Therefore, cytokines are considered to be potent locally acting regulators of immune and inflammatory action.

In certain pathological situations, particularly autoimmune diseases, chronic inflammatory diseases, and some leukemias, the production of cytokines is disregulated. Overproduction of certain cytokines results in a spillover from the local context into the surrounding tissue fluid and into the general circulation. Over the past 3 years, some of the cytokines, in particular tumor necrosis factor (TNF), IL-1, and IL-6, have been shown to have systemic effects reminiscent of the activities of hormones. For example, IL-1 produced at a site of inflammation can affect the temperature control center of the CNS and can stimulate the pituitary-adrenal axis. Thus, if the cytokines are capable of hormone-like actions, perhaps they should also possess a system of checks and balances characteristic of endocrine systems. The need for a means of regulation of cytokine activity becomes more apparent when you consider the fact that cytokines induce the production of other cytokines. TNF induces IL-1 production. Both TNF and IL-1 induce production of granulocytes (G-), monocytes (M-), and granulocyte-monocyte colony-stimulating factor (GM-CSF), as well as IL-6. IL-1 induces production of IL-1, just to cite a few well-documented examples. The runaway production of cytokines might result in the physiological equivalent of the meltdown of a nuclear reactor without the appropriate system of controls.

Our knowledge of inhibitors of cytokines lags behind our knowledge of the cytokines themselves. The molecular cloning of the DNA sequence for, and the production of, a cytokine by recombinant microorganisms usually precedes the full characterization of a cytokine inhibitor. Recombinant cytokines, free of other contaminating factors, have allowed the elucidation of the function of these molecules. From this comes a knowledge of the mechanism of action of the cytokine and an appreciation of its role in a particular pathology. This knowledge then points out the potential value of an inhibitor. Thus, inhibitors of cytokine action have evolved from a nuisance and a source of assay artifacts to potentially useful pharmaceutical tools.

One can imagine a variety of different ways to block the activity of a cytokine. Any step along the path from induction of synthesis to final response of the target cell is a legitimate target. In the following discussion of cytokine inhibitors, it will be clear that a number of different strategies have been employed, and that, in the future, we hope to develop others. Most of this discussion will be focused on inhibitors of IL-1, IL-2, and TNFα, since this is where most of current effort is concentrated, but it will conclude with some hints of more general mechanisms of inhibition.

INHIBITORS OF INTERLEUKIN-1

The study of IL-1 inhibitors has been confounded by the pleiotropic activity of IL-1 itself. At least a dozen assays for IL-1 have appeared in the literature, but the most widely used assay has been the lymphocyte activating factor (LAF) assay. In the LAF assay, a suboptimal fixed concentration of a lectin, PHA, and a variable concentration of IL-1 activate murine lymphocytes to produce IL-2 (and IL-2 receptors). The IL-2 response is measured as incorporation of ^{3}H-thymidine into DNA in proliferating cells. Cell line-based analogues of the LAF assays have also been established.[1] The advantages of the LAF assay are sensitivity and realism; it is based on primary cells that would respond to IL-1 *in vivo*. When used in a study of IL-1 inhibitors, it must be used with the knowledge that the final readout, ^{3}H-thymidine incorporation, is the result of a cascade of events. Interference with PHA, with IL-2 production or action, or with general mechanisms of cellular proliferation will give the same result as interference with IL-1. The first inhibitors discussed in this section could have been viewed as IL-1 inhibitors; indeed, several of them were, but later studies showed that they acted at some other level.

GLIOBLASTOMA FACTOR

In 1985, Schwytzer and Fontana[2] characterized an inhibitor of IL-1- driven T-cell proliferation produced by a glioblastoma cell line from the collection of de Tribollet. This factor is a protein around 90 to 100 kDa. Further work on this molecule led to purification, to homogeneity, amino acid sequencing, and molecular cloning.[3,4] The resulting cDNA sequence showed a 71% homology with the sequence of transforming growth factor β (TGFβ). The glioblastoma factor was the second member of a blossoming family of TGFβs and is now commonly referred to as TGFβ2. TGFβ1 and 2 both suppress lymphocyte proliferation.[5,6,6a]

CONTRA IL-1

Scala et al. in Oppenheim's group described an inhibitor of IL-1 in the supernatants of an Epstein-Barr virus (EBV) -transformed B-cell line, ROHA-9.[7] This inhibitor, called "Contra IL-1", is a protein around 95 kDa that blocks lectin-stimulated thymocyte proliferation as well as antigen- stimulated T-cell proliferation. This factor is likely to be similar to the glioblastoma factor.[8]

ELLNER FACTOR

Ellner's group has reported the presence of an inhibitor of IL-1 in the supernatant of U937 cells.[9] U937 spontaneously produced a protein with a molecular weight around 85,000 that inhibits both IL-1 and IL-2 activity in the murine thymocyte proliferation assay. Further characterization of this factor showed that it interferes with IL-2 production and expression of IL-2 receptor.[9a]

CMV INDUCED IL-1 INHIBITOR

In 1985, Rodgers et al.[10] reported the induction of synthesis of an IL-1 inhibitor by macrophages or by the monocytic cell line U937 when these cells are infected by cytomegalovirus. CMV infection is known to cause a syndrome of immune suppression, and inhibition of IL-1 might be one aspect. The CMV IL-1 inhibitor had a molecular weight of approximately 95,000 and blocked the LAF assay. Further work showed that the inhibitory activity in this assay was due to mycoplasma contamination of the virus preparation.[11]

This is an appropriate point to bring up the problem of mycoplasma. Many cell lines are contaminated with mycoplasma, and this contamination has numerous consequences. For example, the presence or absence of mycoplasma affects the ability of the A431 cell line to transcribe the IL-1β gene.[12] Mycoplasma also has a much more insidious effect: it secretes a thymidine kinase of around 95,000 molecular weight and other nucleotide modifying enzymes.[13,14] Phosphorylation of ^{3}H-thymidine prevents its uptake by cells. This gives a false-negative result when you measure cell proliferation by the usual thymidine incorporation assay.

ROSENSTREICH'S IL-1 INHIBITOR

In 1984, Liao et al. in Rosenstreich's group described an inhibitor of IL-1 in the urine of patients with fever.[15] This factor has a molecular weight around 35,000 and inhibited the standard LAF assay. It appeared to be IL-1 specific in the sense that it did not block the IL-2 or IL-4-driven proliferation of T-cell lines.[16,16a] Paradoxically, this inhibitor stimulated, rather than blocked, the IL-1-driven proliferation of fibroblasts. It was also shown to be distinct from uromodulin, another urinary inhibitor of IL-1 (see below).[17] More recently, this protein has been purified to homogeneity DNase I.[18] Amino terminal sequence analysis reveals a striking homology to DNase 1. The purified protein has DNA-degrading activity, and commercially obtained preparations of DNase 1 had an antiproliferative effect.

α-MELANOCYTE STIMULATING HORMONE

α-MSH, a thirteen amino acid peptide, which is the same as the first thirteen residues of adrenocorticotropin hormone (ACTH), was shown several years ago to be an inhibitor of IL-1-induced fever.[19] Indeed, as early as 1949[20,21] ACTH itself had been shown to be an antipyretic. This area has been reviewed recently by Lipton.[19] Several different physiological activities of IL-1 are inhibited by α-MSH, including fever, neutrophilia, and the acute phase response when α-MSH is given centrally. This prompted investigations of the effects of α-MSH on various *in vitro* activities of IL-1. Cannon et al.[22] have shown that α-MSH blocks the thymocyte proliferation assay and the production of PGE_2 by fibroblasts treated with IL-1. Other investigators[23,24] have reported different results *in vitro*; thus, this area of IL-1 action requires more attention. Nevertheless, α-MSH appears to play a physiological role in the regulation of IL-1 activity.

UROMODULIN

In 1985, Muchmore and Decker reported the presence of a 85,000 molecular weight inhibitor of IL-1 in the urine of pregnant women[25] and called this factor uromodulin. Pregnant women are known to be immunosuppressed to a certain extent, and various attempts had been made to identify suppressive factors in their urine. Uromodulin, once sequenced and cloned,[26,27] proved to be the Tamm-Horsfall protein, a major urinary glycoprotein originally described in the early 1950s. The gene for uromodulin appears to be expressed exclusively in kidney tubule epithelial cells. The amino acid sequence translated from the cDNA specifies a polypeptide of around 62,000 molecular weight, is very rich in cysteine, and contains eight N-linked glycosylation sites. Uromodulin appears to exert its activity via carbohydrate side chains. Digestion of the protein with proteinase K followed by gel filtration yields an active, low-molecular-weight carbohydrate-rich active fraction. Similarly, removal of the carbohydrate side chains with *N*-glycanase results in partitioning of activity with the carbohydrate fraction.[28] A variety of hypotheses have been put forward for the physiological role of Tamm-Horsfall/uromodulin, but none are entirely satisfactory. Although uromodulin blocks the LAF assay and an antigen (tetanus toxoid) driven T-cell proliferation assay, both of which are IL-1-dependent, the activity of uromodulin in these systems is likely to be due to an interaction with the lectins in the assays.[29] Uromodulin was also shown to bind to IL-1 (as well as to TNFα and IL-2)[25,30,30a] in a solid phase assay. However, it was subsequently shown that uromodulin bound preferentially to denatured cytokines and to other denatured proteins.[31] Thus, the notion of a role for uromodulin as a trap for cytokines requires further investigation.

Points in favor of the renal cytokine trap hypothesis include the following. Frozen sections of kidney incubated in solutions of IL-1 reveal, by immunohistology using anti-IL-1 sera, IL-1 binding to tubule epithelial cells exclusively in the anatomical regions where uromodulin is detected with anti-uromodulin sera.[27] Radioiodinated IL-1 injected into the rat is found on whole body section autoradiography exclusively in the kidney (not in the bladder).[32] Taken together, these pieces of circumstantial evidence implicate uromodulin in some sort of interaction with IL-1 and probably other cytokines in the kidney. This seems to make a certain amount of sense. Circulating IL-1 has a number of deleterious effects.[33] Removal of IL-1 by filtration would be one efficient means of limiting the concentration of IL-1 in the blood.

CKS 17

A number of avian and murine retroviral infections are linked to immunosuppression of the host, and the immunosuppression is caused in part by an envelope protein p15E.[34] A number of p15E molecules as well as the gp21E envelope glycoproteins of human T-lymphotropic virus-I (HTLV-I) and HTLV-II contain a conserved 17 amino acid sequence. This sequence, as a synthetic peptide coupled to bovine serum albumin (BSA), has been shown to inhibit several aspects of IL-1 activity.[35-38] Recently, the peptide-BSA complex was shown to inhibit IL-1 action by blocking the transduction of the signal generated subsequent to IL-1 binding to its receptor. In particular, CKS 17-BSA inhibits protein kinase C, which is implicated in the post-receptor action of IL-1.[39]

SUBMANDIBULAR GLAND IL-1 INHIBITOR

Extracts of submandibular glands of rats have been shown to inhibit *in vitro* lymphocyte proliferation.[40] Partial purification of the inhibitory factor by gel filtration and chromatofocusing indicate a molecule with a molecular weight between 50 and 90,000 and a pI around 4.5. This molecule does not inhibit IL-2, but it does inhibit IL-1 and does so in a competitive manner.

NEUTROPHIL IL-1 INHIBITOR

Tiku et al.[41,41a] have shown that human polymorphonuclear leukocytes (PMN) make both IL-1 and a competitive inhibitor of IL-1-driven lymphocyte proliferation. The inhibitor is specific for IL-1 in that it does not block IL-2 stimulation of thymocyte proliferation. This factor is a protein and behaves as two separate molecular weight species on gel filtration: one around 70,000 and the other greater than 160,000.

INFLUENZA VIRUS AND RESPIRATORY SYNOVIAL VIRUS INDUCED IL-1 INHIBITOR

Macrophages infected with either influenza virus or with respiratory synovial virus (RSV) produce both IL-1 and an inhibitor of the IL-1-driven LAF assay. The inhibitor, which has a molecular weight around 99,000, does not inhibit IL-2 and is therefore not a general inhibitor of cell proliferation.[42]

LOW MOLECULAR WEIGHT IL-1 INHIBITOR

Adherent macrophages treated with LPS have been shown to produce both IL-1 and a low molecular weight IL-1 inhibitor.[43] The inhibitor has a molecular weight between 5,000 and 9,000 and a pI around 5. Immature T-cells (thymocytes) IL-1-driven proliferation is blocked by this factor; however, mature peripheral T-cells are not sensitive to it. The inhibitor does not block the standard CTLL-2 IL-2 assay. Inhibition of IL-1 receptor binding may be the mode of action, since the inhibitor must be added to cultures before or at the same time as IL-1. This group has subsequently shown that mononuclear cells from AIDS patients produce both more IL-1 and 20-times more of the low-molecular-weight inhibitor without stimulation.[44] This increase in the production of an IL-1 inhibitor may account for part of the immunosuppression seen in AIDS.

IL-1 BINDING PROTEINS

Preliminary reports have appeared that describe IL-1-binding proteins in serum, plasma, synovial fluid, or cultured cell supernatants.

α2 MACROGLOBULIN

In 1987 M. Teodorescu[45] reported that α2 macroglobulin, a large tetrameric serum protein, bound IL-1β in a reversible manner. More recent studies show two modes of binding: one easily reversible and another tighter, presumably disulfide-binding interaction.[46] G. Mazzei[47] has found that commercial fetal calf serum contains large amounts (+/- 5 ng/ml) of IL-1. This IL-1 is predominantly IL-1α, can be recovered by passage of serum over an anti-IL-1α affinity column followed by acid elution, and is then biologically active. The same unfractionated fetal calf serum on its own had no detectable IL-1 activity and must therefore be either accompanied by large amounts of a countervailing inhibitor or bound in an inactive form.

DUFF'S INHIBITOR

J. Symons and J. Eastgate[48] in G. Duff's laboratory, have isolated an IL-1-binding inhibitor from serum and synovial fluid of rheumatoid arthritis patients. The serum and synovial fluid inhibitors appear to be the same molecule. This inhibitor is a protein of around 45,000 molecular weight that binds preferentially to IL-1β and inhibits its activity. Current efforts to purify this molecule and to determine its amino acid sequence will resolve the question of whether this is a soluble form of the IL-1 receptor.

COZZOLINO'S INHIBITOR

M. Torcia and F. Cozzolino have detected an inhibitor of IL-1 that also acts by binding to IL-1 itself. This molecule appears to be distinct from the IL-1β-binding factor from Duff's laboratory in that it binds specifically to IL-1α. The source of this inhibitor is human T-cells from tumor-draining lymph nodes selected by panning for the OKM1 (helper suppressors) surface marker. This inhibitor can be removed and recovered from culture supernatants and shows a molecular weight of about 50,000. Again, sequencing the protein and isolating a corresponding cDNA will determine whether this is a soluble form of the IL-1 receptor.[49]

KERATINOCYTE INHIBITORS

Skin has been shown by a number of criteria to be a rich source of IL-1. Although the true quantities of IL-1 present are difficult to estimate, the work in the literature shows that skin contains considerable amounts of IL-1 and that it is in an inactive form. Western blotting studies of IL-1 from heel callus scrapings show a +/- 57,000 molecular weight molecule that reacts specifically with polyclonal sera raised against recombinant IL-1.[50]

Schwarz et al. in Luger's laboratory have isolated an inhibitor of IL-1 from UV-irradiated epidermal cells and from the PAM-212 keratinocyte line. This factor has a molecular weight around 40,000 and a pI of 8.8. It appears to be a competitive inhibitor, since the inhibition can be overcome by increasing amounts of IL-1.[57] It is tempting to speculate that the keratinocyte inhibitor might be the molecule complexed with IL-1 in skin extracts. The competitive inhibition and molecular weight are consistent with this hypothesis.

Keratinocytes have also yielded a second inhibitor of thymocyte activation. Walsh et al.[52] have found an inhibitor of the LAF assay in serum-free supernatants of gingival keratinocytes. This molecule has a molecular weight around 97,000 and also inhibits IL-2-stimulated thymocyte proliferation. The size and activity profile suggest that this is a member of the TGFβ2-like group of factors.

MACROPHAGE-DERIVED IL-1 INHIBITORS

Macrophages, the major source of IL-1, have been shown to produce IL-1 inhibitors, by several groups of investigators. If one accepts the need of a mechanism for damping the activity of IL-1, what better cell source than the source of IL-1 itself? One autodamping feedback system has been well characterized. Macrophages stimulated to produce IL-1 also produce prostaglandin E_2 (PGE_2). PGE_2 in turn acts on the macrophage to turn down the expression of IL-1. Interference in this negative feedback loop by blockers of cyclooxygenase, such as aspirin, indomethacin, or ibuprofen, results in a prolonged period of IL-1 production.[53] IL-1 has also been shown to turn off production of IL-1 as well as turn it on.[54,55] Therefore, if these two control mechanisms are orchestrated by the macrophage, there might be others as well.

Several apparently different molecules with IL-1 inhibitory activity have been described in macrophage supernatants or in the urine of patients with macrophage-related disorders.

P388D1 FACTOR

Two different IL-1 inhibitors have been found in supernatants of the murine monocytic cell line P388D1. Nishihara et al.[56] have described a 160,000 molecular weight factor that inhibits IL-1-driven thymocyte proliferation, but does not block IL-2 activity. The production of this factor is induced by treatment of the cells with lipopolysaccharide (LPS) over the course of 72 hr. Isono and Kumagai[57] find IL-1 inhibitory activity at a molecular weight between 40,00 and 60,00 and at two pIs (5.3 and 6.0) in the supernatants of unstimulated P388D1 cells. This inhibitor also appears to be specific for IL-1, as it does not inhibit lymphocyte proliferation stimulated by IL-2, IL-3, or IL-4.

M20

Barak and Treves[58] find a 52,000 molecular weight IL-1 inhibitor in supernatants of the human myelomonocytic cell line M20. It inhibits IL-1-driven thymocyte proliferation, but does not inhibit IL-2 activity. It also does not inhibit growth of a number of other cell types, including human fibroblasts and monoblastic leukemia cells. Recently, this molecule has been shown to bind to the IL-1 receptor.[59]

DAYER'S INHIBITOR

The most thoroughly characterized IL-1 inhibitor in the literature is the urinary IL-1 inhibitor first reported by Balavoine et al. from Dayer's group.[60] A patient with acute monoblastic leukemia produced an activity in his urine that blocked the LAF assay as well as the "mononuclear cell factor" (MCF) assay. The MCF assay measures PGE_2 and collagenase production by primary cultures of human fibroblasts (foreskin or synovial) stimulated by IL-1. This activity was subsequently shown to be a protein about 25,000 molecular weight with a pI of approximately 4.5.[61] Inhibition is strictly limited to IL-1, both α and β forms. It does not inhibit IL-2. More importantly, it does not inhibit $TNF\alpha$ in assays where IL-1 and $TNF\alpha$ have similar activities. Seckinger et al.[62] showed that partially purified urinary IL-1 inhibitor blocked binding of radiolabeled IL-1α to the murine EL4.6.1 thymoma line. Preincubation of the EL4.6.1 cells with inhibitor gave a greater degree of inhibition than did preincubation with the IL-1, thus suggesting that the inhibitor acted at the level of the receptor.

Recently, we have purified the urinary IL-1 inhibitor to homogeneity.[47a] The purified inhibitor is a single polypeptide with a molecular weight of 26,000 on SDS-PAGE. Treatment of the protein with endoglycosidase F to remove N-linked carbohydrate reduced the molecular weight to 24,000; thus, the inhibitor is probably a glycoprotein with one carbohydrate moiety. Purified inhibitor radioiodinated by the Bolton-Hunter method binds to EL4.6.1 cells, and the bound material remains a 26,000 molecular weight protein. Deglycosylation does not affect binding to EL4.6.1 cells. Binding of both native and deglycosylated radiolabeled inhibitor is competed for by both IL-1α and IL-1β. Conversely, unlabeled inhibitor prevents binding of 17,000 molecular weight iodinated IL-1. This strongly suggests that IL-1 and the urinary IL-1 inhibitor compete for a single site on the surface of EL4.6.1 cells. Chemical cross-linking of radioiodinated IL-1α, IL-1β, and purified urinary inhibitor, each independently, to the surface of EL4.6.1 cells yields in each case a complex of a size consistent with each ligand binding to a 80,000 molecular weight species, the size of the major component of the IL-1 receptor.[47]

Since urine is essentially a selected filtrate of blood, proteins found in the urine could in principle come from anywhere in the body. Several clues pointed to the macrophage as the source of the IL-1 inhibitor found in febrile urine. These included other reports of IL-1 inhibitors from macrophage-like cells (see above and below), the disease, AML-M5, of the patient in whose urine the inhibitor was originally found, and the appearance of a similar activity in arthritic patients.[62a,b] We have examined a number of monocytic cell lines for their ability to make the IL-1 inhibitor and found that several, including HL-60, AML-193, and H161.29, can produce it. However, untreated cells produce little if any inhibitor activity. Only when cells are induced to differentiate with phorbol ester and stimulated with human GM-CSF will they produce measurable levels of inhibitor. The inhibitor produced by these cells has the same biological activities and biophysical characteristics as the protein purified from urine. Production of a competitive inhibitor of IL-1 is apparently not a function of transformed cells. Dayer's group[63] and our group[64] have found that long-term cultures of adherent macrophages also produce this activity.

AREND'S INHIBITOR

In 1985 Arend's group[65] reported the production of both IL-1 and an inhibitor of IL-1 by macrophages that had been stimulated with immune complex. Both the LAF assay and the MCF assay are inhibited by this molecule. The molecular weight determined by gel filtration was about 22,000. More recently a scaled up method of production was described along with details of the kinetics of production. Macrophages were culture on IgG-coated plastic dishes. Inhibitor accumulated over a period of 72 hr to a value of about 20 units per ml.[66] Finally, Janson and Arend report that addition of GM-CSF to macrophage cultures enhances production of inhibitor.[67]

Large-scale production of the IL-1 inhibitor has been achieved using macrophages cultures on IgG-coated surfaces, and the inhibitor has been purified from the culture supernatant. The inhibitor appears at two different molecular weight species, a 22.5 kDa glycosylated molecule and a 17 kDa nonglycosylated protein. Peptide sequence information was used to design oligonucleotide probes for the cloning of the cDNA for the IL-1 inhibitor. The genomic sequence has also been cloned, and it shows organizational homology with the IL-1β gene, but it lacks the sequence encoding the long "presequence" found on IL-1α and IL-1β and has instead a classical signal sequence for secretion.[67-69] Although there exists a difference in molecular weight (26 to 24 kDa for Dayer's molecule versus 22.5 to 17 kDa for Arend's molecule) as judged by SDS gel electrophoresis, taken together the available information suggests that Arend's and Dayer's inhibitors are closely related molecules.

INHIBITORS OF INTERLEUKIN-2

Interleukin-2, or T-cell growth factor, has been available as a recombinant protein for a long time. The cDNA for IL-2 was cloned in 1982,[70] and the cDNA for one of the components of the IL-2 receptor, the Tac antigen, was cloned in 1984.[71] Monoclonal antibodies against IL-2 and the Tac antigen were first reported in 1981.[72] Thus, the basic tools for investigation of the range and mechanism of IL-2 action have been available for quite a while. A wealth of information has been generated through *in vitro* experimentation and through numerous clinical trials with

IL-2. One outcome of this work has been the definition of several clinical conditions where there exists a deficit in the function of IL-2. This in turn led to the investigation of the nature of the problem and to the discovery of factors that inhibit the action of IL-2.

The IL-2 inhibitors reported in the literature to date can be divided into two groups: those that inhibit IL-2 production or IL-2 responsiveness via an interference at the level of intracellular signal transduction, and those that act by interfering with IL-2 action outside the cell.

INHIBITORS OF IL-2 PRODUCTION OR RESPONSIVENESS

T-cells, when activated, produce and use IL-2 as a growth factor. Therefore, an immediate means of breaking this cycle would be to limit the production of IL-2 and/ or to reduce the cell's capacity to respond to IL-2. As is the case for IL-1, PGE_2 also serves as a negative regulator of IL-2 production.[73] In addition to this well-characterized mechanism, the fatty acid precursors of PGE_2, dihomogamma linoleic acid and arachidonic acid, are able to inhibit IL-2 production via a pathway independent of cyclooxygenase and PGE_2.[74] Glucocorticoids, one of the most widely used agents for the treatment of autoimmune and inflammatory disorders, have been shown by Redondo et al.[75] to block IL-2 activity. Since inhibitors of protein synthesis prevent this blockade, the effect is thought to be mediated by a glucocorticoid-induced protein.

Although the combination of IL-2 plus IL-4 is now used for the maintenance of T-cell clones, IL-4 itself is also an inhibitor of IL-2 in some contexts. Nagler et al.[76] and Spits et al.[77] have shown that IL-4 inhibits the induction of IL-2-activated killer cells. The mechanism for this inhibition is at present unknown, but IL-4 appears to act early along the pathway to killer cell activation. Llorente et al.[78] have reported that IL-4 blocks the helper effect of IL-2 on antibody production by antigen-activated B-cells. Again, the mechanism is unknown. These reports do, however, emphasize the complexity of cytokine networks.

A number of the above-mentioned investigations of clinically observed IL-2 deficits have turned up inhibitors of IL-2 production. Burton et al.,[79] studying T-cell and NK-cell disfunction in chronic malignant B-cell leukemias, found that malignant B-cells shed or secreted a low-molecular-weight factor (< or = 5000 molecular weight), which blocked phytohemagglutinin (PHA) -stimulated T-cell proliferation and IL-2 production. This factor is inactivated by treatment with neuramidase and is thus proposed to be either a ganglioside or a sialated glucopeptide.

Kobayashi et al.[80] have found that mice with pulmonary hypersensitivity granulomas have in their sera a suppressor of T-cell function. This factor appears to be a protein between 12 to 67,000 molecular weight and acts by suppressing IL-2 production by immune lymph node cells. Curiously, the granulomas themselves do not contain the suppressive factor.

As mentioned above in the section on IL-1 inhibitors, a conserved 17 amino acid peptide, CKS-17, from a highly conserved region of the p15E envelope protein of immunosuppressive retroviruses, inhibits certain aspects of IL-1 action. Gottlieb et al.[39] have shown recently that CKS-17, as a BSA conjugate, in fact blocks IL-2 production by murine thymoma cells treated with either IL-1 or with phorbolester. The blockade would appear to be at the level of protein kinase C, since soluble PKC from the thymoma cells was inhibited by CKS-17-BSA.

SERUM AND SYNOVIAL INHIBITORS OF IL-2

Deficiencies in IL-2 action have been described in the synovial lymphoid nodules in rheumatoid arthritis and in solid tumors. Inhibitors of IL-2 action have been found in synovial fluid and in serum of these patients. This area has been reviewed recently by Kuchary and Goodwin.[81] Two clinical studies that have appeared in the interim reinforce the pathophysiological significance of soluble IL-2 receptors.

In a study of a large number of rheumatoid arthritis patients, Symons et al.,[82] in Duff's laboratory, found elevated levels of soluble IL-2 receptor in both serum and synovial fluid. Mononuclear cells purified from synovial fluid spontaneously released large amounts of soluble IL-2 receptor, whereas peripheral blood mononuclear cells required stimulation with a mitogen to attain comparable levels. Serum and mononuclear cell culture supernatant soluble IL-2 receptor had a molecular weight of approximately 40,000, while synovial fluid sIL-2r was found at 100,000 molecular weight. The 40,000 molecular weight component was the Tac small component of the IL-2 receptor. The 100,000 molecular weight component contained the Tac antigen presumably complexed with the larger second chain of the IL-2 receptor. In a serial study of rheumatoid arthritis patients, serum sIL-2r levels correlated strongly with severity of disease. Somewhat surprisingly, serum sIL-2r levels dropped several days before a clinically measurable improvement. The sIL-2r levels rose again several days before exacerbation of disease. Synovial fluid levels of sIL-2r correlated strongly with levels of IL-1, suggesting a coparticipation of one cytokine and an inhibitor of another cytokine in the pathophysiology of the disease. This leads to the counterintuitive conclusion that IL-2 might be of benefit in rheumatoid arthritis.

Immunosuppression is a frequent complication of major burns. In a study of IL-2 receptors in burn patients Teodorczyk-Injeyan et al.[83] found that the number of cells expressing the IL-2 receptor was reduced. The reduction was moderate and transient in patients who recovered from their burns. In patients who died from their burns, the reduction in cellular IL-2r expression was more severe and was permanent. Serum levels of sIL-2r were elevated in all cases, and when patients became immunosuppressed, the levels of serum sIL-2r rose further. Patients who survived their burns had a less severe elevation of sIL-2r than did the nonsurvivors. The sIL-2r measured by ELISA was able to inhibit exogenous IL-2 *in vitro*. This suggests that sIL-2r participates in the immunosuppression of burn patients.

These and other studies of this type raise the question of the origin, mechanism of production, and molecular nature of the soluble IL-2 receptor. The biological fluids analyzed contained a significant capacity to inhibit IL-2. Much of the literature now available is based on ELISA measurements of the levels of the Tac antigen, the 40,000 molecular weight "low affinity" chain of the IL-2 receptor. This protein, by itself, does bind to and inhibit the activity of IL-2. However, the quantity of soluble Tac required to account for the inhibition observed is quite large. The recent cloning of the DNA sequence for the second chain of the IL-2 receptor should provide the reagents necessary to answer the question of whether Tac alone is a physiological inhibitor of IL-2. In either case, Tac alone or in combination with something else, the source and mechanism of production remain open questions. Is soluble receptor simply a byproduct of lymphocyte death? Is the receptor shaved off of the surface of cells by a specific protease?[84] Is there a system for secretion of soluble receptor: perhaps an mRNA lacking a cytoplasmic tail?[85]

NONRECEPTOR INHIBITORS OF IL-2

In addition to the inhibitors of IL-2 that are clearly some form of soluble IL-2 receptor, there have been reports of other inhibitors that appear to act by different mechanisms. Most of these factors are still at an early stage of classification.

Smith et al.[86,86a] have found an IL-2 inhibitor in normal sera and in sera and synovial fluid of patients with inflammatory joint diseases. This factor would appear to act by reducing IL-2 production by peripheral mononuclear cells. It appears not to bind to either IL-2 or the IL-2 receptor. Inhibition can be overcome by increasing amounts of IL-2.

Emery et al.,[87,87a,b] reexamining the question of the presence of IL-2 activity in rheumatoid synovial fluid, found no IL-2 activity in the standard CTLL-2 assay. Most rheumatoid arthritis (RA) patients, as well as patients with other articular diseases, had an IL-2 inhibitor in their synovial fluid. Curiously, the level of inhibition of RA synovial fluid was overall lower than in patients with other articular diseases. The preliminary characterization, heat lability and ammonium sulfate precipitability, is consistent with this activity being a protein. With the caveat that patient selection can greatly affect results in this kind of study, one would surmise from the relative levels of inhibitor between patient groups here (RA lower than "other articular diseases") and the sIL- 2R levels found by Symons et al.[82] (RA high, osterarthritis low to none) that the inhibitor found by Emery et al. might not be a SIL-2R.

Over the past three years, M. Ziff's group has been pursuing an inhibitor of IL-2 produced by macrophages and by monocytic cell lines. This IL-2 inhibitor is similar to, or the same as, the one originally described by Krakauer.[88] Krakauer's IL-2 inhibitor was originally found in the supernatant of the monocytic cell line THP-1 stimulated with silica. It inhibits both IL-2 response and production and has a molecular weight around 60,000 to 70,000, as determined by gel filtration. Miossec et al.[89] and Kashiwado et al.[90,91] have found this inhibitor in the synovial fluid of rheumatoid arthritis patients and in the supernatants of LPS-stimulated macrophages and human umbilical endothelial cells. This molecule has an apparent molecular weight around 130,000 to 150,000 on gel filtration and an apparent pI at three values from 4.7 to 7.0. Despite the apparent difference in molecular weight between the THP-1-derived molecule of Krakauer and the synovial fluid-, endothelial cell-, and macrophage-derived factors from Ziff's group, all of these IL-2 inhibitors are neutralized by polyclonal sera raised against the THP-1-derived molecule. Thus, these are likely to be a series of similar, if not identical, molecules. Arend and Dayer have prepared a review on this topic.[92]

TNF INHIBITORS

Tumor necrosis factor α shares many activities with IL-1, including the ability to induce the production of other cytokines.[93] Regulation of TNFα activity by negative feedback is, as in the case of IL-1, probably an important means of controlling systemic levels of cytokine activity. To date there would appear to be one inhibitor of TNF. The first report of this activity by Seckinger et al.[94] in Dayer's group described a 40,000 to 60,000 molecular weight protein with a pI around 5.8 from the urine of patients with fever. This inhibitor was specific for TNF in that it blocked binding of radiolabeled TNFα to L929 target cells and hence blocked TNFα-mediated cytotoxicity. It did not block binding of radiolabeled IL-1 to EL4.6.1 target cells. This paper was followed by a report from Peetre et al.[95] in Olsson's group describing a TNFα binding protein in serum and urine from patients undergoing regular hemodialysis. Preliminary characterization indicated a molecular weight around 50,000. This inhibitor also blocked TNFα-mediated growth inhibition of a tumor cell line, H1-60-10 cells in this case.

This protein has been purified to homogeneity either by standard biochemical techniques or by affinity chromatography on immobilized TNFα.[96,98] N terminal amino acid sequencing yields the same sequence for all three preparations (DSVCPQGKYIHPQCNSI); thus, all three groups are working on the same protein. Since this inhibitor of TNFα acts by binding to TNF itself, it is tempting to speculate that this may be a soluble form of a component of the TNF receptor.

Indeed, this has been shown by Engelmann et al.[99] to be the case. Surprisingly, there appear to be two different TNF receptors, both of which are present in urine as soluble, truncated molecules. Both sequences have been cloned, and preliminary results suggest the two genes are differentially expressed on different cell types.

ANTIBODIES AGAINST CYTOKINES

Another means of inhibiting cytokine action *in vitro* is through the use of specific neutralizing antibodies. Indeed, such antibodies have been crucial in the dissection of cytokine pathways. The importance of anticytokine antibodies extends beyond the microtiter plate. Svenson et al.[100] and Fomsgaard, Svenson, and Bendtzen[101] have found neutralizing antibodies to IL-1α and TNFα in the sera of both normal and diseased individuals. In a longitudinal study of patients with septic shock, anti-TNFα antibodies appeared early in the course of the disease, peaked during the first week, and returned to low levels between d 9 to 20. This time course of antibody presence implies that anti-TNF antibodies may play a role in the regulation of TNF activity. Although these studies of anti-IL-1 and anti-TNF are the most advanced, one is tempted to predict that other anticytokine antibodies with pathophysiological function will come to light as the tools for their discovery become available.

SOLUBLE RECEPTORS FOR CYTOKINES

In the discussion of inhibitors of IL-2, soluble forms of the IL-2 receptor hold the place of honor. Recent developments indicate the soluble receptors for other cytokines may also exist and may play roles in the regulation of cytokine action. For example, the TNF inhibitor described by Seckinger, Olsson, and Wallach is a soluble TNF receptor. The IL-1 inhibitors of Duff's group[102] and of Cozzolino and Torcia[103] could be different soluble forms of the IL-1 receptor. The information to confirm or refute this assertion will come from molecular biology. One intriguing finding[104] of a soluble IL-1 receptor from murine fibroblasts would imply that there is more than one IL-1 receptor coding sequence. Two further examples of soluble cytokine inhibitors have been reported by Novick et al.[105] Urine from normal donors, when passed over columns of immobilized IL-6 and γ-interferon, yields soluble forms of the IL-6 receptor and the γ-IFN receptor. In addition, there appears to be a soluble form of the murine receptor for IL-4. Beckman et al.[106] found a cDNA for a soluble form of the IL-4 receptor from a subclone of the classic CTLL-2 thymoma line. The soluble molecule blocks the action of IL-4 on IgE induction. Thus, it would appear that soluble receptors for cytokines in bodily fluids are a general phenomenon. The elucidation of the source, mechanism of solubilization, and the appearance or disappearance of soluble receptors in disease should indicate new directions for the use of cytokines and their inhibitors as pharmacological agents.

SUMMARY

Natural inhibitors of cytokines come in many forms and flavors. So far we have seen soluble receptors, antibodies, cytokine-binding proteins, competitors for receptor binding, and agents that reduce production and/or response to cytokines. Other mechanisms may be lurking over the horizon. At the moment, the picture of the network of cytokine inhibitors is incomplete: even more incomplete than the pictures (four-color, centerfold, or otherwise) of cytokine networks themselves. What is becoming obvious, however, is the notion of a finely tuned system of cells communicating with each other via an expanding family of (mostly) soluble cytokines. This communication is modulated by naturally occurring inhibitors. Understanding the mechanisms that nature has developed to control cytokine action will permit a more rational approach to therapy through use of cytokine inhibitors, both natural and synthetic, and through more enlightened use of the cytokines themselves.

ACKNOWLEDGMENTS

I would like to thank Magali Leemann for her help, patience, and good humor during the production of this manuscript. I would also like to thank Penny Shaw for her help with on-line literature searches. Finally, I must thank all of my friends and colleagues for access to their unpublished data.

REFERENCES

1. Gearing, A.J., C.R. Bird, A. Bristow, S. Poole, and R. Thorpe, A simple sensitive bioassay for interleukin-1 which is unresponsive to 10(3) U/ml of interleukin-2. *J. Immunol. Methods,* 99(1), 7–11, 1987.
2. Schwyzer, M. and A. Fontana, Partial purification and biochemical characterization of a T cell suppressor factor produced by human glioblastoma cells. *J. Immunol.,* 134(2), 1003–1009, 1985.
3. Siepl, C., S. Bodmer, K. Frei, H.R. MacDonald, R. De Martin, E. Hofer, and A. Fontana, The glioblastoma-derived T cell suppressor factor/transforming growth factor-beta 2 inhibits T cell growth without affecting the interaction of interleukin 2 with its receptor. *Eur. J. Immunol.,* 18(4), 593–600, 1988.
4. De Martin, R., B. Haendler, R. Hofer-Warbinek, H. Gaugitsch, M. Wrann, H. Schlusener, J.M. Seifert, S. Bodmer, A. Fontana, and E. Hofer, Complementary DNA for human glioblastoma-derived T cell suppressor factor, a novel member of the transforming growth factor-beta gene family. *EMBO J.* 6(12), 3673–3677, 1987.
5. Wahl, S.M., D.A. Hunt, H.L. Wong, S. Dougherty, N. McCartney-Francis, L.M. Wahl, L. Ellingsworth, J.A. Schmidt, G. Hall, A.B. Roberts, et al. Transforming growth factor-beta is a potent immunosuppressive agent that inhibits IL-1-dependent lymphocyte proliferation. *J. Immunol.,* 140(9), 3026–3032, 1988.
6. Ellingsworth, L.R., D. Nakayama, P. Segarini, J. Dasch, P. Carrillo, and W. Waegell, Transforming growth factors-beta are equipotent growth inhibitors of interleukin-1-induced thymocyte proliferation. *Cell. Immunol.,* 114(1), 41–54, 1988.
6a. Kuppner, M.C., M.F. Hamou, S. Bodmer, A. Fontana, and N. de Tribolet, The glioblastoma-derived T-cell suppressor factor/transforming growth factor beta 2 inhibits the generation of lymphokine-activated killer (LAK) cells. *Int. J. Cancer,* 42(4), 562–567, 1988.
7. Scala, G., Y.D. Kuang, R.E. Hall, A.V. Muchmore, and J.J. Oppenheim, Accessory cell function of human B cells. I. Production of both interleukin 1-like activity and an interleukin 1 inhibitory factor by an EBV-transformed human B cell line. *J. Exp. Med.,* 159(6), 1637–1652, 1984.
8. Oppenheim, J., personal communication.

9. Fujiwara, H. and J.J. Ellner, Spontaneous production of a suppressor factor by the human macrophage-like cell line U937. I. Suppression of interleukin 1, interleukin 2, and mitogen-induced blastogenesis in mouse thymocytes. *J. Immunol.*, 136(1), 181–185, 1986.

9a. Fujiwara, H., Z. Toossi, K. Ohnishi, K. Edmonds, and J.J. Ellner, Spontaneous production of a suppressor factor by a human macrophage-like cell line U937. II. Suppression of antigen- and mitogen-induced blastogenesis, IL 2 production and IL 2 receptor expression in T lymphocytes. *J. Immunol.*, 138(1), 197–203, 1987.

10. Rodgers, B.C., D.M. Scott, and J. Mundin, Sissons, J.G. Monocyte-derived inhibitor of interleukin 1 induced by human cytomegalovirus. *J. Virol.*, 55(3), 527–532, 1985.

11. Scott, D., personal communication.

12. Demczuk, S., C. Baumberger, B. Mach, and J.M. Dayer, Differential effects of *in vitro* mycoplasma infection on interleukin-1 alpha and beta mRNA expression in U937 and A431 cells. *J. Biol. Chem.*, 263(26), 13039–13045, 1988.

13. Neale, G.A., A. Mitchell, and L.R. Finch, Enzymes of pyrimidine deoxyribonucleotide metabolism in *Mycoplasma mycoides* subsp. *mycoides*. *J. Bacteriol.*, 156(3), 1001–1005, 1983.

14. Charron, J. and Y. Langelier, Analysis of deoxycytidine (cD) deaminase activity in herpes simplex virus-infected or HSV TK-transformed cells: association with mycoplasma contamination, but not with virus infection. *J. Gen. Virol.*, 57, 245–250, 1981.

15. Liao, Z., R.S. Grimshaw, and D.L. Rosenstreich, Identification of a specific interleukin 1 inhibitor in the urine of febrile patients. *J. Exp. Med.*, 159(1), 126–136, 1984.

16. Liao, Z., A. Haimovitz, Y. Chen, J. Chan, and D.L. Rosenstreich, Characterization of a human interleukin 1 inhibitor. *J. Immunol.*, 134(6), 3882–3886, 1985.

16a. Rosenstreich, D.L., A. Haimovitz, K.M. Brown, and Z. Liao. *Lymphokines*, 14, 63–89, 1987.

17. Brown, K.M., A.V. Muchmore, and D.L. Rosenstreich, Uromodulin, an immunosuppressive protein derived from pregnancy urine, is an inhibitor of interleukin 1. *Proc. Natl. Acad. Sci. U.S.A.*, 83(23), 9119–9123, 1986.

18. Rosenstreich, D.L., J.H. Tu, P.R. Kinkade, I. Maurer-Fogy, J. Kahn, R.W. Barton, and P.R. Farina, A human urine-derived interleukin 1 inhibitor. Homology with deoxyribonuclease I. *J. Exp. Med.*, 168(5), 1767–7179, 1988.

19. Lipton, J.M. Neuropeptide α-melanocyte-stimulating hormone in control of fever, the acute phase response, and inflammation. *Neuroimmune Networks: Physiol. Dis.*, 243–250, 1989.

20. Hench, P.S., E.C. Kendall, C.H. Slocumb, and H.F. Polley, The effect of a hormone of the adrenal cortex (17-hydroxy-11-dehydrocorticosterone:compound E) and of pituitary adrenocorticotropic hormone on rheumatoid arthritis, *Proc. Staff Meet. Mayo Clinic,* 24:181–197, 1949.

21. Kass, E.H. and M. Finland, Effects of ACTH on induced fever, *N. Engl. J. Med.*, 243:693–695, 1950.

22. Cannon, J.G., J.B. Tatro, S. Reichlin, and C.A. Dinarello, Alpha melanocyte stimulating hormone inhibits immunostimulatory and inflammatory actions of interleukin 1. *J. Immunol.*, 137(7), 2232–2236, 1986.

23. Robertson, B.A., L.C. Gahring, and R.A. Daynes, Neuropeptide regulation of interleukin-1 activities. Capacity of alpha-melanocyte stimulating hormone to inhibit interleukin-1-inducible responses *in vivo* and *in vitro* exhibits target cell selectivity. *Inflammation*, 10(4), 371–385, 1986.

24. Robertson, B., K. Dostal, and R.A. Daynes, Neuropeptide regulation of inflammatory and immunologic responses. The capacity of alpha-melanocyte- stimulating hormone to inhibit tumor necrosis factor and IL-1-inducible biologic responses. *J. Immunol.*, 140(12), 4300–4307, 1988.
25. Muchmore, A.V. and J.M. Decker, Uromodulin: a unique 85-kilodalton immunosuppressive glycoprotein isolated from urine of pregnant women. *Science*, 229, 479–481, 1985.
26. Pennica, D., W.J. Kohr, W.J. Kuang, D. Glaister, B.B. Aggarwal, E.Y. Chen, and D.V. Goeddel, Identification of human uromodulin as the Tamm- Horsfall urinary glycoprotein. *Science*, 236, 83–88, 1987.
27. Hession, C., J.M. Decker, A.P. Sherblom, S. Kumar, C.C. Yue, R.J. Mattaliano, R. Tizard, E. Kawashima, U. Schmeissner, S. Heletky, et al., Uromodulin (Tamm-Horsfall glycoprotein): a renal ligand for lymphokines. *Science*, 237, 1479, 1987.
28. Sherblom, A.P., J.M. Decker, and A.V. Muchmore, The lectin-like interaction between recombinant tumor necrosis factor and uromodulin. *J. Biol. Chem.*, 263(11), 5418–5424, 1988.
29. Moonen, P. and K. Williamson, Bioassay for interleukin-1 inhibitors. *J. Immunol. Methods*, 102(2), 283–284, 1987.
30. Muchmore, A.V. and J.M. Decker, Uromodulin. An immunosuppressive 85-kilodalton glycoprotein isolated from human pregnancy urine is a high affinity ligand for recombinant interleukin 1 alpha. *J. Biol. Chem.*, 261(29), 13404–13407, 1986.
30a. Sherblom, A.P, N. Sathyamoorthy, J.M. Decker, and A.V. Muchmore, IL-2, a lectin with specificity for high mannose glycopeptides. *J. Immunol.*, 143(3), 939–944, 1989.
31. Moonen, P., R. Gaffner, and P. Wingfield, Native cytokines do not bind to uromodulin (Tamm-Horsfall glycoprotein). *FEBS Lett.*, 226(2), 314–318, 1988.
32. Poole, S. personal communication.
33. Dinarello, C.A. Interleukin-1 and its biologically related cytokines. *Adv. Immunol.*, 44, 153–205, 1989.
34. Copelan, E.A., J.J. Rinehart, M. Lewis, L. Mathes, R. Olsen, and A. Sagone, The mechanism of retrovirus suppression of human T cell proliferation *in vitro*, *J. Immunol.*, 131(4), 2017–20, 1983.
35. Kleinerman, E.S, L.B. Lachman, R.D. Knowles, R. Snyderman, and G.J. Cianciolo, A synthetic peptide homologous to the envelope proteins of retroviruses inhibits monocyte-mediated killing by inactivating interleukin 1. *J. Immunol.*, 139(7), 2329–2337, 1987.
36. Harris, D.T., G.J. Cianciolo, R. Snyderman, S. Argov, and H.S. Koren, Inhibition of human natural killer cell activity by a synthetic peptide homologous to a conserved region in the retroviral protein, p15E. *J. Immunol.*, 138(3), 889–894, 1987.
37. Wang, J.M., Z.G. Chen, G.J. Cianciolo, R. Snyderman, F. Breviario, E. Dejana, and A. Mantovani, Production of a retroviral P15E-related chemotaxis inhibitor by IL-1-treated endothelial cells. A possible negative feedback in the regulation of the vascular response to monokines. *J. Immunol.*, 142(6), 2012–2017, 1989.
38. Cianciolo, G.J., T.D. Copeland, S. Oroszlan, and R. Snyderman, Inhibition of lymphocyte proliferation by a synthetic peptide homologous to retroviral envelope proteins. *Science*, 230, 453–455, 1985.
39. Gottlieb, R.A., W.J. Lennarz, R.D. Knowles, G.J. Cianciolo, C.A. Dinarello, L.B. Lachman, and E.S. Kleinerman, E.S. Synthetic peptide corresponding to a conserved domain of the retroviral protein p15E blocks IL-1 mediated signal transduction. *J. Immunol.*, 142(12), 4321–4328, 1989.

40. Kemp, A., L. Mellow, and E. Sabbadini, Inhibition of interleukin 1 activity by a factor in submandibular glands of rats. *J. Immunol.*, 137(7), 2245–2251, 1986.
41. Tiku, K., M.L. Tiku, S. Liu, and J.L. Skosey, Normal human neutrophils are a source of a specific interleukin 1 inhibitor. *J. Immunol.*, 136(10), 3686–3692, 1986.
41a. Tiku, K., M.L. Tiku, and J.L. Skosey, Interleukin 1 production by human polymorphonuclear neutrophils. *J. Immunol.*, 136(10), 3677–3685, 1986.
42. Roberts, N.J., Jr., A.H. Prill, and T.N. Mann, Interleukin 1 and interleukin 1 inhibitor production by human macrophages exposed to influenza virus or respiratory syncytial virus. Respiratory syncytial virus is a potent inducer of inhibitor activity. *J. Exp. Med.*, 163(3), 511–519, 1986.
43. Berman, M.A., C.I. Sandborg, B.S. Calabia, B.S. Andrews, and G.I. Friou, Studies of an interleukin 1 inhibitor: characterization and clinical significance. *Clin. Exp. Immunol.*, 64(1), 136–145, 1986.
44. Berman, M.A., C.I. Sandborg, B.S. Calabia, B.S. Andrews, and G.J. Friou, Interleukin 1 inhibitor masks high interleukin 1 production in acquired immunodeficiency syndrome (AIDS). *Clin. Immunol. Immunopathol.*, 42(1), 133–140, 1987.
45. Teodorescu, M., J.L. Skosey, C. Schlesinger, and J. Waltman, Covalent disulfide binding of IL-1 to alpha-2 macroglobulin alpha-2 M. International workshop on monokines and other non-lymphocytic cytokines, Hilton Head Island, South Carolina, USA. *J. Leukocyte Biol.*, 42(5), 604, 1987.
46. Teodorescu, M., personal communication.
47. Mazzei, G.J., L. Winger, A. Osen-Sand, L.M. Bernasconi, W. Benotto, and A.R. Shaw, Interleukin-1α, a major component of fetal calf serum: artifacts in IL-1 bioassays, in *The Physiological and Pathological Effects of Cytokines*, C.A. Dinarello, et al., Eds., Wiley-Liss, New York, 1990.
47a. Mazzei, G.J. and A.R. Shaw, Purification and characterization of a 26-kDa competitive inhibitor of interleukin 1, *Eur. J. Immunol.*, 20(3), 683–689, 1990.
48. Symons, J. and J. Eastgate, in press.
49. Torcia, M and F. Cozzolino, personal communication.
50. Didierjean, L., D. Salomon, Y. Merot, G. Siegenthaler, A. Shaw, J.M. Dayer, and J.H. Saurat, Localization and characterization of the interleukin 1 immunoreactive pool (IL-1 alpha and beta forms) in normal human epidermis. *J. Invest. Dermatol.*, 92(6), 809–816, 1989.
51. Schwarz, T., A. Urbanska, F. Gschnait, and T.A. Luger, UV-irradiated epidermal cells produce a specific inhibitor of interleukin 1 activity. *J. Immunol.*, 138(5), 1457–1463, 1987.
52. Walsh, L.J., P.E. Lander, G.J. Seymour, and R.N. Powell, Isolation and purification of ILS, an interleukin 1 inhibitor produced by human gingival epithelial cells. *Clin. Exp. Immunol.*, 68(2), 366–374, 1987.
53. Endres, S., J.G. Cannon, R. Ghorbani, R.A. Dempsay, S.D. Sisson, G. Lonnemann, J. van der Meer, S.M. Wolff, and C.A. Dinarello, *In vitro* production of IL-1β, IL-1α, TNF and IL-2 in healthy subjects: distribution, effect of cyclooxygenase inhibition and evidence of gene regulation. *Eur. J. Immunol.*, 19(12), 2327–2333, 1989.
54. Dinarello, C.A., T. Ikejima, S.J. Warner, S.F. Orencole, G. Lonnemann, J.G. Cannon, and P. Libby, P. Interleukin 1 induces interleukin 1. I. Induction of circulating interleukin 1 in rabbits in vivo and in human mononuclear cells *in vitro*. *J. Immunol.*, 19(2), 261–265, 1989.
55. Manson, J.C., J.A. Symons, F.S. Di Giovine, S. Poole, and G.W. Duff, Autoregulation of interleukin 1 production. *Eur. J. Immunol.*, 19(2), 261–265, 1989.

56. Nishihara, T., T. Koga, and S. Hamada, Production of an interleukin-1 inhibitor by cell line P388D1 murine macrophages stimulated with *Haemophilus actinomycetemcomitans* lipopolysaccharide. *Infect. Immun.*, 56(11), 2801–2807, 1988.
57. Isono, N. and K. Kumagai, Production of interleukin-1 inhibitors by the murine macrophage cell line P388D1 which produces interleukin-1. *Dermatologica*, 179(1), 134, 1989.
58. Barak, V., A.J. Treves, P. Yanai, M. Halperin, D. Wasserman, S. Biran, and S. Braun, Interleukin 1 inhibitory activity secreted by a human myelomonocytic cell line (M20). *Eur. J. Immunol.*, 16(11), 1449–1452, 1986.
59. Barak, V., D. Perritt, P. Yanai, M. Halperin, and A.J. Treves, Regulation of immune responses and inflammation by an IL-1 inhibitor. *Cytokine*, 1(1), 89, 1989.
60. Balavoine, J.F., B. de Rochemonteix, K. Williamson, P. Seckinger, A. Cruchaud, and J.-M. Dayer, Prostaglandin E2 and collagenase production by fibroblasts and synovial cells is regulated by urine-derived human interleukin 1 and inhibitor(s). *J. Clin. Invest.*, 78(4), 1120–1124, 1986.
61. Seckinger, P., K. Williamson, J.F. Balavoine, B. Mach, G. Mazzei, A. Shaw, and J.M. Dayer, A urine inhibitor of interleukin 1 activity affects both interleukin 1 alpha and 1 beta but not tumor necrosis factor alpha. *J. Immunol.*, 139(5), 1541–1545, 1987a.
62. Seckinger, P., J.W. Lowenthal, K. Williamson, J.M. Dayer, and H.R. MacDonald, A urine inhibitor of interleukin 1 activity that blocks ligand binding. *J. Immunol.*, 139(5), 1546–1549, 1987b.
62a. Prieur, A.M, M.T. Kaufmann, C. Griscelli, and J.M. Dayer, Specific interleukin-1 inhibitor in serum and urine of children with systemic juvenile chronic arthritis. *Lancet*, 2, 1240–1242, 1987.
62b. Roux-Lombard, P., C. Modoux, and J.M. Dayer, Inhibitors of IL-1 and TNFα activities in synovial fluids and cultured synovial fluid cell supernatants. *Calif. Tissue Int.*, 42, S (A47), 1988.
63. Roux-Lombard, P., C. Modoux, and J.M. Dayer, Production of interleukin-1 (IL-1) and a specific IL-1 inhibitor during human monocyte- macrophage differentiation. Influence of granulocyte-monocyte colony- stimulating factor (GM-CSF). *Cytokine*, 1, 45–51, 1989.
64. Shields et al., in press.
65. Arend, W.P., F.G. Joslin, and R.J. Massoni, Effects of immune complexes on production by human monocytes of interleukin 1 or an interleukin 1 inhibitor. *J. Immunol.*, 134(6), 3868–3875, 1985.
66. Arend, W.P., F.G. Joslin, R.C. Thompson, and C.H. Hannum, An IL-1 inhibitor from human monocytes. Production and characterization of biologic properties. *J. Immunol.*, 143(6), 1851–1858, 1989.
67. Janson, R.W. and W.P. Arend, The effects of GM-CSF on LPS-induced IL-1β vd IL-1 inhibitor (IL-1i). *Cytokine*, 1(1), 91, 1989.
68. Eisenberg, S.P., C.H. Hannum, R.J. Evans, W.P. Arend, M.T. Brewer, and R.C. Thompson, Cloning and expression of IL-1i, a human interleukin-1 inhibitor. *Cytokine*, 1(1), 75, 1989.
69. Hannum, C.H., W.P. Arend, C.J. Wilcox, D.J. Dripps, F.G. Joslin, and R.C. Thompson, Purification and characterization of a human monocyte IL-1 inhibitor (IL-1i). *Cytokine*, 1(1), 91, 1989.
70. Taniguchi, T., H. Matsui, T. Fujita, C. Takaoka, N. Kashima, R. Yoshimoto, and J. Hamuro, Structure and expression of a cloned cDNA for human interleukin-2. *Nature*, 302, 305–310, 1983.

71. Nikaido, T., A. Shimizu, N. Ishida, H. Sabe, K. Teshigawara, M. Maeda, T. Uchiyama, J. Yodoi, and T. Honjo, Molecular cloning of cDNA encoding human interleukin-2 receptor. *Nature,* 311, 631–635, 1984.
72. Uchiyama, T., S. Broder, and T.A. Waldmann, A monoclonal antibody (anti-Tac) reactive with activated and functionally mature human T cells. I. Production of anti-Tac monoclonal antibody and distribution of Tac (+) cells. *J. Immunol.,* 126(4), 1393–1397, 1981.
73. Santoli, D., R.B. Zurier, Z. Liao, A. Haimovitz, Y. Chen, J. Chan, and D.L. Rosenstreich, Characterization of a human interleukin 1 inhibitor. *J. Immunol.,* 134(6), 3882–3886, 1985.
74. Santoli, D. and R.B. Zurier, Prostaglandin E precursor fatty acids inhibit human IL-2 production by a prostaglandin E-independent mechanism. *J. Immunol.,* 143(4), 1303–1309, 1989.
75. Redondo, J.M., M. Fresno, and A. Lopez-Rivas, Inhibition of interleukin 2- induced proliferation of cloned murine T cells by glucocorticoids. Possible involvement of an inhibitory protein. *Eur. J. Immunol.,* 18(10), 1555–1559, 1988.
76. Nagler, A., L.L. Lanier, and J.H. Phillips, The effects of IL-4 on human natural killer cells. A potent regulator of IL-2 activation and proliferation. *J. Immunol.,* 141(7), 2349–2351, 1988.
77. Spits, H., H. Yssel, X. Paliard, R. Kastelein, C. Figdor, and J.E. de Vries, IL-4 inhibits IL-2-mediated induction of human lymphokine-activated killer cells, but not the generation of antigen-specific cytotoxic T lymphocytes in mixed leukocyte cultures. *J. Immunol.,* 141(1), 29–36, 1988.
78. Llorente, L., M.C. Crevon, S. Karray, T. Defrance, J. Banchereau, and P. Galanaud, Interleukin (IL) 4 counteracts the helper effect of IL2 on antigen-activated human B cells. *Eur. J. Immunol.,* 19(4), 765–769, 1989.
79. Burton, J.D., C.H. Weitz, and N.E. Kay, Malignant chronic lymphocytic leukemia B cells elaborate soluble factors that down regulate T cell and NK function. *Am. J. Hematol.,* 30, 61–67, 1989.
80. Kobayashi, K., C. Allred, and T. Yoshida, Suppression of interleukin 2 production by sera obtained from hypersensitivity granuloma-bearing mice with defective T cell-mediated immune responses. *Immunobiology,* 178(4–5), 329–339, 1989.
81. Kucharz, E.J. and J.S. Goodwin, Minireview. Serum Inhibitors of interleukin-2. *Life Sci.,* 42, 1485–1491, 1988.
82. Symons, J.A., N.C. Wood, F.S. Di Giovine, and G.W. Duff, Soluble IL-2 receptor in rheumatoid arthritis. Correlation with disease activity, IL-1 and IL-2 inhibition. *J. Immunol.,* 141(8), 2612–2618, 1988.
83. Teodorczyk-Injeyan, J.A., B.G. Sparkes, G.B. Mills, R.E. Falk, and W.J. Peters, Increase of serum interleukin 2 receptor level in thermally injured patients. *Clin. Immunol. Immunopatho.,* 51, 205–215, 1989.
84. Herrmann, T., O. Josimovic-Alasevic, A. Mouzaki, and T. Diamantstein, Demonstration of two distinct forms of released low affinity-type rat IL- 2 receptors. *Immunology,* 66(3), 384–387, 1989.
85. Treiger, B.F., W.J. Leonard, P. Svetlik, L.A. Rubin, D.L. Nelson, and W.C. Greene, A secreted form of the human interleukin 2 receptor encoded by an "anchor minus" cDNA. *J. Immunol.,* 136(11), 4099–4105, 1986.
86. Smith, M.D., D.R. Haynes, and P.J. Roberts-Thomson, Interleukin 2 and interleukin 2 inhibitors in human serum and synovial fluid. I. Characterization of the inhibitor and its mechanism of action. *J. Rheumatol.,* 16(2), 149–157, 1989.

86a. Smith, M.D. and P.J. Roberts-Thomson, Interleukin 2 and interleukin 2 inhibitors in human serum and synovial fluid. II. Mitogenic stimulation, interleukin 2 production and interleukin 2 receptor expression in rheumatoid arthritis, psoriatic arthritis and Reiter's syndrome. *J. Rheumatol.*, 16(7), 897–903, 1989.
87. Emery, P., N. Wood, K. Gentry, A. Stockman, I.R. Mackay, and O. Bernard, High-affinity interleukin-2 receptors on blood lymphocytes are decreased during active rheumatoid arthritis. *Arthritis Rheum.*, 31(9), 1176–1181, 1988.
87a. Emery, P. and I.R. Mackay, Interleukin-2 inhibitor and measurement of interleukin-2 in synovial fluid of patients with rheumatoid arthritis (letter). *Arthritis Rheum.*, 31(12), 1590–1591, 1988.
87b. Emery, P., K.C. Gentry, A. Kelso, and I.R. Mackay, Interleukin 2 inhibitor in synovial fluid. *Clin. Exp. Immunol.*, 72(1), 60–66, 1988.
88. Krakauer, T. Biochemical characterization of interleukin 1 from a human monocytic cell line. *J. Leukoc. Biol.*, 37(5), 511–518, 1985.
89. Miossec, P., T. Kashiwado, and M. Ziff, Inhibitor of interleukin-2 in rheumatoid synovial fluid. *Arthritis Rheum.*, 30(2), 121–129, 1987.
90. Kashiwado, T., P. Miossec, N. Oppenheimer-Marks, and M. Ziff, Inhibitor of interleukin-2 synthesis and response in rheumatoid synovial fluid. *Arthritis Rheum.*, 30(12), 1339–1347, 1987.
91. Kashiwado, T., N. Oppenheimer-Marks, and M. Ziff, T cell inhibitor secreted by macrophages and endothelial cells. *Clin. Immunol. Immunopathol.*, 53, 137–150, 1989.
92. Arend, W.P. and J.-M. Dayer, Cytokines and cytokine inhibitors or antagonists in rheumatoid arthritis. *Arthritis Rheum.*, in press 1990.
93. Le, J. and J. Vilcek, Tumor necrosis factor and interleukin 1: cytokines with multiple overlapping biological activities. *Lab. Invest.*, 56(3), 234–248, 1987.
94. Seckinger, P., S. Isaaz, and J.M. Dayer, A human inhibitor of tumor necrosis factor alpha. *J. Exp. Med.*, 167(4), 1511–1516, 1988.
95. Peetre, C., H. Thysell, A. Grubb, and I. Olsson, A tumor necrosis factor binding protein is present in human biological fluids. *Eur. J. Haematol.*, 41(5), 414–419, 1988.
96. Olsson, I., M. Lantz, E. Nilsson, C. Peetre, H. Thysell, A. Grubb, and G. Adolf, Isolation and characterization of a tumor necrosis factor binding protein from urine. *Eur. J. Haematol.*, 42(3), 270–275, 1989.
97. Seckinger, P., S. Isaaz, and J.M. Dayer, Purification and biologic characterization of a specific tumor necrosis factor alpha inhibitor. *J. Biol. Chem.*, 264(20), 11966–11973, 1989.
98. Seckinger, P et al., in press.
99. Engelmann, H., D. Aderka, M. Rubinstein, D. Rotman, and D. Wallach, A tumor necrosis factor-binding protein purified to homogeneity from human urine protects cells from tumor necrosis factor toxicity. *J. Biol. Chem.*, 264(20), 11974–11980, 1989.
100. Svenson, M., L.K. Poulsen, A. Fomsgaard, and K. Bendtzen, IgG autoantibodies against interleukin 1 alpha in sera of normal individuals. *Scand. J. Immunol.*, 29(4), 489–492, 1989.
101. Fomsgaard, A., M. Svenson, and K. Bendtzen, Auto-antibodies to tumour necrosis factor alpha in healthy humans and patients with inflammatory diseases and gram-negative bacterial infections. *Scand. J. Immunol.*, 30(2), 219–223, 1989.
102. Symons, J. et al., in press.
103. Cozzolino, F. and M. Torcia, personal communication.

104. Tominaga, S. I., A putative protein of a growth specific cDNA from BALB/c-3T3 cells is highly similar to the extracellular portion of mouse interleukin 1 receptor. *FEBS Lett.*, 258(2), 301–304, 1989.
105. Novick, D., H. Engelmann, D. Wallach, and M. Rubinstein, Soluble cytokine receptors are present in normal human urine. *J. Exp. Med.*, 170(4), 1409–1414, 1989.
106. Beckmann, M.P., B. Mosley, C.J. March, R. Idzerda, T. Vanden Bos, D. Friend, A. Alpert, J. Wignall, S. Gimpel, J. Jackson, J. Sims, D. Anderson, C. Smith, B. Gallis, M. Widmer, D. Cosman, and L. Park, Molecular cloning and characterization of multiple forms of the murine interleukin-4 receptor. *Cytokine,* 1(1), 148, 1989.

3. Low Molecular Weight Inhibitors of Interleukin-1

FRANCIS J. PERSICO

INTRODUCTION

Cytokines are a series of polypeptides produced by a variety of cells with an extraordinarily diverse and yet often overlapping spectrum of activities. Complicating our understanding of how these molecules function is the multitude of macro- and microenvironments in which they act. In the intact host, they are influenced by factors including, but not limited to, endocrine and local hormonal elements, products of metabolism, access to target sites regulated by diffusion rates through tissues, vascularization, physiological status, and other elements. Such complexity offers exceptional opportunities for interdisciplinary study by scientists with different interests, but also presents an almost impenetrable maze to others wishing to delineate those elements primarily responsible for the pathophysiology of a particular disease and to design strategies for therapeutic intervention. In this respect, pure compounds of defined structure with known activity and properties that can reliably be delivered to experimental targets can be valuable tools as well as clinical advances.

One problem actively being considered relates to the relative degree, if any, to which interleukin-1 (IL-1) contributes to human arthritis. The majority of activities assigned to IL-1 are "proinflammatory" in nature, and most can be associated with one or more facets of the complex immunopathology of this prevalent, chronic disease. There is certainly circumstantial evidence in the sense that IL-1 levels are elevated in the synovial fluids of patients with arthritis. On this basis it would seem rational to consider developing novel approaches for treating arthritis by inhibiting the synthesis or activity of this molecule. However, many of the functions of IL-1 are also exhibited by other cytokines, notably tumor necrosis factor (TNF) and interleukin-6 (IL-6), and these molecules are also elevated during inflammation. In fact, some of the functions attributed to IL-1 are secondary to its ability to induce or exhibit synergy with these and other cytokines. Furthermore, consideration must be given to the potential consequences of interfering with the function of a molecule evidently so vital to mobilization of the host response to pathogenic microorganisms. Because of the significance of these issues and the potential for using easily delivered inhibitors of IL-1 to contribute to their resolution, this chapter will concentrate on reviewing various classes of compounds claimed to affect the synthesis or response to this cytokine, primarily within the context of arthritic disease.

NONSTEROID ANTI-INFLAMMATORY DRUGS (NSAIDS)

One of the original terms for what was ultimately identified as IL-1 was endogenous or leukocytic pyrogen,[1-3] and it is now well accepted that IL-1 is one substance capable of elevating body temperature by acting on the hypothalamus. Aspirin and NSAIDs suppress fever and therefore antagonize the effect of IL-1, but the mechanism is indirect; prostaglandins synthesized in the brain in response to IL-1 are responsible for the fever,[2,4] and aspirin-like NSAIDs inhibit prostaglandin synthesis,[5] thereby suppressing the hyperthermic response. Similar indirect attenuation of the response to IL-1 as a consequence of alterations in arachidonic acid metabolism can occur in a variety of tissues. IL-1 activates phospholipase A_2[6,7] to deacylate membrane phospholipids, making arachidonic acid available for cyclooxygenation in a variety of cells,[8] including fibroblasts and synoviocytes,[9-11] chondrocytes,[12] endothelial cells,[13] and tumor cell lines.[14] The products of the arachidonic acid cascade initiate a number of biological responses that must be considered in attempting to determine whether or not an agent known to affect eicosanoid metabolism also directly influences cytokine activity.

Indomethacin was shown to reduce the production of plasminogen activator[11] and enhance DNA synthesis[15] in synovial cells activated with IL-1. In both studies, exogenous PGE_2 nullified inhibition by indomethacin. However, caution must be used in generalizing from the results of such *in vitro* studies, since they can be highly dependent on the choice of assay and cell line. In this respect, IL-1 induction of collagenase was not attenuated by indomethacin in the studies of Mochan et al.,[11] despite the inhibitory effect on plasminogen activator. Using bovine catabolin (IL-1) to elicit cartilage degradation, Rainsford[16] found that concentrations of NSAIDs likely to be acheived *in vivo* had no effect on glycosaminoglycan release, and similar observations were made by Arner et al.[17] using IL-1 purified from human monocytes.

The studies of Couchman and Sheppeard[18] supported the observation that catabolin(IL-1)-induced cartilage degeneration was not affected by cyclooxygenase (CO) inhibitors, but found that a number of NSAIDs shared the common property of suppressing catabolin production by synovial tissue. Inconsistent results relative to a previous communication by the same group[19] claiming that aspirin and clozic failed to inhibit the elaboration of IL-1 were attributed to differences in preparation of the compounds for assay. In contrast, Brandwein[20] found that indomethacin caused slight stimulation of IL-1 production by mouse peritoneal macrophages. He noted that prostaglandin E_2 (PGE_2), cyclic adenosine monophosphate (AMP), and agents that elevated cAMP suppressed IL-1 synthesis, particularly when added early in the course of macrophage activation before IL-1 was evident in the supernatants. IL-1 and PGE_2 were secreted concurrently in response to lipopolysaccharide (LPS), and the concentration of prostanoid thus generated (ca. 30 nM) was sufficient to cause weak inhibition of IL-1 (the IC_{50} for PGE_2 was 50 nM). Consistent with this, slight augmentation of IL-1 production was observed when PGE synthesis was inhibited with indomethacin. Kunkel et al.[21] reported stimulation of LPS-induced IL-1 synthesis by indomethacin, piroxicam, and ibuprofen, but others[3,22] have failed to observe NSAID augmentation of IL-1 synthesis. In a careful study, Otterness et al.[23] observed augmented thymocyte proliferation at low dilutions of IL-1-containing macrophage-conditioned supernatants from cultures exposed to indomethacin and piroxicam, but as the culture supernatants were diluted further, the enhanced responses were diminished until they were identical to untreated control values. As a consequence, when the data were calculated as units, there was no evidence of increased production of IL-1 in response to the prostaglandin synthesis inhibitors. Using antibody, it was demonstrated that enhancement of thymocyte proliferation by the NSAIDs was due to elimination of the antiproliferative PGE_2 from LPS-treated macrophage supernatants, but there was no evidence of increased IL-1 synthesis when stringent, bioassay-independent criteria were employed (i.e., competition for IL-1 binding to membrane sites or western blotting with anti-IL-1 antibody).

Thus, consistent evidence for a direct effect on IL-1 synthesis by CO inhibitors is lacking; however, indirect regulation by prostaglandins is possible under some *in vitro* conditions, dependent on the cell type and stimulus. IL-1 clearly stimulates prostaglandin synthesis in cells and has been shown to induce synthesis of the CO enzyme.[24-26] Kunkel et al.[21,27] suggested that PGE_2 synthesis induced by IL-1 could serve as a negative regulator of IL-1 production. However, due to the nonspecific nature of the bioassay (LAF) generally used to quantitate IL-1, these studies could not adequately address the possibility that IL-1 induced the synthesis of a second cytokine, such as IL-6, also capable of providing the proliferative comitogenic signal to lectin-treated thymocytes.[28,29] Dinarello et al.[30] observed that IL-1 could induce it's own expression in human mononuclear cells, but found that higher cytokine concentrations (100 ng/ml) were less effective. Diminution of the response at high cytokine concentrations was shown to be the result of PGE_2 synthesis and was abrogated by treating the cells with 1 µg/ml indomethacin. IL-1 was unequivocally identified by radioimmunoassay and Northern transfer as well as by bioassay. PGE-regulated IL-1 autoamplification loops have also been proposed to function in large granular lymphocytes,[31] human vascular smooth muscle,[32] and endothelial cells,[33] and human dermal fibroblasts.[34] In these studies, increased production of IL-1 appeared to be at least partially dependent on new protein synthesis. Control was not at the level of transcription, since Knudsen et al.[35] reported that PGE_2 had little effect on the synthesis of IL-1 mRNA and Warner et al.[33] showed that indomethacin failed to influence synthesis of IL-1 gene transcripts in endothelial cells. Instead, it was speculated that PGE_2 suppressed IL-1 synthesis by modifying a posttranscriptional event regulated by intracellular cAMP levels.[35] However, using human dermal fibroblasts, Mauviel et al.[34] provided data indicating that indomethacin prevented increased transcription of the IL-1β gene in response to either IL-1α or IL-1β and suggested that regulation of IL-1 autoamplification loops can vary in different cell types.

The question of whether CO inhibitors can directly regulate IL-1 activity is a critical consideration for the development of novel treatments for human arthritis. Since NSAIDs provide only palliative relief for arthritis and have little effect on the underlying disease process, the demonstration that they directly influence the synthesis or suppress the activity of IL-1 would severely compromise arguments proposing IL-1 to be a major factor in the pathogenesis of human arthritis. As noted above, most studies to address this question employed tissue culture, and due to a number of factors, no consensus has emerged. In any event, the relevance of *in vitro* results to the complexities of whole animal physiology or pathology can always be questioned. Indeed, the results with at least one *in vitro* experimental paradigm have been indicted as being an artifact of culture conditions; decreased pH as a consequence of prolonged incubation of synovial cells is responsible for the apparent inhibition of IL-1 production by NSAIDs.[36] The *in vivo* data available to date suggest that the aspirin-like CO inhibitors inhibit responses to IL-1 such as fever, shock,[37] or natriuresis,[38] by virtue of suppressing prostaglandin synthesis, rather than any direct effect on IL-1 activity per se. There is no convincing *in vivo* evidence of IL-1 synthesis being affected by NSAIDs. Splenic macrophages from rats with adjuvant arthritis display an increased capacity to synthesize IL-1 in response to LPS. In general, treatment of such animals with NSAIDs causes substantial improvement in the inflammatory condition, but minor or no effects on their enhanced IL-1 synthetic capability.[39-41]

STEROIDS

In a review of the anti-inflammatory and immunosuppressive properties of corticosteroids, Fauci[42] commented that the literature could justify the notion that ". . . virtually every aspect, phase, and cell type involved in immunologic and inflammatory reactivity can be modulated to a greater or lesser degree by glucocorticosteroids." In view of this, it is not surprising that steroids have been reported to affect the synthesis and response to IL-1.

Dillard and Bodel[43] observed that the production of endogenous pyrogen by human leukocytes exposed to heat-killed staphylococcal organisms or etiocholanalone was reduced by exposing the cells to hydrocortisone. This observation was subsequently confirmed by the studies of Unanue and Snyder,[44] which demonstrated that release of IL-1 from mouse macrophages stimulated with LPS was diminished in a dose-related fashion by hydrocortisone in concentrations ranging from 10^{-6} to 10^{-8} *M*. Since then, this phenomenon has been confirmed and extended by a number of laboratories.[45-48] These studies used bioassays to measure the production of IL-1, but by using procedures such as dialysis to remove the drug prior to assay for cytokine production, care was taken to minimize the chances that the results were due to steroid influences on the bioassay to measure IL-1. Our laboratory, which routinely uses lymphocyte activation (LAF or some other comitogenic assay) to quantitate IL-1 produced from LPS-stimulated macrophage populations, restricts exposure of the cells to drug to a 4- to 18-hr period following stimulation with LPS. Drug-containing supernatants are decanted, the cells washed, and fresh, drug-free, LPS-free medium added for an additional 24 hr of incubation before harvesting the cytokine-rich supernatants. ^{3}H-dexamethasone was used to establish that the steroid was removed by the washing procedure, yet an initial 4-hr exposure was sufficient to significantly inhibit the production of IL-1.[49]

Tracey et al.[48] employed a complex multistep protocol to demonstrate that nontoxic concentrations of hydrocortisone diminished the proliferative response of T-cells to IL-1 and IL-2. These investigators suggested that it is unlikely that corticosteroids inhibit responses to IL-1 (or IL-2) by functioning as receptor antagonists; IL-1 and corticosteroids do not compete with each other in radioligand binding studies. In fact, despite the immunosuppressive nature of the response to these compounds, there are paradoxical reports suggesting an increase in the expression of functional lymphoid IL-1 receptors after exposure to glucocorticoids.[50,51] Thus, the mechanism by which these compounds oppose the effects of IL-1 is not apparent, but in view of the characteristics of glucocorticoids in general, it is likely to involve nuclear events.

Glucocorticoid receptors (members of the steroid/thyroid hormone receptor superfamily reviewed by Evans[52]) differ from most receptors by nature of being intracellular rather than cell surface molecules. These receptors, once activated by ligand, translocate to the nucleus, attach to specific chromosomal sites,[53] and regulate the transcription of selected target genes in a specific fashion. The steroid receptor-ligand complex is known to exert both positive and negative sequence-specific control,[54] but often down regulates the gene transcription. Glucocorticoids decrease the transcription of the a_1-fetoprotein gene in liver,[55] the proopiomelanocortin gene in pituitary tumor cells,[56] the synthesis of type 1 procollagen message in fibroblasts,[57] and transcripts for a number of proteolytic enzymes, including stromelysin,[58] collagenase[59] and plasminogen activator from rabbit synoviocytes,[60] and mammary carcinoma.[61]

Antipodal responses to corticosteroids and IL-1 are often the consequence of differential regulation of specific genes. For example, in contrast to the above-mentioned suppression of transcription of a number of genes coding for proteolytic enzymes, IL-1β caused a dose-dependent increase in stromelysin mRNA levels,[58] and both α and β forms of IL-1 increased transcription of the urokinase-type plasminogen activator gene[60] in synovial fibroblasts. Lee and colleagues[62] recently demonstrated that hydrocortisone (10 μM) and dexamethasone (100 nM) directly suppressed transcription of the IL-1β gene induced by LPS/PMA (lipopolysaccharide/phorbol 12-myristate 13-acetate) in U-937 promonocytic human leukemic cells. Dexamethasone prevented induction of IL-1 mRNA synthesis, and by means of a nuclear run-off assay, it was demonstrated that incorporation of [α-^{32}P]UTP into nascent IL-1 message was specifically suppressed without significant effects on *c-fos*, *c-myc*, or actin nuclear RNAs. These observations were confirmed by Nishida et al.[63] Thus, dexamethasone had a direct effect on IL-1 transcription. Lee et al.[62] also showed that corticosteroids could selectively decrease IL-1 mRNA stability. In the presence of actinomycin D and cyclohexamide to inhibit further transcription and translation, the degradation of IL-1, but not *c-fos*, mRNA was accelerated by exposure to dexamethasone. Increased messenger lability in the presence of dexamethasone was not simply due to the possession of a particular A+U-rich sequence common to many cytokine messages,[64] since both *c-fos* and IL-1 messages also possess this sequence. Thus, glucocorticoids decreased production of the IL-1 gene product both by suppressing transcription and by accelerating hydrolysis of the IL-1 mRNA. Allison and Lee[65] proposed that the ability of glucocorticoids to inhibit IL-1 is a major element contributing to their anti-inflammatory and immunosuppressive properties.

Mutual antagonism between IL-1 and glucocorticoids can be demonstrated *in vivo*. For example, dexamethasone administration to LPS-treated mice reduced the level of IL-1 in serum,[66] while dosing mice with IL-1 decreased the expression of cytosolic glucocorticoid receptors in the liver.[67] In animals with chronic degenerative joint disease, corticosteroids decrease the excessive production of IL-1; Johnson et al.[40] showed that prednisolone reduced the enhanced LPS-induced secretion of IL-1 exhibited by splenic macrophages derived from arthritic rats. This correlated with a reduction in the inflammtory response. It differed from the results obtained with the CO inhibitor, indomethacin, which reduced inflammation, but not the production of IL-1. In similar studies, Connolly et al.[68] noted that in addition to alleviating inflammation and reducing IL-1 levels, dexamethasone decreased production of acute phase reactants, causing them to speculate that suppression of IL-1 contributed to the anti-inflammatory activity of steroids in the adjuvant arthritic rat.

Several studies suggest that the mutually opposing effects of IL-1 and glucocorticoids function as a homeostatic neuroimmunoendocrine feedback mechanism.[67,69] Although some disagreement exists regarding the details and specificity of the cascade of events comprising this response,[70] the essence is as follows. IL-1 (but not TNF, IL-2, or IFN$_\gamma$[69]) induces the secretion of adrenocorticotropic (ACTH) and other hormones[71,72] from the pituitary gland. This response is due to IL-1 stimulation of responsive neurons,[73] possibly including those with fibers containing immunoreactive IL-1 located in specific hypothalamic areas,[74] to activate central neuroendocrine pathways leading to the release of corticotropin-releasing factor.[73,75] Corticotropin-releasing factor causes ACTH to be produced by the pituitary and to stimulate the adrenal gland to produce glucocorticoids, which can reciprocally antagonize the production and activity of IL-1.

INHIBITORS OF THE CYCLOOXYGENASE AND LIPOXYGENASE (CO/LO) PATHWAYS OF ARCHIDONIC ACID METABOLISM

Dinarello et al.[2] noted that BW 755C and 5,8,11,14-eicosatetraynoic acid (ETYA), two compounds that inhibit oxidation of arachidonic acid catalyzed by both CO and LO enzymes, suppressed the enhanced comitogenic proliferation of thymocytes exhibited in the presence of IL-1. However, compounds such as ibuprofen, which only inhibit cycloxygenation, had no effect on IL-1-induced thymocyte proliferation at concentrations that inhibited PGE synthesis. On this basis, it was suggested that products derived from the lipoxygenation of arachidonic acid mediated responses to IL-1. In agreement with this report, Farrar and Humes[76] used cell lines specifically responsive to IL-1 and IL-2 to show that activation of the LO pathway and production of IL-2 in response to IL-1 was sensitive to inhibition by the CO/LO inhibitors BW 755C, ETYA, and nordihyroguaiaretic acid (NDGA), but not indomethacin. Using procedures not directly dependent on cellular proliferation, other investigators have implicated products of the LO pathway as mediating the response to IL-1;[77,78] thus, this phenomenon is not simply an artifact of a particular bioassay. However, failure to fully restore the response to interleukins compromised by inhibitors of lipoxygenation by providing metabolites of the LO pathway,[76,79] and the complex nature of processes such as lymphocyte activation[80] induced by IL-1, make it likely that other factors are involved. Furthermore, as demonstrated by Liu et al.,[81] using novel, selective 5-LO inhibitors, inhibition of lipoxygenation does not necessarily result in suppression of lymphocyte activation. These investigators concluded that LO products did not appear to be obligatory mediators of the response to IL-1.

Dinarello et al.[82] showed that BW755C and ETYA inhibited the production of IL-1 from human monocytes when the compounds were added prior to activation of the cells by several stimuli. Treatment with these inhibitors after cellular activation was ineffective. Indirect evidence suggesting a role for products of the LO pathway early in the sequence of events leading to the production of IL-1 came from the observation that the CO inhibitor ibuprofen suppressed PGE_2 synthesis, but had no effect on IL-1 production. The studies of Brandwein[20] indicated that ETYA inhibited the synthesis of IL-1; but in contrast to the observation of Dinarello and colleagues, BW755C enhanced its secretion. PGE_2 suppressed cytokine production, but neither exogenous LTB_4 nor activation of lipoxygenation by treatment of the cells with the calcium ionophore, A23187, provoked IL-1 secretion, leading Brandwein to suggest that LO metabolites were not involved in the regulation of IL-1 synthesis. Kunkel et al.[27] presented evidence that leukotrienes C_4 (LTC_4) and B_4 (LTB_4) augmented the production of IL-1 by resident murine macrophages and that the response to LTC_4 could be blocked by the peptidoleukotriene receptor antagonist, FPL 55712. He suggested that control of macrophage IL-1 production *in vitro* was primarily by means of PGE, but LO metabolites could "fine-tune" the system.

The discrepancies between these reports may be accounted for by a variety of factors, such as the type, source, condition, and state of activation of the cells, culture conditions, time of addition, concentration, and length of exposure to compounds. For example, Dinarello and colleagues[82] noted that time of addition of the inhibitor to the culture was critical and used compound concentrations in the 100 μM range, whereas the maximum compound concentration employed by Brandwein[20] was 10 μM, and he found preincubation to be of no consequence. Additionally, it is worth emphasizing that with the exception of the work of Liu et al.,[81] the compounds used in these studies were not pure LO inhibitors, but also inhibited the CO enzyme. It may be that in order to inhibit IL-1, coordinate inhibition of both CO and LO pathways, similar to that attained with steroids, is required and that the ability to achieve this is not only dependent on the relative potency of the compound against both enzymes, but also the nature and intensity of the activating stimulus. Using a series of potent, pure LO inhibitors, our laboratory has not observed suppression of IL-1 synthesis.[83] Thus, it is not clear that LO products play an obligatory role in the regulation of IL-1 synthesis. However, several compounds reported to suppress both pathways of arachidonate metabolism have been observed to down regulate IL-1.

CP-66,248 (Figure 1) is currently being evaluated for efficacy in the treatment of arthritis in humans.[84] In preclinical studies, it was observed to inhibit the 5-LO with IC_{50} (concentration of compound required to inhibit the response by 50%) values of 7 to 13 μM.[85] It also inhibited the CO enzyme ($IC_{50} = 0.7$ μM), and consistent with this, it was active in animal models traditionally used to study the aspirin-like NSAIDs.[85,86] Reports implicating products of the arachidonic acid cascade with regulation of IL-1 synthesis prompted investigation of the effect of CP-66,248 on the production of IL-1 by mouse peritoneal macrophages. It was found that concentrations of CP-66,248 as low as 1 $\mu g/ml$ prevented the 34 kD pro-IL-1_α molecule from being made intracellularly, indicating that it truly inhibited synthesis rather than simply suppressed secretion. Global inhibition of protein synthesis was not observed at compound concentrations that inhibited the production of IL-1 in response to either LPS or zymosan.[87] As noted above, using similar assay conditions, Otterness et al.[23] had previously provided evidence that CO inhibition had no effect on IL-1 synthesis. However, this does not necessarily mean that suppression of cytokine production by CP-66,248 is dependent on its ability to inhibit LO activity. In fact, it is more potent an inhibitor of IL-1 synthesis ($IC_{50} = 3$ μM) than of the 5-LO enzyme.

CP-66,248 has also been demonstrated to inhibit IL-1 production from human cells in culture. McDonald et al.[88] reported that in contrast to pure prostaglandin synthetase inhibitors (indomethacin, naproxen, and piroxicam), CP-66,248 added coincidentally or shortly following cellular stimulation decreased the production of IL-1 from human peripheral blood monocytes. In addition to its effects on IL-1 synthesis, these investigators presented evidence to show that CP-66,248 also inhibited cytokine function. It was shown to diminish the comitogenic response of thymocytes and proliferation of the cloned, IL-1-sensitive murine T-cell line, D10.G4.1, in response to recombinant human IL-1.

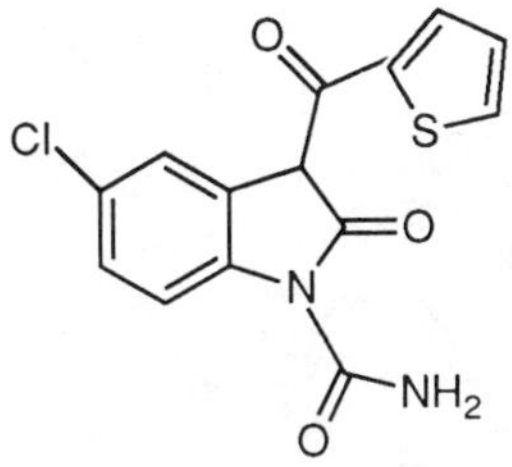

Figure 1. Chemical structure of CP-66,248.

Clinically, McDonald and colleagues[88a] used the D10.G4 cell line and immunoblotting to quantitate IL-1 levels in synovial fluids of patients with rheumatoid arthritis, and demonstrated a reduction following treatment with CP-66,248 for 7 d. Samples from these patients spiked with recombinant human IL-1 (rhIL-1) exhibited full biological activity, so this response was not due to an effect of the drug on any natural, endogenous inhibitor in the synovial effluents. Although the possibility exists that CP-66,248 also influences the activity of other cytokines, such as TNF or IL-6, which can produce similar biological responses or which potentiate the activity of IL-1,[29,89,90] it is evident that one component of the biological activity is indeed IL-1, since activity in synovial fluids was blocked by antibody specific for IL-1β.

Patients with both rheumatoid[84] and osteoarthritis[91] treated with CP-66,248 exhibited clinical improvement. As noted above, CP-66,248, like the aspirin-like NSAIDs, is an inhibitor of the CO enzyme; therefore, it was not surprising that several of the clinical parameters that improved (e.g., number of painful joints, grip strength, etc.) were those that would respond to treatment with NSAIDs. However, in addition CP-66,248 reduced the circulating levels of the acute phase reactant, C-reactive protein (CRP). The ability to effect a decrease in CRP or other acute phase proteins is not a property common to CO inhibitors[92] and, in fact, was not observed with a comparative NSAID in these studies.[86] The capacity to reduce elevated CRP levels in patients has been suggested as discriminating between drugs providing only symptomatic relief and those capable of modifying the progression of rheumatoid arthritis.[93,94] This implies that CP-66,248 may possess unusual therapeutic properties. However, as noted by Otterness et al.,[86] it is difficult to define the mechanism by which this compound brings about a reduction in CRP levels. IL-1 induces IL-6 in a number of systems;[95,96] CRP and other acute phase glycoproteins are produced from the liver presumably in response to IL-6 or IL-1 acting in concert with IL-6.[97-99] Thus, it is possible that the effect of CP-66,248 on CRP levels is a direct consequence of its capacity to decrease synthesis of IL-1 (or possibly IL-6), rather than a reflection of resolution of the disease. In any case, it remains to be established that the novel antiarthritic properties of this compound, if any, are due to suppression of IL-1 activity, rather than some other aspect of its pharmacology.[86]

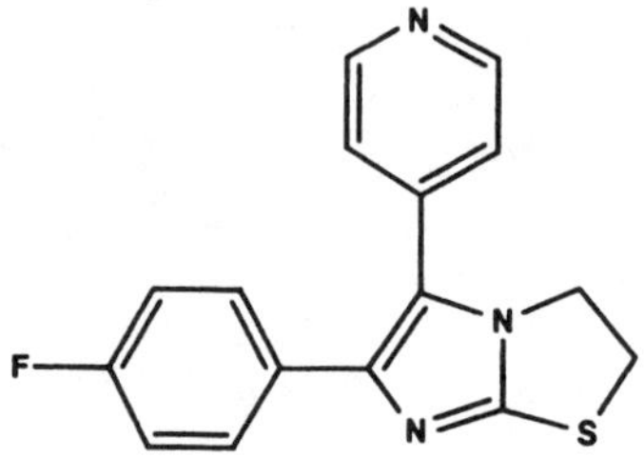

A. SK&F 86002

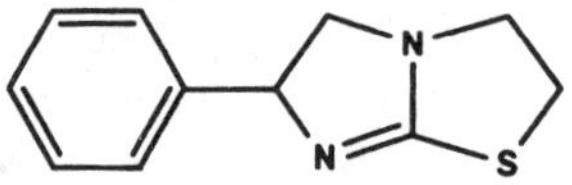

B. Levamisole

Figure 2. Chemical structures of SK&F 86002 and levamisole.

SK&F 86002 (Figure 2) is another compound that impedes arachidonic acid metabolism by both LO and CO pathways[100] and also diminishes the production of IL-1. However, Lee et al.[22] claimed that inhibition of IL-1 and oxidation of arachidonate were distinct properties. In their studies, independent of stimulus, SK&F 86002 reduced the production of cell-associated and secreted IL-1 from human peripheral blood monocytes. Inhibition was reversible and optimal with continuous exposure of the cells to the drug. However, the IC_{50} for inhibition of IL-1 was 1.3 µM, while that for inhibition of LTC_4 (LO pathway) and PGE_2 (CO pathway) was 13.3 µM and 0.3 µM, respectively. Ibuprofen also inhibited PGE_2 synthesis (IC_{50} = 1.2 µM), but at concentrations as high as 100 µM had no effect on IL-1 synthesis. BW 755C decreased IL-1 synthesis, but the concentration required was approximately nine-times greater than that necessary to suppress oxidation of arachidonic acid. The IC_{50} for lipoxygenation of arachidonate by SK&F 86002 (13.3 µM) and BW 755C (8.8 µM) were approximately equivalent, yet SK&F 86002 was nearly two orders of magnitude more potent as an inhibitor of IL-1 synthesis. Only NDGA inhibited IL-1 and arachidonate metabolism with approximately equivalent potencies, but the concentrations required to do so were nearly toxic.

As expected with inhibitors of prostaglandin synthesis, SK&F 86002 was effective in animal models of arthritis and inflammatory disease,[101] but in addition, it suppressed the enhanced *ex vivo* production of IL-1 exhibited by splenocytes obtained from adjuvant arthritic rats. The latter is not a property of aspirin-like CO inhibitors, but since SK&F 86002 shares with them the ability to inhibit the synthesis of prostaglandins, it is difficult to determine whether this novel attribute contributes to alleviation of the disease in a model such as adjuvant arthritis, which responds so well to treatment with NSAIDs.

There is some structural similarity between SK&F 86002 and levamisole (Figure 2), a drug with immunoregulatory properties clinically tested for the treatment of arthritis,[102,103] cancer,[104] and other conditions. Despite the structural resemblance, levamisole does not exhibit pharmacologic properties similar to SK&F 86002; it does not inhibit arachidonate metabolism, nor does it suppress IL-1 synthesis. In fact, our laboratory has demonstrated that levamisole has quite the opposite effect; it enhances the LPS-stimulated production of IL-1 from $P388D_1$ and mouse adherent peritoneal cell populations *in vitro* and *ex vivo*.[105] Thus, these two structurally related compounds both affect IL-1 production, but in opposite ways. It remains to be established whether there are common elements to the mechanism(s) by which they act to influence the synthesis of IL-1. For example, they may differentially regulate a common genetic element, resulting in an alteration of transcription or translation, or they may occupy the same receptor, but activate different signal transduction pathways. Whatever the mechanism, the fact that both compounds influence IL-1, but only SK&F 86002 affects arachidonic acid metabolism, is consistent with proposals suggesting that the two events are independent of each other.

Preliminary studies have indicated that E-5110, a recently described inhibitor of both LO and CO enzymes, and its hydroxylated metabolite (Figure 3), can cause concentration-dependent suppression of IL-1 generation from rat leukocytes[106] and human monocytes.[107] Using bioassays (LAF, pyrogen, and collagenase production from rat synoviocytes) as well as an ELISA for IL-1, it was determined that synthesis of intracellular and secreted IL-1 was diminished by treatment with E-5110. Suppression of IL-1 synthesis was selective; E-5110 had no effect on nonspecific protein synthesis at concentrations that inhibited IL-1 (IC_{50} = 1.2 μM). However, like Lee et al.,[22] these investigators concluded that inhibition of IL-1 generation was independent of eicosanoid synthesis. In this respect, BW-755C and the specific 5-LO inhibitor, AA861, failed to cause significant suppression of IL-1 synthesis at concentrations up to 10 μM, and indomethacin and piroxicam were also inactive.

OTHER DRUGS USED TO TREAT ARTHRITIS

Gold compounds have been used for decades to treat arthritis, but despite the variety of hypotheses that have been advanced to account for their activity (for review, see reference 109), none have been universally accepted.[110] In view of the variety of proinflammatory activities linked to IL-1 and similar cytokines, as well as reports that gold modulates the function of a major cellular source of such

A. E-5110

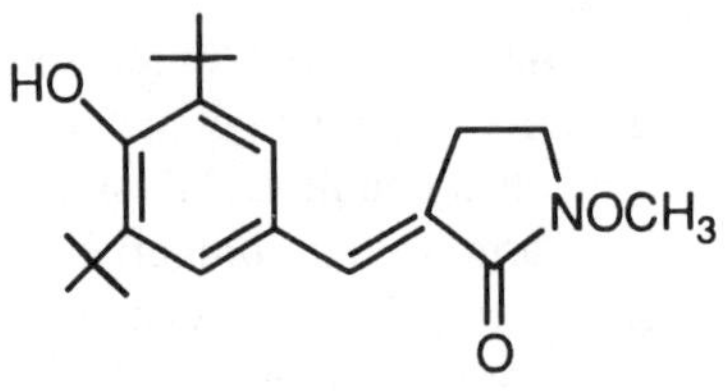

B. Metabolite

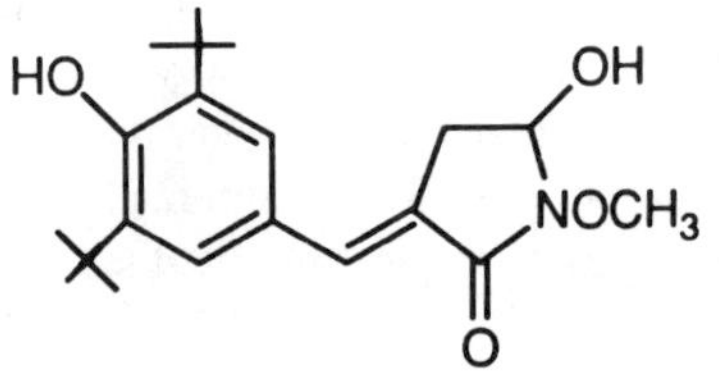

Figure 3. Chemical structures of E-5110 and its hydroxylated metabolite.

products, the macrophage/monocyte,[111-113] it is not surprising that gold has been studied in the context of regulation of IL-1 activity. However, there is no consensus regarding the capacity of gold compounds to influence the synthesis or antagonize the activity of IL-1 *in vitro*.

Griswold et al.[114] initially observed that pretreatment of adherent peritoneal cells with auranofin for 1 hr and then washing the monolayers free of drug before adding LPS caused a reduction in the quantity of IL-1 subsequently produced 48 hr later. The studies of Barrett and Lewis[115] agreed; auranofin inhibited baseline LPS- and phytohemagglutinin (PHA) -induction of IL-1 from adherent human synovial cell cultures ($IC_{50} < 1$ µg/ml). However, these investigators reported that the parenteral gold preparations, aurothiomalate and aurothioglucose, only suppressed IL-1 release when it was stimulated by PHA. At variance with the latter, Drakes et al.[116] used the thymocyte comitogenesis (LAF) assay to test for IL-1 and found that aurothiomalate suppressed both production of and response to the IL-1-like activity found in supernatants from LPS-treated human peripheral blood monocytes. In contrast, Matsubara et al.[117] claimed that neither aurothiomalate nor auranofin affected the secretion of the cytokine from human monocytes treated with LPS, but clinically relevant (<1 µg/ml) concentrations of both inhibited IL-1 induced fibroblast proliferation. However, they, as well as several of the above investigators, noted that suppression of proliferation was not unique to cytokine-stimulated events, but rather a consequence of the general capacity of these drugs to inhibit cellular proliferation in certain circumstances. These investigators suggested that gold may nonspecifically counteract the proliferative response of synovial fibroblasts and suppress the formation of pannus (invasive hyperplastic synovium characteristic of rheumatoid arthritis), but concluded that there was no evidence for a direct antagonism of either synthesis or response to IL-1 by gold salts.

Thus, the *in vitro* effects of gold on events mediated by IL-1 remain controversial. In part, this may reflect the individual spectrum of nonspecific activities intrinsic to each gold compound and the relative concentrations at which these activities become evident using nonselective bioassays. Additionally, there is a critical need to control for drug carry-over in many of the *in vitro* systems. *In vivo*, Lee et al.[118] reported that prophylactic administration of auranofin to arthritic rats reduced the disease-associated elevation of IL-1 synthesis and restored the production of IL-2 by spleen cells. Treatment of rats with established adjuvant arthritis resulted in normalization of IL-1 levels, but IL-2 production remained depressed. Identical treatments of nonarthritic animals had no effect on interleukin levels, and neither prophylactic nor therapeutic dosing altered the depressed IL-3 levels. However, the partial restoration of normal interleukin levels by auranofin was associated with only minor improvements in inflammation, whereas indomethacin, while having no effect on interleukin levels, clearly suppressed joint inflammation. The authors concluded that in this experimental model of arthritis, abnormalities in interleukin production are more sensitive to the disease-modifying properties of auranofin than those lesions responsive to inhibition of prostaglandin synthesis. It is of obvious significance to determine which of these sets of properties is more closely associated with the underlying pathophysiology of rheumatoid arthritis in humans. Since the effects of the aspirin-like NSAIDs, including indomethacin, in human arthritis are primarily palliative, whereas gold has been suggested to modify the progression of the disease, it is tempting to surmise that normalization of interleukin levels will ultimately be of greater value.

The antimalarial compounds, chloroquine and hydroxychloroquine, are accepted treatments for rheumatoid arthritis.[119] In a preliminary study of several classes of antiarthritic and analgesic compounds, only hydroxychloroquine prevented IL-1-induced proteoglycan degradation at reasonable concentrations.[16] This observation prompted additional study by Rainsford[120] of a series of quinolines and related compounds. Of these, mefloquine (Figure 4) was the most potent inhibitor of cartilage degradation by porcine catabolin, being about twice as potent as chloroquine and hydroxychloroquine (IC_{50}s approximately 50 to 60 μM). When rhIL-1$_\alpha$ was used to provoke proteoglycan degradation, rather than the pig cytokine preparation, mefloquine was equivalent to chloroquine and hydroxychloroquine. This suggested to the author that there might be subtle distinctions in drug or interleukin receptors in bovine cartilage,[120] but it may have been that minor contaminants in the porcine catabolin preparation interacted with the drugs or receptors. Skotnicki et al.[121]

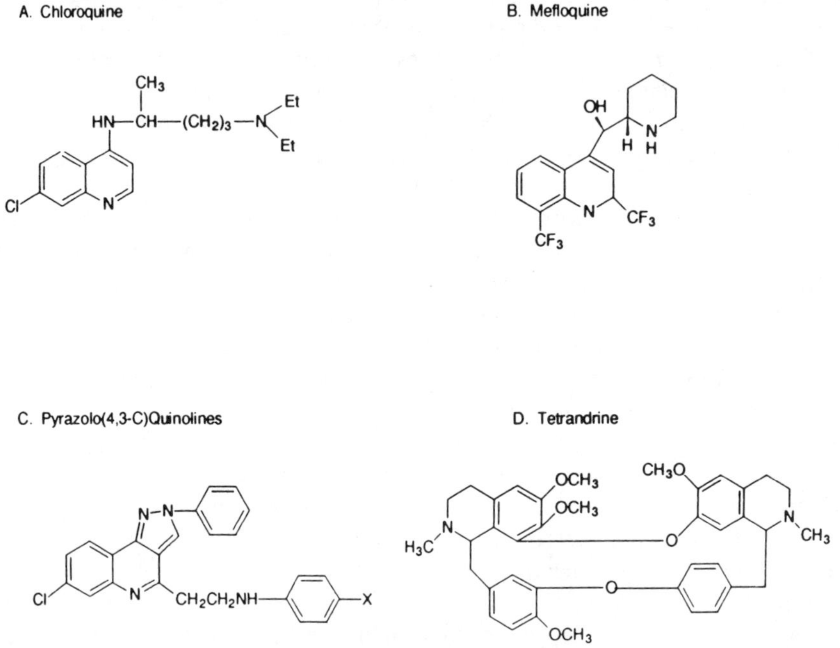

Figure 4. Quinolines. Chemical structures of (A) the antimalarial chloroquine; (B) mefloquine; (C) pyrazolo(4,3-C)quinolines claimed in Skotnicki et al.,[121] where when X = CO_2Et, the compound is designated WY-48988, and when X = CN, the compound is WY-48989; and (D) the isoquinoline, tetrandrine.

disclosed a series of pyrazolo-quinolines (Figure 4), which by virtue of their ability to inhibit IL-1 were proposed as potentially useful in the treatment of inflammation and diseases associated with enzymatic destruction of tissues. The bisbenzylisoquinoline, tetrandrine (Figure 4), which affects a number of monocyte/macrophage functions, has also been reported by Seow et al.[122] to inhibit the production of IL-1 from human monocytes. It remains to be established whether antagonism of cytokine-induced cartilage destruction by antimalarial compounds contributes to their antiarthritic effects in humans. In the adjuvant arthritic rat, chloroquine failed to influence the excessive production of IL-1 by either splenic macrophages or cells from peritoneal exudates.[102]

Cyclosporin A (CsA), an immunosuppressive fungal metabolite that has had a major impact on the field of organ transplantation, is presently being used experimentally to treat arthritis.[123,124] A number of studies (for reviews see references 125 and 126) have indicated that this cyclic undecapeptide appears to prevent the activation of T-cells by mechanisms involving the production and response to IL-2. Dos Reis and Shevach[127] performed studies suggesting that CsA might affect IL-2 by interfering with the response of cells that produce IL-2 to IL-1. Palacios[128,129] observed that IL-2-producing lymphocytes lost the ability to respond to IL-1 after treatment with CsA, while Bunjes et al.,[130] reported that increasing concentrations of exogenous IL-1 could abrogate CsA suppression of proliferation, but not IL-2 secretion in response to Con A. Using mixing experiments in which only one cell population was treated with CsA, Palacios[129] suggested that CsA decreased the production of IL-1 not by a direct effect on macophages, but in some fashion involving $CD4^+$ lymphocytes. However, CsA can also directly affect macrophages in the absence of T-cell mediation.[131] Bendtzen and Dinarello[132] demonstrated that active, but not inactive, forms of CsA can antagonize and displace IL-1 binding to T-cells. Thus, the *in vitro* evidence suggests that CsA can both antagonize and prevent the production of IL-1. Nevertheless, these effects have not been demonstrated to be specific and do not mean that CsA interacts with the IL-1 receptor in a selective manner; similar observations have been made with respect to displacement of prolactin[133,134] and other immunoregulatory molecules from binding sites on T-cells.

In vivo, oral dosing with this fungal metabolite resulted in alterations in IL-1 production in experimentally induced adjuvant arthritic rats. Connolly et al.[68] reported that the oral administration of CsA to rats at doses of 3 and 5 mg/kg/d prevented the development of inflammation, reduced the production of thymocyte comitogenic activity by spleen cells, and reduced the levels of plasma CRP and fibronectin. As noted above, normalization of the acute phase response can either be a consequence of alleviating the disease in these animals or a reflection of the drug's ability to reduce IL-1 levels. NSAIDs are very effective therapy for adjuvant arthritis and, consequently, have been noted to reduce the concentration of acute phase reactants in this animal model.[135,136] However, in this study, both aspirin (maximum dose, 200 mg/kg/d) and phenylbutazone (maximum dose, 30 mg/kg/d) alleviated inflammation, but consistent with their inablility to suppress the production of IL-1 by splenocytes, neither influenced the acute phase response.

The use of low doses of the folic acid antagonist methotrexate, a drug used extensively to treat leukemia and other types of cancer,[137] has recently been demonstrated to possess considerable promise for the treatment of rheumatoid arthritis.[138] Methotrexate alleviated joint inflammation in experimental models of arthritis in rats and was shown by Hu et al.[139] to suppress Ia antigen expression and the synthesis of IL-1 by peritoneal macrophages obtained from arthritic animals. It did not inhibit the synthesis of IL-1 by resident peritoneal macrophages from nonarthritic rats challenged *in vitro* with LPS. This caused speculation that the effects of methotrexate on macrophage interleukin synthesis might be indirect, possibly a consequence of activating T-cells to produce cytokines (e.g., interferon and colony stimulating factors) known to regulate macrophage function. Irrespective of mechanism, the fact that dosing with methotrexate reduced IL-1 levels and Ia antigen expression suggested that suppression of macrophage function might be responsible for the effectiveness of methotrexate in this animal model. Whether this may also be true in humans is not known.

The ability of methotrexate to suppress IL-1 production in animal models of arthritis has been confirmed in other studies.[40,68,140] In fact, Johnson et al.[40] reported that in their study, methotrexate uniquely inhibited all the indices of macrophage activation measured — IL-1 and PGE_2 production, cyanine dye accumulation — as well as suppressing inflammation and the influx of Ia^+ macrophages into synovial tissues. Penicillamine was inactive, indomethacin only inhibited paw edema, and the effects of prednisolone were limited to alleviation of swelling and suppression of IL-1 production; methotrexate was the only compound tested that suppressed all parameters of disease.

PEPTIDES

Several endogenous, naturally occurring inhibitors have been described. It would appear that these molecules serve to limit or down regulate IL-1 activity, particularly in those circumstances in which it is produced in quantity. These inhibitors are discussed elsewhere in this volume (Chapter 2) and will not be covered here.

Lipton and co-workers[141,142] reported that α-melanocyte stimulating hormone (α-MSH) and its analogue, [Nle^4, D-Phe^7]-α-MSH, antagonized the development of fever in response to IL-1 in rabbits. These data support the existence of neuroimmune feedback loops mentioned above. IL-1 stimulates secretion of ACTH by the pituitary, and both ACTH and α-MSH (which corresponds to the first thirteen amino acids of ACTH) oppose the activities of IL-1. *In vitro,* Cannon et al.[143] observed that concentrations of α-MSH as low as 10^{-11} M inhibited the comitogenic property of IL-1 with thymocytes and the production of PGE_2 from fibroblasts, but [Nle^4,D-Phe^7]-α-MSH did not share this activity. Unfortunately, Daynes et al.[144] were unable to confirm the *in vitro* data and concluded that neither α-MSH nor [Nle^4,D-Phe^7]-α-MSH influenced PGE_2 production or LAF activity. However, both peptides reduced a number of *in vivo* responses to exogenous IL-1, including pyrexia, production of the acute phase protein, serum amyloid-P (SAP), and the delayed cutaneous hypersensitivity response to dinitrofluorobenzene. Thus, products of the proopiomelanocortin gene, the transcription of which can be regulated by IL-1, can exert neuroendocrine influences on the expression of at least some of the *in vivo* immunological and inflammatory responses to IL-1. This is not receptor-mediated antagonism. Similar observations have been made by Robertson et al.[145] demonstrating that α-MSH inhibited responses to exogenous TNF; TNF and IL-1 are not extensively homologous and utilize distinct receptors. It is unlikely that α-MSH would possess sufficient affinity for both to antagonize the endogenous ligands and also the selectivity to trigger its own physiologic receptor. Finally, it has been shown by Cannon et al.[143] that α-MSH does not affect the synthesis or secretion of IL-1.

CKS-17 is a peptide that exhibits immunosuppressive properties when coupled to albumin. It corresponds to the first 16 amino acids[146] of a highly conserved 26-amino acid sequence found in the transmembrane envelop proteins (P15E) of retroviruses isolated from a number of species. P15E-like antigens are expressed by human tumors and may participate in subversion of the host antitumor immune response. In the process of studying inhibition by CKS-17 of *in vitro* monocyte-mediated antitumor cytotoxic responses, Kleinerman et al.[147] noted that the immunosuppressive peptide prevented both α and β forms of IL-1 from lysing A375 tumor cells. Despite the high peptide concentration (10^{-5} *M*) required for activity, inhibition was not due to monocyte toxicity and appeared somewhat specific, since when bound to albumin, the peptides neurotensin and CS-2 (partially homologous to CKS-17) failed to block lysis of tumor cells by IL-1. Furthermore, CKS-17-albumin appeared to possess selectivity for responses to IL-1. It had no effect on the TNF-mediated lysis of L929 cells or the activity of IL-3 on FDCP-1 cells, but did inhibit proliferative response of several cell lines to IL-1. CKS-17 did not alter synthesis or secretion of IL-1. It was proposed by these investigators that CKS-17 modified monocyte function by inactivating exogenous IL-1. Bakouche et al.[148] extended these studies by showing that CKS-17 could also suppress responses to IL-1 bound to membranes.

IL-1 and several peptide analogues derived from amino acids 187-204 of IL-1β were reported by Ferreira et al.[149] to be hyperalgesic following systemic injection in the rat. The tripeptide, Lys-Pro-Thr, was thought to be a partial agonist, since doses insufficient to elicit hyperalgesia could suppress nociceptive responses to IL-1. Substitution of Pro with its D-isomer yielded a peptide (Lys-D-Pro-Thr) that lacked intrinsic agonist activity, but retained the capability to attenuate hyperalgesia elicited by the injection of IL-1β or carrageenan. Suppression of hyperalgesia could be overcome by increasing the dose of IL-1. Specificity was inferred from the observation that IL-1$_\alpha$, which lacks the Lys-Pro-Thr sequence, is also weakly hyperalgesic; yet IL-1$_\alpha$ hyperalgesia was resistant to attenuation by Lys-D-Pro-Thr, indicating specificity. The possibility that Lys-D-Pro-Thr reduces the hyperalgesic response to IL-1 indirectly, like NSAIDs, by interfering with prostaglandin synthesis is unlikely. At concentrations up to 200 µg/ml, it had no effect on basal, IL-1, or LPS-stimulated production of PGE_2 from human peripheral blood mononuclear cells *in vitro* and was not antipyretic in rabbits. Lys-D-Pro-Thr has been proposed by Ferreira and colleagues[149] as prototypical of a new class of peripheral analgesics: drugs that antagonize IL-1.

SOME MISCELLANEOUS INHIBITORS

Roche et al.[150] reported that the quinolone antibiotics, pefloxacin and ciprofloxacin (Figure 5), inhibited the secretion of IL-1 from LPS-stimulated monocytes, but did not affect cellular levels. Unfortunately, the concentrations required for suppression of cytokine release were above the plasma concentrations (10 µg/ml) normally achieved with therapeutic doses of the antibiotics.

Sprirogermanium is a metal-containing, cytotoxic compound demonstrated by DiMartino et al.[39] to suppress inflammation in the adjuvant arthritic rat. It also corrected the enhanced synthesis of IL-1 exhibited by these animals *ex vivo* but did

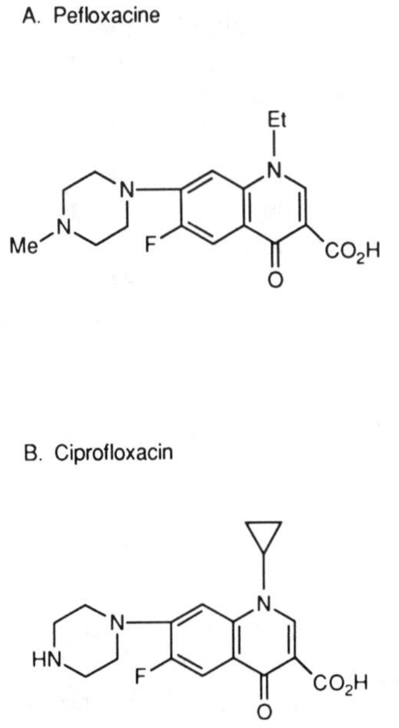

Figure 5. Structures of the antibiotics pefloxacine and ciprofloxacin.

not normalize abnormalities in the production of IL-2 and IL-3. The immunosuppressive activity of spirogermanium in this animal model was described as being generally mild, and the authors speculated that the antiarthritic activity displayed in this study may derive from suppression of macrophage function.

In view of the polypeptide nature of the interleukins, it is not surprising that agents that interfere with protein synthesis nonspecifically inhibit their production. Similarly, since many of the responses to IL-1 are dependent on gene activation and new enzyme synthesis, compounds which interfere with transcription or translation will decrease such responses. For example, Rainsford[16] observed that the intercalator echinomycin was one of the few compounds that reduced catabolin-induced cartilage degeneration, and more recently[36] observed that actinomycin and doxorubicin, inhibitors of transcription, had similar properties.

Bender et al.[151] claimed that compounds of the general structure shown in Figure 6 inhibited the production of IL-1 from human monocytes under conditions where ibuprofen was inactive and the CO/LO inhibitors phenidone and NDGA were only marginally active.

Several investigators have been studying the effects of second messengers and compounds affecting signal transduction in order to define the pathways activated by IL-1. For a discussion of the results of these efforts, please refer to Chapter 4 in this volume.

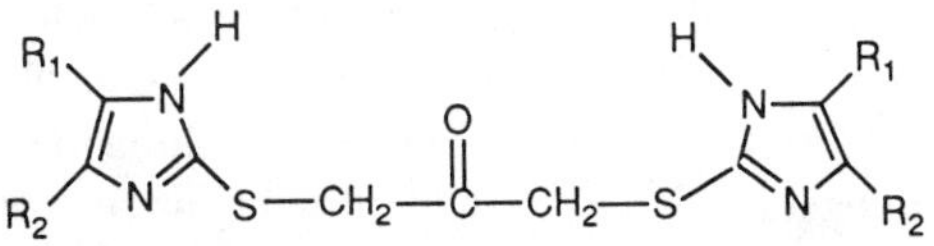

Figure 6. Generic formula of interleukin-1 inhibitors disclosed in Bender et al.[151] R_1 or R_2 is 4-pyridyl, and the other is a monohalo-substituted phenyl.

FUTURE APPROACHES

As noted in the introduction, it is difficult to predict or dissect the consequences of ablating the function of a particular cytokine in biological systems. The use of compounds that specifically inhibit selected cytokines is one approach to trying to answer such questions. However, the observation that a compound modulates the synthesis or response to IL-1 does not exclude the strong likelihood that it also modifies the activity of other cytokines. The compounds discussed in this chapter were not designed as specific cytokine inhibitors, and therefore, they possess other pharmacologic properties that can complicate interpretations of the mechanism responsible for their effects in complex biological systems. One way to achieve the desired selectivity would be to use specific anti-cytokine antibodies, but this approach is primarily restricted to neutralization of cell surface or secreted peptides and is of limited use in evaluating intracellular cytokine synthesis. One additional method that shows considerable promise for future studies is the use of antisense oligo- or polynucleotides. This approach requires knowledge of the base sequence of the molecule of interest, but thereafter offers considerable specificity, versatility, and apparently wide applicability. Conceptually, it should be possible not only to inhibit the production of a particular cytokine, but also to prevent the expression of its receptor and to determine the biological consequences. The use of antisense polynucleotides to answer complex questions is not yet common, but application of this technology is increasing. An example of its power and selectivity is the elegant study of Harel-Bellan et al.[152] who used antisense oligonucleotides to IL-2 and IL-4 to differentiate between Th1 and Th2 T helper cell subsets. Recently our laboratory has used a 21-base antisense oligonucleotide sequence to inhibit the production of IL-1 from mouse thioglycollate-elicited macrophages stimulated with LPS. We plan to do the same with antisense oligonucleotides to IL-6 and TNF to evaluate the relative contribution of individual cytokines and various combinations ("cocktails") to macrophage-mediated immunological and inflammatory processes. The primary problem with this approach at the present time is the difficulty in applying this technology to studies in whole animals. Using transgenic mice, it has been demonstrated by Katsuki et al.[153] that antisense genes can produce responses *in vivo*, but such procedures are not routine. Nevertheless, advances in the stabilization of these oligomers are being made, and perhaps a combination of methods, such as increasing cellular uptake and resistance to degradation by chemical modification of the oligonucleotide,[154,155] packaging in liposomes, or binding to some other protective or targeting macromolecule,[156] may provide the means for the application of this technology to study complex physiological or pathological models. It is even conceivable that the development of advances in methods for administration of these molecules to living organisms may ultimately be the basis for rational development of an entirely new class of highly specific therapeutic agents[157] that could be used singly or in various combinations designed to meet the needs of treating a specific disease.

NOTE ADDED IN PROOF

Since this manuscript was prepared, there have been additional reports of compounds with the capacity to inhibit IL-1. In this respect, IX 207-887 is somewhat unusual since it was shown to suppress the release of IL-1 from human monocytes and mouse peritoneal macrophages with minimal effects on IL-1 synthesis (intracellular levels).[158] Suppression of IL-1 was demonstrated using biological and immunological (RIA/ELISA) techniques. Inhibition of release was relatively selective; suppression of TNF, IL-6, and lysozyme secretion was reduced or absent relative to the effect on IL-1. The concentrations used in these *in vitro* studies were within the therapeutic range achieved in humans and animals, and IX 207-887 was reported to be active *in vivo* in experimental animal models and in human rheumatoid arthritis. It is not a CO inhibitor.

In vitro, inhibition of human monocyte TNF_α and IL-6 with the appropriate 18-base antisense oligonucleotide fragments has been reported.[159] The corresponding sense fragments were devoid of activity. The use of antisense technology is becoming quite widespread; a new journal devoted to the topic (*Antisense Research and Development,* Mary Ann Liebert, Inc., New York) has recently been introduced.

REFERENCES

1. Dinarello, C.A. and S.M. Wolff, The molecular basis of fever in humans. *Am. J. Med.,* 72, 799, 1982.
2. Dinarello, C.A., S.O. Marnoy, and J.L. Rossenwasser, Role of arachidonate metabolism in the immunoregulatory function of human leukocytic pyrogen/lymphocyte-activating factor/interleukin 1. *J. Immunol.,* 130, 890, 1983.
3. Dinarello, C.A., Interleukin-1. *Rev. Infect. Dis.,* 61, 51, 1984.
4. Dinarello, C.A. and H.A. Bernheim, The ability of human leukocytic pyrogen to stimulate brain prostaglandin synthesis *in vitro. J. Neurochem.,* 37, 702, 1981.

5. Flower, R.J. and J.R. Vane, Inhibition of prostaglandin biosynthesis. *Biochem. Pharmacol.*, 23, 1439, 1974.
6. Chang, J., S.C. Gilman, and A.E. Lewis, Interleukin-1 activates phospholipase A_2 in rabbit chorndrocytes: a possible signal for IL-1 action. *J. Immunol.*, 136, 1283, 1986.
7. Godfrey, R.W., W.J. Johnson, and S.T. Hoffstein, Interleukin-1 stimulation of phospholipase activity in rat synovial fibroblasts. Possible regulation by cyclooxygenase products. *Arthr. Rheum.*, 31, 1421, 1988.
8. Dinarello, C.A., An update on human interleukin-1: from molecular biology to clinical relevance, *J. Clin. Immunol.*, 5, 287, 1985.
9. Dayer, J.-M., D.R. Robinson, and S.M. Krane, Prostaglandin production by rheumatoid synovial cells: stimulation by a human lymphocyte factor. *Science*, 195, 181, 1977.
10. Balavoine, J.-F., B. de Rochemonteix, K. Williamson, P. Seckinger, A. Cruchaud, and J.-M. Dayer, Prostaglandin E_2 and collagenase production by fibroblasts and synovial cells is regulated by urine-derived human interleukin 1 and inhibitor(s). *J. Clin. Invest.*, 78, 1120, 1986.
11. Mochan, E., J. Uhl, and R. Newton, Evidence that interleukin-1 induction of synovial cell plasminogen activator is mediated via prostaglandin E_2 and cAMP. *Arthr. Rheum.*, 29, 1078, 1986.
12. Sung, K., D. Mendelow, H.I. Georgescu, and C.H. Evans, Characterization of chondrocyte activation in response to cytokines synthesized by a synovial cell line. *Biochim. Biophys. Acta*, 971, 148, 1988.
13. Rossi, V., R. Breviario, P. Ghezzi, E. Dejani, and A. Mantovani, Prostacyclin synthesis induced in vascular cells by interleukin-1. *Science*, 229, 174, 1985.
14. Last-Barney, K., C.A. Homon, R.B. Faanes, and V.J. Merluzzi, Synergistic and overlapping activities of tumor necrosis factor-alpha and IL-1. *J. Immunol.*, 141, 527, 1988.
15. Butler, D.M., D.S. Piccoli, P.H. Hart, and J.A. Hamilton, Stimulation of human synovial fibroblast DNA synthesis by recombinant human cytokines. *J. Rheumatol.*, 15, 1463, 1988.
16. Rainsford, K.D., Preliminary investigations on the pharmacological control of catabolin-induced cartilage destruction *in vitro*. *Agents Actions*, 16, 55, 1985.
17. Arner, E.C., L.R. Darnell, M.A. Pratta, R.C. Newton, N.R. Ackerman, and W. Galbraith, Effect of anti-inflammatory drugs on interleukin-1-induced cartilage degradation. *Agents Actions*, 21, 334–336, 1987.
18. Couchman, K.G. and H. Sheppeard, The effect of anti-rheumatic drugs on factors from porcine synovium inducing chondrocyte mediated cartilage degradation. *Agents Actions*, 19, 116, 1986.
19. Sheppeard, H., L.M.C. Pilsworth, B. Hazleman, and J.T. Dingle, Effects of antirheumatoid drugs on the production and action of porcine catabolin. *Ann. Rheum. Dis.*, 41, 463, 1982.
20. Brandwein, S.R., Regulation of interleukin 1 production by mouse peritoneal macrophages: effects of arachidonic acid metabolites, cyclic nucleotides and interferons. *J. Biol. Chem.*, 261, 8624, 1986.
21. Kunkel, S.L., S.W. Chensue, and S. Phan, Prostaglandins as endogenous mediators of interleukin 1 production. *J. Immunol.*, 136, 186, 1986.
22. Lee, J.C., D.E. Griswold, B. Votta, and N. Hanna, Inhibition of monocyte IL-1 production by the anti-inflammatory compound, SK&F 86002. *Int. J. Immunopharmacol.*, 7, 835, 1988.

23. Otterness, I.G., M.L. Bliven, J.D. Eskra, M. Reinke, and D.C. Hanson, The pharmacologic regulation of interleukin-1 production: the role of prostaglandins. *Cell. Immunol.,* 114, 385, 1988b.
24. Raz, A., A. Wyche, N. Siegel, and P. Needleman, Regulation of fibroblast cyclooxygenase by interleukin-1. *J. Biol. Chem.,* 263, 3022, 1988.
25. Burch, R.M., J.R. Connor, and J. Axelrod, Interleukin 1 amplifies receptor-mediated activation of phospholipase A_2 in 3T3 fibroblasts. *Proc. Natl. Acad. Sci. U.S.A.,* 85, 6306, 1988.
26. Korn, J.H., E. Downie, G.J. Roth, and S.-Y. Ho, Synergy of interleukin 1 (IL-1) with arachidonic acid and A23187 in stimulating PGE synthesis in human fibroblasts: IL-1 stimulates fibroblast cyclooxygenase. *Clin. Immunol. Immunopathol.,* 50, 196, 1989.
27. Kunkel, S.L., S.W. Chensue, M. Spengler, and J. Geer, Effect of arachidonic acid metabolites and their metabolic inhibitors on interleukin-1 production. *Prog. Leukoc. Biol.,* 2, 297–307, 1985.
28. Van Damme, J., J. Van Beeuman, B. Decock, J. Van Snick, M. De Ley, and A. Billiau, Separation and comparison of two monokines with LAF activity (interleukin-1Beta and hybridoma growth factor): identification of leucocyte-derived HGF as interleukin-6. *J. Immunol.,* 140, 1534, 1988.
29. Lotz, M., F. Jirik, P. Kabouridis, C. Tsourkas, T. Hirano, T. Kishimoto, and D. Carson, B cell stimulating factor 2/interleukin 6 is a costimulant for human thymocytes and T lymphocytes. *J. Exp. Med.,* 167, 1253, 1988.
30. Dinarello, C.A., T. Ikejima, S.J.C. Warner, S.F. Orencole, G. Lonnemann, J.G. Cannon, and P. Libby, Interleukin 1 induces interleukin 1. I. Induction of circulating interleukin 1 in rabbits in vivo and humans in vitro. *J. Immunol.,* 139, 1902, 1987.
31. Herman, J. and A.R. Rabson, Prostaglandin E_2 depresses natural cytotoxicity by inhibiting interleukin-1 production by large granular lymphocytes. *Clin. Exp. Immunol.,* 57, 380, 1984.
32. Warner, S.J., K.R. Auger, and P. Libby, Interleukin 1 induces interleukin 1. II. Recombinant human interleukin 1 induces interleukin 1 production by adult human vascular endothelial cells. *J. Immunol.,* 139, 1911, 1987a.
33. Warner, S.J.C., K.R. Auger, and P. Libby, Human interleukin 1 induces interleukin 1 gene expression in human vascular smooth muscle cells. *J. Exp. Med.,* 165, 1316, 1987b.
34. Mauviel, A., N. Temime, D. Charron, G. Loyau, and J.-P. Pujol, Interleukin-1 alpha and beta induce interleukin-1 beta gene expression in human dermal fibroblasts. *Biochem. Biophys. Res. Commun.,* 156, 1209, 1988.
35. Knudsen, P.J., C.A. Dinarello, and T.B. Strom, Prostaglandins posttranscriptionally inhibit monocyte expression of interleukin 1 activity by increasing intracellular cyclic adenosine monophosphate. *J. Immunol.,* 137, 3189, 1986.
36. Rainsford, K.D., Actions of nonsteroidal anti-inflammatory drugs on the functions of lymphocytes. *Agents Actions Suppl.,* 24, 54, 1988.
37. Okusawa, S., J.A. Gelfand, T. Ikejima, R.J. Connolly, and C.A. Dinarello, Interleukin 1 produces a shock-like state in rabbits. Synergism with tumor necrosis factor and the effect of cyclooxygenase inhibition. *J. Clin. Invest.,* 81, 1162, 1988.
38. Beasley, D., C.A. Dinarello, and J.G. Cannon, Interleukin-1 induces natriuresis in conscious rats: role of renal prostaglandins. *Kidney Int.,* 33, 1059, 1988.
39. DiMartino M.J., J.C. Lee, A.M. Badger, K.M. Muirhead, C.K. Mirabelli, and N. Hanna, Antiarthritic and immunoregulatory activity of spirogermanium. *J. Pharm. Exp. Ther.,* 236, 103, 1986.

40. Johnson, W.J., M.J. DiMartino, P.C. Meunier, K.A. Muirhead, and N. Hanna, Methotrexate inhibits macrophage activation as well as vascular and cellular inflammatory events in rat adjuvant induced arthritis. *J. Rheumatol.*, 15, 745, 1988.
41. Connolly, K.M., V.J. Stecher, E. Danis, D.J. Pruden, and T. LaBrie, Alteration of interleukin-1 production and the acute phase response following medication of adjuvant arthritic rats with cyclosporin-A or methotrexate. *Int. J. Immunopharmacol.*, 10, 717, 1988.
42. Fauci, A.S., Mechanisms of the immunosuppressive and anti-inflammatory effects of glucocorticosteroids. *J. Immunopharmacol.*, 1, 1–25, 1978.
43. Dillard, G.M. and P. Bodel, Studies on steroid fever. II. Pyrogenic and anti-pyrogenic activity of some endogenous steroids of man. *J. Clin. Invest.*, 49, 2418, 1970.
44. Snyder, D.S. and E.R. Unanue, Corticosteroids inhibit murine macrophage Ia expression and interleukin 1 production. *J. Immunol.*, 129, 1803, 1982.
45. Stosic-Grujicic, S. and M.M. Simic, Modulation of interleukin 1 production by activated macrophages: in vitro action of hydrocortisone, colchicine, and cytochalasin. *B. Cell. Immunol.*, 69, 235, 1982.
46. Draber, P., Mechanism of the inhibitory effect of hydrocortisone on conconavalin A-induced activation of thymus cells. Role of interleukins. *Int. J. Immunopharmacol.*, 4, 401, 1982.
47. Gery, I., W.R. Benjamin, and R.B. Nussenblatt, Cyclosporin selectively inhibits certain mitotic responses of thymocytes. *Transplant. Proc.*, 15, 2311, 1983.
48. Tracey, D.E., M.M. Hardee, K.A. Richard, and J.W. Paslay, Pharmacological inhibition of interleukin-1 activity on T cells by hydrocortisone, cyclosporin, prostaglandins, and cyclic nucleotides. *Immunopharmacology*, 15, 47, 1988.
49. Stim, T., E. Kimball, and F. Persico, unpublished observation.
50. Akahoshi, T., J.J. Oppenheim, and K. Matsushima, Induction of high affinity functional receptors for interleukin 1 by glucocorticoid hormones on human peripheral blood lymphocytes. *Lymphokine Res.*, 6, 1240, 1987.
51. Matsushima, K., Y. Kobayashi, T.D. Copeland, T. Akahoshi, and J.J. Oppenheim, Phosphorylation of a cytosolic 65-kDa protein induced by interleukin 1 in glucocorticoid pretreated normal human peripheral blood mononuclear leukocytes. *J. Arthr. Immunol.*, 139, 3367, 1987.
52. Evans, R.M., The steroid and thyroid hormone receptor superfamily. *Science*, 240, 889, 1988.
53. Schule, R., M. Muller, C. Kaltschmidt, and R. Renkawitz, Many transcription factors interact synergistically with steroid receptors. *Science*, 242, 1418, 1988.
54. Oro, A.E., S.M. Hollenberg, and R.M. Evans, Transcriptional inhibition by a glucocorticoid receptor-beta-galactosidase fusion protein. *Cell*, 55, 1109, 1988.
55. Guertin, M., P. Baril, J. Bartkowiak, A. Anderson, and L. Belanger, Rapid suppression of alpha$_1$-fetoprotein gene transcription by dexamethasone in developing rat liver. *Biochemistry*, 22, 4296, 1983.
56. Brown, S.L., L.R. Smith, and J.E. Blalock, Interleukin 1 and interleukin 2 enhance proopiomelanocortin gene expression in pituitary cells. *J. Immunol.*, 139, 3181, 1987.
57. Cockayne, D., K.M. Sterling, Jr., S. Shull, K.P. Mintz, S. Illeyne, and K.R. Gutreono, Glucocorticoids decrease the synthesis of type I procollagen mRNAs. *Biochemistry*, 25, 3209, 1986.
58. Frisch, S.M. and H.E. Ruley, Transcription from the stromolysin promoter is induced by interleukin-1 and repressed by dexamethasone. *J. Biol. Chem.*, 262, 16300, 1987.

59. Brinckerhoff, C.E., I.M. Plucinska, L.A. Sheldon, and G.T. O'Connor, Half-life of synovial cell collagenase mRNA is modulated by phorbol myristate acetate but not by all-trans-retinoic acid or dexamethasone. *Biochemistry*, 25, 6378, 1986.
60. Vitti, G. and J.A. Hamilton, Modulation of urokinase-type plasminogen activator messenger RNA levels in human synovial fibroblasts by interleukin-1, retinoic acid, and a glucocorticoid. *Arthr. Rheum.*, 31, 1046, 1988.
61. Busso, N., D. Belin, C. Failly-Crepin, and J.-D. Vassalli, Glucocorticoid modulation of plasminogen activators and of one of their inhibitors in the human mammary carcinoma cell line MDA-MB-231. *Cancer Res.*, 47, 364, 1987.
62. Lee, S.W., Tsou, A.-P., H. Chan, J. Thomas, K. Petrie, E.M. Eugui, and A.C. Allison, Glucocorticoids selectively inhibit the transcription of the interleukin 1_{beta} gene and decrease the stability of interleukin 1_{beta} mRNA. *Proc. Natl. Acad. Sci. U.S.A.*, 85, 1204, 1988.
63. Nishida, T., M. Takano, T. Kawakami, N. Nishino, S. Nakai, and Y. Hirai, The transcription of the interleukin 1-beta gene is induced with PMA and inhibited with dexamethasone in U937 cells. *Biochem. Biophys. Res. Commun.*, 156, 269, 1988.
64. Beutler, B., P. Thompson, J. Keyes, K. Hagerty, and D. Crawford, Assay of a ribonuclease that preferentially hydrolyses mRNAs containing cytokine derived UA-rich instability sequences. *Biochem. Biophys. Res. Commun.*, 152, 973, 1988.
65. Allison, A.C. and S.W. Lee, Pro-inflammatory and catabolic effects of interleukin-1 and theri antagonism by glucocorticoids. *Agents Actions Suppl.*, 24, 207, 1988.
66. Staruch, M.J. and D.D. Wood, Reduction of serum interleukin-like activity after treatment with dexamethasone. *J. Leukoc. Biol.*, 37, 193, 1985.
67. Hill, M.R., R.D. Stith, and R.E. McCallum, Interleukin 1, a regulatory role in glucocorticoid-regulated hepatic metabolism. *J. Immunol.*, 137, 858, 1986.
68. Connolly, K.M., V.J. Stecher, T. LaBrie, and C. Fluno, Modulation of the acute phase response and in vitro measurement of interleukin-1 activity following administration of dexamethasone to adjuvant-arthritic rats. *Immunopharmacology*, 15, 133, 1988.
69. Besedovsky, H., A. Del Rey, E. Sorkin, and C.A. Dinarello, Immunoregulatory feedback between interleukin-1 and glucocorticoid hormones. *Science*, 233, 652, 1986.
70. Lumpkin M.D., The regulation of ACTH secretion by IL-1. *Science*, 238, 452, 1987.
71. Bernton E.W., J.E. Beach, J.W. Holliday, R.C. Smallridge, and H.G. Fein, Release of multiple hormones by a direct action of interleukin-1 on pituitary cells. *Science*, 238, 519, 1987.
72. Larrick, J.W., Native interleukin 1 inhibitors. *Immunol. Today*, 10, 61, 1989.
73. Berkenbosch, F., J. Van Oers, A. Del Rey, F. Tilders, and H. Besedovsky, Corticotropin-releasing factor-producing neurons in the rat activated by IL-1. *Science*, 238, 452, 1987.
74. Breder, C.D., C.A. Dinarello, and C.B. Safer, Interleukin-1 immunoreactive innervation of the human hypothalamus. *Science*, 240, 321, 1988.
75. Sapolskiy, R., C. Rivier, G. Yamamoto, P. Plotsky, and W. Vale, Interleukin-1 stimulates the secretion of hypothalamic corticotropin-releasing factor. *Science*, 238, 522, 1987.
76. Farrar, W.L. and J.L. Humes, The role of arachidonic acid metabolism in the activities of interleukin 1 and 2. *J. Immunol.*, 135, 1153, 1985.
77. Smith, R.J., B.J. Bowman, and S.C. Speziale, Interleukin-1 stimulates granule exocytosis from human neutrophils. *Prog. Leukoc. Biol.*, 2, 31, 1985.
78. Ulich, T.R., J. Del Castillo, M. Keys, G.A. Granger, and R.-X. Ni, Kinetics and mechanisms of recombinant human interleukin 1 and tumor necrosis factor-alpha-induced changes in circulating numbers of neutrophils and lymphocytes. *J. Immunol.*, 139, 3406, 1987.

79. Schultz, R.M. and M.G. Altom, Modulation of interleukin-1 activity on murine thymocytes by various inhibitors of arachidonic acid oxygenation. *Immunopharmacol. Immunotoxicol.*, 10, 21, 1988.
80. Crabtree, G.R., Contingent genetic regulatory events in T lymphocyte activation, *Science*, 243, 355, 1989.
81. Liu, D.S., F.Y. Liew, and J. Rhodes, Immunoregulatory properties of novel specific inhibitors of 5-lipoxygenase. *Immunopharmacology* 17, 1, 1989.
82. Dinarello, C.A., I. Bishai, L.J. Rosenwasser, and F. Coceani, The influence of lipoxygenase inhibitors on the *in vitro* production of human leukocytic pyrogen and lymphocyte activating factor (interleukin-1). *Int. J. Immunopharmacol.*, 6, 43, 1984.
83. Persico, F.J. et al., unpublished observations.
84. Katz, P., A.P. Borger, and L.D. Loose, Evaluation of CP-66,248 [5-chloro-2,3-dihydro-2-oxo-3(2-thienylcarbonyl)-indole-1-carboxamide] in rheumatoid arthritis (RA). *Arthr. Rheum.*, 31, S52, 1988.
85. Carty, T.J., H.J. Showell, L.D. Loose, and S.B. Kadin, Inhibition of both 5-lipoxygenase (5-LO) and cyclooxygenase (CO) pathways of arachidonic acid (AA) metabolism by CP-66,248. *Arthr. Rheum.*, 31, S89, 1988.
86. Otterness, I.G., T.J. Carty, and L.D. Loose, CP-66,248, A new drug for arthritis, in *Therapeutic Approaches to Inflammatory Diseases*, A.J. Lewis, N.R. Ackerman. Eds., Elsevier, New York, in press.
87. Otterness, I.G., M.L. Bliven, J.T. Downs, and D.C. Hanson, Effects of CP-66,248 on IL-1 synthesis by murine peritoneal macrophages. *Arthr. Rheum.*, 31, S90, 1988a.
88. McDonald, B., L. Loose, and L.J. Rosenwasser, The influence of a novel arachidonate inhibitor, CP66,248 on the production and activity of human monocyte IL-1. *Arthr. Rheum.*, 31, S17, 1988.
88a. McDonald, B., L. Loose, and L.J. Rosenwasser, Synovial fluid IL-1 in RA patients receiving a novel arachidonate inhibitor CP66,248, *Arthr. Rheum.*, 31, A88, 1988.
89. Le, J., G. Fredrickson, L.F.L. Reis, T. Diamanstein, T. Hirano, T. Kishimoto, and J. Vilcek, Interleukin 2-dependent and interleukin 2-independent pathways of regulation of thymocyte function by interleukin 6. *Proc. Natl. Acad. Sci. U.S.A.*, 85, 8643, 1988.
90. Wong, G.G. and S.C. Clark, Multiple actions of interleukin 6 within a cytokine network. *Immunol. Today*, 9, 137, 1988.
91. Davis, J.S., L. Loose, and A.P. Borger, Clinical efficacy of CP-66,248 [5-chloro-2,3-dihydro-2-oxo-3(2-thienylcarbonyl)-indole-1-carboxamide] in osteoarthritis (OA). *Arthr. Rheum.*, 31, S72, 1988.
92. McConkey, B., R.A. Crockson, A.P. Crockson, and A.R. Wilkinson, The effects of some anti-inflammatory drugs on the acute-phase proteins in rheumatoid arthritis. *Q. J. Med.*, New Series XLII, 785, 1973.
93. Amos, R.S., T.J. Constable, R.A. Crockson, A.P. Crockson, and B. McConkey, Rheumatoid arthritis: relation of serum C-reactive protein and erythrocyte sedimentation rates to radiographic changes. *Brit. Med. J.*, 1, 195, 1977.
94. Wright, V. and R. Amos, Do drugs change the course of rheumatoid arthritis? *Brit. Med. J.*, 280, 964, 1980.
95. Oppenheim, J.J., K. Matsushima, T. Yoshimura, and E.J. Leonard, The activities of cytokines are pleiotropic and interdependent. *Immunol. Lett.*, 16, 179, 1987.
96. Zhang, Y., J.-X. Lin, Y.K. Yip, and J. Vilcek, Enhancement of cAMP levels and of protein kinase activity by tumor necrosis factor and interleukin 1 in human fibroblasts: role in the induction of interleukin 6. *Proc. Natl. Acad. Sci. U.S.A.*, 85, 6802, 1988.

97. Gauldie, J., C. Richards, D. Harnish, P. Lansdorp, and H. Baumann, Interferon beta$_2$/ B-cell stimulatory factor type 2 shares identity with monocyte-derived hepatocyte-stimulating factor and regulates the major acute phase protein response in liver cells. *J. Immunol.*, 84, 7251, 1987.
98. Moshage, H.J., H.M.J. Roelofs, J.F. van Pelt, B.P.C. Hazenberg, M.A. van Leeuwen, P.C. Limburg, L.A. Aarden, and S.H. Yap, The effect of interleukin-1, interleukin-6 and its interrelationship on the synthesis of serum amyloid A and C-reactive protein in primary cultures of adult human hepatocytes. *Biochem. Biophys. Res. Commun.*, 155, 112, 1988.
99. Ganapathi, M.K., D. Schultz, A. Mackiewicz, D. Samols, S. Hu, A. Brabenec, S.S. Macintyre, and I. Kushner, Heterogeneous nature of the acute phase response. Differentail regulation of human serum amyloid A, C-reactive protein, and other acute phase proteins by cytokines in Hep 3B cells. *J. Immunol.*, 141, 564, 1988.
100. Griswold, D.E., P.J. Marshall, E.F. Webb, R. Godfrey, J. Newton, M.J. DiMartino, H.M. Sarau, J.G. Gleason, G. Poste, and N. Hanna, SK&F 86002, A structurally novel anti-inflammatory agent that inhibits lipoxygenase- and cyclooxygenase-mediated metabolism of arachidonic acid. *Biochem. Pharmacol.*, 36, 3463, 1987.
101. DiMartino, M.J., D.E. Griswold, B.A. Berkowitz, G. Poste, and N. Hanna, Pharmacologic characterization of the anti-inflammatory profiles of a new dual inhibitor of lipoxygenase and cyclooxygenase. *Agents Actions*, 20, 113, 1987.
102. Veys, E., H. Mielants, J. Symoens, G. Vetter, E.C. Huskisson, J. Scott, D.D. Felix-Davies, B. Wilkinson, M. Rosenthal, T.L. Vischer, and J.C. Gerster, A multicentre randomized double-blind study comparing two dosages of levamisole in rheumatoid arthritis. *J. Rheumatol.*, 5 (suppl. 4), 5, 1972.
103. Huskisson, E.C., The place of levamisole in the armamentarium for rheumatoid arthritis. *J. Rheumatol.*, 5 (suppl. 4), 149, 1972.
104. Amery, W.K. and H. Verhaegen, Effects of levamisole treatment in cancer patients. *J. Rheumatol.*, 5 (suppl. 4), 123, 1972.
105. Kimball, E.S., M.C. Clark, C.R. Schneider, and F.J. Persico, Enhancement of *in vitro* lipopolysaccharide-stimulated interleukin-1 production by levamisole. *Clin. Immunol. Immunopathol.*, 58, 385, 1991.
106. Shirota, H., S. Kobayashi, K. Terato, Y. Sakuma, K. Yamada. H. Ikuta, Y. Yamagishi, I. Yamatsu, and K. Katayama, Effect of the novel non-steroidal antiinflammatory agent N-Methoxy-3-(3,5-di-tert-butyl-4-hydroxybenzylidene)pyrrolidin-2-one on in vitro generation of some inflammatory mediators. *Arz. Forsch./Drug Res.*, 37, 936, 1987.
107. Shirota, H., M. Goto, R. Hashida, I. Yamatsu, and K. Katayama, Inhibitory effects of E-5110 on interleukin-1 generation from human monocytes, presented at Fourth International Conference of Inflammation Research Association, 1988.
108. Lee, J.C., D.E. Griswold, B. Votta, and N. Hanna, Inhibition of monocyte IL-1 production by the anti-inflammatory compound, SK&F 86002. *Int. J. Immunopharmacol.*, 7, 835, 1988.
109. Leibfarth, J.H. and R.H. Persellin, Review: mechanisms of action of gold. *Agents Actions*, 11, 458, 1981.
110. Crooke, S.T., R.M. Snyder, T.R. Butt, D.J. Ecker, H.S. Allaudeen, B. Monia, and C. Mirabelli, Cellular and molecular pharmacology of auranofin and related gold complexes. *Biochem. Pharmacol.*, 35, 3423, 1986.
111. Persellin, R.H. and M. Ziff, The effect of gold salt on lysosomal enzymes of the peritoneal macrophage. *Arthr. Rheum.*, 9, 57, 1966.

112. Lipsky, P. and M. Ziff, Inhibition of antigen- and mitogen-induced human lymphocyte proliferation by gold compounds. *J. Clin. Invest.*, 59, 455, 1977.
113. Lipsky, P., K. Ugai, and M. Ziff, Alterations in human monocyte structure and function induced by incubation with gold sodium thiomalate. *J. Rheumatol.*, 5 (suppl.), 130, 1979.
114. Griswold, D.E., J.C. Lee, G. Poste, and N. Hanna, Modulation of macrophage-lymphocyte interactions by the antiarthritic compound, auranofin. *J. Rheumatol.*, 12, 490, 1985.
115. Barrett, M.L. and G.P. Lewis, Unique properties of auranofin as a potential antirheumatic drug. *Agents Actions*, 19, 109, 1986.
116. Drakes, M.L., M. Harth, S.B. Galsworthy, and G.A. McCain, Effects of gold on the production and response to human interleukin-1. *J. Rheumatol.*, 14, 1123, 1987.
117. Matsubara, T., Y. Saegusa, and K. Hirohata, Low-dose gold compounds inhibit fibroblast proliferation and do not affect interleukin-1 secretion by macrophages. *Arthr. Rheum.*, 31, 1272, 1988.
118. Lee, J.C., M.J. DiMartino, B.J. Votta, and N. Hanna, Effect of auranofin treatment on aberrant splenic interleukin production in adjuvant arthritic rats. *J. Immunol.*, 139, 3268, 1987.
119. Rynes, R., Antimalarials, in *Textbook of Rheumatology*, 2nd ed., W.N. Kelley, E.D. Harris, S. Ruddy, and C. Sledge, Eds., W.B. Saunders, Philadelphia, 1985, 774–788.
120. Rainsford, K.D., Effects of antimalarial drugs on interleukin 1-induced cartilage proteoglycan degradation in-vitro. *J. Pharm. Pharmacol.*, 38, 829, 1986.
121. Skotnicki, J.S., S.C. Gilman, B.A. Steinbaugh, and J.H. Musser, Pyrazolo(4,3-c)quinolines, U.S. Patent No. 4748246, 1988.
122. Seow, W.K., A. Ferrante, L. Si-Ying, and Y.H. Thong, Suppression of human interleukin 1 production by the plant alkaloid tetrandrine. *Clin. Exp. Immunol.*, 75, 47, 1989.
123. Ruddy, S., The management of rheumatoid arthritis, in *Textbook of Rheumatology*, 2nd ed., W.N. Kelley, E.D. Harris, S. Ruddy, and C. Sledge, Eds., W.B. Saunders, Philadelphia, 1985, 979–992.
124. Tugwell, P., C. Bombardier, M. Gent, K. Bennett, D. Ludwin, E. Grace, W.W. Buchanan, W.G. Bensen, N. Bellamy, G.F. Murphy, and B. von Graffenried, Low dose cyclosporine in rheumatoid arthritis: a pilot study. *J. Rheumatol.*, 14, 1108, 1987.
125. Borel, J.F., H.U. Gubler, P.C. Hierstand, and R.M. Wenger, Immunopharmacological properties of cyclosporine (Sandimmune®) and (Val^2)-dihydrocyclosporine and their prospect in chronic inflammation. *Adv. Inflamm. Res.*, 2, 277, 1986.
126. Sternberg, E.S. and C.W. Parker, Pharmacologic aspects of lymphocyte regulation, in *The Lymphocyte — Structure and Function*, 2nd ed., J.J. Marchalonis, Ed., Marcel Dekker, New York, 1987, 1–54.
127. Dos Reis, G.A. and E.M. Shevach, Effect of cyclosporin A on T cell function *in vitro*. The mechanism of suppression of T cell proliferation depends on the nature of the T cell stimulus as well as the differentiation state of the responding cells. *J. Immunol.*, 129, 2360, 1982.
128. Palacios, R., HLA-DR antigens render interleukin-2-producer T lymphocytes sensitive to interleukin-1. *Scand. J. Immunol.*, 14, 321, 1981.
129. Palacios, R., Conconavlin A triggers T lymphocytes by directly interacting with their receptors for activation. *J. Immunol.*, 128, 337, 1982.
130. Bunjes, D., C. Hardt, M. Rollinghoff, and H. Wagner, Cyclosporin A mediates immunosuppression of primary cytotoxic T-cell responses by impairing the release of interleukin 1 and interleukin 2. *Eur. J. Immunol.*, 11, 657, 1981.

131. Manca, F., A. Kunkl, and F. Celada, Inhibition of the accessory function of murine macrophages *in vitro* by cyclosporin. *Transplantation,* 39, 644, 1985.
132. Bendtzen, K. and C.A. Dinarello, Mechanism of action of cyclosporin A. Effect on T-cell binding of interleukin-1 and antagonizing effect of insulin. *Scand. J. Immunol.,* 20, 43, 1984.
133. Russell, D.H., R. Kibler, L. Matrisian, D.F. Larson, B. Poulos, and B. Magun, Prolactin receptors on human T and B lymphocytes: antagonism of prolactin binding by cyclosporine. *J. Immunol.,* 134, 3027, 1985.
134. Hiestand, P.C., P. Mekler, R. Nordmann, A. Grieder, and C. Permmongkol, Prolactin as a modulator of lymphocyte responsiveness provides a possible mechanism of action for cyclosporin. *Proc. Natl. Acad. Sci. U.S.A.,* 83, 2599, 1986.
135. Billingham, M.E.J., Models of arthritis and the search for anti-arthritic drugs. *Pharmacol. Ther.,* 21, 389, 1983.
136. Persico, F.J. and T.B. Stim, The erythrocyte sedimentation rate in rodents, in *Handbook of Animal Models for the Rheumatic Diseases, Vol. II,* R.A. Greenwald and H.S. Diamond, Eds., CRC Press, Boca Raton, FL, 1988, 165–179.
137. Johns, D.G. and J.R. Bertino, Folic antagonists, in *Cancer Medicine,* J.F. Holland and E. Frei, III, Eds., Lea and Febiger, Philadelphia, 1973, 739–754.
138. Wilkins and Watson, 1982.
139. Hu, S.-K., Y. Mitcho, A.L. Oronsky, and S.S. Kerwar, Studies on the effect of methotrexate on macrophage function. *J. Rheumatol.,* 15, 206, 1988.
140. DiMartino, M.J., W.J. Johnson, B. Votta, and N. Hanna, Effect of antiarthritic drugs on the enhanced interleukin-1 (IL-1) production by macrophages from adjuvant-induced arthritic (AA) rats. *Agents Actions,* 21, 348, 1987.
141. Lipton, J.M., Antagonism of IL-1 fever by the neuropeptide alpha-MSH. *Prog. Leukoc. Biol.,* 2, 121, 1985.
142. Holdeman, J. and J.M. Lipton, Antipyretic activity of a potent alpha-MSH analog. *Peptides,* 6, 273, 1985.
143. Cannon, J.G., J.B. Tatro, S. Reichlin, and C.A. Dinarello, Alpha melanocyte stimulating hormone inhibits immunostimulatory and inflammatory actions of interleukin 1. *J. Immunol.,* 137, 2232, 1986.
144. Daynes, R.A., B.A. Robertson, B.-H. Cho, D.K. Burnham, and R. Newton, Alpha-melanocyte-stimulating hormone exhibits target cell selectivity in its capacity to affect interleukin 1-inducible responses in vivo and in vitro. *J. Immunol.,* 139, 103, 1987.
145. Robertson, B., K. Dostol, and R.A. Daynes, Neuropeptide regulation of inflammatory and immunologic responses. The capacity of alpha-melanocyte-stimulating hormone to inhibit tumor necrosis factor and IL-1-inducible biologic responses. *J. Immunol.,* 140, 4300, 1988.
146. Cianciolo, G.J., T.D. Copeland, S. Oroszlan, and R. Snyderman, Inhibition of lymphocyte proliferation by a synthetic peptide homologous to envelope proteins of retroviruses. *Science,* 230, 453, 1986.
147. Kleinerman, E.S., L.B. Lachman, R.D. Knowles, R. Snyderman, and G.J. Cianciolo, A synthetic peptide homologous to the envelope proteins of retroviruses inhibits monocyte-mediated killing by inactivation interleukin 1. *J. Immunol.,* 139, 2329, 1987.
148. Bakouche, O., L.B. Lachman, R.D. Knowles, and E.S. Kleinerman, Cytotoxic liposomes: membrane interleukin 1 presented in multilamellar vesicles. *Lymphokine Res.,* 7, 445, 1988.

149. Ferreira, S.H., B.B. Lorenzetti, A.F. Bristow, and S. Poole, Interleukin-1-beta as a potent hyperalgesic agent antagonized by a tripeptide analogue. *Nature,* 334, 698, 1988.
150. Roche, Y., M. Fay, and M.-A. Gougerot-Pocidalo, Effects of quinolones on interleukin-1 production in vitro by human monocytes, *Immunopharmacology,* 13, 99, 1987.
151. Bender, P.E., D.E. Griswold, N. Hanna, and J.C. Lee, Inhibition of interleukin-1 production by monocytes and/or macrophages, U.S. Patent No. 4778806, 1988.
152. Harel-Bellan, A., S. Duram, K. Muegge, A.K. Abbas, and W.L. Farrar, Specific inhibition of lymphokine biosynthesis and autocrine growth using antisense oligonucleotides in Th1 and Th2 helper T cell clones, *J. Exp. Med.,* 168, 2309, 1988.
153. Katsuki, M., M. Sato, M. Kimura, M. Yokoyama, K. Kobayashi, and T. Nomura, Conversion of normal behavior to shiverer by myelin basic protein antisense cDNA in transgenic mice. *Science,* 241, 593, 1988.
154. Agrawal, S., J. Goodchild, M.P. Civeira, A.H. Thornton, P.S. Sarin, and P.C. Zamicnik, Oligodeoxynucleoside phosphoramidates and phosphorothioates as inhibitors of human immunodeficiency virus. *Proc. Natl. Acad. Sci. U.S.A.,* 85, 7079, 1988.
155. Sarin, P.S., S. Agrawal, M.P. Civeira, J. Goodchild, T. Ikeuchi, and P.C. Zamecnik, Inhibition of acquired immunodeficiency syndrome virus by oligodeoxynucleoside methylphosphonates. *Proc. Natl. Acad. Sci. U.S.A.,* 85, 7448, 1988.
156. Wu, G.Y. and C.H. Wu, Receptor-mediated gene delivery and expression *in vivo. J. Biol. Chem.,* 263, 14621, 1988.
157. Loose-Mitchell, D.S., Antisense nucleic acids as a potential class of pharmaceutical agents. *TIPS,* 9, 45, 1988.
158. Schnyder, J., P. Bollinger, and T. Payne, Inhibition of interleukin-1 release by IX 207-887, *Agents Actions,* 30, 350, 1990.
159. Zucali, J.R. and J. Moreb, Use of antisense oligonucleotides to inhibit monocyte derived cytokine, *Prog. Leukoc. Biol.,* 10A, 315, 1990.

4. Control of IL-1 and TNFα Production at the Level of Second Messenger Pathways

ELIZABETH J. KOVACS

INTRODUCTION

A number of macrophage functions have been ascribed to their ability to produce immune mediators (cytokines).[1-4] The best documented of these macrophage-derived cytokines (or monokines) are interleukin-1 (IL-1) α, IL-1 β, and tumor necrosis factor (TNF) α. While unstimulated macrophages do not produce appreciable levels of these mediators, stimulation with a variety of agents, including lipopolysaccharide (LPS), triggers the expression of IL-1 α, IL-1 β, and TNF-α genes.[5-8] While a great deal of information has been accumulated in recent years about the agents capable of triggering the production of IL-1 and TNF genes, the molecular mechanisms responsible for this control remain unclear. Studies from several laboratories have indicated that several second messenger pathways may be involved in triggering the production of these cytokines. It is the intent of this chapter to provide an overview of the regulation of expression of IL-1α, IL-1 β, and TNF-α genes. In addition, it will concentrate on the second messenger pathways involved in their control. Since the vast majority of studies on the regulation of these genes have been performed in monocytes and macrophages, the discussion will be confined to cells of that lineage.

IL-1 AND TNFα GENES AND PROTEINS

IL-1 α and β

There are at least two species of IL-1, designated IL-1α and IL-1β, which differ by isoelectric points.[9] The cDNAs for human and murine IL-1 have been cloned[10-12] and the protein sequences shown to have 26% amino acid homology.[13] Both species of IL-1 are synthesized as precursor proteins of about 270 amino acids, which are cleaved to produce the secreted forms that are about 17 kd. While IL-1 was originally described as a product of monocytes and macrophages, it is now clear that a wide variety of cells, both of hematopoietic and nonhematopoietic origin, are capable of synthesizing and secreting IL-1 (for a review see reference 2).

TNFα

TNFα was first described by Carswell et al.[14] as a factor from endotoxin-treated animals, which produced hemorrhagic necrosis of tumors *in vivo* and was cytotoxic for certain types of tumor cells *in vitro*. The cloning of human TNFα cDNA revealed that, like IL-1 α and β, TNFα is produced as a precursor protein. In the case of human TNFα, the precursor is 230 amino acids.[15-17] The mature protein is 17.3 kd (157 amino acids) and is nonglycosylated. There are two cysteine residues that probably form an intramolecular disulfide bond.[18]

TNFα is both structurally and functionally related to lymphotoxin (TNFβ).[15] There is about 30% amino acids homology between TNFα and LT with two highly conserved regions.

CONTROL OF IL-1 AND TNFα GENE EXPRESSION IN MONOCYTES

Regulation of IL-1α and β mRNA Expression

Numerous reports have documented the induction of expression of IL-1 genes in monocytes and macrophages following treatment with LPS.[19-22] Most reports agree that in the absence of stimulation, macrophages do not express detectable levels of cytokine mRNAs. However, IL-1 genes are rapidly induced after treatment with LPS.[19,23] Matsushima et al.[19] showed that IL-1β mRNA expression is triggered as soon as 1 hr after LPS stimulation, reaches a peak at 6 hr, and remains elevated for at least 16 hr.

Early reports suggested that in the human the β species of IL-1 was the primary one expressed in LPS-treated monocytes[12] and that the ratio of IL-1β to IL-1α was roughly 10:1. Furthermore, others suggested from isoelectric focusing of murine IL-1, that the ratio of IL-1β to IL-1α was reversed (i.e., 90% IL-1α and 10% IL-1β). However, following LPS treatment of mouse peritoneal macrophages, the mRNA levels of both species of IL-1 are expressed in nearly equal amounts.[22] The inequality between the levels of protein and mRNA (higher IL-1α protein in murine macrophages than IL-1β, yet equal amounts at the mRNA level) might suggest that the relative rates of translation of the two messages differ.

The Role of Adherence in the Induction of IL-1β mRNA Expression

As one might predict from the morphological changes initiated by adherence of monocytes to tissue culture plastic, the expression of IL-1 α and β mRNAs transiently elevated. Fuhlbrigge et al.[20] reported that concurrent induction of IL-1α and β mRNAs was observed within 1 hr of initiation of culture of murine peritoneal macrophages. Unlike peptone-elicited peritoneal exudate cells from CBA/J mice reported by Fuhlbrigge et al.,[20] we found that thioglycolate elicited macrophages from C57BL/6N mice do not spontaneously express detectable levels of IL-1α or IL-1β mRNAs in response to adherence to tissue culture plastic.[22] This difference could easily be explained by the different forms of elicitation of cells or, possibly, by the strains of mice used in the studies.

Induction of IL-1 mRNA Expression by IL-2

Recent data demonstrate that IL-1β mRNA expression can be added to the growing list of activities and functions induced in monocytes and macrophages by IL-2.[23,24] These activities include the induction of cytotoxicity,[25] the production of monokines capable of inducing acute phase proteins,[26] the production of IL-1 protein,[27] and following pretreatment with interferon-γ (IFN-γ), the expression of IL-1β mRNA.[28]

The culture of the mixed population of peripheral blood leukocytes (PBL) with IL-2 leads to the induction of expression of several cytokine genes.[29] The first set of cytokine genes to be induced after IL-2 treatment of PBL are IL-1α and IL-1β, peaking at 2 hr. The rapid kinetics of activation suggests that the induction of IL-1 gene expression in PBL by IL-2 was direct. Although IL-1β mRNA can be produced by many of the subsets of cells within the PBL, including macrophages,[19,20,24] T-cells,[30] B-cells,[31] and large granular lymphocytes,[32] several pieces of evidence suggest that monocytes could be induced by IL-2 to express IL-1 genes. These include (1) the extremely rapid kinetics of expression (within 1 hr), (2) the extremely high level of IL-1 mRNA produced, (3) the absence of detectable levels of other IL-2-inducible cytokines, such as IFN-γ (within 6 hr), and (4) the demonstration of the presence of IL-2 receptors on macrophages.[33,34]

The kinetics of IL-1β mRNA expression in response to IL-2 were similar to that of LPS.[23] IL-1β mRNA was not detectable in untreated cells, was elevated within 1 hr, and reached a peak at 6 hr. Differences between LPS and IL-2-induced expression of IL-1 were only observed after 6 hr. IL-1β mRNA levels remained elevated in monocytes treated with LPS and rapidly declined in cells stimulated with IL-2. Because of the transient expression of IL-1 genes following adherence to plastic,[20] these studies were conducted with monocytes that were "rested" (incubated in medium without IL-1-inducing agents) overnight, until the adherence-induced expression subsided.

Regulation of TNFα mRNA Expression

Like IL-1, TNFα can be induced following treatment with a variety of stimuli, including LPS, IFN-γ, and phorbol esters.[4,35-38] According to Beutler et al.,[6] the production of TNFα is controlled at two distinct levels. In peritoneal macrophages and peripheral blood monocytes there are small pools of TNFα mRNA that are maintained in an untranslated state. The addition of LPS (or other inducing agents) triggers (within minutes) an acceleration of gene transcription and the translation of the "stored" mRNA as well as the newly synthesized mRNA. After a few hours the synthesis of TNFα returns to pretreatment levels even in the continued presence of LPS. If TNFα mRNA is constitutively produced, then the level of that production must be very low. In contrast, other laboratories find that murine peritoneal macrophages fail to express TNFα mRNA in the absence of stimulation.[22,39]

The LPS resistant C3H/HeJ mouse represents a good model for the examination of cytokine gene expression. Beutler et al.[6] have shown that in this strain of mouse, LPS fails to induce translation of existing TNFα mRNA. It is possible to circumvent this resistance by the administration of IFN-γ.[40] While unable to induce TNFα production alone, IFN-γ is capable of relieving the translational block in C3H/HeJ mice.

The kinetics of TNFα mRNA expression following treatment of macrophages with LPS is more rapid than the expression in response to IL-2. Kunkel et al.[22] show that TNFα mRNA peaks at 90 min after LPS treatment. In contrast, a recent report demonstrated that the accumulation of TNFα mRNA following treatment with IL-2 was delayed in comparison to treatment with LPS.[41] They claim that the delayed induction of TNFα mRNA may be a result of the requirement for *de novo* synthesis of IL-2 receptor prior to IL-2 stimulation of TNFα gene expression.

Arachidonic Acid Metabolites and Cytokine Production

Arachidonic acid metabolism can be divided into the cyclooxygenase pathway and the lipoxygenase pathway. The role of prostaglandins, products of the cyclooxygenase pathway, in the activation of macrophages has been reviewed elsewhere.[42] Several laboratories have demonstrated the role of prostaglandin E_2 (PGE_2) in the regulation of production of TNFα[43-46] and IL-1.[47]

Studies have reported the coincidental production of cytokines and PGE_2, suggesting a possible autocrine or paracrine role for arachidonate metabolites in the control of IL-1 and TNFα gene expression.[48-49] Some reports indicate that the IL-1 and TNFα production increases in a linear fashion with increasing concentration of PGE_2; however, others have shown suppressive effects of PGE_2.[50] TNFα mRNA expression may exhibit a biphasic response to PGE_2 treatment. PGE_2 apparently suppresses cytokine production by macrophages, yet indomethacin, an inhibitor of the cyclooxygenase pathway, enhances both the transcription and synthesis of TNFα and IL-1.[45,49,51]

Horiguchi et al.[46] compared the level of TNFα transcription in HL-60 cells (a line that is capable of differentiating toward the monocyte lineage following treatment with phorbol esters). In their studies, indomethacin did not alter the rate of TNFα transcription induced by phorbol esters; however, ketoconazole, an inhibitor of 5-lipoxygenase, increased TNFα mRNA levels. Moreover, leukotriene B4, a metabolite of 5-lipoxygenase, also, increased TNFα mRNA. These data suggest that phorbol esters increase TNFα mRNA expression through metabolites of arachidonic acid and that leukotriene B4 and PGE2, or their metabolites, play a role in modulating its expression. Differences between the reports of Horiguchi et al. and those described above could be a function of the type of cells (a tumor cell line vs. primary culture monocytes and macrophages) utilized in the studies.

A number of reports demonstrate that the production of IL-1 is enhanced by agents that inhibit PGE2 levels.[38,48,52,53] Like TNFα production, the activity of IL-1 triggered by LPS, or LPS and IFN-γ, was reduced by PGE2 and stimulated by indomethacin.[51]

Requirement For One or Two Signals For the Induction of Cytokine Gene Expression

Several reports (reviewed elsewhere, references 54 and 55) suggest that macrophage activation requires two stages, which can be satisfied by priming and triggering signals. While the kinetics of induction of expression of IL-1β mRNA following treatment with LPS and IL-2 are similar[23] (see above), the onset of induction of TNFα mRNA expression in human monocytes by IL-2 was delayed with respect to induction by LPS, as described above.[41] Strieter and co-workers[41] suggest that the delayed onset in IL-2-treated monocytes may reflect a difference in signal transduction pathways involved in LPS and IL-2 activation of cells or the requirement for *de novo* production of IL-2 receptor prior to the establishment of IL-2 responsiveness. These data suggest that the two signals could be (1) an inducer of cell surface IL-2 receptor and (2) IL-2.

Herrmann et al.[28] report that two signals are required for the induction of IL-1β gene expression. They showed that the sequential treatment of monocytes with IFN-γ followed by IL-2 is required for IL-1β mRNA expression. In contrast, our data and those of Strieter et al.[41] agree about the requirement for only one signal, IL-2, for the induction of expression of cytokine mRNAs. Herrmann and co-workers[28] state that it is possible to induce IL-1β mRNA after treatment with high levels of IL-2 ($\geq$ 500 units/ml). Similar high levels were needed for the induction of TNFα mRNA expression.[41] Our studies show that monocytes can be induced to express IL-1β mRNA in the presence of low levels of IL-2 ($\leq$ 100 units/ml).[23] As described above, the overnight "resting" of monocytes, allowing for the adherence-induced expression of IL-1β mRNA to subside, may have satisfied the requirement for the "priming" signal described. Further studies involving monocytes purified by elutriation and cultured in vessels in which cells cannot attach will be required to determine the relative roles of cell adherence to substrate and IFN-γ as a "priming" signal.

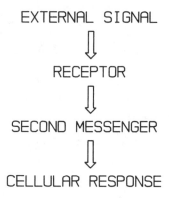

Figure 1. Flow chart of the general scheme of signal transduction.

AU-Rich Regions and mRNA Stability

It is clear that the steady-state level of a given mRNA is dependent on two factors, namely the synthesis and the turnover of that mRNA. The rates of synthesis and turnover of eukaryotic mRNAs vary widely. A set of inflammatory and immune mediators, as well as protooncogenes (*c-fos* and *c-myc*), have been described as having AT rich regions in the 3' untranslated DNA that may be involved in mRNA stability.[56,57] Further studies are required to identify the level(s) of mRNA control at which the inhibitors are acting.

REGULATION OF IL-1 AND TNFα GENES BY AGONISTS AND ANTAGONISTS OF SECOND MESSENGER PATHWAYS

Overview of Second Messenger Pathways

The activation of cellular genes by exogenous stimuli requires the transmission of the extracellular signal across the plasma membrane and cytoplasm to the nucleus. In eukaryotic systems, this is accomplished by second messenger pathways (Figure 1). External signaling is mediated by cell surface receptors and the signal is transferred by second messengers.

In spite of the large amount of potential stimuli, the number of intracellular signals is very small. Two major signal pathways have been well documented. One utilizes cyclic adenosine monophosphate (cAMP) (Figure 2), and the other, which includes calcium (Ca^{2+}), employs diacylglycerol (DAG) and inositol triphosphate (IP_3) (Figure 3). Both pathways are similar in that they are activated by association of a ligand and a cell surface receptor, and guanosine triphosphate (GTP) binding proteins (referred to as G proteins) are involved in the transmembrane transduction of the signal. Furthermore, the intracellular activation of a second messenger path-

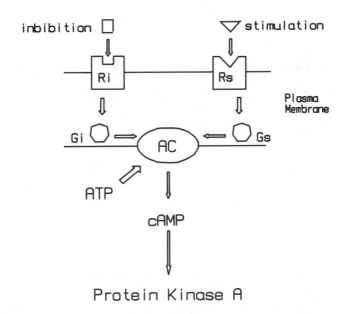

Figure 2. Signal transduction pathway through cyclic AMP (cAMP) and protein kinase A. External signals (either stimulatory or inhibitory) arrive through cell surface receptors, referred to as Rs (for stimulatory signals) and Ri (for inhibitory signals). These signals are processed through G proteins (specific for stimulatory and inhibitory). The signals either activate or inactivate adenylate cyclase (AC), which in turn converts ATP (adenosine triphosphate) into cAMP (cyclic adenosine monophosphate). Through the binding of cAMP to a regulatory component of protein kinase A, the kinase enzymatic activity is turned on, resulting in a cellular response.

way can lead to the amplification of the signal triggered by the external stimulus (for a review, see reference 58).

Protein Kinase C (PKc), Calcium, and Calmodulin Kinase

The involvement of second messenger pathways in the control of macrophage function is not unique to the expression of cytokine genes (for a review, see reference 59). Phorbol myristate acetate (PMA) has been used as a PKc activator. A number of early PMA-inducible activation events have been observed. These include the phosphorylation of proteins,[60] the secretion of metabolites of arachidonic acid,[55] and the expression of early genes, such as c-fos and c-myc.[61,62]

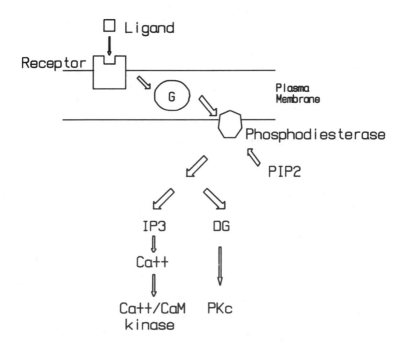

Figure 3. Signal transduction pathway through CaM kinase and PKc. External signals arrive through the interaction of a ligand with a cell surface receptor. The signal is processed a through GTP binding protein (G protein) to activate phosphodiesterase. Phosphodiesterase mediates the conversion of PIP_2 (phosphatidyl-inositol 4,5-biphosphate) into IP_3 (inositol triphosphate) and DG (diacylglycerol). DG activates PKc (protein kinase C). IP_3 in turn mobilizes cellular stores of Ca^{++}, which binds to the regulatory Ca^{++} binding protein, calmodulin (CaM). The interaction of CaM with Ca^{++} activates Ca^{++}/CaM kinase.

The molecular activation of macrophages following their interaction with LPS remains unresolved;[63,64] however, a number of activities of LPS-treated macrophages are identical to those initiated by PMA, suggesting that some LPS-induced functions are mediated by PKc. Katakami et al.[65] demonstrated that both LPS and phorbol-12,13-dibutyrate, a PKc-activating phorbol ester, induced IL-1 production, further implicating the involvement of a PKc-dependent pathway in the triggering of IL-1 synthesis. However, recent studies have shown that the activation of PKc alone is not sufficient to account for the complete functional response of macrophages to LPS.[66] In view of this heterogeneity of the effect of LPS, studies were designed to determine the nature of the pathway(s) by which LPS induces gene expression for IL-1 and TNFα.[22,67] These studies in the mouse demonstrated that the use of inhibitors of either PKc or calmodulin-dependent (CaM) kinase could block the induction of IL-1α and β gene expression induced by LPS.[22] In contrast, only PKc-dependent pathways were required for LPS induction of TNFα mRNA expression. These studies demonstrate that inhibitors of second messenger pathways can specifically block the signal transduction leading to the expression of IL-1α, IL-1β, and TNFα genes and biological activities. Furthermore, they revealed that all cytokine genes are not coordinately regulated. Macrophages are capable of activating the expression of one cytokine gene without concomitant expression of another. Surprisingly, it was found that, unlike IL-1 and TNFα mRNAs, TGFβ mRNA is constitutively expressed in murine peritoneal macrophages.[22] The steady state levels of TGFβ mRNA were neither induced nor suppressed by treatment with LPS or inhibitors of PKc and CaM kinase. The constitutive expression of TGFβ observed in these studies may, in part, be a result of the state of activation of a selected subset of macrophages or related to the species examined. In the mouse, for example, thioglycolate-elicited macrophages express TGFβ constitutively;[22] resident peritoneal macrophages and blood monocytes may not. Further investigation will clarify this point.

Since the kinetics of induction of IL-1β mRNA expression following treatment of human monocytes with LPS and IL-2 were similar, as shown above,[23] it seemed likely that the similar second messenger pathways might be involved in the transduction of signal(s) initiated by those two inducing agents. Data identical to those of LPS induction of IL-1β mRNA expression in murine peritoneal macrophages were found in human peripheral blood monocytes. IL-1β mRNA expression induced by LPS could be inhibited in a dose-dependent manner by a PKc inhibitor, 1-(5-isoquinolinesulfonyl)-2-methylpiperazine hydrochloride (H7) (Figure 4). Similar data were obtained for an inhibitor of CaM kinase, N-(6-amino-hexyl)-5-chloro-1-napthalenesulfonamide hydrochloride (W7) (data not shown). Using H7, it was determined that a PKc-dependent pathway is involved in the transduction of signals initiated by IL-2 (Figure 5). In contrast, agents that block CaM kinase were used to show that that pathway is not involved in the induction of IL-1β by IL-2 (data not shown, but these data are summarized in Figure 6).[23]

As with any other inhibitors, it is not possible to rule out that alterations of biochemical pathways other than PKc or CaM may contribute to the effect of H7 or W7, respectively. The levels of H7 and W7 used to block cytokine gene expression in these studies were identical to those reported to inhibit PKc-inducible *c-fos* expression in murine macrophages,[68] and they were much lower than the doses of H7 at which protein kinase A may be inhibited.[69] Furthermore, the data with retinal, which was shown to be selective for PKc,[70] further confirm the involvement of PKc in the signal transduction pathway triggered by LPS leading to IL-1 production. Studies with other inhibitors with overlapping activities, including N-(2-guanidinoethyl)-5-isoquinolinesulfonamide (HT1004), help support these data. At the doses tested, H7 blocks both PKc and cAMP, and HT1004 blocks cAMP and cGMP (and not PKc).[69] Since HT1004 does not inhibit the production of IL-1 or TNFα, it suggests that the PKc pathway, and not cAMP, is involved in LPS-induced cytokine production.

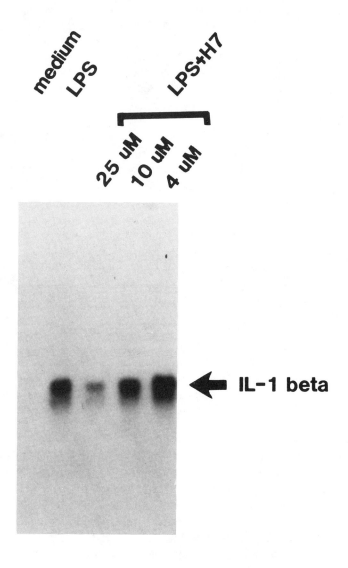

Figure 4. Northern blot analysis of IL-1β mRNA expression in human monocytes. Monocytes were treated with or without LPS for 6 hr in the presence or absence of an inhibitor of PKc (H7). Note the dose-dependent inhibition of IL-1β mRNA expression with increasing concentrations of inhibitor.

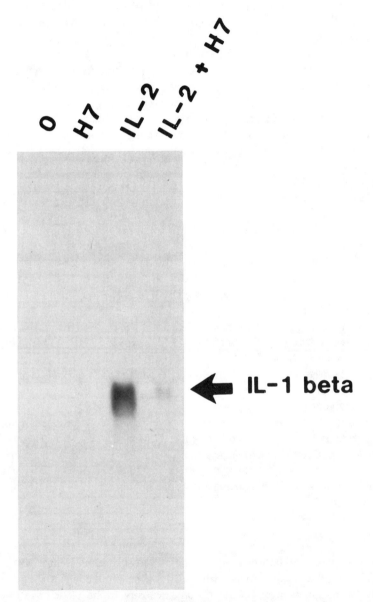

Figure 5. Northern blot analysis of IL-1β mRNA expression in IL-2-treated human monocytes. Monocytes were treated with or without IL-2 for 6 hr in the presence or absence of an inhibitor of PKc (H7). Monocytes treated with IL-2 alone express high levels of IL-1β mRNA. In contrast, monocytes treated with IL-2 and H7 express minimal levels of IL-1β mRNA.

Inhibitors of CaM kinase are not absolutely specific for this set of enzymes; however, the use of several inhibitors with overlapping specificities has enabled investigators to successfully study second messenger pathways. At the concentrations used by Kovacs et al.,[22] trifluoperazine dichloride (TFP) has been shown to completely inhibit CaM-dependent enzymes *in vitro*;[71,72] however, additional CaM-independent pathways may also be affected.[71-73] Van Belle demonstrated that calmidazolum was 500 times as effective as TFP at inhibiting brain phosphodiesterase by calmidazolum.[74] Calmidazolum is also a potent inhibitor of Ca^{2+}-transport ATPase, as well as other calmodulin-related functions in red blood cells. Wright and co-workers found calmidazolum to be the most reproducible and effective inhibitor of macrophage stimulation by LPS and macrophage-activating factor.[75] They demonstrated that W7 was also an effective antagonist of activation by these agents.

The inability of CaM kinase inhibitors, W7, N-(6-aminobutyl)-5-chloro-2-napthalenesulfonamide (W13), calmidazolum, and TFP to block TNFα production by LPS provides a clear indication that CaM-dependent kinases are *not* involved in the production of TNFα by LPS. Moreover, these data further support the lack of general cytotoxicity of W7 under the experimental conditions utilized. It is of interest that the inhibitors of CaM kinase block LPS-induced IL-1β mRNA expression, but enhance IL-2-induced expression (increasing it by over 40%).[23] Studies were designed to determine whether the enhanced expression of IL-1β by IL-2 and CaM kinase inhibitors over the level induced by IL-2 alone is a function of the rate of gene transcription, mRNA stability, or a combination of the two.

In addition to activation of PKc and CaM kinase, LPS has been reported to initiate the breakdown of phosphatidylinositol-4,5-biphosphate (PIP_2) into inositoltriphosphate and diacylglycerol,[60,76] which (Figure 3) is one of the steps leading to PKc and CaM kinase activation. Furthermore, a specific set of proteins is phosphorylated in macrophages in response to treatment with LPS,[60] further implicating a PKc-like enzyme activity.

Several lines of evidence suggest that the level of intracellular calcium plays a role in macrophage activation, in addition to the CaM kinase data shown above. For example, the induction of macrophage tumoricidal activity is calcium dependent.[75] Inhibitors of calcium fluxes block the effects of LPS and IFN-γ,[66,77] and the physiological effects of these inducing agents can be mimicked, in part, by calcium ionophore A23187.[66,78]

Cyclic Nucleotides

There is a direct link between prostaglandins and second messenger pathways. When PGE_2 receptors are activated, the result is the production of cyclic AMP (cAMP), a second messenger generated from adenosine triphosphate (ATP) by the enzyme adenylate cyclase (see Figure 2). Agents that cause an elevation of cAMP, including dibuteryl cAMP and 8-bromo cAMP, inhibit TNF production.[65,79] These data provide a mechanism by which the effects of PGE_2 can be achieved. Renz et

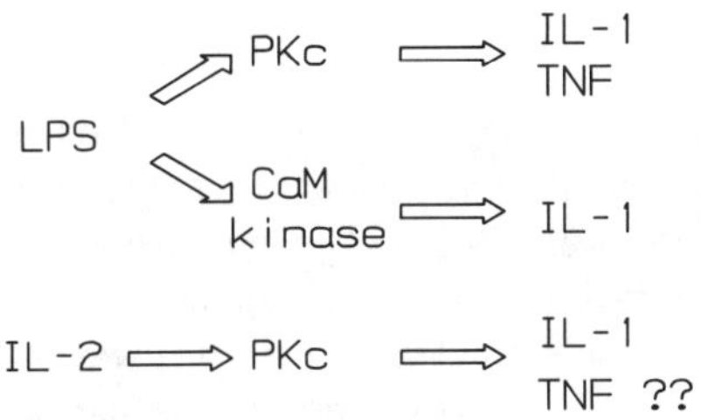

```
           ┌─→ PKc        ⟹  IL-1
           │                  TNF
    LPS ───┤
           └─→ CaM
               kinase     ⟹  IL-1

    IL-2  ⟹  PKc         ⟹  IL-1
                              TNF  ??
```

Figure 6. Summary of second messenger pathways involved in the control of IL-1 and TNFα mRNA expression in monocytes following treatment with LPS and IL-2.

al.[45] demonstrated that (1) agents that increase intracellular cAMP levels suppress TNFα production and (2) the elevation of cGMP stimulates TNFα production. It is of interest to note that both *in vitro* and *in vivo* toxic effects of TNFα can be blocked by inhibitors of the cyclooxygenase pathway.[44,80] Presumably the production of IL-1 by prostaglandins is also mediated through a cAMP-dependent pathway.

POTENTIAL THERAPEUTIC INTERVENTION WITH INHIBITORS OF SECOND MESSENGER PATHWAY INHIBITORS

Because of the numerous interactions of second messenger pathways with one another, it is clear that inhibition of one pathway may have profound effects on cellular mechanisms other than those that are desired. As a consequence of this, it is difficult to fathom the clinical use of these inhibitors to combat disease. However, TFP, a CaM kinase inhibitor, has been used *in vivo* to block inflammatory cell activities in an animal model of lung injury.[81] In addition, inhibitors of arachidonic acid metabolism, indomethacin, ibuprofen,[80] and SK&F 86002,[82] have been used to block the toxic effects of TNFα *in vivo,* suggesting that inhibition of cAMP would result in the inhibition of TNF production. These data suggest that, in spite of the overlap of second messenger pathways, it is possible to use inhibitors *in vivo* to block biological function.

The results of the studies using H7 and W7 to block PKc and CaM kinase, respectively, have both immediate practical and long-term value. Agents like W7 and TFP can be used to block the production of IL-1, but not TNF-α. This will allow us not only to selectively suppress a subset of the functions of differentiated macrophages, which may be beneficial clinically, but also to further expand our knowledge of the signals controlling the expression of these pluripotent mediators.

SUMMARY

The studies reported herein suggest that multiple second messenger pathways are involved in the transduction of signal(s) that lead to the induction of IL-1 and TNFα gene expression. The use of inhibitors of second messenger pathways in the regulation of cytokine gene expression is in its infancy. Since there are distinct, but overlapping signals for these second messenger pathways, it is a difficult system to dissect. It is of critical importance to determine mechanisms by which the expression of cytokine genes can be controlled; the compilation of a list of agents capable of inhibiting the synthesis of these potent mediators will enable us to selectively block their production.

ACKNOWLEDGMENTS

This work was sponsored by grants from the Potts Foundation, BRSG, and the Office of Naval Research.

REFERENCES

1. Degliantoni, G., M. Murphy, M. Kobayashi, M.K. Francis, B. Perussia, and G. Trinchieri, Natural killer (NK) cell-derived hematopoietic colony-inhibiting factor: relationship with tumor necrosis factor and synergism with immune interferon. *J. Exp. Med.,* 162, 1512, 1985.
2. Oppenheim, J.J., E.J. Kovacs, K. Matsushima, and S.K. Durum, There's more than one interleukin-1. *Immunol. Today,* 7, 45, 1986.
3. Rubin, B.Y., S.L. Anderson, S.A. Sullivan, B.D. Williamson, E.A. Carswell, and L.J. Old, Nonhematopoietic cells selected for resistance to tumor necrosis factor produce tumor necrosis factor. *J. Exp. Med.,* 164, 1350, 1986.
4. Beutler, B. and A. Cerami, Cachectin and tumor necrosis factor: two sides of the same coin. *Nature,* 320, 584, 1986.
5. Gery, I., R.K. Gershon, B.H. Waksman, Potentiation of the thymocyte response to mitogens. *J. Exp. Med.,* 136:128 1972.
6. Beutler, B., N. Krochin, I.W. Milsark, C. Luedke, and A. Cerami, Control of cachectin (tumor necrosis factor) synthesis: mechanisms of endotoxin resistance. *Science,* 232:977, 1986a.
7. Kornbluth, R.S. and T.S. Eddington, Tumor necrosis factor production by human monocytes is a regulated event: induction of TNFα-mediated cellular cytotoxicity by endotoxin. *J. Immunol.,* 137:2585, 1986.
8. Gifford, G.E. and M.-L. Lohmann-Matthes, The requirement for the continual presence of LPS for the production of TNFα by thioglycolate induced peritoneal murine macrophages. *Int. J. Cancer,* 38:135, 1986.
9. Cameron, P.M., G.A. Limjuco, J. Chin, L. Silberstein, and J.A. Schmidt, Purification to homogeneity and amino acid sequence analysis of two anionic species of human interleukin-1. *J. Exp. Med.,* 146:237, 1986.

10. Lomedicio, P.T., U. Gubler, C.P. Hellmann, M. Dukovich, J.G. Giri, Y.-C.E. Pan, K. Collier, R. Semionow, A.O. Chua, and S.B. Mizel, Cloning and expression of murine interleukin-1 cDNA in *Escherichia coli. Nature,* 312:458, 1984.
11. Auron, P.E., A.C. Webb, L.J. Rosenwasser, S.F. Mucci, A. Rich, S.M. Wolff, and C.A. Dinarello, Nucleotide sequence of human monocyte interleukin 1 precursor cDNA. *Proc. Natl. Acad. Sci. U.S.A.,* 81:7907, 1984.
12. March, C.J., B. Mosley, A. Larsen, D.P. Cerretti, G. Braedt, V. Price, S. Gillis, C.S. Henney, S.R. Kronheim, K. Grabstein, P.J. Conlon, T.P. Hopp, and D. Cosman, Cloning, sequence and expression of two distinct human interleukin-1 complementary DNAs. *Nature,* 315:641, 1985.
13. Gubler, U., A.O. Chua, A.S. Stern, C.P. Hellerman, M.P. Vitek, T.M. Dechiara, W.R. Benjamin, K.C. Collier, M. Dukovich, P.C. Familletti, C. Fiedler-Nagy, J. Jenson, K. Kaffka, PL. Kilian, D. Stremlo, B.H. Wittreichm, D. Woehle, S.B. Mizel, and P.T. Lomedico. *J. Immunol.,* 136:2492, 1986.
14. Carswell, E.A., L.J. Old, R.L. Kassel, S. Green, N. Fiore, and B. Williamson, An endotoxin induced serum factor that causes necrosis of tumors. *Proc. Natl. Acad. Sci. U.S.A.,* 72:3666, 1975.
15. Pennica, D., G.E. Nedwin, J.S. Haflick, P.H. Seeburg, R. Derynk, M.A. Pallidino, and D.V. Goeddel, Human tumor necrosis factor: precursor structure, expression and homology to lymphotoxin. *Nature,* 312:724, 1984.
16. Wang, A.M., A.A. Creasy, M.B. Ladner, L.S. Lin, J. Strickler, J.N. Van Arsdell, R. Yamamoto, and D.F. Mark, Molecular cloning of the complementary DNA from human tumor necrosis factor. *Science,* 228:149, 1985.
17. Shirai, T., H. Yamaguchim, H. Ito, C.W. Todd, and R.B. Wallace, Cloning and expression in *Escherichia coli* of the gene for human tumor necrosis factor. *Nature,* 313:803, 1985.
18. Aggarwal B.B., W.J. Kohr, P.E. Hass, B. Moffat, S.A. Spencer, W.J. Henzel, T.S. Bringman, G.E. Nedwin, D.V. Goeddel, and R.N. Harkins, Human tumor necrosis factor: production, purification, and characterization. *J. Biol. Chem.,* 260:2345, 1985.
19. Matsushima, K., M. Taguchi, E.J. Kovacs, H.A. Young, and J.J. Oppenheim, Intracellular localization of human monocyte associated interleukin 1 (IL 1) activity and release of biologically active IL-1 by trypsin and plasmin. *J. Immunol.,* 136:2883, 1986.
20. Fuhlbrigge, R.C., D.D. Chaplin, J.-M. Kiely, and E.R. Unanue, Regulation of interleukin 1 gene expression by adherence and lipopolysaccharide. *J. Immunol.,* 138:3799, 1987.
21. Fenton, M.J., B.D. Clarke, K.L. Collins, A.C. Webb, A. Rich, and P.E. Auron, Transcriptional regulation of the human prointerleukin-1 β gene. *J. Immunol.,* 138:3972, 1987.
22. Kovacs, E.J., H.A. Young, D. Radzioch, and L. Varesio, Differential biochemical pathways involved in the expression of interleukin-1 (IL-1) and tumor necrosis factor (TNF) mRNAs in mouse peritoneal macrophages. *J. Immunol.,* 141:3031, 1988b.
23. Kovacs, E.J., B. Brock, L. Varesio, and H.A. Young. 1989b. IL-2 induction of IL-1β mRNA expression in monocytes: regulation by agents which block second messenger pathways, submitted, 1989b.
24. Kovacs, E.J., H.A. Young, and F.W. Ruscetti, Interleukin-2 induction of interleukin-1 mRNA expression in human monocytes. *J. Leukoc. Biol.,* 42:368, 1987.
25. Malkovsky, M., B. Loveland, M. North, G.L. Asherson, L. Gao, P. Ward, and W. Fiers, Recombinant interleukin-2 directly augments the cytotoxicity of human monocytes. *Nature,* 325:262, 1987.

26. Mier, J.W., C.A. Dinarello, M.B. Atkins, P.I. Punsal, and D.H. Perlmutter, Regulation of hepatocyte acute phase protein synthesis by products of interleukin-2 (IL-2) stimulated human peripheral blood mononuclear cells. *J. Immunol.*, 139:1268, 1988.
27. Numerof, R.P., F.R. Aronson, and J.W. Meir, IL-2 stimulates the production of IL-1α and IL-1β by human peripheral blood mononuclear cells. *J. Immunol.*, 141:4250, 1988.
28. Herrmann, F., S.A. Cannistra, A. Lindemann, D. Blohm, A. Rambaldi, R.H. Mertelsmann, and J.D. Griffin, Functional consequences of monocyte IL-2 receptor expression: Induction of IL-1β secretion by IFN-gamma and IL-2. *J. Immunol.*, 142:139, 1989.
29. Kovacs, E.J., S.K. Beckner, D.L. Longo, L. Varesio, and H.A. Young, Cytokine gene expression during the generation of lymphokine-activated killer cells: early induction of interleukin-1β by interleukin-2. *Cancer Res.*, 49:940, 1989a.
30. Tartakovsky, B., E.J. Kovacs, L. Takacs, and S.K. Durum, T cell clone producing an IL-1 like activity following stimulation by antigen-producing B cells. *J. Immunol.*, 1986.
31. Matsushima, K., A. Procopio, H. Abe, G. Scala, J.R. Ortaldo, J.J. Oppenheim, Production of interleukin-1 by normal human peripheral blood B lymphocytes. *J. Immunol.*, 135:1132, 1985.
32. Galli, M.C., S. Peppoloni, E.J. Kovacs, H.A. Young, C.W. Reynolds, and J.R. Ortaldo, IL-1 gene expression in human CD3-large granular lymphocytes. *Fed. Proc.*, 46:787, 1987.
33. Holter, W., R. Grunow, H. Stockinger, and W. Knapp, Recombinant interferon-gamma induces interleukin-2 receptors on human peripheral blood monocytes. *J. Immunol.*, 136:2171, 1986.
34. Rambaldi, A., D.C. Young, F. Herrmann, S.A. Cannistra, and J.D. Griffin, Interferon-gamma induces expression of the interleukin-2 receptor gene in human monocytes. *Eur. J. Immunol.*, 17:153, 1987.
35. Beutler, B., J. Mahoney, N. Le Trang, P. Pekala, and A. Cerami, Purification of cachectin, a lipoprotein lipase suppressing hormone secreted by endotoxin-induced RAW 264.7 cells. *J. Exp. Med.*, 161:984, 1985.
36. Old, L.J., Tumor necrosis factor (TNF). *Science*, 230:630, 1985.
37. Le, J. and J. Vilcek, Biology of disease: tumor necrosis factor and interleukin-1: cytokines with multiple overlapping biological activities. *Lab. Invest.*, 56:234, 1987.
38. Kunkel, S.L., M. Spengler, M.A. May, R. Spengler, J. Larrick, and D. Remick, Prostaglandin E_2 regulates macrophage derived tumor necrosis factor gene expression. *J. Biol. Chem.*, 253:5380, 1988c.
39. Kunkel, S.L., M. Spengler, G. Kwon, M.A. May, and D.G. Remick, Production and regulation of tumor necrosis factor-α: a cellular and molecular analysis. *Methods Achiev. Exp. Pathol.*, 13:240, 1988b.
40. Beutler, B., V. Tkacenko, I. Milsark, N. Krochin, and A. Cerami, Effect of γ interferon on cachectin expression by mononuclear phagocytes, *J. Exp. Med.*, 164:1791, 1986b.
41. Strieter, R.M., D.G. Remick, P.A. Ward, R.N. Spengler, J.P. Lynch, III, J. Larrick, and S.L. Kunkel, Cellular and molecular regulation of tumor necrosis factor α production by pentoxifyline. *Biochem. Biophys. Res. Commun.*, 155:1230, 1988.
42. Zwilling, B.S. and L.B. Justement, Prostaglandin regulation of macrophage function, in *Macrophages and Cancer*, G.H. Heppner and A.M. Fulton, Eds., CRC Press, Boca Raton, 1988, 62.
43. Kunkel, S.L., W.E. Scales, R. Spengler, M. Spengler, and J. Larrick, Dynamics and regulation of tumor necrosis factor-α, interleukin-1α and interleukin-1β gene expression by arachidonate metabolites, in *Monokines and Other Non-Lymphoid Cytokines*, M.C. Podwanda, J.J. Oppenheim, M. Kluger, and Dinarello, Eds., Alan R. Liss, New York, 1988a, 61.

44. Lehmann, V., B. Benninghoff, and W. Droge, Tumor necrosis factor induced activation of peritoneal macrophages is regulated by prostaglandin E_2 and cAMP. *J. Immunol.*, 141:587, 1988.
45. Renz, A., J.-H. Gong, A. Schmidt, M. Nain, and D. Gemsa, Release of tumor necrosis factor α from macrophages: enhancement and suppression are dose dependently regulated by prostaglandin E_2 and cyclic nucleotides. *J. Immunol.*, 141:2388, 1988.
46. Horiguchi, J., D. Spriggs, K. Imamura, R. Stone, R. Leubbers, and D. Kufe, Role of arachidonic acid metabolism in transcriptional induction of tumor necrosis factor gene expression by phorbol ester. *Mol. Cell. Biol.*, 9:252, 1989.
47. Otterness, I.G., M.L. Bliven, J.D. Eskra, M. Reinke, and D.C. Hanson, The pharmacologic regulation of interleukin-1 production: the role of prostaglandins. *Cell. Immunol.*, 114:385–397, 1988.
48. Kunkel, S.L., S.W. Chensue, and S.H. Phan, Prostaglandins as endogenous mediators of interleukin 1 production. *J. Immunol.*, 136:182, 1986a.
49. Kunkel, S.L., R.C. Wiggins, S.W. Chensue, and J. Larrick, Regulation of macrophage tumor necrosis factor production by prostaglandin E_2. *Biochem. Biophys. Res. Commun.*, 137:404, 1986b.
50. Remick, D.G., J.W. Larrick, and S.L. Kunkel, Tumor necrosis factor induced alterations in circulating leukocyte populations. *Biochem. Biophys. Res. Commun.*, 3141:818, 1987.
51. Hart, P.H., G.A. Whitty, D.S. Piccoli, and J.A. Hamilton, Control by IFN-gamma and PGE_2 of TNFα and IL-1 production by human monocytes. *Immunology*, 66:376, 1988.
52. Oppenheim, J.J., W.J. Koopman, L.M Wahl, and S.F. Dougherty, Prostaglandin E_2 rather than lymphocyte activating factor produced by activated human mononuclear cells stimulates increases in murine thymocyte cyclic AMP. *Cell. Immunol.*, 49:64, 1980.
53. Kunkel, S.L. and S.W. Chensue, Arachidonic acid metabolites regulate interleukin-1 production. *Biochem. Biophys. Res. Commun.*, 128:892, 1985.
54. Adams, D.O. and T.A. Hamilton, The cell biology of macrophage activation. *Annu. Rev. Immunol.*, 2:283, 1984.
55. Adams, D.O. and T.A. Hamilton, Molecular transduction mechanisms by which IFN gamma and other signals regulate macrophage development. *Immunol. Rev.*, 97:5, 1987.
56. Caput, D., B. Beutler, K. Hartog, R. Thayer, S. Brown-Shimer, and A. Cerami, Identification of a common nucleotide sequence in the 3'-untranslated region of mRNA molecules specifying inflammatory mediators. *Proc. Natl. Acad. Sci. U.S.A.*, 83:1670, 1986.
57. Shaw, G. and R. Keamen, A conserved AU sequence from the 3" untranslated region of GM-CSF mRNA that mediates selective mRNA degradation. *Cell*, 46:659-667, 1986.
58. Berridge, M.J., The molecular basis of communication within the cell. *Sci. Am.*, 254:143, 1985.
59. Hamilton, T.A. and D.O. Adams, Molecular mechanisms of signal transduction in macrophages. *Immunol. Today*, 8:151, 1987.
60. Weiel, J.E., T.A. Hamilton, and D.O. Adams, LPS induces altered phosphate labeling of proteins in murine peritoneal macrophages. *J. Immunol.*, 136:3012, 1986.
61. Radzioch, D., B. Bottazzi, and L. Varesio, Augmentation of *c-fos* mRNA expression by activators of protein kinase C in fresh, terminally differentiated resting macrophages. *Molec. Cell. Biol.*, 7:595, 1986.
62. Introna, M., T.A. Hamilton, R.E. Kaufman, D.O. Adams, and R.C. Bast, Jr, Treatment of murine peritoneal macrophages with bacterial lipopolysaccharide alters expression of *c-fos* and *c-myc* oncogenes. *J. Immunol.*, 137:2711, 1986.

63. Jacobs, D.M., Structural features of binding of lipopolysaccharide to murine lymphocytes. *Rev. Infect. Dis.*, 6:501, 1984.
64. Morrison, D.C. and J.A. Rudbach, Endotoxin-cell membrane interactions leading to transmembrane signalling. *Contemp. Top. Mol. Immunol.*, 8:187, 1981.
65. Katakami, Y., Y. Nakao, T. Koizumi, N. Katakami, R. Ogawa, and T. Fujita, Regulation of tumor necrosis factor production by mouse peritoneal macrophages: the role of cellular cyclic AMP. *Immunology*, 64:719, 1988.
66. Sommers, S.D., J.E. Weiel, T.A. Hamilton, and D.O. Adams, Biochemical mechanisms of macrophage activation: Phorbol esters and calcium ionophore act synergistically to prime macrophages for tumor cell destruction. *J. Immunol.*, 136:4199, 1986.
67. Kovacs, E.J., D. Radzioch, H.A. Young, and L. Varesio, Control of LPS induced interleukin-1 and tumor necrosis factor mRNAs expression by inhibitors of second messenger pathways in murine macrophages, in *Monokines and Other Non-Lymphoid Cytokines*, M.C. Podwanda, J.J. Oppenheim, M. Kluger, and C.A. Dinarello, Eds., Alan R. Liss, New York, 1988a, 55–60.
68. Radzioch, D. and L. Varesio, Protein kinase C inhibitors block the activation of macrophages by IFN-β but not by IFN-gamma. *J. Immunol.*, 140:1259, 1988.
69. Hidaka, H., M. Inagaki, S. Kawamato, and Y. Sasaki, Isoquinolinesulfonamides, novel and potent inhibitors of cyclic nucleotide dependent protein kinase and protein kinase C. *Biochemistry*, 236:5036, 1984.
70. Taffet, S.M., A.R.L. Greenfield, and M.K. Haddox, Retinal inhibits TPA activated, calcium-dependent, phospholipid-dependent protein kinase ("C" kinase). *Biochem. Biophys. Res. Commun.*, 114:1194, 1983.
71. Weiss, B., W. Prozaikeck, M. Cimino, M.S. Barnette, and T.L. Wallace, Pharmacological regulation of calmodulin. *Ann. N.Y. Acad. Sci.*, 365:319, 1980.
72. Roufogalis, B.D., Specificity of trifluoperazine and related phenothiazines for calcium-binding proteins, in *Calcium and Cell Function*, W.Y. Cheung, Ed., Academic Press, New York, 1982, 129.
73. Luthra, M.G., Trifluoperazine inhibition of calmodulin-sensitive Ca^{2+}-ATPase and calmodulin insensitive $(Na^+ + K^+)$ and Mg^{2+}-ATPase activities of human and rat blood cells. *Biochim. Biophys. Acta*, 962:271, 1982.
74. Van Belle, H., R24 571: a potent inhibitor of calmodulin activated enzymes. *Cell Calcium*, 2:483, 1981.
75. Wright, B., I. Zeidman, R. Greig, and G. Poste, Inhibition of macrophage activation by calcium channel blockers and calmodulin antagonists. *Cell. Immunol.*, 95:46, 1985.
76. Prpic, V., J.E. Weiel, S.D. Sommers, J. DiGuiseppi, S.D. Gonias, S.V. Pixxo, T.A. Hamilton, B. Herman, and D.O. Adams, Effects of bacterial lipopolysaccharide on the hydrolysis of phophatidylinositol 4,5-bis phosphate in murine peritoneal macrophages. *J. Immunol.*, 139:526, 1987.
77. Hamilton, T.A., J.A. Rigsbee, W.A. Scott, and D.O. Adams, Gamma interferon enhances the secretion of arachidonic acid metabolites from murine macrophages stimulated with phorbol derivatives. *J. Immunol.*, 134:2631, 1985.
78. Weiel, J.E., T.A. Hamilton, and D.O. Adams, Biochemical models of transferrin receptors on murine peritoneal macrophages following treatment *in vitro* with PMA or A23187. *J. Immunol.*, 135:293, 1985.
79. Taffet, S., Regulation of TNF gene expression by lipopolysaccharide and interferon. *J. Leukoc. Biol.*, 42:542, 1987.

80. Kettelhut, I.S., W. Fiers and A.L. Goldberg, The toxic effects of tumor necrosis factor in vivo and their prevention by cyclooxygenase inhibitors. *Proc. Natl. Acad. Sci. U.S.A.,* 84:4273, 1987.
81. Nakashima, J.M., D.M. Hyde, and S.N. Giri, Effects of a calmodulin inhibitor of bleomycin-induced lung inflammation in hamsters. *Am. J. Pathol.,* 124:528, 1986.
82. Badger, A.M., D. Olivera, J.E. Talmadge, and N. Hanna, Protective effect of SK&F 86002, a novel dual inhibitor of arachidonic acid metabolism, in murine models of endotoxin shock: inhibition of tumor necrosis factor as a possible mechanism of action. *Circ. Shock,* 27:51, 1989.

5. Cytokines and Growth Factors in Arthritic Diseases: Mechanisms of Cell Proliferation and Matrix Degradation in Rheumatoid Arthritis

CONSTANCE E. BRINCKERHOFF AND
ANNE M. DELANY

INTRODUCTION

Other than rheumatoid arthritis, few nonmalignant diseases have so much potential for tissue invasion and destruction. In this disease, the synovial tissue lining the joint organizes into a mass that infiltrates and degrades articular cartilage, tendons, and bone (reviewed in references 1–3). In nondiseased individuals, synovial tissue consists of a thin membrane of only two or three cell layers, comprised principally of fibroblast-like synovial cells and rare resident macrophages. On the other hand, rheumatoid synovial tissue consists of a mixture of cell types: immune T- and B-cells, monocyte/macrophages, polymorphonuclear leucocytes, and the fibroblast-like cells with their rampant proliferative ability. With the exception of the fibroblasts, most of these cells have been recruited to the rheumatoid joint in response to inflammatory stimuli that occur as part of the pathology of this disease.

Although the etiology of rheumatoid arthritis is not clear, it is suspected that an unknown antigen, such as a bacterium, virus, or mycoplasma, is deposited in the joints as a consequence of a systemic infection.[2] Normally, this antigen is cleared and no disease arises; however, in genetically susceptible individuals, the antigen elicits an acute inflammatory/foreign body response in which some autologous tissue damage occurs. This, in turn, develops into an (auto)immune response and eventually leads to a chronic inflammatory and immunologic reaction within the synovial lining of the joint (Figure 1). Thus, there is a *potpourri* of activated cell types, and the cytokines they produce continuously fuel the proliferative and destructive ability of the synovial fibroblasts.

Many of these cytokines are produced by the monocytes/macrophages and immune cells that are recruited to the inflamed joint. They function as paracrine proteins, stimulating the synovial fibroblast-like cells. Other cytokines act as autocrines, i.e., proteins that are produced by and act on the synovial cells, themselves. The net effect is that these cells proliferate vigorously and degrade their extracellular matrix.[1-3] This degradation is accomplished principally through the excessive production of prostaglandin E_2 (PGE_2), which increases levels of bone collagenase,[1-3] and the neutral metalloproteinases, collagenase, and stromelysin.

Collagenase is synthesized as a single polypeptide, preprocollagenase, Mr 59K.[4] The enzyme is a neutral metalloproteinase: it is active at neutral pH; Zn is an integral part of its structure; and Ca^{++} is required for activity.[1-3] Evidence suggests that the same collagenase gene product is produced by fibroblasts,[5-7] macrophages,[8] and endothelial cells[9] and that the enzyme is secreted immediately after synthesis.[10] In contrast, neutrophil collagenase, which is a distinct gene product,[107] remains stored in granules within the cell and is released only upon extrusion of the granules.[11,12]

Collagenase is secreted in an inactive form, procollagenase, with a predominant species, Mr 57K, and a minor amount of glycosylated enzyme, Mr 61K.[10,14] Conversion of procollagenase to active enzyme is accomplished proteolytically by serine proteases, such as plasmin, kallikrein, or trypsin, or by organic mercurial compounds.[1-3] However, a second metalloproteinase that is closely related to collagenase may work in conjunction with the serine proteases or the mercurials to maximally activate collagenase.[13-16] This related metalloproteinase, originally termed activator because of its ability to activate latent collagenase,[13] was subsequently found to be identical to stromelysin (Mr 55K).[7,17-20] In addition to its ability to activate latent collagenase, stromelysin degrades noncollagenous components of the extracellular matrix, e.g., proteoglycans, laminin, and fibronectin, as well as type IV collagen.[15,16,18-22] Since stromelysin/activator is also secreted in latent form, it is likely that proteases, such as trypsin or plasmin, or the organomercurial compounds, first activate prostromelysin/proactivator, which then goes on to activate procollagenase.

Thus, collagenase has the singular ability to initiate breakdown of interstitial collagen, the body's most abundant protein, while stromelysin degrades noncollagenous matrix. These two proteinases, then, can act in concert to digest nearly all connective tissue macromolecules. Other cells present in the rheumatoid joint, such as monocytes and polymorphonuclear leucocytes, also secrete collagenase. However, compared to the fibroblast-like rheumatoid synovial cells, the amounts produced by these cells are small, thereby singling out the synovial cells as a major source of collagenase and stromelysin.[1-3] Because these two enzymes have been so strongly implicated in the degradation of the extracellular matrix in arthritic diseases and because expression of these enzymes can be modulated by cytokines, a discussion of mechanisms controlling their expression will be a major focus of this chapter.

We will first summarize some of those cytokines that have been shown to affect synovial cell (fibroblast) growth and enzyme production (Table 1). We will then

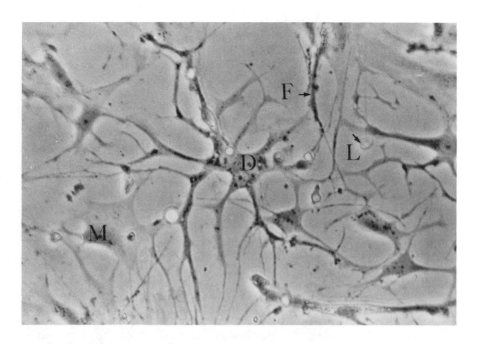

Figure 1A. Modulation of synovial cell function by cytokines. Rheumatoid synovium was received from the operating room at Mary Hitchcock Memorial Hospital at the time of synovectomy. The tissue was dissociated into a single cell suspension[56,89-93] and plated in Dulbecco's modified Eagle's medium with 10% fetal bovine serum. (A) Phase-contrast photomicrograph of primary culture of rheumatoid synovial cells. (Magnification × 100.)

choose several lymphokines/cytokines whose actions have been studied in some detail, and we will use the data gathered with them to begin to ask important questions about mechanism: do various cytokines modulate cell proliferation and metalloproteinase production by transcriptional or posttranscriptional mechanisms? Is the expression of genes responsible for the proliferative and destructive capabilities of rheumatoid synovial cells associated or linked in any way? Can we block or interrupt the expression of these genes? If so, by what mechanism?

Several years ago, Sporn and Harris[23] advanced the concept of nonmalignant proliferative diseases in which underlying biochemical pathways, rather than descriptive pathology, form the basis for disease categories. They emphasized the idea that common biochemical mechanisms controlling excessive proliferation and turnover of extracellular matrix are significantly involved in the pathogenesis of cancer and of nonmalignant proliferative diseases, including rheumatoid arthritis, scleroderma, and psoriasis. Furthermore, they postulated an important role for polypeptide hormones and hormone-like substances (cytokines) and for proteinases as mediators of these biochemical pathways. Here, as we describe the effects of various cytokines on fibroblast behavior, we support this hypothesis. By document-

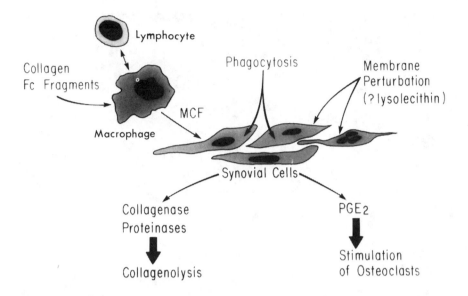

Figure 1B. Diagram of cell-cell interactions in primary cultures of rheumatoid synovial macrophage (M); lymphocyte (L) (T or B cell); fibroblast (F); dendritic cell (D) (probably derived from fibroblasts.[56,89-93]). MCF = mononuclear cell factor/Interleukin-1.[57,58]

ing the biochemical and morphological effects of these cytokines on fibroblast proliferation and invasiveness, the similarities between proliferating synovium and invading malignant tissue become increasingly evident. The heightened expression of protooncogenes that occurs in rheumatoid arthritis as a result of cytokine stimulation of synovial cells, and the subsequent biochemical and phenotypic similarity to overt malignancy, are central themes of this chapter.

EFFECTS OF PHORBOL MYRISTATE ACETATE ON SYNOVIAL CELLS AND ANTAGONISM BY RETINOIDS AND GLUCOCORTICOIDS

The tumor promoter phorbol myristate acetate (PMA) is a potent activator of protein kinase C (reviewed in 24), which is important in triggering a series of intracellular events that can lead to cell proliferation and enhanced production of enzymes. By directly activating this enzyme, PMA bypasses the usual signal/transduction pathway in the cell. Normally, cytokines stimulate increases in diacylglycerol and inositol triphosphate in the cell membrane, which then, as second messengers, activate protein kinase C.

PMA is also a potent inducer of metalloproteinases and is probably among the best studied of all compounds that induce these enzymes. Originally known for its inflammatory properties,[24,25] the possibility arose that these properties and its

Table 1. Modulation of Fibroblast Proliferation and Metalloproteinase Production

Agent	Proliferation	Metalloproteinase Production	References
Cytokines:			
phorbol myristate acetate	↑	↑	17,21,24,26,27, 29,30,35,39,40, 43–50
interleukin 1	↑	↑	28,31,48,53–55, 57–60,64–66,68, 70,72
tumor necrosis factor	↑	↑	61,63,67–69,71
epidermal growth factor	↑	↑	36–40,48,50,69, 83–86, 89–91
transforming growth factor β	↑↓	↑↓	82–91
platelet-derived growth factor	↑	↑	50,69,70,79,81,90
Crystals:			
monosodium urate monohydrate	?	↑	27,32,75
basic calcium phosphate	↑	↑	77–81
Pharmacologic agents:			
glucocorticoid hormones	↑	↓	17,29,30,42,48, 56,64,89,91
retinoids	↓	↓	17,29,30,42,89–91, 93–95

ability to promote tumor formation were associated with enhanced proteinase production. To test this hypothesis, Weinstein and Wigler[25] showed that PMA induced plasminogen activator, a serine protease often linked with malignant tissue.

Shortly thereafter, in an attempt to help build the argument that rheumatoid synovial tissue shared proliferative and invasive characteristics with malignant tissue, we added PMA to cultures of quiescent rabbit synovial fibroblasts.[26] We found that nmolar amounts of this inflammatory agent dramatically increased expression in normal synovial fibroblasts of three hallmarks of rheumatoid synovium: cell proliferation, collagenase production, and prostaglandin E_2 (PGE_2) synthesis. Thus, phorbol-treated rabbit synovial fibroblasts mimicked primary cultures of rheumatoid synovial cells, and this observation has provided us with a valuable model system for studying mechanisms controlling the behavior of rheumatoid synovial cells.

To develop this model system, we cultured rabbit synovial fibroblasts from the synovium lining the knee joints of young (4 to 6 week old) healthy New Zealand White rabbits.[26,27] These cells are readily available, grow well in culture, and are amenable to experimental manipulation. Collagenase synthesis in these cells can be both induced[17,26-30] and repressed.[17,29,20] The cells can be induced by treatment with a variety of agents, including Interleukin-1 (IL-1),[28,31] PMA,[17,26-30] crystals of monosodium urate monohydrate,[4,27,32] and heat shock[33] (see also Table 1). Alternatively, they can be repressed by the addition of glucocorticoid hormones or vitamin A analogues (retinoids).[17,29,30] It is important to point out that although aspirin and the nonsteroidal anti-inflammatory drugs (NSAIDs) abolish PGE_2 production, they do not reduce collagenase synthesis,[26] thus dissociating these two mediators of matrix destruction.

The earliest experiments to understand metalloproteinase expression focused on collagenase. This is because, in contrast to many assays designed to measure activity of other metalloproteinases, the ^{14}C-collagen fibril assay specifically detected collagenase activity.[1-3,26] This allowed us to follow accurately the ability of various compounds to modulate collagenase production. Since treatment with PMA greatly increased collagenase and since this increase could be blocked with α amanitin,[27] an inhibitor of RNA polymerase II, we speculated that PMA might be acting transcriptionally to stimulate collagenase. To begin testing this hypothesis, we used mRNA from PMA-stimulated cells to isolate a partial cDNA clone for collagenase.[32] Out of only 167 recombinant transformants, four represented collagenase, thus suggesting that collagenase mRNA was a major gene product in induced fibroblasts. With this cDNA clone, we measured steady state levels of collagenase mRNA in cells after addition of urate crystals.[32] Collagenase mRNA (~2.1 kb) was first detected by 4 to 5 hr, and it increased steadily over the next 24 to 36 hr. At the same time, collagenase protein, measured by Western blotting with monospecific polyclonal antibody, increased in the culture medium.[32] These findings supported data from other experiments, which showed (a) that collagenase is rapidly synthesized and secreted into the culture medium[10] and (b) that there is a direct correlation between the levels of collagenase mRNA and collagenase protein.[27,30]

To demonstrate clearly that an increase in steady state levels of mRNA resulted from increased transcription, we treated cultures of synovial fibroblasts with PMA in the presence of ^3H-uridine to label newly synthesized collagenase mRNA.[30] Over an 18-hr period, we detected increasing amounts of collagenase mRNA, and by 18 hr, it represented ~2% of the mRNA population. The biosynthesis of new ^3H-uridine-labeled collagenase mRNA shows that PMA induced transcription of the collagenase gene, and the abundance of the mRNA once again emphasizes that collagenase is a major gene product of these PMA-treated cells.

By using pulse-chase experiments, we explored a possible role for mRNA stabilization in this system.[30] For these studies, collagenase ^3H-mRNA was induced by PMA in the presence of ^3H-uridine. After 18 hr, radioactivity was removed and the cells were chased with nonradioactive medium containing 5 mM uridine and 2.5 mM cytidine. At intervals during the chase period, cultures were terminated and RNA extracted. The amount of ^3H-mRNA present was determined by hybridization to collagenase cDNA spotted onto nitrocellulose filters. As a control for changes in the levels of steady state collagenase mRNA, poly A+ RNA was spotted onto nitrocellulose filters and hybridized with ^{32}P-cDNA for collagenase. The decay of total poly A+ RNA in PMA vs. control cells showed that the half-life of the poly A+ RNA was ~24 hr and that this was unaffected by PMA. However, the half-life of collagenase mRNA varied greatly: in one experiment where the chase period was carried out in the presence of PMA, the half-life was 38 hr, while in a second experiment where PMA was absent during the chase, the half-life was 12 hr.[30]

To try to explain this discrepancy, we measured simultaneously the amount of steady state collagenase mRNA in the cells during the chase periods in experiment 1 and in experiment 2.[30] We found that at time zero, the level of total collagenase mRNA in experiment 1 was nearly twice that of experiment 2. Furthermore, the total collagenase mRNA in experiment 1 remained high during the chase period in the presence of PMA, while collagenase mRNA in experiment 2 declined rapidly in a chase that did not contain PMA. Thus, the half-life of collagenase mRNA was directly correlated with the level of collagenase mRNA induced in the cells: the greater the level of induction, the longer the half-life.

Reasons for this variation in PMA-induced half-life are not fully understood. One possible explanation may lie in the nucleic acid sequences found in the 3′ untranslated region of the collagenase mRNA. The sequence, ATTTA, is present in the 3′ untranslated region of mRNAs for certain transiently expressed inflammatory proteins and oncogene products and has been implicated in the destabilization of mRNA.[34] However, in the presence of PMA, the stability of mRNAs containing this sequence is increased. It is thought that PMA treatment causes a change in the ATTTA-mediated degradation pathway, resulting in a relative stability of mRNAs containing this sequence.[34] Of potential importance is the fact that this sequence is repeated three times in the 3′ untranslated region of the rabbit[7,34] and human collagenase genes,[6,7,18] once in the 3′ untranslated region of the rabbit activator/stromelysin gene,[7,17] and twice in rat transin (stromelysin) mRNA.[36,37] The fact that treatment of rabbit synovial fibroblasts with PMA stabilizes the collagenase message suggests that this sequence may, indeed, have a functional role.

As mentioned earlier, collagenase synthesis can be repressed by retinoids and by glucocorticoids. This suppression is reflected by the fact that continued treatment (up to 60 hr) of collagenase-producing cells with retinoic acid or dexamethasone results in a reduction in steady state collagenase mRNA in the cells and collagenase protein in the culture medium.[17,29,30] We wanted to determine whether these compounds were influencing the half-life of collagenase mRNA.[30] Therefore, as part of the pulse-chase experiments described above, cells that had been induced with PMA and radiolabeled with ^{3}H-uridine were exposed to all-*trans*-retinoic acid (10^{-6} M) or dexamethasone (10^{-7} M) during the chase period. We found that neither compound affected the half-life of collagenase mRNA, thus suggesting that retinoids and steroids act at a transcriptional level to depress collagenase mRNA levels.

While these experiments were being performed, a number of laboratories were cloning metalloproteinase genes. First, the full-length human collagenase gene was simultaneously cloned by a number of investigators,[5-7] and full-length cDNA clones for rabbit collagenase became available.[7,35] Sequence analysis of these cDNAs revealed that rabbit and human collagenase were ~50% homologous and that the same collagenase was found in fibroblasts throughout the body.

Second, the genes for other metalloproteinases were cloned: rabbit[7,17,38] and human stromelysin[7,18,19] and rat transin/stromelysin.[36,37] Both rabbit and human stromelysin were ~50% homologous with collagenase,[5-7] but they were ~75 to 80% homologous with transin, a mRNA isolated from rat fibroblasts.[7,17-19,36,37] The very high degree of homology between stromelysin and transin suggested that transin was the rat analogue of stromelysin. This was particularly exciting because the rat transin gene was expressed in fibroblasts treated with growth factors, such as epidermal growth factor (EGF), or transformed with oncogenes, such as *h-ras* or *src* or polyoma virus.[37,38] Thus, for the first time a direct link was established between oncogenes and expression of metalloproteinases. This observation only strengthened the hypothesis of Sporn and Harris[23] that the phenotype of nonmalignant proliferative disease can biochemically mimic that of genotypically transformed tumor cells.

Third, we isolated and characterized a gene for rabbit collagenase.[35] We partially sequenced this gene and found that it is encoded by 9.1 kb of DNA containing 9 introns and 10 exons. The size of the rabbit collagenase gene and its intron/exon structure resemble significantly that of several other members of the metalloproteinase family: rat transin/stromelysin,[40] rat transin/stromelysin 2,[40] human collagenase,[41] human stromelysin,[39] and human stromelysin 2.[39] The homology among these genes substantiates the hypothesis that they evolved from a common ancestor gene family.[7,17-20,35] The conservation in cDNA and protein sequences suggests that common regulatory sequences might also be conserved and that some of these genes might be coordinately expressed.

To investigate this possibility, we used slot-blot analysis to determine whether PMA could coordinately induce both collagenase and stromelysin mRNAs. Stromelysin mRNA (1.9 kb), like collagenase, is greatly increased in rabbit synovial fibroblasts treated with PMA for varying periods.[17] To determine coordinate induction, blots were hybridized with the appropriate cDNA clones for collagenase or stromelysin. We found that mRNAs for both collagenase and stromelysin were low at time 0. They began to increase between 2.5 and 5 hr and continued to do so until at least 30 hr. Other investigators have also reported the coordinate induction of collagenase and stromelysin by PMA.[38] It is important to point out that coordinate expression is not limited to phorbol esters. For example, IL-1[31] and heat shock (42°C for 3 hr or 45°C for 1 hr)[33] increase both collagenase and stromelysin, and other agents may do so as well. However, it should be pointed out that the response of collagenase and stromelysin genes is not always coordinate. For example, EGF, *src*, or *h-ras* induces stromelysin in rat fibroblasts, but not collagenase.[37] One important area for future study will be an investigation of mechanisms controlling the differential expression of these genes in rat cells, as well as in fibroblasts from other species.

We tested whether collagenase and stromelysin mRNAs could be coordinately suppressed by all-*trans*-retinoic acid and dexamethasone.[17] In this experiment, cells were stimulated with PMA for 24 hr to induce both mRNAs. Then, the PMA was removed (time 0) and all-*trans*-retinoic acid ($10^{-6} M$) or dexamethasone ($10^{-7} M$) was added. Cultures were terminated at selected times, and the level of mRNA for collagenase and stromelysin was determined by slot-blot analysis. Both drugs substantially decreased mRNAs (by ~60%) for both enzymes by 24 hr, while mRNA in the untreated control cultures remained relatively constant throughout the experiment.[17]

To begin to address the question of regulatory sequences located in the 5' flanking DNA of these genes,[35,42] we sequenced 1.2 kb of DNA of the rabbit collagenase gene that is 5' to the start site of transcription. We analyzed this DNA for the presence of putative regulatory elements.[35,42] Beginning at −32 bp from the transcription start site is the sequence, 5'-TATAAA-3',[35] a TATA box implicated in the accurate initiation of transcription of other mRNAs. At −78 to −70 bp is the sequence, 5'-ATGAGTCAG-3', which has been termed the phorbol ester response element[35,43-48] and which is important for induction of collagenase following exposure to PMA.

Recent work by a number of investigators suggests that it is the interaction of the protooncogene products c-JUN and c-FOS with this phorbol response element that increases transcription of the collagenase gene.[42-47] Initially, the complex nuclear transcription factors, AP-1, was identified as binding to the phorbol element, thus implicating this protein as a major player in activating metalloproteinase genes.[43,44,47] Soon it was documented that AP-1 was (probably) the product of the protooncogene c-JUN and c-FOS.[45,49] Other investigators found that anti-FOS antisera precipitated a complex of c-FOS and several other associated proteins, one of which was similar to c-JUN.[49] This finding led the authors to conclude that c-FOS is a transacting factor that interacts with c-JUN, but does not directly bind to the DNA.

Studies by two different groups demonstrated that c-FOS plays a major role in the activation of collagenase and stromelysin genes by protooncogenes, growth factors, and phorbol esters.[46,50] Inactivation of *c-fos* transcripts using antisense *fos* RNA abolished *fos*, *mos*, *ras*, *src*, and phorbol ester-induced collagenase gene transcription.[46] Another group found that the use of antisense *fos* RNA reduced or prevented induction of rat stromelysin by PMA or platelet-derived growth factor (PDGF).[50] However, EGF stimulated stromelysin despite an equivalent inhibition of *fos* levels. Since the phorbol response element was implicated in induction with both PDGF and EGF,[50] the authors postulate the existence of both *fos*-dependent and *fos*-independent pathways in regulating metalloproteinase gene expression. These findings, once again, link the expression of oncogenes to the expression of metalloproteinases.

Thus, one model holds that a complex of at least two proteins, the AP-1 complex of the *c-jun* gene product and the *c-fos* gene product, bind to the phorbol element.[51,52] The interaction of these proteins with the DNA may be enhanced by protein-protein interactions that increase their affinity for the DNA. It has been suggested that, in some instances, overproduction of *c-jun* may result in the formation of homodimers that can compensate for the absence of FOS protein and induce gene transcription of the collagenase gene independently of FOS.[51,52]

There is still much to be learned. What, for example, induces overproduction of the AP-1 complex? Is expression of the two protooncogenes *c-jun* and *c-fos* necessary and sufficient for metalloproteinase induction by other agents? Are differences in the expression of the AP-1 complex responsible for the variation in the responsiveness of our out-bred population of rabbit fibroblasts to PMA? How do the agents that inhibit metalloproteinase synthesis (i.e., retinoids and glucocorticoids) block JUN and FOS? Thus, although we have learned a great deal about the molecular mechanisms by which PMA induces metalloproteinase synthesis, we need to integrate this knowledge with the biology and physiology of fibroblasts during normal development and remodeling and during disease states such as rheumatoid arthritis and tumor invasion.

EFFECTS OF INTERLEUKIN-1 AND TUMOR NECROSIS FACTOR ON SYNOVIAL CELLS

Interleukin-1 (IL-1) is probably one of the most studied and well-characterized lymphokines (reviewed in reference 53). While macrophages are presumed to be a major source of IL-1,[53] fibroblasts have also been reported to produce an IL-1-like molecule.[54,55] The two forms of Il-1, α and β, share only ~30% homology, yet both interact with cells via the same receptor. Both forms are translated as precursor polypeptides (31 kDa), and despite the fact that IL-1 is found in the extracellular space, it does not contain a signal peptide. The generation of mature IL-1 (17.5 kDa) and smaller peptides occurs via serine proteases that cleave at the NH_2 terminal end of the protein.

Confluent fibroblast monolayers contain between 5,000 and 15,000 IL-1 receptors. Presumably, receptor-ligand interaction is an important first step in the signal/transduction pathway leading to expression of two of the effects of IL-1 on these cells: increased proliferation and induction of metalloproteinase synthesis. The ability of IL-1 to influence synovial cell function was first noted in 1976 when Dayer et al.[56] found that monolayer cultures of dissociated synovial cells taken from rheumatoid patients contained a mixture of cell types that grew rapidly and that "spontaneously" secreted large quantities of collagenase and PGE_2 into the culture medium. Soon after, they reported that supernatants from monocytes contained a factor, Mr approximately 12,000 Da, that was capable of inducing synovial fibroblasts to produce PGE_2 and collagenase.[57] They termed the protein mononuclear cell factor (MCF). Subsequent work by this group demonstrated that MCF was the same as interleukin-l (IL-1),[58] and most recently, Dayer et al.[59] and others[60] have shown that recombinant IL-1 can elicit collagenase and PGE_2 production by synovial cells, as well as enhance fibroblast growth. It is important to point out that IL-1 is exceedingly potent: pmol concentrations of either IL-1 α or IL-1 β will induce collagenase and PGE_2 and will stimulate cell proliferation.[59,60] Consequently, very small quantities of IL-1 can have a relatively large impact on the pathophysiology of a disease process such as rheumatoid arthritis.

Tumor necrosis factor (TNF) is another inflammatory mediator that can influence the pathophysiology of rheumatoid arthritis.[61-63] It has a Mr of ~17 K, and like IL-1, it is produced by macrophages. TNF also shares many activities with IL-1, including stimulation of bone resorption, PGE_2 and collagenase induction, and mitogenicity for fibroblasts.[59-67]

These common biologic activities suggest that perhaps common intracellular pathways may be utilized by both compounds as they increase cell growth and stimulate metalloproteinase synthesis. In earlier studies, IL-1 and TNF-induced proliferation was measured by incorporation of ^{3}H-thymidine, cell number, and protein content. The increases were two- to fivefold above controls and required 24 to 48 hr to be seen.[59-67] More recent studies by Lin and Vilcek[68] on IL-1 and TNF-induced proliferation of human fibroblasts have been directed towards mechanisms. They found that the increase in cell proliferation triggered by both compounds was preceded by a transient increase in steady state mRNA levels for *c-fos* and *c-myc*. Protooncogene mRNA levels were enhanced within 20 min of IL-1 (1 ng/ml) treatment or TNFα (30 ng/ml). Both these mRNAs peaked at 30 min and returned to basal levels (*c-myc*) or became undetectable (*c-fos*) by 60 or 90 min. The presence of cycloheximide did not prevent the response, and this suggested that the increases in *c-myc* and *c-fos* was a primary response not requiring the synthesis of intermediary proteins.

More recent data extend these findings and begin to suggest that after an initial similarity, divergent pathways may ultimately be needed to stimulate cell proliferation.[69,70] Vilcek and colleagues note that at 1 ng/ml, TNF was only weakly mitogenic for growth-arrested fibroblasts and that its potency was enhanced two- to fourfold by EGF or PDGF. However, antiserum to either EGF or PDGF failed to block mitogenic action of TNF, and this led the authors to conclude that autocrine PDGF or EGF probably does not play a significant role in the mitogenicity of TNF. Rather, these growth factors may be acting indirectly to enhance the effects of TNF.[69]

In contrast, another report directly implicates PDGF in the mitogenicity of IL-1.[70] These investigators argued that IL-1 receptor-mediated ligand internalization was too slow to account directly for an increase in *c-fos* and *c-myc* and the subsequent increase in cell proliferation. They began by analyzing the kinetics of ^{3}H-thymidine incorporation by cultures of human fibroblasts treated with either IL-1-α or β or with PDGF. They found that the first increase in ^{3}H-thymidine incorporation in response to isoforms of PDGF (PDGF-AA, PDGF-BB, or PDGF-AB) was seen at 16 hr and was maximal by 24 hr. On the other hand, with either IL-1-α or β, an increase in ^{3}H-thymidine incorporation was not seen until 24 hr and was maximal at 36 hr. They concluded that the delay observed with IL-1 is consistent with the interpretation that IL-1 stimulates the expression and translation of a biologically active molecule that induces the mitogenic stimulation. They went on to show that IL-1-induced proliferation was due to the induction and release of an isoform of PDGF, PDGF-AA: IL-1 transiently increased expression of the PDGF-A chain mRNA, with a maximal increase 2 hr after addition of Il-1. This was followed by appearance of PDGF protein in the culture medium by 4 hr and the subsequent down regulation of the binding of PDGF-AA protein to the cell surface. Importantly, antibodies to PDGF completely inhibited the mitogenicity of Il-1. Thus, these data indicate that the mitogenic action of IL-1 is mediated indirectly via the induction of proteins (PDGF-AA) that act as autocrines.

The parallel between IL-1 and TNF can be extended to include mechanisms controlling the induction of metalloproteinase synthesis in fibroblasts. The ability of both these compounds to increase collagenase production has been known for some time, and results of recent studies indicate that, similar to phorbol ester-mediated induction of collagenase, IL-1 and TNF also act at a transcriptional level.[71,72] Using transcriptional run-off assays with nuclei isolated from human fibroblasts treated with TNFα, Angel, Karin and co-workers showed a fourfold stimulation in transcription of the collagenase gene.[71] Furthermore, this stimulation was preceded by a TNFα-induced prolonged induction of *c-jun* mRNA, beginning at 30 min and lasting at least 6 hr. The prolonged expression of *c-jun* contrasts with the transient expression seen when cells were treated with phorbol ester and provides an example of a physiologically important mediator that can induce transcription of *c-jun*. The authors also found that TNF elicited a rapid and transient increase in *c-fos*, which was detected at 1 hr. Therefore, TNFα stimulates expression of both *c-jun* and *c-fos* genes whose products interact to stimulate transcription of AP-1 responsive genes, e.g., collagenase and stromelysin.

IL-1 also elicits a rapid rise in *c-jun* mRNA.[72] Treatment of human dermal fibroblasts with IL-1 caused an induction of *c-jun*, which was maximal 1 hr after stimulation. In contrast to TNF, *c-jun* levels quickly declined, returning to baseline levels between 3 and 6 hr. Collagenase mRNA began to appear 2 hr after IL-1 addition and continued to rise for at least 24 hr. This time course of increasing collagenase mRNA is very similar to that seen when rabbit synovial cells are induced with either crystals of monosodium urate monohydrate,[32] PMA,[17] or heat shock,[33] and this suggests that similar pathways of induction may be operational.

Both IL-1 and TNF-α appear to mediate an increase in collagenase via an increase in *c-jun* and *c-fos*, which then interact with the phorbol responsive sequence located at the 5' end of the collagenase and stromelysin genes. However, as with IL-1 and TNF-induced cell proliferation, there are differences in the induction process, possibly suggesting alternative pathways of signal/transduction. Nonetheless, the results to date indicate an important role for the phorbol response element in stimulating metalloproteinase synthesis by demonstrating that physiologically important mediators may act via this element.

HEAT SHOCK AND INORGANIC CRYSTALS AS "CYTOKINES"

Many compounds increase metalloproteinase synthesis and enhance cell proliferation (Table 1). Some of these are not cytokines. An example, which is not listed in Table 1, is heat shock. Recently, we have determined that heat shock of rabbit synovial fibroblasts at 45°C for 1 hr or at 42°C for 3 hr increased mRNAs for collagenase and stromelysin.[33] Similar to induction of metalloproteinases by other agents, these mRNAs were detected by ~3 hr following initiation of the stimulus. Concomitant treatment with PMA at 10^{-8} or 10^{-9} *M* and heat shock was not additive or synergistic, but all-*trans*-retinoic acid (10^{-6} *M*), added just prior to heat shock, was antagonistic. Although we do not know the mechanisms controlling the heat shock response, analysis of the 5' end of the rabbit collagenase and stromelysin genes reveals several putative heat shock elements. Present at −78 is a putative heat shock element, 5'-C<u>A</u>TGAAATT<u>G</u>CAA<u>C</u>-3',[33] which shares 11 base pairs (3 mismatches are underlined) with the heat shock element, 5'-CTNGAANNTTCNAG-3'[73] Perhaps these sequences play a role in increasing transcription of these genes following exposure to higher temperature.

In searching for an explanation as to why heat shock should increase metalloproteinase levels, we note that increases in these enzymes have long been associated with stress, whether induced by cell fusion,[74] by phagocytosis of debris,[27,32,75] or by treatment with agents that rearrange the cytoskeleton (i.e., cytochalasin B).[76] Finding that exposure of synovial fibroblasts to elevated temperatures will increase mRNA for collagenase and stromelysin broadens our definition of the classical heat shock response and implies a possible role for heat shock in chronic inflammatory states such as rheumatoid arthritis.

The stimulation of collagenase and cell proliferation by inorganic crystals, such as monosodium urate monohydrate[27,32,75] and basic calcium phosphate (BCP) crystals (e.g., hydroxyapatite, octacalcium phosphate, tricalcium phosphate, and calcium urate),[77-81] also requires us to broaden our definition of cytokines (see Table 1). These crystals have been implicated in the pathophysiology of arthritic diseases: crystals of monosodium urate monohydrate help to mediate the destructive component of gouty arthritis,[75] and deposition of BCP crystals in joints is accompanied by symptoms of classical arthritis with synovitis and degenerative joint disease.[77-81] The interesting question here is the mechanism by which the effects of these crystals mimic the classical cytokines. With both sodium urate monohydrate and BCP crystals, the initial stimulus is physical ingestion, i.e., phagocytosis. Following that, the signal/transduction pathway leading to increased levels of collagenase mRNA by either type of crystal is not known, but may perhaps involve some of the mechanisms discussed above.

Evidence is accumulating that BCP crystals are mitogenic.[79-81] Most of this work has been carried out by Cheung, McCarty, et al., and in their early studies, they found that BCP crystals could substitute for PDGF as a competence factor in initiating cell proliferation.[79] Subsequently, they determined that BCP crystals induced the expression of both *c-myc* and *c-fos* protooncogenes with a time course similar to PDGF.[81] Treatment of growth-arrested BALB/3T3 cells with either BCP crystals or PDGF resulted in maximal accumulation of *c-fos* mRNA at 30 min and maximal induction of *c-myc* by 1 hr with maximal levels by 3 hr. This is the first demonstration that BCP crystals can activate protooncogene expression in cultured cells.

Initially, their hypothesis was that BCP crystals were solubilized in the acidic environment of secondary lysosomes after endocytosis, as free Ca^{++} appeared in the extracellular medium within 2 hr.[79-81] Ca^{++} in transit from crystal dissolution either diffused or was actively pumped from the lysosome, raising the intracellular Ca^{++} concentration. As a consequence, many Ca^{++}-activated processes, such as microtubule depolymerization and alterations in phospholipid metabolism (including protein phosphorylation and protease activation), occurred as part of a program leading eventually to increased DNA synthesis. However, they have recently concluded that intracellular dissolution probably does not proceed fast enough to account for the rapid induction in *c-myc* and *c-fos* mRNAs. Rather, they believe that it is likely that dissolution and elevation of intracellular Ca^{++} are required for the later events occurring during the stages of the cell cycle.[81] Nonetheless, the intriguing fact remains that this Ca^{++}-dependent proliferative cycle, usually activated by growth factors and the products of oncogenes, can be initiated by the phagocytosis of inorganic crystals.

EFFECTS OF TGFβ ON SYNOVIAL CELLS

Transforming growth factor β (TGFβ; Mr 25 K) is a multifunctional growth factor with many pleiotropic effects, including inhibition or enhancement of cell proliferation, chemotaxis of fibroblasts and monocytes, and modulation of the extracellular matrix (reviewed in reference 82). Originally isolated from tumor cells, TGFβ was subsequently found in nearly all normal tissues. It is particularly plentiful in platelets, but it is also present in substantial quantities in mesenchymal cells, where it plays an important role in the remodeling of connective tissue that occurs during inflammation, bone resorption, and wound healing.[82]

Several laboratories have studied the effect of TGFβ on metalloproteinase synthesis,[83-87] and with one exception,[83] the general consensus is that TGFβ suppresses the production of both collagenase and stromelysin.[84-87] Chua et al.[83] reported that TGFβ increased the synthesis of procollagenase in neonatal fibroblasts. Other investigators,[84-87] using rat and human fibroblasts, found that TGFβ antagonized collagenase mRNA synthesis that was either constitutive or EGF induced. The mechanism appears to be transcriptional. The overall decrease in collagenase activity is amplified by the fact that TGFβ also increases the production of TIMP (Tissue Inhibitor of Metalloproteinases),[84,87] a 25 kDa protein that complexes with active forms of metalloproteinases and inhibits them.[1-3]

Reasons for the reported increase in collagenase following TGFβ treatment are not entirely clear. Perhaps they are related to the strain of cells used or to unknown factors in the culture media. These possibilities are potentially important because they emphasize the fact that there is a constant agonist/antagonist interaction among the numerous factors and cytokines that are present. This interaction may be an ever-changing dynamic that can influence the eventual outcome of the disease process, depending upon the relative concentrations and activities of the stimulating and suppressive factors. Consequently, the complement of cytokines within a tissue is crucial in determining whether the delicate balance of gene regulation is shifted towards extracellular matrix deposition or destruction, or towards enhanced or reduced cell proliferation.

The recent documentation of elevated levels of active and latent forms of TGFβ in synovial fluid of patients with rheumatoid arthritis[88] raises the possibility that this peptide may participate in the initiation or maintenance of synovitis. Using monolayer cultures of human skin and rabbit synovial fibroblasts, we have explored a potential role for TGFβ in the pathophysiology of rheumatoid arthritis.[89] When

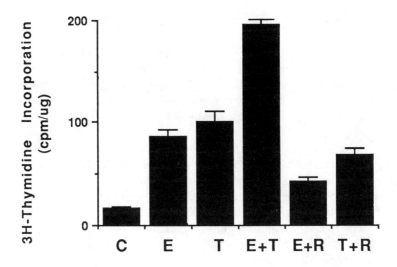

Figure 2. Effect of TGFβ and EGF on ³H-thymidine incorporation by human foreskin fibroblasts. Monolayer cultures of human foreskin fibroblasts (passage 9) in Dulbecco's modified Eagle's medium with 10% fetal bovine serum were treated with EGF (10 ng/ml; Collaborative Research, Bedford, MA) or TGFβ (2.5 ng/ml; Collaborative Research), either alone or in combination, or in the presence of all-*trans*-retinoic acid (10^{-6} *M*) for 72 hr. The cultures were then pulse-labeled with ³H-thymidine (25 μCi/ml in 10^{-4} *M* thymidine) for 2 hr.[89] Cell proliferation was measured as cpm ³H-thymidine incorporated per μg cell protein.[89]

purified TGFβ was added to rabbit fibroblasts, it either had no effect or was slightly growth suppressive.[89] However, TGFβ was mitogenic when added in conjunction with EGF. A 24- to 48-hr treatment with EGF (2 ng/ml) gave a three- to fivefold stimulation of ³H-thymidine incorporation. Addition of TGFβ augmented this stimulation at least twofold, and treatment with all-*trans*-retinoic acid antagonized the mitogenicity of both TGFβ and EGF. These findings can be extended to include human skin fibroblasts cultured as monolayers in 10% serum (Figure 2). In this experiment, nanogram concentrations of either TGFβ or EGF were mitogenic, and combined treatment resulted in a further enhancement of cell proliferation. As with the rabbit fibroblasts, mitogenicity could be antagonized by all-*trans*-retinoic acid (10^{-6} *M*).

These data contrast sharply to a report documenting the growth-inhibitory properties of TGFβ on rheumatoid synovial cells.[90] The investigators show that both human rheumatoid synovial cells and synoviocytes taken from rats with experimentally induced arthritis grew as colonies in soft agar in the presence of 20% serum or PDGF. TGFβ, as well as all-*trans*-retinoic acid, inhibited cell growth, while EGF, IL-1, TNF-α, and γ-interferon had no effect.

These data are difficult to reconcile. The actions of TGFβ differ dramatically perhaps because of different culture conditions (monolayer vs. agarose culture), because of the presence of other (unknown) growth factors, or because of subtle differences in the cells. However, these results serve to emphasize the fact that TGFβ is, indeed, multifunctional,[82] and its behavior may well depend upon its environment. In reality, these growth factors probably do not function in isolation, but rather as part of a mixture. Consequently, it is the relative concentrations of each growth factor and cytokine present within a microenvironment that will determine cell behavior at any given time. Therefore, even though TGFβ has the potential to actually inhibit cell proliferation [82,90] and hence impede the rheumatoid lesion, in the presence of other factors it may have exactly the opposite effect.

In addition to studying the effects of TGFβ on cell proliferation, we studied its ability to affect cell morphology, and we correlated these changes with the organization of actin filaments within the cells (Figure 3).[91] Control fibroblasts were flat, with a very well-organized pattern of actin filaments. Treatment with EGF changed the cell shape and actin pattern only slightly, while treatment with TGFβ, in combination with EGF, resulted in a piling up of cells into foci. This was accompanied by a disruption in the pattern of actin filaments, a finding usually associated with malignant transformation. Note that the morphology and pattern of actin filaments seen in primary cultures of human rheumatoid synovial cells greatly resembled that seen in the cultures of rabbit fibroblasts treated with TGFβ. Thus, primary cultures of rheumatoid synovial cells mimic malignant cells in their morphology and in their proliferative ability.

Many of these primary cultures of rheumatoid tissue are derived from patients of advanced years in whom fibroblasts have only a limited number of cell doublings remaining. Nonetheless, when these cells were placed in primary culture, not only did they grow rapidly, but they were not contact inhibited and they piled up.[56,89-93] These cells appeared transformed. However, this transformation was phenotypic rather than genotypic, since when the cells were passaged and the population of cells became exclusively fibroblast-like, the rate of proliferation became commensurate with the age of the patient: cultures derived from older patients grew more slowly than did cultures from younger patients.[56, 89-93]

It is reasonable to speculate that the interaction among the various cell types and cytokines that are present in inflamed rheumatoid synovium is largely responsible for the extraordinary proliferative capacity of this tissue. Studies by Wilder and colleagues[90,93] in the streptococcal cell-wall model of arthritis in rats support this speculation. Using *in situ* immunochemical and histochemical stains, electron microscopy, *in vitro* culture and staining techniques, and cell transfer studies to athymic nude mice,[93] they confirmed that the most abundant cell in the synovium of athritic rats is an immature, spindle-shaped mesenchymal cell that exhibited characteristics generally associated with transformed cells. In addition, they found that, like the human rheumatoid adherent synovial cells, the synovial cells from the arthritic rats lost the characteristics of the transformed phenotype on extended culture.[90,93] This suggests that it is highly unlikely that the transformed phenotype was virally

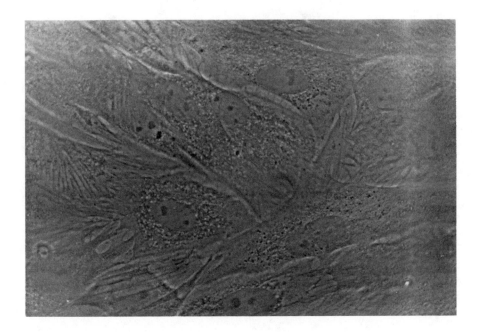

A

Figure 3. Effect of transforming growth factors on synovial cell morphology and actin filaments and antagonism by all-*trans*-retinoic acid. Confluent cultures of rabbit synovial fibroblasts in Dulbecco's modified Eagle's medium with 10% fetal bovine serum were grown on glass coverslips and treated with partially purified EGF and TGFβ at a final concentration of 200 µg protein per ml or with EGF at 5 ng/ml for 48 hr. Selected cultures were treated with EGF and TGFβ and with all-*trans*-retinoic acid (10^{-6} M). Also cultured on coverslips in 10% fetal bovine serum for 5 d were primary human rheumatoid cells.[56,89-91] To fix the cells, cultures were washed twice in phosphate buffered saline and treated with a 3.7% formaldehyde solution in buffered saline for 10 min. The coverslips were washed twice in buffered saline and extracted with a solution of acetone at 4°C for 5 min. After a quick water wash, actin filaments were stained with NBD-phallocidin (Molecular Probes, Plano, TX) at 60 units stain per ml phosphate buffered saline for 30 min at room temperature. The coverslips were washed rapidly twice, mounted cell-side down on a glass slide, and the cells visualized by phase-contrast and immunufluorescence microscopy with a 40 X oil-immersion objective. (A, B) Control rabbit synovial fibroblasts: (A) phase contrast; (B) phallocidin-stained. (C, D) Rabbit synovial fibroblasts treated with EGF: (C) phase contrast; (D) phallocidin-stained. (E, F) Rabbit synovial fibroblasts treated with EGF and TGFβ: (E) phase contrast; (F) phallocidin-stained. (G, H) Primary culture of human rheumatoid synovial cells: (G) phase contrast; (H) phallocidin-stained. (I, J) Rabbit synovial fibroblasts treated with EGF and TGFβ and all-*trans*-retinoic acid: (I) phase contrast; (J) phallocidin-stained. (From Brinckerhoff, C.E. et al., in *Retinoids, Differentiation and Disease*, Ciba Foundation Symposium, 113, 1985. © John Wiley & Sons. With permission.)

B

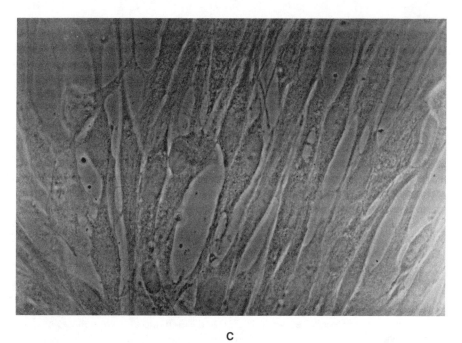

C

Figure 3. *(Continued.)*

D

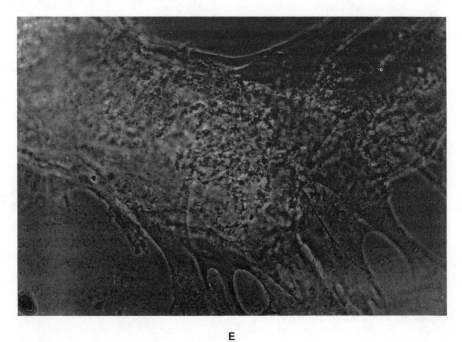

E

Figure 3. *(Continued.)*

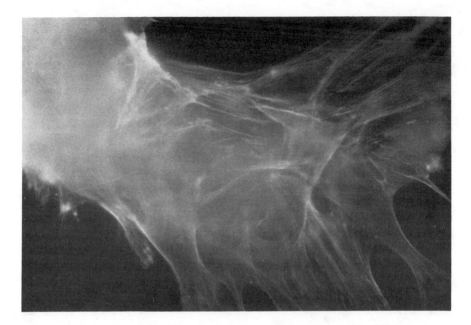

F

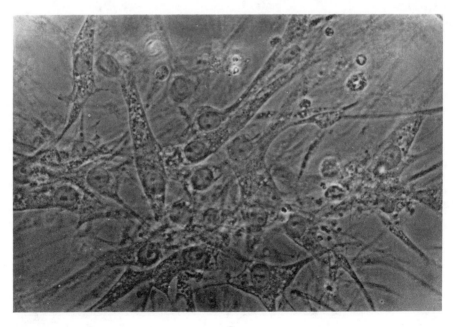

G

Figure 3. *(Continued.)*

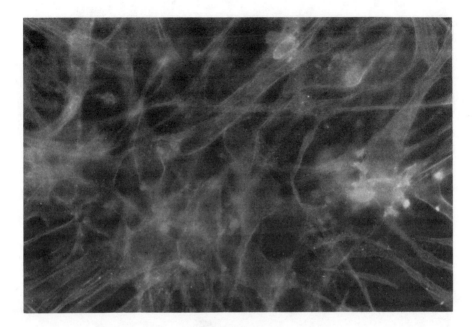

H

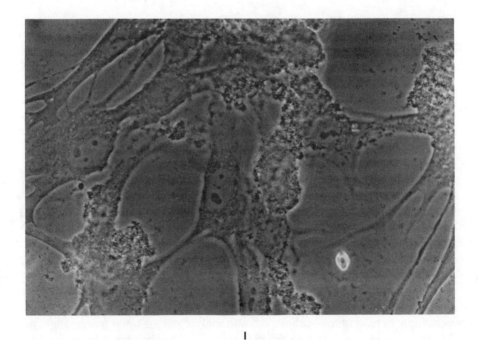

I

Figure 3. *(Continued.)*

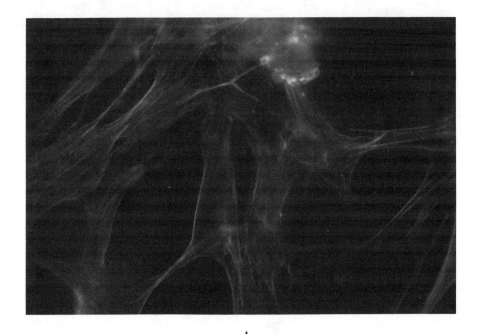

J

Figure 3. *(Continued.)*

induced. The critical question then remains: what paracrine growth factors drive the synovial cells to acquire phenotypic characteristics associated with malignancy?

Indeed, the presence of these factors in rheumatoid synovium may have been responsible for the ability of rheumatoid tissue to survive for a limited time when implanted subcutaneously in nude mice.[92] When a single cell suspension of normal synovial fibroblasts was injected into nude mice, the cells disappeared from the injection site within a few days, while dissociated rheumatoid synovial cells organized themselves into a mass remarkably similar to rheumatoid tissue. For at least 3 to 4 weeks, the cells remained as a distinct lump at the injection site and retained their ability to synthesize collagenase and PGE_2. We can speculate that the proliferative ability of primary rheumatoid synovial cells in nude mice may be mediated by growth factors and cytokines, some of which may be the products of protooncogenes. Eventually, when the cells producing these cytokines died off, so did the mass of proliferating synovial cells.

We wanted to document the association in primary rheumatoid synovium between increased expression of the protooncogenes mediating proliferation and the metalloproteinases mediating tissue invasion and destruction. For these studies, we used Northern blots of total cellular RNA to analyze mRNA taken from cultures of rheumatoid synovium from five different patients. The cells were cultured for 4 to

10 d in 10% fetal bovine serum. Passaged cultures of rheumatoid synovial cells were used as controls, since they have characteristics of normal fibroblasts.[56] Figure 4 shows a Northern blot probed with cDNA for human collagenase (Panel A), stromelysin (Panel B), TGFβ (Panel C), or *v-sis*, which codes for a truncated form of the β chain of PDGF (Panel D). The figure shows that collagenase and stromelysin are major gene products of these primary synovial cells, and that metalloproteinase expression is coordinate. In some cultures, increased expression of metalloproteinases and protooncogenes mediating proliferation (TGFβ) is linked (patient 3), while in others they are not (patient 4), thus making it difficult to determine whether TGFβ has a suppressive or stimulatory role. Interestingly, one culture showed increased expression of protooncogenes TGFβ and PDGF-*v-sis*, but not increased metalloproteinase mRNA. These findings reflect the heterogeneous cell types present in primary cultures of rheumatoid synovium, as well as variations in the time of culture. However, the data do suggest that the proliferative response of fibroblasts to cytokines may not be necessarily linked to the destructive response mediated by metalloproteinases.

It is important to point out that the failure to demonstrate expression of a particular mRNA does not mean that the gene is not expressed or the protein it encodes has no role in rheumatoid disease. While this interpretation may be correct, it is also possible that the mRNA in question is not expressed in a particular specimen at the time the RNA is harvested. Furthermore, negative data can result from the fact that an mRNA is expressed at such low levels that it can be detected only by using poly A+ RNA rather than total cellular RNA.

To see if the proliferative response can be dissociated from the destructive response, we cultured human foreskin fibroblasts with EGF (10 ng/ml) along with dexamethasone (10^{-7} M) or all-*trans*-retinoic acid (10^{-6} M) for 24 hr. Whole cell RNA was isolated and RNA slot-blot analysis was used to quantitate levels of the cell-cycle-dependent protooncogene, human *c-myc* mRNA (Figure 5, Panel A). Here, steady state RNA levels of *c-myc* are used to monitor proliferation in EGF-treated cells vs. time-matched controls. We found that EGF increased *c-myc* levels twofold over the control. Importantly, dexamethasone further augmented this EGF-induced proliferative response, while in contrast, retinoic acid slightly antagonized it.

When the RNA slot-blot was probed with the human collagenase cDNA (Panel B), we found that EGF induced collagenase eightfold over the control. The timeframe for induction of *c-myc* and collagenase mRNAs by EGF is different, suggesting distinct mechanisms. As expected, both dexamethasone and retinoic acid reduced collagenase mRNA: dexamethasone by 90% and retinoic acid by 40%.[17,29,30] Thus, dexamethasone differentially affected the proliferative and destructive responses of the cells, while retinoic acid decreased both. However, the time frame for each of the retinoid-mediated effects was distinct, suggesting distinct mechnaisms. Taken together, these findings imply that the proliferative and destructive potential of fibroblasts is not linked, even though expression of protooncogenes may be required for both.

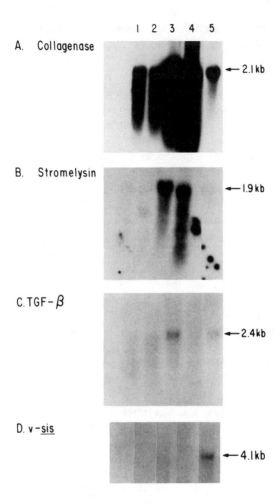

Figure 4. Northern blot analysis of whole cell RNA from primary cultures of human rheumatoid synovial cells. Primary cultures were established[56] from synovial tissue of 5 patients received from the operating room of Mary Hitchcock Memorial Hospital. The cells were cultured from 4 to 10 d in Dulbecco's modified Eagles' medium with 10% fetal bovine serum. Whole cell RNA was harvested,[30] and 20 μg RNA was loaded in each lane of a formadehyde/agarose gel[6] for Nothern blot analysis. Blots of RNA from primary (1°) or passaged control (C) cells were probed with ^{32}P-oligolabeled[6] cDNA probes for (A) human collagenase (3 d exposure), (B) human stromelysin (3 d exposure), (C) human TGFβ (3 d exposure), or (D) v-sis (6 d exposure). Each lane represents synovium from a different patient.

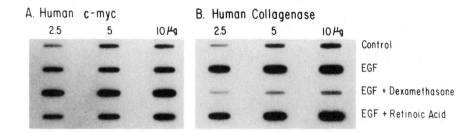

Figure 5. Slot-blot analysis of human *c-myc* and collagenase mRNAs. Whole cell RNA was isolated[30] from confluent cultures of human foreskin fibroblasts (passage 10) that were cultured for 24 hr in Dulbecco's modified Eagle's medium with 10% fetal bovine serum alone, or with EGF (10 ng/ml) in the presence or absence of all-*trans*-retinoic acid (10^{-6} *M*) or dexamethasone (10^{-7} *M*). Slot blots represent 2.5, 5, and 10 μg whole cell RNA loaded, and were probed with ^{32}P-oligolabeled cDNAs[6] for human *c-myc* (Panel A), or collagenase (Panel B).

PHARMACOLOGIC MODULATION OF SYNOVIAL CELL FUNCTION: MECHANISMS CONTROLLING THE EFFECTS OF GLUCOCORTICOIDS AND RETINOIDS

Despite the fact that cytokines may "naturally" down regulate the proliferative and destructive behavior of cells, pharmacologic intervention is probably the most effective way to modulate synovial cell function. The NSAIDs and the glucocorticoids are often used in the treatment of rheumatoid arthritis.[2] A third class of compounds, the vitamin A analogues, or retinoids, have not yet been used as therapeutic agents in patients, but they are effective in treating animal models of arthritis.[90,92-95] All three classes of compounds suppress inflammation and PGE_2 production,[26,29,56,90,93-95] and by doing so, they may interrupt the chronic inflammatory cascade that culminates in the production of collagenase and PGE_2. They may, therefore, indirectly decrease joint destruction. As mentioned earlier, the retinoids and the glucocorticoids also suppress metalloproteinase production by acting directly on the synovial cells.[17,29,30] Since they are able to suppress the inflammatory aspect of the disease as well as the proliferative/destructive component, these observations suggest that retinoids and steroids may be acting at several steps in the pathogenic pathways, but exactly where (and how) is not known.

We do, however, have some information on how these two classes of compounds affect collagenase gene expression. Of potential significance is the fact that retinoids and glucocorticoids act synergistically: at concentrations of 10^{-10} *M* neither compound alone suppressed collagenase synthesis, but combined treatment resulted in a 50% decrease in collagenase production.[29] This finding may have important therapeutic implications by allowing effective target-specific therapy with doses of compounds that are below the threshold for eliciting undesirable side effects. The synergism also implies that the effects of retinoids and steroids may be mediated by different transcriptional mechanisms.

To begin to study mechanisms of transcriptional control, we isolated fragments of DNA flanking the 5' end of the rabbit collagenase gene (Figure 6).[42] The largest fragment of DNA that we have isolated thus far extends 5' a length of 1.2 kb from the start site of transcription. Smaller fragments (0.9 kb, 0.6 kb, 0.4 kb, and 0.25 kb) were derived by deleting portions of DNA from the 5' end of the 1.2 kb fragment. All contained the putative phorbol ester-responsive element located at –72 bp.[35] These segments of DNA were ligated to the promoterless bacterial chloramphenicol acetyl transferase (CAT) gene contained in the plasmid pSVOCAT,[96] and the ability of these fragments to drive transcription was assessed by measuring CAT enzyme activity in rabbit synovial fibroblasts that were transiently transfected with each of the various plasmid constructs. The response of these DNA fragments to inducers and inhibitors of collagenase gene expression was measured after treating transfected cells with PMA, dexamethasone, or all-*trans*-retinoic acid

As expected, each of the CAT constructs directed low levels of CAT enzyme activity in unstimulated control cells, but CAT activity increased 14- to 40-fold in cells treated with PMA for 24 hr. The fact that even the most proximal DNA fragment (0.24 kb) conferred phorbol ester inducibility is consistent with the expectation that the short sequence with 89% sequence identity to the human AP-1 binding site[35,43,44,47] is responsible for this induction.

We also found that each of the promoter fragments was sensitive to dexamethasone. Treatment of transfected cells for 24 hr with 10^{-7} M dexamethasone resulted in an approximate three- to fivefold decrease in PMA-induced CAT enzyme activity for the 1.2 and 0.9 kb DNA fragments, whereas shorter DNA fragments conferred a reduced, but still measurable, inhibition of CAT activity. These findings suggest that glucocorticoid regulation probably involves multiple discrete upstream DNA elements and that a plasmid that contains just 180 bp of 5' flanking DNA (0.24 CAT) is sufficient to confer at least a partial inhibitory response by dexamethasone. These data are similar to recent results obtained by Sakai et al.[97] in studies using the bovine prolactin gene promoter. The negative glucocorticoid regulatory DNA sequences (negative glucocorticoid regulatory elements, nGREs) bound by the glucocorticoid receptor described so far appear to be poorly conserved, but related to positive GREs, and a "consensus" sequence has recently been reported.[98] In the case of the prolactin promoter[97] and the human glycoprotein hormone α-subunit promoter,[99] it has been proposed that negative regulation occurs by the binding of the glucocorticoid receptor to a sequence(s) that overlaps the binding site of a transcription factor required for mRNA synthesis. Decreased CAT activity following treatment with dexamethasone would then result from competition for binding to the mutually exclusive sites. Consistent with this model, we have observed sequences located at approximately –105, –195, –210, –425, and –670 bp with respect to the start site of transcription with similarity to the nGRE consensus sequence.[100-102]

We also tested each of the collagenase-CAT constructs for the ability to respond to retinoic acid.[42] The 1.2, 0.9, and 0.24 kb CAT constructs were repressed 20 to 40% below the PMA-induced control. This degree of repression was modest, but compared favorably to the approximately 50% reduction in collagenase mRNA

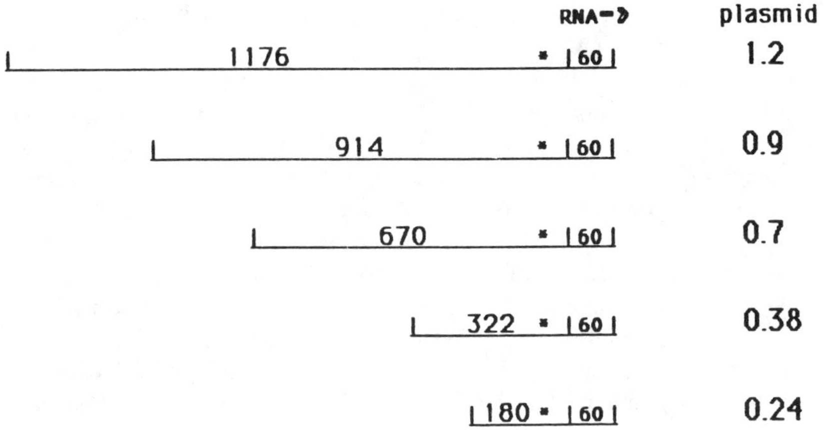

		RNA→	plasmid
⊢————— 1176 ——————	* |60|	1.2
⊢————— 914 —————	* |60|	0.9
⊢———— 670 ————	* |60|	0.7
⊢— 322 —	* |60|	0.38
|180 —	|60|	0.24

* homology with human collagenase phorbol ester responsive element

Figure 6. Chimeric constructs containing DNA flanking the 5′ end of the rabbit collagenase gene. Fragments of the 5′ flanking DNA of the rabbit collagenase gene[35] were ligated to the promoterless bacterial gene, chloramphenicol acetyltransferase (CAT), contained in the reporter plasmid pSVOCAT (965). The sizes of the DNA fragments in base pairs are indicated. The largest DNA fragment is about 1200 base pairs, and the other fragments are 5′ deletions of this piece. The start site of transcription begins at the arrow and proceeds rightwards through the 60 base pairs of untranslated leader sequence. The 5′ deletions of the 1200 base pair fragment were obtained by cleaving the DNA at convenient restriction sites, blunting the ends where necessary with the Klenow fragment fo DNA polymerase I, ligating Hind III linkers to the blunt ends, and then cloning the adapted Hind III fragments into the unique Hind III site of pSVOCAT.[42,96] (From Brinckerhoff, C.E. and D.T. Auble, in *Structure, Molecular Biology, and Pathology of Collagen,* Vol. 580, New York Academy of Science, 1990, 355–374. With permission.)

levels measured under similar conditions.[30] Results with the 0.7 and 0.38 CAT constructs indicated loss of suppression (or even stimulation) of CAT activity in the presence of both retinoic acid and PMA. Importantly, in the same experiment, measurement of collagenase gene expression via collagen fibril assay showed that the endogenous gene was functioning properly. Presently, we favor a model that in the presence of the –0.38 to –0.7 kb fragment of DNA, additional upstream DNA is required to manifest retinoic acid repression. It is also possible that other mechanisms, such as a decrease in transcript elongation by retinoic acid,[103] contributes to suppression of collagenase gene expression.

One conclusion to be drawn from the data with retinoic acid obtained thus far is that a relatively small fragment of DNA (0.24 kb, which contains 0.18 kb of 5′

flanking DNA) appears to be as responsive to retinoic acid as the largest fragment (1.2 kb) tested. This fragment of DNA contains the octanucleotide sequence TCAGGCTA, which is identical in six out of eight positions to the 5′ half palindrome of the reported consensus thyroid hormone-retinoic acid response element.[98] This sequence is a candidate for the negative retinoic acid response element (nRRE) if the inhibitory effects of retinoic acid are mediated by the direct binding of a retinoic acid receptor (RAR)[104,105] to a site within the collagenase promoter. Another possibility involves a 15 base pair sequence with similarity to the estrogen response element (ERE),[98] which overlaps the putative PMA-responsive AP-1 binding site. The thyroid hormone receptor (which shares binding specificity with the retinoic acid receptor) binds to the ERE, but does not activate transcription.[106] A similar mechanism applied to the retinoic acid receptor might involve inhibition of transcription by binding of the RAR to the ERE-like sequence, preventing binding of activators to the PMA-responsive element. Consistent with the data obtained with the thyroid hormone and the ERE, the RAR would presumably fail to stimulate transcription of the collagenase gene because of the nature of the ERE-like site.

SUMMARY AND CONCLUSIONS

In this chapter, we have discussed some of the possible mechanisms by which cytokines modulate fibroblast proliferation and production of the metalloproteinases, collagenase and stromelysin. In discussing the metalloproteinases, we have used studies with phorbol ester-stimulated and EGF-stimulated rabbit and human fibroblasts, respectively, as models. These studies have provided considerable insights into the cellular and molecular biology involved in increasing metalloproteinase production. The most important conclusions are (a) transcription of new mRNA for these enzymes occurs via interaction of at least two protooncogene products (c-JUN and c-FOS) with a specific nucleotide sequence located in the DNA flanking the 5′ end of the metalloproteinase genes, and (b) transcription can be antagonized by retinoids and steroids that may mediate their effects, at least in part, through specific DNA sequences also located within the 5′ flanking DNA.

Our knowledge of the mechanisms by which phorbol esters induce metalloproteinases may help us to understand metalloproteinase induction by physiologically important compounds, such as IL-1 and TNF. Indeed, it appears that IL-1 and TNF induce collagenase and stromelysin via mechanisms similar to those described for phorbol esters. A major goal of future work will be determining whether this mechanism represents a final common pathway to increase the synthesis of these enzymes in response to a wide variety of agents.

Similarly, in discussing the proliferative ability of fibroblasts, we have again emphasized the ability of cytokines to stimulate protooncogenes, such as *c-fos* and *c-myc*. Of particular interest is the fact that phagocytosis of inorganic crystals of basic calcium phosphate, by triggering a rise in intracellular Ca^{++}, can induce a proliferative response in fibroblasts that is usually associated only with classical lymphokines and cytokines.

The role of TGFβ in modulating synovial cell function is pivotal and controversial. Although this cytokine has the ability to suppress growth and metalloproteinase synthesis, in combination with other cytokines, it may stimulate cell proliferation and even enhance matrix degradation. Thus, the microenvironment in which TGFβ interacts with varying concentrations of a mixture of cytokines may determine ultimately the behavior of this multifunctional growth factor.

There is one additional point to be made that extends the implications of lymphokine-mediated induction of metalloproteinases beyond the confines of connective tissue disease and into the areas of cancer cell biology and metastasis. There may be a battery of genes that can be induced or suppressed in unison by a variety of mediators and cytokines, such as those discussed here, thus permitting a concerted modulation of the connective tissue matrix. The fact that we are able to establish a direct link between proteinases commonly thought to be involved mainly in connective tissue metabolism and those proteinases influenced by oncogene expression and involved in cancer metastasis provides firm biochemical support for the theory that nonmalignant proliferative diseases share many features usually attributed solely to malignancy. Most certainly, increased understanding of mechanisms controlling the the ability of cytokines to modulate expression of oncogene products and synovial cell function will further our knowledge of the pathophysiology of rheumatoid arthritis. However, these findings may well be applicable to our knowledge of cancer metastasis and tumor cell invasion.

ACKNOWLEDGMENTS

This research was supported by grants from USPHS-NTH (AR-26599), from the Council for Tobacco Research, from Merck, Sharp and Dohme, and from The RGK Foundation (Austin, TX).

REFERENCES

1. Harris, E.D., Jr., H.G. Welgus, and S.M. Krane, Regulation of the mammalian collagenases. *Coll. Relat. Res.*, 4:493–512, 1984.
2. Harris, E.D. Jr., Pathogenesis of rheumatoid arthritis, in *Textbook of Rheumatology*, W.N. Kelley, E.D. Harris, Jr., S. Ruddy, C.B. Sledge, Eds., W.B. Saunders, Philadelphia, 1985, 886–914.
3. Woolley, D.E. and J.M. Evanson, Eds., *Collagenase in Normal and Pathological Connective Tissues*, J. Wiley & Sons, New York, 1980.
4. Nagase, H., R.C. Jackson, C.E. Brinckerhoff, C.A. Vater, and E.D. Harris, Jr., A precursor form of latent collagenase produced in a cell-free system with mRNA from rabbit synovial cells. *J. Biol. Chem.*, 256:11951–11954, 1981.
5. Goldberg, G.L., S.M. Wilhelm, A. Kronberger, E.A. Bauer, G.A. Grant, and A.Z. Eisen, Human fibroblast collagenase. *J. Biol. Chem.*, 261:6600–6605, 1986.

6. Brinckerhoff, C.E., P.L. Ruby, S.D. Austin, M.E. Fini, and H.D. White, Molecular cloning of human synovial cell collagenase and selection of a single gene from genomic DNA. *J. Clin. Invest.*, 79:542–546, 1987.
7. Whitham, S.E., G. Murphy, P. Angel, H-J. Rahmsdorf, B.J. Smith, A. Lyons, T.J.R. Harris, J.J. Reynolds, P. Herrlich, and A.P. Docherty, Comparisons of human stromelysin and collagenase by cloning and sequence analysis. *Biochem. J.*, 240:913–916, 1986.
8. Campbell, E.J., J.D. Cury, C.J. Lazarus, and H.G. Welgus, Monocyte procollagenase and tissue inhibitor of metalloproteinases. *J. Biol. Chem.*, 262:15862–15868, 1987.
9. Herron, G.S., Z. Werb, K. Dwyer, and M.J. Banda, Secretion of metalloproteinases by stimulated capillary endothelial cells. *J. Biol. Chem.*, 261:2810–2813, 1986.
10. Nagase, H., C.E. Brinckerhoff, C.A. Vater, and E.D. Harris, Jr., Biosynthesis and secretion of procollagenase by rabbit synovial fibroblasts. *Biochem. J.*, 214:281–288, 1983.
11. Hasty, K.A., M.S. HIbbs, A.H. Kang, and C.L. Mainardi, Heterogeneity among human collagenases demonstrated by monoclonal antibody that selectively recognizes and inhibits human neutrophil collagenase. *J. Exp. Med.*, 156:379–390, 1984.
12. Weiss, S.J., Tissue destruction by neutrophils. *N. Engl. J. Med.*, 320:365–376, 1989.
13. Vater, C.A., H. Nagase, and E.D. Harris, Jr., Purification of an endogenous activator of procollagenase from rabbit synovial fibroblast culture medium. *J. Biol. Chem.*, 258:9374–9382, 1983.
14. Vater, C.A., H. Nagase, and E.D. Harris, Jr., Proactivator-dependent activation of prollagenase induced by treatement with EGTA. *Biochem. J.*, 237:853–858, 1986.
15. Murphy, G., M.I. Crockett, P.E. Stephens, B.J. Smith, and A.J.P. Docherty, Stromelysin is an activator of prollagenase. A study with natural and recombinant enzymes. *Biochem. J.*, 248:265–268, 1987.
16. Ito, A and H. Nagase, Evidence that human rheumatoid synovial matrix metalloproteinase 3 is an endogenous activator of procollagenase. *Arch. Biochem. Biophys.*, 267:211–216, 1988.
17. Fini, M.E., M.J. Karmilowicz, P.L. Ruby, A.M. Beeman, K.A. Borges, and C.E. Brinckerhoff, Cloning of a complementary DNA for rabbit proactivator. A metalloproteinase that activates synovial cell collagenase, shares homology with stromelysin and transin, and is coordinately regulated with collagenase. *Arthr. Rheum.*, 30:1255–1264, 1987.
18. Wilhelm S.M., I.E. Collier, A. Kronberger, A.Z. Eisen, B.L. Marmer, G.A. Grant, E.A. Bauer, and G.I. Goldberg, Human fibroblast stromelysin: structure, glycoslyation, substrate specificity, and differential expression in normal and tumorigenic cells. *Proc. Natl. Acad. Sci. U.S.A.*, 84:6725–6729, 1987.
19. Saus, J., S. Quinones, Y. Otani, H. Nagase, E.D. Harris, Jr., and M.K. Kurkinen, The complete primary structure of human matrix matalloproteinase 3 (MMP-3): Identity with stromelysin. *J. Biol. Chem.*, 263:6742–6745, 1988.
20. Murphy, G., H. Nagase, and C.E. Brinckerhoff, Relationship of procollagenase activator, stromelysin, and matrix metalloproteinase 3. *Coll. Relat. Res.*, 8:389–391, 1988.
21. Chin, J.R., G. Murphy, and Z. Werb, Stromelysin: a connective tissue degrading metallo-endopepetidase secreted by stimulated rabbit synovial fibroblasts in parallel with collagenase. *J. Biol. Chem.*, 260:12367–12376, 1985.
22. Okada, Y., H. Nagase, and E.D. Harris, Jr, A metalloproteinase from human rheumatoid synovial fibroblasts that digests connective tissue matrix components. *J. Biol. Chem.*, 261:14245–14255, 1986.

23. Sporn, M.B. and E.D. Harris, Jr., Proliferative diseases. *Am. J. Med.*, 70:1231–1236, 1981.
24. Blumberg, P., Protein kinase C as the receptor for phorbol ester tumor promoters. *Cancer Res.*, 48:1–8, 1988.
25. Weinstein, I.B. and M. Wigler, Cell culture studies provide new information on tumor promoters. *Nature.*, 270:559–560, 1977.
26. Brinckerhoff, C.E., R.M. McMillan, J.V. Fahey, and E.D. Harris, Jr., Collagenase production by synovial fibroblasts treated with phorbol myristate acetate. *Arthr. Rheum.*, 22:1109–1116, 1979.
27. Brinckerhoff, C.E., R.H. Gross, H. Nagase, L.A. Sheldon, R.C. Jackson, and E.D. Harris, Jr., Increased level of translatable collagenase mRNA in rabbit synovial fibroblasts treated with phorbol myristate acetate or crystals of monosodium urate monohydrate. *Biochemistry*, 21:2674–2679, 1982.
28. Brinckerhoff, C.E. and T.I. Mitchell, Autocrine control of collagenase synthesis by synovial fibroblasts. *J. Cellular Physiol.*, 136:72–80, 1988.
29. Brinckerhoff, C.E. and E.D. Harris, Jr., Modulation by retinoic acid and corticosteroids of collagenase production by rabbit synovial fibroblasts treated with phorbol myristate acetate or polyethylene glycol. *Biochim. Biophys. Acta*, 677:424–432, 1981.
30. Brinckerhoff, C.E., I.M. Plucinska, L.A. Sheldon, and G.T. O'Connor, Half-life of synovial cell collagenase mRNA is modulated by phorbol myristate acetate but not by all-*trans*-retinoic acid or dexamethasone. *Biochemistry*, 25:6378–6384, 1986.
31. Murphy, G., R.M. Hembry, and J.J. Reynolds, Characterization of a specific antiserum to rabbit stromelysin and demonstration of the synthesis of collagenase and stromelysin by stimulated rabbit articular chondrocytes. *Coll. Relat. Res.*, 6:351–364, 1986.
32. Gross, R.H., L.S. Sheldon, C.F. Fletcher, and C.E. Brinckerhoff, Isolation of a collagenase cDNA clone and measurement of changing collagenase mRNA levels during induction in rabbit synovial fibroblasts. *Proc. Natl. Acad. Sci. U.S.A.*, 81:1919–1985, 1984.
33. Vance, B.A., C.G. Kowalski, and C.E. Brinckerhoff, Heat shock of rabbit synovial fibroblasts increases expression of mRNAs for two metalloproteinases, collagenase and stromelysin. *J. Cell Biol.*, in press, 1989.
34. Shaw, G. and R. Kamin, A conserved AU sequence from the 3' untranslated region of GM-CSF mRNA mediates selective mRNA degradation. *Cell,* 46:659–667.35, 1986.
35. Fini, M.E., I.M. Plucinska, A.M. Mayer, R.H. Gross, and C.E. Brinckerhoff, A gene for rabbit synovial cell collagenase: member of a family of metalloproteinases that degrade the connective tissue matrix. *Biochemistry*, 29:6156–6165, 1987.
36. Matrisian, L.M., N. Glaichentraus, M.C. Gesnel, and R. Breathnach, Epidermal growth factor and oncogenes induce transcription of the same cellular mRNA in rat fibroblasts. *EMBO J.*, 4:1435–1440, 1985.
37. Matrisian, L.M., P. Leroy, C. Ruhlmann, M.C. Gesnel, and R. Breathnach, Isolation of the oncogene and epidermal growth-factor induced transin gene: complex control in rat fibroblasts. *Mol Cell. Biol.*, 6:1679–1686, 1986.
38. Frisch, S.M., E.J. Clark, and Z. Werb, Coordinate regulation of stromelysin and collagenase genes determined with cDNA probes. *Proc. Natl. Acad. Sci. U.S.A.*, 84:2600–2604, 1987.
39. Sirum, K.L. and C.E. Brinckerhoff, Cloning of the genes for human stromelysin and stromelysin 2: differential expression in rheumatoid synovial fibroblasts. *Biochemistry*, 28:8691–8698, 1989.

40. Breathnach, R., L.M. Matrisian, M.C. Gesnel, A. Staub, and P. Leroy, Sequences coding for part of oncogene-induced transin are highly conserved in a related rat gene. *Nucleic Acids Res.*, 15:1139–1151, 1987.
41. Collier, I.E., J. Smith, A. Kronberger, E.A. Bauer, S.M. Wilhelm, A.Z. Eisen, and G.I. Goldberg, The structure of the human skin fibroblast collagenase gene. *J. Biol. Chem.*, 263:10711–10713, 1988.
42. Brinckerhoff, C.E. and D.T. Auble, Regulation of collagenase gene expression in synovial cells, in: *Structure, Molecular Biology and Pathology of Collagen*, Vol. 580, New York Academy of Science, 1990, 355–374.
43. Angel, P., M. Imagawa, R. Chiu, B. Stein, R.J., Imbra, H.J. Rahmsdorf, C. Jonat, P. Herrlich, and M. Karin, Phorbol-ester inducible genes contain a common sis element recognized by a TPA-modulated trans-acting factor. *Cell*, 49:729–739, 1987.
44. Lee, W., P. Mitchell, and R.J. Tjian, Purified transcription factor AP-1 interacts with TPA-inducible enhancer elements. *Cell*, 49:741–752, 1987.
45. Angel, P., E.A. Allegretto, S.T. Okino, K.H. Jattori, W.J. Boyle, T. Hunter, and M. Karin, Oncogene *jun* encodes a sequence-specific transactivator similar to AP-1. *Nature*, 332:166–171, 1988.
46. Schonthal, A., P. Herrlich, H.H. Rahmsdorf, and H. Ponta, Requirement for *fos* expression in the transcriptional activation of collagenase by other oncogenes and phorbol esters. *Cell*, 54:325–334, 1988.
47. Angel, P., I. Baumann, B. Stein, H. Delius, H.J. Rahmsdorf, and P. Herrlich, 12-0-tetradecanoyl-phorbol-13-acetate induction of the human collagenase gene is mediated by an inducible enhancer element located in the 5'-flanking region. *Mol. Cell. Biol.*, 7:2256–2266, 1987.
48. Frisch, S.M. and H.E. Ruley, Transcription from the stromelysin promoter is induced by interleukin-1 and repressed by dexamethasone. *J. Biol. Chem.*, 262:16300–16304, 1987.
49. Chiu, R., W.J. Boyle, J. Meek, T. Smeal, T. Hunter, and M. Karin, The c-FOS protein interacts with *c-jun*/AP-1 to stimulate transcription of AP-1 responsive genes. *Cell*, 54:541–552, 1988.
50. Kerr, L.D., J.T. Holt, and L.M. Matrisian, Growth factors regulate transin gene expression by *c-fos* dependent and *c-fos* independent pathways. *Science*, 242:1424–1427, 1988.
51. Nakabeppu, Y., K. Ryder, and D. Nathans, DNA binding activities of three murine *jun* proteins: stimulation by *fos*. *Cell*, 55:907–915, 1988.
52. Halazonetis, T.D., K. Georgopoulos, M.E. Greenberg, and P. Leder, c-Jun dimerizes with itself and with c-fos, forming complexes of different DNA affinities. *Cell*, 55:917–924, 1988.
53. Dinarello, C., Biology of interleukin-1. *FASEB J.*, 2:108–115, 1988.
54. Saklatvala, J., V.A. Curry, and S.J. Sarsfeld, Purification to homogeneity of pig leukocyte catabolin, a protein that causes cartilage resorption in vivo. *Biochem. J.*, 215:385–392, 1983.
55. Dalton, B., J.R. Connor, and W.J. Johnson, Interleukin-1 induces interleukin 1 alpha and interleukin 1 β gene expression in synovial fibroblasts and peripheral blood monocytes. *Arthr. Rheum.*, 32:279–287, 1989.
56. Dayer, J.-M., S.M. Krane, R.G.G. Russell, and D.R. Robinson, Production of collagenase and prostaglandins by isolated adherent synovial cells. *Proc. Natl. Acad. Sci. U.S.A.*, 73:945–949, 1976.

57. Dayer, J.-M., L. Bread, L. Chess, and S.M. Krane, Participation of monocyte-macrophages and lymphocytes in the production of a factor that stimulates collagenase and prostaglandin release by rheumatoid synovial cells. *J. Clin. Invest.*, 64:1386–1392, 1979.
58. Mizel, S.B., J.-M. Dayer, S.M. Krane, and S.E. Mergenhagen, Stimulation of rheumatoid cell collagenase and prostaglandin by partially purified lymphoctye-activating factor (interleukin-1). *Proc. Natl. Acad. Sci. U.S.A.*, 78:2474–2478, 1981.
59. Dayer, J.-M., B. deRochemonteix, B. Burrus, S. Demczuk, and C.A. Dinarello, Human recombinant interleukin-1 stimulates collagenase and prostaglandin E_2 production by human synovial cells. *J. Clin. Invest.*, 77:645–648, 1986.
60. Rupp, E.A., P.M. Cameron, C.S. Ranawat, J.A. Schmidt, and E.K. Bayne, Specific bioactivities of monocyte-derived interleukin 1 alpha and interleukin 1 β are similar to each other on cultured murine thymocytes and on cultured human connective tissue cells. *J. Clin. Invest.*, 78:836–939, 1986.
61. Dayer, J.-M., B. Beutler, and A. Cerami, Cachectin/tumor necrosis factor stimulates collagenase and prostaglandin E_2 production by human synovial cells and dermal fibroblasts. *J. Exp. Med.*, 162:2163–2168, 1985.
62. Aggarwal, B.B., W.J. Kohrn, P.E. Hass, B. Moffat, S.A. Spencer, et al., Human tumor necrosis factor. *J. Biol. Chem.*, 260:2345–2354, 1985.
63. Saklatvala, J., Tumor necrosis factor alpha stimulates resorption and inhibits synthesis of proteoglycans in cartilage. *Nature*, 322:547–549, 1986.
64. Korn, J. H., P.V. Haluska, and E.C. LeRoy, Mononuclear cell modulation of connective tissue function: suppression of fibroblast growth by stimulation of endogenous prostaglandin production. *J. Clin. Invest.*, 65:543–554, 1984.
65. Postlewaite, A.E., L.B. Lachman, and A.H. Kang, Induction of fibroblast proliferation by interleukin-1 derived from human monocyte leukemic cells. *Arthr. Rheum.*, 27:995–1001, 1984.
66. Postlewaite, A.E., R. Raghow, G.P. Stricklin, H. Poppleton, J.M. Seyer, and A.H. Kang, Modulation of fibroblast functions by interleukin-1: increased steady state accumulation of type I procollagen messenger RNAs and stimulation of other functions but not chemotaxis by human recombinant interleukin 1 alpha and β. *J. Cell. Biology.*, 106:311–318, 1988.
67. Sugarman, B.J., B.B. Aggarwal, P.E. Hass, I.S., Figari, M.A. Palladino, and H.M. Shepard, Recombinant human tumor necrosis factor. *Science*, 230:943–945, 1985.
68. Lin, J.Y. and J. Vilcek, Tumor necrosis factor and interleukin-1 cause a rapid and transient stimulation of *c-fos* and *c-myc* mRNA levels in human fibroblasts. *J. Biol. Chem*, 262:11,908–11911, 1987.
69. Palombella, V.J., J. Mendelsohn, and J. Vilcek, Mitogenic action of tumor necrosis factor in human fibroblasts: interaction with epidermal growth factor and platelet derived growth factor. *J. Cell. Physiol.*, 243:393–396, 1988.
70. Raines, E.W., S.K. Dower, and R. Ross, Interleukin 1 mitogenic activity for fibroblasts and smooth cells is due to PDGF-AA. *Science*, 243:93–398, 1989.
71. Brenner, D.A., M. O'Hara, P. Angel, M. Chojkier, and M. Karin, Prolonged activation of jun and collagenase genes by tumor necrosis factor alpha. *Nature*, 337:661–663, 1989.
72. Conca, W., P.B. Kaplan, and S.M. Krane, Increases in levels of procollagenase mRNA in cultured fibroblasts induced by human recombinant interleukin 1 β or serum follow *c-jun* expression and are dependent on new protein synthesis. *J. Clin. Invest.*, 83:1753–1757, 1989.

73. Pelham, H.R.B., A regulatory upstream promoter element in the Drosophila HSP-70 heat shock gene. *Cell*, 30:517–528, 1982.
74. Brinckerhoff, C.E. and E.J. Harris, Jr., Collagenase production by cultures containing multinucleated cells derived from synovial fibroblasts. *Arthr. Rheum.*, 21:745–753, 1978.
75. McMillan, R.M., C.A Vater, P. Hasselbacher, J. Hahn, and E.D. Harris, Jr., Induction of collagenase and prostaglandin synthesis in synovial fibroblasts treated with monosodium urate crystals. *J. Pharm. Pharmacol.*, 33:382–383, 1981.
76. Harris, E.D. Jr., J.J. Reynolds, and Z. Werb, Cytochalasin B increases collagenase production by cells in vitro. *Nature*, 257:243–244, 1975.
77. Cheung, H.S., P.B. Halverson, and D.J. McCarty, Release of collagenase, neutral protease and prostaglandins from cultured mammalian synovial cells by hydroxyapatite and calcium pyrophosphate dihydrate crystals. *Arthr. Rheum.*, 24:1338:1344, 1981.
78. Dayer., J.-M., V. Evequoz, C. Grob-Zavadil, M.D. Grypnpas, P.T. Cheng, et al., Effect of synthetic calcium pyrophosphate and hydroxyapatite crystals on the interaction of human blood mononuclear cells with chondrocytes, synovial cells and fibroblasts. *Arthr. Rheum.*, 30:1372–1381, 1987.
79. Cheung, H.S., J.J. Van Wyk, W.E. Russell, and D.J. McCarty, Mitogenic activity of hydroxyapatite: requirement for somatomedin C. *J. Cell. Physiol.*, 128:143–148, 1986.
80. Rothenberg, R.J. and H.S. Cheung, Rabbit synoviocyte inositol phospholipid metabolism is stimulated by hydroxyapatite crystals. *Am. J. Physiol.*, 254 (Cell Physiol. 23): C4554–C4559, 1988.
81. Cheung, H.S., P.G. Mitchell, and W.J. Pledger, Induction of expression of *c-fos* and *c-myc* protooncogenes by basic calcium phosphate crystal: effect of β-interferon. *Cancer Res.*, 49:134–138, 1989.
82. Sporn, M.B. and A.B. Roberts, Peptide growth factors are multifunctional. *Nature*, 332:217–219, 1988.
83. Chua, C.C., D.E. Geiman, G.H. Keller, and R.L. Ladda, Induction of collagenase secretion in human fibroblast cultures by growth promoting factors. *J. Biol. Chem.*, 260:5213–5216, 1985.
84. Edwards, D.R., G. Murphy, J.J. Reynolds, S.E. Whitham, A.J.P. Docherty, P. Angel, and J.K. Heath, Transforming growth factor beta modulates the expression of collagenase and metalloproteinase inhibitor. *EMBO J.*, 6:1899–1904, 1987.
85. Machida, C.M., L.L. Muldoon, K.D. Rodland, and B. Magun, Transcriptional modulation of transin gene expression by epidermal growth factor and transforming growth factor β. *Mol. Cell. Biol.*, 8:2479–2483, 1988.
86. Kerr, L.D., N.E. Olashaw, and C.M. Matrisian, Transforming growth factor β1 and cAMP inhibit transcription of epidermal growth factor and oncogene-induced transin RNA. *J. Biol. Chem.*, 263:16999–17005, 1988.
87. Overall, C.M., J.L. Wrana, and J. Sodek, Independent regulation of collagenase, 72 kd progelatinase and metalloendoproteinase inhibitor expression in human fibroblasts by transforming growth factor β. *J. Biol. Chem.*, 264:1860–1869, 1989.
88. Fava, R., N. Olsen, J. Keska-Oja, H. Moses, and T. Pincus, Active and latent forms of transforming growth factor β activity in synovial effusions. *J. Exp. Med.*, 169:291–296, 1989.
89. Brinckerhoff, C.E., Morphologic and mitogenic responses of rabbit synovial fibroblasts to transforming growth factor β require transforming growth factor alpha or epidermal growth factor. *Arthr. Rheum.*, 26:1370–1379, 1983.

90. Lafyatis, R., E.F. Remmers, A.B. Roberts, D.E. Yocum, M.B. Sporn, and R.L. Wilder, Anchorage-independent growth of synoviocytes from arthritis and normal joints. *J. Clin. Invest.*, 83:12167–12176, 1989.
91. Brinckerhoff, C.E., L.A. Sheldon, M.C. Benoit, D.R. Burgess, and R.L. Wilder, Effect of retinoids on rheumatoid arthritis, a proliferative and invasive non-malignant disease, in *Retinoids, Differentiation and Disease*, Ciba Foundation Symposium, 113, Pitman Press, London, 1985, 191–207.
92. Brinckerhoff, C.E. and E.D. Harris, Jr., Survival of human rheumatoid synovium implanted into nude mice. *Am J. Pathol.*, 103:411–418, 1981.
93. Yocum, D.E., R. Lafyatis, E.F. Remmers, H.R. Schumacher, and R.L. Wilder, Hyperplastic synoviocytes from rats with streptococcal cell wall-induced arthritis exhibit a transformed phenotype that is thymic dependent and retinoid inhibitable. *Am. J. Pathol.*, 132:38–48, 1988.
94. Brinckerhoff, C.E., J.W. Coffey, and A.C. Sullivan, Inflammation and collagenase production in rats with adjuvant arthritis reduced with 13-*cis*-retinoic acid. *Science*, 221:756–758, 1983.
95. Haraoui, B., R.L. Wilder, J.B. Allen, M.B. Sporn, R.K. Helfgott, and C.E. Brinckerhoff, Dose-dependent suppression by the synthetic retinoid 4-hydroxyphenylretinamide of streptococcal cell-wall induced arthritis in rats. *Int. J. Immunopharmacol.*, 7:903–916, 1985.
96. Gorman, C.M., L.F. Moffat, and B.H. Howard, Recombinant genomes which express chloramphenicol acetyltransferase in mammalian cells. *Mol. Cell. Biol.*, 2:1044–1051, 1982.
97. Sakai, D.D., S. Helms, J. Carlstedt-Duke, J.-A. Gustafsson, F.M. Rottman, and K.R. Yamamoto, Hormone-mediated repression: a negative glucocorticoid response element from the bovine prolactin gene. *Genes Dev.*, 2:1144–1154, 1988.
98. Beato, M., Gene regulation by steroid hormones. *Cell*, 56:335–344, 1989.
99. Akerblom, I.E., E.P. Slater, M. Beato, J.D. Baxter, and P.L. Mellon, Negative regulation by glucocorticoids through interference with a cAMP responsive enhancer. *Science*, 241:350–353, 1988.
100. Bentley, D.L. and M. Groudine, A block to elongation is largely responsible for decreased trasncription of c-myc in differentiated HL60 cells. *Nature*, 321:702–706, 1986.
101. Rosenfeld, P.J. and T.J. Kelly, Purification of nuclear factor I by DNA recognition site affinity chromatography. *J. Biol. Chem.*, 261:1398–1408, 1986.
102. Chodosh, L.A., A.S. Baldwin, R.W. Carthew, and P.A. Sharp, Human CCAAT-binding proteins have heterologous subunits. *Cell*, 53:11–24, 1988.
103. Rossi, P., G. Karsenty, A.B. Roberts, N.S. Roche, M.B. Sporn, and B. deCrombrugghe, A nuclear factor I binding site mediates the transcriptional activation of a type I collagen promoter by transforming growth factor-beta. *Cell*, 52:405–414, 1988.
104. Giguere, V., E.S. Ong, P. Segui, and R.M. Evans, Identification of a receptor for the morphogen retinoic acid. *Nature*, 330:624–629, 1987.
105. Petkovich, M., N.J. Brand, A. Krust, and P. Chambon, A human retinoic acid receptor which belongs to the family of nuclear receptors. *Nature*, 330:444–450, 1987.
106. Glass, C.K., J.M. Holloway, O.V. Devary, and M.G. Rosenfeld, The thyroid hormone receptor binds with opposite transcriptional effects to a common sequence motif in thyroid hormone and estrogen response elements. *Cell*, 54:313–323, 1988.
107. Hasty, K.A., T.F. Pourmotabbed, G.I. Goldberg, J.P. Thompson, D.G. Spinella, R.M. Stevens, and C.L. Mainardi, Human neutrophil collagenase. A distinct gene product with homology to other matrix metalloproteinases. *J. Biol. Chem.*, 265:11421–11424, 1990.

6. Cytokines and Bone Metabolism: Resorption and Formation

JOSEPH A. LORENZO

INTRODUCTION

Bone is a complex organ with multiple functions. It forms the body's structural framework, it is a storehouse for calcium, and it is the site of hematopoiesis.[1] To accomplish these roles, bone contains a variety of cell types. Osteoblasts and osteoclasts are the cells in bone that form and maintain the skeleton. Osteoblasts produce a collagen-rich organic matrix that calcifies.[2] Osteoclasts are multinuclear giant cells that resorb calcified bone.[3] In addition, bone contains the hematopoietic and immune cells of the bone marrow.

It has become apparent that a number of local factors (cytokines) are produced by hematopoietic and immune cells.[4] Cytokines appear to regulate cellular function through paracrine and autocrine mechanisms. Many cytokines are now known to have powerful effects on osteoblasts and osteoclasts.[2,3] In addition, osteoblasts have now been shown to produce cytokines.[5-9] Hence, interactions among the various cell types in bone through cytokines may play an important role in the normal homeostasis of both the skeleton and the bone marrow.

The skeleton is a dynamic organ that is constantly reforming.[10] During childhood this remodeling process is necessary to allow bones to reshape and grow. However, after closure of the ephyseal plates, when linear growth stops, remodeling continues and appears to cause the changes in bone mass that have been documented with aging. From birth until the second or third decade of life, bone mass increases.[11] Thereafter, the skeleton is in negative calcium balance, and bone mass is lost. In women after menopause, when ovarian hormone production ceases, this loss of bone is accelerated and may lead to the clinical syndrome of osteoporosis.[11] Various pathologic states are also associated with accelerated bone loss. Inflammatory processes, such as rheumatoid arthritis or osteomyelitis, may produce localized areas of bone destruction.[12,13] Tumors of various organs can also affect the skeleton. Ectopic hormones, which cause the hypercalcemia of malignancy syndrome, may be produced by tumors.[14,15] In addition, local factors that stimulate osteoclasts to resorb and cause either local or generalized osteopenia may be produced by metastatic tumor cells in bone or by marrow cell malignancies.[14,15]

In general, all states of decreased bone mass result from an imbalance between the closely coupled bone forming and bone resorbing mechanisms. The initial step in bone remodeling is believed to be activation of the calcified matrix by lining cells.[10] These cells are adjacent to the mineralized surface of bone and are thought to originate from osteoblasts. They appear to produce enzymes that initiate the degradation of the organic elements in bone.[16] Degradative enzymes may prepare the calcified matrix for resorption by osteoclasts.[17] Within 24 h after the activation of lining cells, osteoclasts are found in increased numbers on the calcified bone surface.[18] This event signals the beginning of the resorption phase. Osteoclasts are the primary bone resorbing cells, and increases in their number correlate directly with accelerated rates of calcium and organic matrix release from the skeleton.[3] Osteoclasts appear to efficiently perform this function because they can isolate the calcified skeletal surface from the extracellular space.[19] This allows them to produce an optimal microenvironment in the resorption area that has a low pH and a high lysosomal enzyme concentration.[20] After resorption is completed, the osteoclasts detach from the bone and the reversal phase begins.[21] During this phase, macrophages may attach to the bone. In a few weeks osteoblasts migrate to the old resorption site, and new bone is deposited during the formation phase. In some pathologic states, like multiple myeloma,[14,15] this process is uncoupled, and resorption proceeds without formation. However, in general, some formation always follows resorption, though not always in equivalent amounts. In osteoporosis the amount of bone gained during the formation phase is less than that lost during resorption, while during linear growth the reverse is true.[11]

ORIGIN OF BONE CELLS

Osteoblasts and osteoclasts were once thought to share a common origin. However, most investigators now believe that they are derived from different precursors. The origin of the osteoblast is the least well known. Owen has proposed that osteoblasts are derived from a common mesenchymal precursor that also differentiates into reticular, fibroblastic, adipocytic, and osteogenic cells.[22] She also believes that the chondroblast and the osteoblast share a common osteoprogenitor. Osteoblasts may exist in different stages. The active osteoblast is defined histologically as a cuboidal cell on the surface of bone at sites of new matrix synthesis.[23] These cells are actively producing a collagen-rich matrix that calcifies extracellularly and forms the skeleton. Lining cells are flat cells found on the majority of bone in areas where remodeling is not occurring. These cells appear to be relatively inactive metabolically. However, they may be important for isolating the bone fluid from the extracellular space and initiating the remodeling cycle.[24] Osteocytes are resting osteoblasts that are encased in calcified matrix. They are in contact with lining cells through dendritic-like cellular processes. However, their function in bone is unknown.

Osteoclast precursors are almost certainly derived from hematopoietic progenitor cells.[3] Walker was the first to show that transplanting normal hematopoietic tissue into animals with a congenital defect in osteoclast development (osteopetrosis) cures their disease.[25] Subsequent studies by others have found similar effects *in vitro*. Burger has shown that osteoclastic progenitors can be removed from fetal bone rudiments by stripping off their outer tissue layer.[26,27] If these stripped bones are cultured, they do not develop osteoclasts. However, if stripped bones are incubated with radiolabeled normal marrow cells, labeled osteoclasts will form. Osteoblasts or another accessory cell appear necessary for the development of mature osteoclasts, since in cultures where normal bone marrow is incubated with devitalized bone explants, osteoclasts do not form.[26,27] Sheven et al. have shown that early hematopoietic progenitors, which can differentiate into a variety of myeloid cells, can substitute for bone marrow cells in this system.[28] Osteoclasts appear to be derived from a pathway that is separate from that of the monocyte, since addition of monocytes to these cultures does not lead to osteoclast formation.[26,27] However, both the point in osteoclast development where its differentiation diverges onto a unique pathway and the factors regulating this process are unknown.

FACTORS REGULATING BONE CELL FUNCTION

Osteoblasts and osteoclasts can respond to a large number of factors. Systemic hormones have been the most extensively studied. However, it has recently become clear that cytokines can also have potent effects. Because of the close association of osteoblasts, osteoclasts, and marrow cells in bone, factors produced locally by one cell type may influence the function of multiple cell types in bone.

SYSTEMIC HORMONES WHICH INFLUENCE BONE

Parathyroid Hormone

Parathyroid hormone (PTH) is the most important hormonal regulator of bone cell function in man. It is an 84 amino acid peptide that is produced by the parathyroid gland in direct response to decreases in the serum ionized calcium. PTH is rapidly metabolized to smaller peptides *in vivo*.[29] However, essentially all its biologic activity is contained in the first 34 amino acids of the N terminus.[30] The purpose of this metabolism is unclear. PTH stimulates bone resorption and new osteoclast formation.[31] However, its effects on bone formation are more complicated and appear to be determined by whether PTH is administered continuously or intermittently. Sustained PTH elevations *in vivo* are associated with increased rates of bone turnover, but only a small net loss of bone.[32] In contrast, continuous PTH treatment of bone *in vitro* causes a marked inhibition of bone collagen synthesis.[2] Intermittent stimulation of bone by PTH in experimental animals[33] or humans[34] increases bone mass and collagen synthesis. The *in vitro* anabolic response of collagen synthesis to intermittent PTH treatment appears to be mediated, at least in part, by local factors, since antibodies to insulin, like growth factor 1 (IGF-1), inhibit it.[35]

In addition, treatment of bone cultures with PTH increases both steady state levels of IGF-1 mRNA and IGF-1 concentrations.[36] PTH also increases the activity of transforming growth factor β (TGF β) in the conditioned medium of bone cultures[37] and the binding of TGF β to its receptors on bone cells.[38] Since TGF β is anabolic for bone,[39,40] this mechanism may also be involved in the growth promoting effects of PTH.

PTH is a potent stimulus of new osteoclast formation,[31] and this response is believed to mediate some of its effects on bone resorption. It appears to increase the rate that preformed osteoclast precursor cells fuse into either new or existing osteoclasts.[31] The cells in bone that contain receptors and initiate a resorptive response to PTH appear to not be osteoclasts. Isolated osteoclast cultures do not respond directly to PTH,[41] and PTH receptors are not found on the surface of osteoclasts.[42] However, if isolated osteoclasts are incubated with conditioned medium from PTH-treated osteoblasts, increased osteoclastic resorption occurs.[41] McSheehy and Chambers have found that a small factor (<1000 kDa) in the CM from PTH-treated osteoblast cultures can stimulate resorption in isolated osteoclast cultures.[41] Hence, a humoral factor produced by osteoblasts or another bone cell may mediate the resorptive effects of PTH. Rodan and Martin have alternatively proposed that the PTH resorptive response is initiated by a change in the shape of the lining cells.[43] They suggest that uncovered bone matrix is chemoattractive for osteoclast precursors, and its exposure initiates osteoclastic recruitment to the calcified bone surface.

1,25(OH)2 Vitamin D

1,25(OH)2 vitamin D, the most active vitamin D metabolite, is produced in the kidney from 25(OH) vitamin D by the enzyme 1 α hydroxylase.[44] The activity of this enzyme is regulated by PTH and by serum phosphate. In addition, $1,25(OH)_2$ vitamin D can feed back to inhibit its own production.[45] While it was originally described as a regulator of bone and calcium metabolism, $1,25(OH)_2$ vitamin D is now known to regulate the differentiation of a variety of cell types, including those of the immune system.[46-50] $1,25(OH)_2$ vitamin D stimulates bone resorption and enhances gut absorption of calcium. It appears to be involved in the differentiation of the osteoclast precursor cell,[51] though it may not be essential, since animals that are deficient in vitamin D produce osteoclasts.[52] As with PTH, osteoblasts or another cell in bone appear to be intermediates in the resorptive response of osteoclasts to $1,25(OH)_2$ vitamin D.[53] Isolated osteoclast cultures do not respond directly to $1,25(OH)_2$ vitamin D. However, when they are cultured with the conditioned medium from $1,25(OH)_2$ vitamin D-stimulated osteoblast cultures, their rate of resorption increases.

In vitro $1,25(OH)_2$ vitamin D inhibits collagen synthesis.[2] However, in vivo it is essential for normal rates of new bone growth. The in vivo effects of $1,25(OH)_2$ vitamin D may result from both its direct effects on osteoblasts and by its actions to increase gut transport of calcium.[44] Activated macrophages can contain the active

1 α hydroxylase enzyme, and increased serum 1,25(OH)$_2$ vitamin D concentrations are sometimes seen in patients with granulomatous diseases.[44]

Calcitonin

The major action of calcitonin appears to be a direct inhibition of osteoclast-mediated bone resorption.[54] However, it also has been reported to have some *in vitro* mitogenic effects on osteoblasts.[55] High-affinity receptors for calcitonin are found in large numbers on the cell membrane of osteoclasts,[56] and these appear to be coupled to an adenyl cyclase. Calcitonin directly inhibits the resorptive activity of isolated osteoclast cultures.[57] The inhibitory effects that calcitonin has on osteoclast function are transient and are termed "escape".[58] This phenomenon may result from the down regulation of receptors for calcitonin on osteoclasts.[59] In contrast to other calcium-regulating hormones, calcitonin does not appear to be essential for normal calcium homeostasis. Abnormalities in skeletal growth and abnormal serum calcium levels are not found in individuals who have either no detectable calcitonin production or hypersecretion of calcitonin from a medullary carcinoma.[54] In addition to osteoclasts, calcitonin responses are also seen in lymphocyte cell lines.[60]

Other Systemic Hormones

Besides parathyroid hormone, 1,25(OH)$_2$ vitamin D, and calcitonin, other hormones may influence the function of osteoblasts and osteoclasts. *In vitro* glucocorticoids inhibit bone resorption and have biphasic effects on bone formation.[2] After 24 h at low concentrations (10 n*M*), cortisol stimulates collagen synthesis in organ culture, while at higher concentrations or after longer times it is inhibitory. The complex nature of the response to cortisol appears partially to result from its effects on both the differentiation of osteoblast precursor cells and its direct inhibitory effects on protein synthesis. *In vivo* cortisol inhibits bone growth, increases bone resorption, and predisposes patients to the clinical syndrome of osteoporosis.[2,11] These effects appear to result from both its direct actions on bone cells and its ability to inhibit gut transport of calcium.

Thyroid hormone can also directly stimulate bone resorption,[61] and its absence during growth is associated with abnormal skeletal development. Sex steroids also have powerful effects on bone. Loss of estrogens during menopause is associated with accelerated bone loss that may lead to the clinical syndrome of osteoporosis.[11] Estrogen receptors have recently been recognized in bone cells,[62,63] and direct effects of estrogens on bone cell function have also been demonstrated.[62] Growth hormone is a powerful stimulus of bone growth, and its effects on bone appear to be mediated through the local production of insulin-like growth factor 1 (IGF-1).[2] *In vitro* IGF-1 is the most potent stimulus of bone collagen synthesis and cell replication known.[64] Production of this agent by bone seems essential for the normal development of the skeleton. In addition, two related polypeptides, IGF-2 and insulin, have similar actions on bone cells.[64]

CYTOKINES AND OTHER LOCAL FACTORS

A large number of locally produced factors are now known to influence the function of both bone-forming and bone-resorbing cells. These factors include cytokines, which are agents that have been identified as mediators of immune cell-cell interactions, and growth factors, which have been identified as mitogens for a variety of cell types. Often, as with the transforming growth factors or interleukin-2, these agents are found to act in both contexts. In order to simplify the nomenclature, I shall use the term *cytokine* to describe all factors that act locally.

Prostaglandins

Prostaglandins are released from a variety of cells and have a large number of activities in bone. Prostaglandins of the E series and in particular PGE_2 are potent stimuli of bone resorption *in vitro*.[65] Paradoxically, they inhibit the resorptive activity of isolated osteoclasts.[57] Hence, their stimulatory effects on resorption appear to be indirect. Prostaglandins may also be important local regulators of bone growth. *In vitro* PGE_2 stimulates collagen and noncollagen protein synthesis at low concentrations (100 nM) and inhibits them at higher concentrations.[66] PGE_2 also has similar effects on DNA synthesis. *In vivo* infusion of prostaglandins is associated with increased rates of new bone formation.[67]

Prostaglandins are produced by bone organ cultures as a response to a variety of local factors. These include interleukin-1,[68-70] tumor necrosis factor,[71] transforming growth factors α and β,[72] epidermal growth factor,[73] and platelet-derived growth factor.[74]

The inhibitory effects that cortisol has on prostaglandin synthesis may mediate some of its actions on bone. Addition of PGE_2 to bone cultures treated with cortisol reverses the inhibitory effects that cortisol has on collagen and DNA synthesis.[75]

Interleukin-1

Interleukin-1 (IL-1) was the first of the polypeptide mediators of immune cell function that was shown to regulate bone resorption[68,69,76-79] and formation.[80-82] IL-1 is a family of two 17 to 18 kDa proteins (IL-1 α and IL-1 β) that are produced by activated monocytes and a variety of other cells.[83,84] IL-1 α has an acidic pI and appears to be the only bioactive form produced by mice.[85,86] IL-1 β has a neutral or basic pI and is the predominant form produced by human cells, though IL-1 α is also found. Both forms of IL-1 have only limited sequence homology. However, each binds to the same cell surface receptor and has similar biologic actions.[87,88] Both IL-1 forms originate as larger precursors that are typically 33 kDa. In activated mouse macrophages, IL-1 β mRNA is found in greater amounts than IL-1 α mRNA.[89-91]

In bone organ culture, IL-1 α and β are the most potent peptide stimulators of bone resorption yet identified.[68,69,76-79] Typically, they are active at concentrations of 10 to 50 pM and greater. Their effects, like those of PTH, appear to be indirect, since IL-1-treated isolated osteoclast cultures do not increase their rates of bone resorption unless they are cocultured with conditioned medium from IL-1-treated osteoblasts.[92] IL-1 β is now known to be the factor in the conditioned medium from phytohemagglutinin-treated human peripheral blood mononuclear cells that stimulates bone resorption, an activity that had previously been labeled osteoclast activating factor (OAF).[68,78] IL-1 is also a potent stimulus of prostaglandin synthesis in bone.[68-70] This effect may account for some of its resorptive activity, since prostaglandins are potent stimulators of resorption.[65] However, IL-1 can also produce prostaglandin-independent resorption.[68,76]

IL-1 has powerful effects on bone formation. In organ cultures its predominant activity appears to be inhibitory.[80-82] However, at low concentrations (0.01 to 1 U/ml) and early time points (24 h), it can have a small stimulatory effect.[81] IL-1 also stimulates DNA synthesis in both bone organ cultures and primary cultures of human bone cells.[80,81,93] The early stimulatory effects of IL-1 on collagen synthesis appear to be mediated by prostaglandins, since they are inhibited by indomethacin.[81] However, the stimulatory effects on DNA synthesis and the inhibitory effects of high concentrations and longer culture periods on collagen synthesis are not blocked by inhibitors of prostaglandin synthesis.[94] In cultures of the osteoblast-like osteosarcoma cell line MC3T3-E1, IL-1 stimulates DNA synthesis and inhibits collagen synthesis.[94,95] IL-1 also has a biphasic inhibitory effect on alkaline phosphatase activity in these cells.[94] This enzyme is believed to be involved in bone mineralization. The effects of IL-1 on alkaline phosphatase are independent of prostaglandin synthesis. Recently, IL-1 was shown to inhibit procollagen gene expression in MC3T3-E1 cells by a transcriptional mechanism.[96] This effect appeared to be mediated by protein kinase C, since it was mimicked by phorbol esters. IL-1 has been reported to stimulate adenylate cyclase in some cell lines.[97,98] However, I have been unable to demonstrate it to have this activity in either bone organ cultures or MC3T3-E1 cells (unpublished observation). In addition, neither forskolin, which stimulates adenyl cyclase directly, nor Dibutryl cAMP, which mimics the action of cAMP, altered the inhibitory effects of IL-1 on collagen synthesis or transcriptional regulation of collagen mRNA in MC3T3-E1 cells.[96] Hence, the major effects of IL-1 on osteoblast function do not appear to be mediated by cAMP.

Prostaglandin synthesis is regulated by a series of enzyme steps. Central to these is prostaglandin H (PGH) synthetase (cyclooxygenase). IL-1 enhances both the production of this protein and the steady state levels of its mRNA in MC3T3-E1 cells.[99] However, additional mechanisms must be involved, since increases in PGH synthetase mRNA do not correlate well with increases in PGE_2 in the medium.

IL-1 is a potent stimulus of bone resorption *in vivo*.[100] Subcutaneous infusions of IL-1 in mice cause hypercalcemia and increase bone resorption. Histologic examination of bone from these animals shows increased osteoclast numbers and resorption surfaces.

IL-1 may be involved in the differentiation of the osteoclast from its hematopoietic progenitor cell. IL-1 is a known differentiating factor for other cells,[101] and it stimulates the production of osteoclast-like multinucleated giant cells in cultured bone marrow.[102]

Primary cultures of cells from newborn mouse calvaria and the osteoblast-like MC3T3-E1 cell line have been found to release an IL-1-like activity into their conditioned medium.[103-105] However, these activities have not yet been completely characterized by specific techniques.

The IL-1 receptor in cell membranes has now been identified and cloned.[106,107] This peptide is an 80 kDa glycoprotein that belongs to the immunoglobulin superfamily. At least one osteoblast-like osteosarcoma cell line, UMR 106, has been demonstrated to have IL-1 receptors.[87]

Production of IL-1 may be involved in the development of osteoporosis, a disease characterized by decreased bone mass and pathologic bone fractures.[11] Increased IL-1 production has been found in monocytes from some patients with high turnover osteoporosis.[108] In addition, one group has found that *in vivo* estrogen treatment, which prevents the development of osteoporosis after menopause, reduced the amount of IL-1 that was released from cultured monocytes.[109] However, a similar study has failed to confirm this finding.[110]

Tumor Necrosis Factor

Like IL-1, tumor necrosis factor (TNF) is a family of two related polypeptides that are the products of separate genes.[111-115] Both TNF α and β bind the same receptor and have similar biologic activities. TNF α, or cachectin, is a 17 to 18 kDa polypeptide that circulates in a trimeric or pentameric form. It is produced principally by macrophages after stimulation with bacterial or viral antigens. TNF β, or lymphotoxin, is a similar-sized peptide that has 35% homology with TNF α in the murine forms. It is principally the product of activated lymphocytes. Like IL-1, TNF α and β have a wide variety of biologic activities. Both are potent stimulators of bone resorption[68,71,79,116,117] and inhibitors of bone collagen synthesis.[80,116,118-120] Typically these effects occur at concentrations of 100 pM. The *in vitro* model that is used to demonstrate bone resorption appears to determine the degree that TNF-mediated resorption is dependent on endogenous prostaglandin synthesis. In newborn mouse calvaria cultures, TNF α stimulates a resorptive response that is completely blocked by inhibitors of prostaglandin synthesis.[71] In contrast, inhibitors of prostaglandin synthesis have only small effects on the resorptive response of fetal rat long bone cultures to TNF.[79] The mechanisms underlying these differences are unknown.

In vivo TNF α increases the serum calcium of mice that are injected with either 5 or 25 μg of recombinant human TNF every 8 h,[71] and similar effects are seen with TNF β.[117] In a more detailed study, CHO cells, which were transfected with the human TNF α gene so that they released high amounts of active peptide, were injected into nude mice.[121] These animals developed tumors and became hypercalcemic within two weeks. Bone histomorphometry demonstrated a tenfold increase in the number of osteoclasts in the bones of these animals compared to controls (animals injected with CHO cells that contained an empty vector). In addition, the percentage of the bone surface undergoing active resorption was similarly increased in animals receiving the TNF-producing cells.

The effects of TNF on resorption appear to be mediated by osteoclasts since they are associated with increases in osteoclast number[121] and are inhibitable by calcitonin.[79] Like IL-1, TNF can also enhance the formation of osteoclast-like cells in bone marrow culture,[102] and therefore, it may regulate osteoclast differentiation. However, as with IL-1, no direct resorptive effects of TNF are seen in isolated osteoclast cultures.[122] Only in the presence of osteoblasts does TNF increase the rate that osteoclasts resorbed bone. Hence, as with most other stimulators, the effects of TNF on resorption appear to be mediated through other cells.

In contrast to the inhibitory effects that continuous treatment with TNF has on collagen synthesis in bone, transient treatment with TNF for 24 h causes a rebound increase in collagen synthesis in primary cultures of osteoblast-like cells from rat calvaria.[119] These results imply that TNF may stimulate the differentiation of osteoblast precursor cells into mature osteoblasts. In these cells, continuous treatment with TNF does not inhibit steady state levels of type 1 collagen mRNA. These results suggest that posttranscriptional regulation of message translation may be involved in the effects that TNF has on collagen synthesis.[119]

In bone organ cultures and in cultures of rat osteoblast-enriched cells and human osteoblast-like cells, TNF stimulates DNA synthesis.[80,118,119,123] However, in an osteoblast-like osteosarcoma cell line, ROS 17/2.8, TNF did not stimulate DNA synthesis, but did inhibit collagen synthesis.[120] Furthermore, addition of hydroxyurea, an inhibitor of DNA synthesis, to primary rat osteoblast-enriched cultures did not alter the inhibitory effects of TNF on collagen synthesis.[119] Hence, the effects that TNF has on cell replication do not appear linked to its effects on collagen synthesis.

TNF β has been implicated as one cause of the effects that hematologic malignancies have on bone.[117] A tumor cell line was established from a patient with multiple myeloma who had developed hypercalcemia and osteolytic bone lesions. This cell line was found to contain detectable steady state levels of TNF β mRNA and to produce TNF β *in vitro*. Furthermore, an antibody to TNF β blocked some, but not all, of the *in vitro* bone resorbing activity that was present in the conditioned medium from these cells and from other established myeloma cell lines.

There are a number of unique characteristics of the bone disease and hypercalcemia that is produced by hematologic malignancies.[14,15] Probably the most important of these is that little or no bone formation is produced in response to the activation of osteoclastic resorption by these tumors. It is likely that cytokines that are produced by these malignancies are responsible for these effects.[117] In contrast, the more common solid tumor hypercalcemia syndrome, which causes high bone turnover, but little osteopenia, appears to be caused by a factor that is related to parathyroid hormone.[124]

Interferon Gamma

Interferon gamma is another regulatory cytokine with a wide variety of biologic activities. Human interferon (INF) gamma is a polypeptide of 17 kDa. However, molecular sizing of circulating INF gamma suggests that it forms multimers, since it is found with sizes ranging from 35 to 70 kDa. In addition, it has at least two potential gycosylation sites that may contribute to its size heterogeneity. Murine INF gamma has only 40% peptide sequence homology with the human form, and it is believed that this difference is responsible for the species specificity of the two peptides.

In bone, INF gamma was first shown to inhibit resorption.[125-129] This effect appears to be more specific for the response to IL-1 and TNF, since lower concentrations of INF gamma inhibit the maximum activity of these factors compared to resorption stimulated by PTH or 1,25(OH)$_2$ vitamin D.[128] The actions of INF gamma on resorption may be mediated by its effects on osteoclast progenitor cells. INF gamma inhibits the ability of 1,25(OH)$_2$ vitamin D, PTH, and IL-1 to stimulate the formation of multinucleated giant cells with osteoclast characteristics in 21 d cultures of human bone marrow.[130] INF gamma is also known to inhibit other hematopoietic functions, including erythropoiesis.[131] However, it does not appear to act directly on osteoclasts, since incubation of 500 U/ml of murine INF gamma with isolated rat osteoclasts did not inhibit their rate of resorption.[132]

The effects of INF gamma on collagen synthesis are also inhibitory. In bone organ culture, INF gamma decreases both collagen and noncollagen protein synthesis and the percentage of total protein synthesis that was collagen.[80] This effect did not appear dependent on prostaglandin synthesis, since indomethacin did not inhibit it. INF gamma also enhances the inhibitory effects that IL-1 and TNF have on collagen and total protein synthesis.[80] The effects of INF gamma on collagen synthesis in the osteoblast-like ROS 17/2.8 cell line are similar to those seen in organ cultures.[120]

INF gamma also inhibits DNA synthesis in bone organ cultures,[80] in human cultured bone cells with osteoblast characteristics,[123] and in the ROS 17/2.8 cells.[120] It also inhibits the stimulatory effects that TNF and IL-1 have on cell replication in bone cultures[80] and that TNF has in human osteoblast-like cells[123] and the ROS 17/2.8 osteosarcoma cell line.[120] Hence, the overall effect of INF gamma on bone resorption and bone formation appears to be inhibitory.

Colony-Stimulating Factors

Colony-stimulating factors (CSF) regulate the development of the myeloid cell system. Four CSFs are known. These are: interleukin 3 (IL-3) (or multi-CSF), granulocyte-macrophage colony-stimulating factor (GM-CSF), granulocyte colony-stimulating factor (G-CSF), and monocyte stimulating factor or (CSF-1).[133,134] IL-3 regulates the development of the earliest stem and progenitor cells. GM-CSF acts on a slightly more differentiated progenitor. G-CSF and M-CSF influence the development of only granulocytic or monocytic precursors, respectively. Osteoclasts appear to be derived from myeloid progenitors.[3] In a system that allows osteoclasts to form in live bone, myeloid progenitor cells were shown to be sufficient to allow osteoclasts to differentiate.[26,27] IL-3 was found to be capable of maintaining the progenitor cells of the osteoclasts for this system.[28] This result implies that the osteoclast originates from an early myeloid progenitor cell, since IL-3 is necessary to maintain that cell *in vitro*. In osteopetrotic rats, which do not form normal osteoclasts, transplants of hematopoietic precursor cells enriched for granulocyte-macrophage colony-forming cells (GM-CFCs) and granulocyte colony forming cells (G-CFCs) cured the disease, but macrophage colony-forming cells (M-CSFs) did not.[135] Similar effects were observed *in vitro*.[136] Hence, the macrophage appears to differentiate along a pathway that is distinct from that of the osteoclast, while the granulocyte may be more closely related.

The role of CSFs in the differentiation and activation of the osteoclast has not yet been clearly defined. A squamous cell carcinoma was found to produce a factor in culture that had bone resorbing and colony stimulating activity. However, when these activities were fractionated, they were found to be distinct.[137] IL-3 and GM-CSF do not stimulate bone resorption in fetal mouse long bone cultures, and in fact, IL-3 is slightly inhibitory.[18] In addition, both appear to decrease the rate that recently replicated nuclei are incorporated into osteoclasts.[18] In other experiments GM-CSF did not increase the rate that progenitor cells developed into osteoclasts *in vitro*.[138] M-CSF has been shown to not alter the development of osteoclasts from marrow cells in mouse bone cultures[139] and to directly inhibit the resorptive activity of isolated murine osteoclasts.[132] In contrast, human GM-CSF and M-CSF enhance the rate that osteoclast-like cells form in baboon bone marrow cultures.[140] Hence, differences may exist between the responses of murine and human osteoclast precursors to their respective CSFs.

Recently, osteoblasts were shown to produce CSFs. Both MC3T3-E1, a murine osteoblast-like osteosarcoma cell line, and primary cultures of osteoblast enriched murine calvaria cells produce M-CSF.[6,141] In addition, lipopolysaccharide (LPS),[6] 1,25(OH)$_2$ vitamin D,[6] IL-1,[142] and TNF[9] have been shown to enhance the production of M-CSF by bone cells. LPS and PTH have also now been shown to stimulate the production of GM-CSF in both primary bone cell cultures that were enriched for osteoblasts and in the rat osteoblast-like osteosarcoma cell line ROS 17/2.8.[5,7,8] LPS has also been shown to stimulate the production of G-CSF by osteoblast-enriched primary bone cell cultures.[5] In addition to the other CSFs, IL-3 has been found in the conditioned medium of LPS-stimulated bone organ cultures.[5]

Osteoblast can also respond to CSFs. Human osteosarcoma cells and human osteoblast-like primary cell cultures increase their rate of DNA synthesis when they are treated with GM-CSF.[143,144] GM-CSF also inhibits the 1,25(OH)$_2$ vitamin D-mediated production of osteocalcin and alkaline phosphatase by human primary osteoblast-like cells.[144] Taken together these results suggest that a complicated paracrine regulatory network exists between osteoblast, osteoclasts, and the bone marrow that is mediated by CSFs.

Growth Factors

Osteoblasts and osteoclasts respond to a large number of growth factors. These include transforming growth factors (TGF) α and β, insulin like growth factors (IGF), platelet-derived growth factor (PDGF), fibroblast growth factors (FGF), and epidermal growth factor (EGF).[145] Of these factors, osteoblasts are found to make TGF β[146,147] and IGF,[36] while bone matrix contains TGF β,[148] PDGF, and FGF.[149] Since hydroxyapatite, the crystal of bone, has a high affinity for a variety of proteins, additional factors may also be present in matrix. It is hypothesized that calcified bone matrix is a storehouse for growth factors and that these factors are released during the resorption process.

Transforming growth factor β is the most abundant of the growth factors yet found in bone martix.[148] It has a wide variety of activities. It stimulates cell replication in osteoblast-like primary cultures with a biphasic dose response curve,[150] but inhibits DNA synthesis in osteosarcoma cell lines.[40,151] It also appears to regulate collagen synthesis.[40,150,151] The effects of TGF β on bone resorption are complex. It stimulates bone resorption in newborn mouse calvaria cultures by a prostaglandin synthesis-dependent mechanism.[72,152] However, in fetal rat long bone cultures, TGF β inhibits bone resorption.[152] The inhibitory effects that TGF β has on resorption may be mediated by its effects on osteoclast formation, since it inhibits the formation of osteoclast-like cells in bone marrow cultures.[153] Hence, different culture systems respond differently to this factor. The mechanism for these differences remains unknown.

TGF β may regulate the responses of osteoblasts to PTH. Treatment of bone cultures with PTH increases the activity of TGF β in the conditioned medium.[37] This effect may represent activation of latent precursor forms of TGF β by osteoclasts. In addition, TGF β also increases the affinity of osteoblast-enriched primary cell cultures for PTH.[38]

Like TGF β, TGF α has effects on bone that differ with the assay systems in which it is tested. In newborn mouse calvaria cultures, it stimulates bone resorption by a mechanism that is dependent on prostaglandin synthesis.[72,154,155] However, in fetal rat long bone cultures, inhibitors of prostaglandin synthesis do not block its resorptive response.[154,155] *In vivo* infusions of TGF α in mice cause serum calcium levels to rise, and these effects are blocked by inhibitors of prostaglandin synthesis.[156] The stimulatory effects that TGF α has on bone resorption may be mediated in part by its effects on osteoclast formation, since in bone marrow cultures it enhances the formation of osteoclast-like cells.[157]

TGF α also stimulates DNA synthesis in fetal rat long bone cultures.[158] The effects of TGF α may be similar to those of EGF, since they are analogues and appear to bind the same receptor.[159,160] In fetal rat long bone cultures, EGF-mediated resorption can be inhibited by agents that block DNA synthesis.[161] EGF is also a potent inhibitor of collagen synthesis in bone cultures.[162] EGF and TGF α have also been found to inhibit PTH-mediated adenyl cyclase in osteoblast-like cells.[163] Recently, TGF α was found to be a product of activated macrophages.[164,165] Hence, it may be involved in the responses that bone has to inflammation and infection.

PDGF stimulates bone resorption in newborn mouse calvaria cultures by a prostaglandin synthesis-dependent mechanism.[74] It also has direct stimulatory effects on bone protein and DNA synthesis.[166] Endothelial cell growth factor (ECGF), a form of acidic fibroblast growth factor, does not stimulate bone resorption *in vitro*.[167] However, it does stimulate bone protein and DNA synthesis. Basic FGF appears to have similar effects, but may be more potent.[168]

Additional Factors

Differentiation inducing factors (DIFs) have been purified from the conditioned medium of concanavalin A (con A)-stimulated murine spleen cell cultures[169] and from the conditioned medium of the murine osteoblast-like osteosarcoma cell line MC3T3-E1.[170] These factors have an M_r of about 70 kDa for the spleen cells and 20 kDa and 50 kDa for the osteosarcoma cells. DIFs stimulate bone resorption *in vitro* and the differentiation of the M1 murine myeloid leukemia cell line. Partially purified DIFs from the MC3T3-E1 cells do not have IL-1 or TNF bioactivity.[170] Hence, they may represent a new regulator of bone cell function. Recently, a 20 kDa peptide leukemia inhibitory factor (LIF) was purified and cloned.[171] This protein is glycosylated and circulates with an M_r of between 40 and 60 kDa. LIF could be one activity that has previously been identified as DIF, since it also stimulates M1 cell differentiation. Conditioned medium from concanavalin A-stimulated murine spleen cells has also been found to directly stimulate bone resorption by isolated osteoclasts.[172] Hence, a factor in these supernatants, which is a source of DIFs,[169] appears to directly stimulate osteoclasts to resorb. In contrast, PTH,[41] $1,25(OH)_2$ vitamin D,[53] IL-1,[92] and TNF[122] all appear to require the production of an intermediate factor by osteoblasts or another cell to stimulate osteoclastic resorption.

Interactions among Factors

A variety of cytokines and hormones have been shown to interact in their effects on bone. IL-1 synergistically augments the *in vitro* resorptive response to PTH,[173] TNF α,[79] and TGF α.[158] The effects of IL-1 on TGF α are mediated by prostaglandin synthesis and can be inhibited by indomethacin, while those to PTH and TNF were not. IL-1 also appeared to block the effects that TGF α has on DNA synthesis.[158] *In vivo* IL-1 synergistically augments the resorptive response to PTH-related peptide (PTHRP).[174] This recently described protein is an analogue of PTH and is believed to mediate the hypercalcemia that is associated with many tumors.[124] PTH may also induce lymphocytes to stimulate bone resorbing factors.[175] INF gamma, as mentioned previously, has inhibitory effects on resorption that are more potent for cytokine-mediated responses.[128] This suggests that it may be a useful drug to treat the hypercalcemia of multiple myeloma and other hematologic malignancies, since these cancers may stimulate resorption through the production of cytokines.[117]

SUMMARY

In this review I have tried to summarize the current knowledge of the effects that cytokines have on bone. It appears that, in addition to their well-known effects on hematopoietic and immune cells, these agents are potent regulators of osteoblast and osteoclast function. Since local cytokine production is probably a normal event in bone, these factors may represent a second level of regulation for skeletal homeostasis. It is likely that cytokines interact with humoral factors in their effects on the skeleton. Hence, the response of bone to a given stimulus probably represents the net responses of a number of local and systemic agents.

In addition, I would like to stress that the marrow elements in bone, which are known to be both a source and target of cytokine action, may be intimately tied to the function of bone's skeletal elements through cytokines. Hence, maintenance of the marrow space or the myelofibrosis that is associated with hyperparathyroidism may result from the combined actions of both bone marrow and osteoblast-produced cytokines.

Much additional research needs to be done to completely define the role that cytokines have on skeletal function. We need to identify the cells in bone that normally produce cytokines and define the mechanisms that regulate the production of these factors. In addition, we need to better understand the roles these factors have in both normal bone homeostasis and bone diseases. It is hoped that by understanding these mechanisms, we will be better able to develop therapies for diseases of both the skeleton and the bone marrow.

REFERENCES

1. Cormack, D.H., *Ham's Histology*, J.B. Lippincott, London, 1989.
2. Raisz, L.G. and B.E. Kream, Regulation of bone formation. *N. Engl. J. Med.*, 309:29–35 and 83–89, 1983.
3. Mundy, G.R. and G.D. Roodman, Osteoclast ontogeny and function, in *Bone and Mineral Research*, W.A. Peck, Ed., Elsevier, Amsterdam, 209–279, 1987.
4. Harrison, L.C. and I.L. Campbell, Cytokines: an expanding network of immunoinflammatory hormones. *Mol. Endocrinol.*, 2:1151–1156, 1988.
5. Felix, R., P.R. Elford, C. Stoerckle, M. Cecchini, A. Wetterwald, U. Trechsel, H. Fleisch, and B.M. Stadler, Production of hemopoietic growth factors by bone tissue and bone cells in culture. *J. Bone Miner. Res.*, 3:27–36, 1988.
6. Elford, P.R., R. Felix, U. Cecchini, U. Trechsel, and H. Fleisch, Murine osteoblastlike cells and the osteogenic cell MC3T3-E1 release a macrophage colony-stimulating activity in culture. *Calcif. Tissue Int.*, 41:151–156, 1987.
7. Horowitz, M.C., D.L. Coleman, P.M. Flood, T.S. Kupper, and R.L. Jilka, Parathyroid hormone and lipopolysaccharide induce murine osteoblast-like cells to secrete a cytokine indistinguishable from granulocyte-macrophage colony-stimulating factor. *J. Clin. Invest.*, 83:149–157, 1989.
8. Weir, E.C., K.L. Insogna, and M.C. Horowitz, Osteoblast-like cells secrete granulocyte-macrophage colony-stimulating factor in response to parathyroid hormone and lipopolysaccharide. *Endocrinology*, 124:899–904, 1989.
9. Sato, K., K. Kasono, Y. Fujii, M. Kawakami, T. Tsushima, and K. Shizume, Tumor necrosis factor (cachectin) stimulates mouse osteoblast-like cells (MC3T3-E1) to produce macrophage-colony stimulating activity and prostaglandin E_2. *Biochem. Biophys. Res. Commun.*, 145:323–329, 1987.

10. Eriksen, E.F., Normal and pathological remodeling of human trabecular bone: three dimensional reconstruction of the remodeling sequence in normals and in metabolic bone disease. *Endocr. Rev.,* 7:379–408, 1986.
11. Riggs, B.L. and L.J., Melton, III, Involutional osteoporosis. *N. Engl. J. Med.,* 314:1676–1686, 1986.
12. Genant, H.K., Radiology of rheumatic diseases, in *Arthritis and Allied Conditions,* D.J. McCarthy, Ed., Lea and Febiger, Philadelphia, 1985, 89.
13. Case records of the Massachusetts General Hospital. Weekly clinicopathological exercises. Case 28-1986. An eight-year-old girl with multiple osteolytic lesions during the preceding six months. *N. Engl. J. Med.,* 315:178–185, 1986.
14. Mundy, G.R., K.J. Ibbotson, and S.M. D'Souza, Tumor products and the hypercalcemia of malignancy. *J. Clin. Invest.,* 76:391–394, 1985.
15. Mundy, G.R., Hypercalcemia of malignancy revisited. *J. Clin. Invest.,* 82:1–6, 1988.
16. Hamilton, J.A., S. Lingelbach, N.C. Partridge, and T.J. Martin, Regulation of plasminogen activator production by bone-resorbing hormones in normal and malignant osteoblasts. *Endocrinology,* 116:2186–2191, 1985.
17. Chambers, T.J., J.A. Darby, and K. Fuller, Mammalian collagenase predisposes bone surfaces to osteoclastic resorption. *Cell Tissue Res.,* 241:671–675, 1985.
18. Lorenzo, J.A., S.L. Sousa, J.M. Fonseca, J.M. Hock, and E.S. Medlock, Colony-stimulating factors regulate the development of multinucleated osteoclasts from recently replicated cells in vitro. *J. Clin. Invest.,* 80:160–164, 1987.
19. Baron, R., L. Neff, P.T. Van, J.R. Nefussi, and A. Vignery, Kinetic and cytochemical identification of osteoclast precursors and their differentiation into multinucleated osteoclasts. *Am. J. Pathol.,* 122:363–378, 1986.
20. Baron, R., L. Neff, D. Louvard, and P.J. Courtoy, Cell-mediated extracellular acidification and bone resorption: evidence for a low pH in resorbing lacunae and localization of a 100-kD lysosomal membrane protein at the osteoclast ruffled border. *J. Cell Biol.,* 101:2210–2222, 1985.
21. Raisz, L.G., Local and systemic factors in the pathogenesis of osteoporosis. *N. Engl. J. Med.,* 318:818–828, 1988.
22. Owen, M., Lineage of osteogenic cells and their relationship to the stromal system, in *Bone and Mineral Research,* W.A. Peck, Ed., Elsevier, Amsterdam, 1985, 1–23.
23. Nijweide, P.J., E.H. Burger, and J.H.M. Feyen, Cells of bone: proliferation, differentiation, and hormonal regulation. *Physiol. Rev.,* 66:855–886, 1986.
24. Talmage, R.V., Morphological and physiological considerations in a new concept of calcium transport in bone. *Am. J. Anat.,* 129:467–476, 1970.
25. Walker, D.G., Control of bone resorption by hematopoietic tissue. *J. Exp. Med.,* 142:651–663, 1975.
26. Burger, E.H., J.W.M. Van Der Meer, and P.J. Nijweide, Osteoclast formation from mononuclear phagocytes: role of bone-forming cells. *J. Cell Biol.,* 99:1901–1906, 1984.
27. Burger, E.H., J.W.M. Van Der Meer, J.C. Van De Gavel, and R. Van Furth, In vitro formation of osteoclasts from long-term culture of bone marrow mononuclear phagocytes. *J. Exp. Med.,* 156:1604–1614, 1982.
28. Scheven, B.A.A., J.W.M. Visser, and P.J. Nijweide, In vitro osteoclast generation from different bone marrow fractions, including a highly enriched haematopoietic stem cell population. *Nature,* 321:79–81, 1986.
29. Canterbury, J.M., L.A. Bricker, G.S. Levey, P.L. Kozlovskis, E. Ruiz, J.E. Zull, and E. Reiss, Metabolism of bovine parathyroid hormone. *J. Clin. Invest.,* 55:1245–1253, 1975.

30. Bader, C.A., J.D. Monet, P. Rivaille, C.M. Gaubert, M.S. Moukhtar, G. Milhaud, and J.L. Funck-Brentano, Comparative in vitro biological activity on 1-34 N-terminal synthetic fragments of human parathyroid hormone on bovine and porcine kidney membranes. *Endocr. Res.,* 3:167–186, 1976.
31. Lorenzo, J.A., L.G. Raisz, and J.M. Hock, DNA synthesis is not necessary for osteoclastic responses to parathyroid hormone in cultured fetal rat long bones. *J. Clin. Invest.,* 72:1924–1929, 1983.
32. Wilson, R., S.D. Rao, M. Kleerekoper, and A.M. Parfitt, Is asymptomatic primary hyperparathyroidism (PHPT) a risk factor for vertebral compression fractures (VCF). *J. Bone Miner. Res.,* 2 (suppl. 1) Abstr. S19, 1987.
33. Gunness-Hey, M. and J.M. Hock, Increased trabecular bone mass in rats treated with human synthetic parathyroid hormone. *Metab. Bone Dis. Rel. Res.,* 5:177–181, 1984.
34. Slovik, D.M., D.I. Rosenthal, S.H. Doppelt, J.T. Potts, Jr., M.A. Daly, J.A. Campbell, and R.M. Neer, Restoration of spinal bone in osteoporotic men by treatment with human parathyroid hormone (1-34) with 1,25-dihydroxyvitamin D. *J. Bone Miner. Res.,* 1 4:377–381, 1986.
35. Canalis, E., M. Centrella, W. Burch, and T.L. McCarthy, Insulin-like growth factor I mediates selective anabolic effects of parathyroid hormone in bone cultures. *J. Clin. Invest.,* 83:60–65, 1989.
36. McCarthy, T.L., M. Centrella, and E. Canalis, Parathyroid hormone enhances the transcript and polypeptide levels of insulin-like growth factor I in osteoblast-enriched cultures from fetal rat bone. *Endocrinology,* 124 3:1247–1253, 1989.
37. Pfeilschifter, J. and G.R. Mundy, Modulation of type beta transforming growth factor activity in bone cultures by osteotropic hormones. *Proc. Natl. Acad. Sci. U.S.A.,* 84:2024–2028, 1987.
38. Centrella, M., T.L. McCarthy, and E. Canalis, Parathyroid hormone modulates transforming growth facter beta activity and binding in osteoblast-enriched cell cultures from fetal rat parietal bone. *Proc. Natl. Acad. Sci. U.S.A.,* 85:5889–5893, 1988.
39. Centrella, M., J. Massague, and E. Canalis, Human platelet-derived transforming growth factor-beta stimulates parameters of bone growth in fetal rat calvariae. *Endocrinology,* 119:2306–2312, 1986.
40. Pfeilschifter, J., S.M. D'Souza, and G.R. Mundy, Effects of transforming growth factor-beta on osteoblastic osteosarcoma cells. *Endocrinology,* 121:212–218, 1987.
41. McSheehy, P.M.J. and T.J. Chambers, Osteoblast-like cells in the presence of parathyroid hormone release soluble factor that stimulates osteoclastic bone resorption. *Endocrinology,* 119:1654–1659, 1986.
42. Rouleau, M.F., J. Mitchell, and D. Goltzman, In vivo distribution of parathyroid hormone receptors in bone: evidence that a predominanat osseous target cell is not the mature osteoblast. *Endocrinology,* 123:187–191, 1988.
43. Rodan, G.A. and T.J. Martin, Rose of osteoclasts in hormonal control of bone resorption — a hypothesis. *Calcif. Tissue Int.,* 33:349–351, 1981.
44. Bell, N.H., Vitamin D-endocrine system. *J. Clin. Invest.,* 76:1–6, 1985.
45. Slatopolsky, E., C. Weerts, J. Thielan, J. Horst, H. Harter, and K.J. Martin, Marked suppression of secondary hyperparathyroidism by intravenous administration of 1,25 dihydroxycholecalciferol in uremic patients. *J. Clin. Invest.,* 74:2136–2143, 1984.
46. Abe, E., C. Miyaura, H. Sakagami, M. Takida, K. Konno, T. Yamazaki, S. Yoshiki, and T. Suda, Differentiation of mouse myeloid leukemia cells induced by 1 alpha, 25-dihydroxyvitamin D3. *Proc. Natl. Acad. Sci. U.S.A.,* 78:4990–4994, 1981.

47. Tsoukas, C.D., D.M. Provvedini, and S.C. Manolagas, 1,25 dihydroxyvitamin D3: a novel immunoregulatory hormone. *Science*, 224:1438–1440, 1984.
48. Rigby, W.F.C., T. Stacy, and M.W. Fanger, Inhibition of T lymphocyte mitogenesis by 1,25 dihydroxyvitamin D3 (calcitriol). *J. Clin. Invest.*, 74:1451–1455, 1984.
49. Amento, E.P., A.K. Bhalla, A.K. Kurnick, J.T. Kardin, T.L. Clemens, S.A. Holick, M.R. Holick, and S.M. Krane. 1884. 1 Alpha 25-dihydroxyvitamin D3 induces maturation of the human monocyte cell line U937, and, in association with a factor from human T lymphocytes, augments production of the monokine, mononuclear cell factor. *J. Clin. Invest.*, 73:731–739, 1984.
50. Provvedini, D.M., C.D. Tsoukas, L.J. Deftos, and S.C. Manolagas, 1,25 dihydroxyvitamin D3 receptors in human leukocytes. *Science*, 221:1181–1183, 1983.
51. Roodman, G.D., K.J. Ibbotson, B.R. MacDonald, T.J. Kuehl, and G.R. Mundy, 1,25-Dihydroxyvitamin D3 causes formation of multinucleated cells with several osteoclast characteristics in cultures of primate marrow. *Proc. Natl. Acad. Sci. U.S.A.*, 82:8213–8217, 1985.
52. Underwood, J.L. and H.F. DeLuca, Vitamin D is not directly necessary for bone growth and mineralization. *Am. J. Physiol.*, 246:E493–E498, 1984.
53. McSheehy, P.M. and T.J. Chambers, 1,25 Dihydroxyvitamin D3 stimulates rat osteoblastic cells to release a soluble factor that increases osteoclastic bone resorption. *J. Clin. Invest.*, 1987:425–429, 1987.
54. Austin, L.A. and H. Heath, Calcitonin: physiology and pathophysiology. *N. Engl. J. Med.*, 304(5):269–278, 1981.
55. Farley, J.R., N.M. Tarbaux, S.L. Hall, T.A. Linkhart, and D.J. Baylink, The anti-bone-resorptive agent calcitonin also acts in vitro to directly increase bone formation and bone cell proliferation. *Endocrinology*, 123:159–167, 1988.
56. Nicholson, G.C., J.M. Moseley, P.M. Sexton, F.A. Mendelsohn, and T.J. Martin, Abundant calcitonin receptors in isolated rat osteoclasts. Biochemical and autoradiographic characterization. *J. Clin. Invest.*, 1986:355–360, 1986.
57. Chambers, T.J., P.M.J. McSheehy, B.M. Thomson, and K. Fuller, The effect of calcium-regulating hormones and prostaglandins on bone resorption by osteoclasts disaggregated from neonatal rabbit bones. *Endocrinology*, 60:234–239, 1985.
58. Wener, J.A., S.J. Gorton, and L.G. Raisz, Escape from inhibition or resorption in cultures of fetal bone treated with calcitonin and parathyroid hormone. *Endocrinology*, 90(3):752–759, 1972.
59. Tashjian, A.H.Jr., D.R. Wright, J.L. Ivey, and A. Pont, Calcitonin binding sites in bone: relationships to biological response and "escape". *Recent Prog. Horm. Res.*, 34:285–334, 1978.
60. Moran, J., W. Hunziker, and J. Fischer, Calcitonin and calcium ionophores: cyclic AMP responses in cells of a human lymphoid line. *Proc. Natl. Acad. Sci. U.S.A.*, 75:3984–3988, 1978.
61. Mundy, G.R., J.L. Shapiro, J.G. Bandelin, E.M. Canalis, and L.G. Raisz, Direct stimulation of bone resorption by thyroid hormones. *J. Clin. Invest.*, 58:529–534, 1976.
62. Komm, B.S., C.M. Terpening, D.J. Benz, K.A. Graeme, A. Gallegos, M. Korg, G.L. Greene, B.W. O'Malley, and M.R. Haussler, Estrogen binding, receptor mRNA, and biologic response in osteoblast-like osteosarcoma cells. *Science*, 241:81–84, 1988.
63. Eriksen, E.F., D.S. Colvard, N.J. Berg, M.L. Graham, K.G. Mann, T.C. Spelsberg, and B.L. Riggs, Evidence of estrogen receptors in normal human osteoblast-like cells. *Science*, 241:84–86, 1988.

64. Canalis, E., T. McCarthy, and M. Centrella, Growth factors and the regulation of bone remodeling. *J. Clin. Invest.*, 81:277–281, 1988.
65. Klein, D.C. and L.G. Raisz, Prostaglandins: stimulation of bone resorption in tissue culture. *Endocrinology*, 86:1436–1440, 1970.
66. Chyun, Y.S. and L.G. Raisz, Stimulation of bone formation by prostaglandin E_2. *Prostaglandins*, 27(1):97–103, 1984.
67. Ueda, K., A. Saita, H. Nakano, et al., Cortical hyperostosis following long-term administration of prostaglandin E_1 in infants with cyanotic congenital heart disease. *J. Pediatr.*, 97:834–836, 1980.
68. Lorenzo, J.A., S.L. Sousa, C. Alander, L.G. Raisz, and C.A. Dinarello, Comparison of the bone-resorbing activity in the supernatants from phytohemagglutinin-stimulated human peripheral blood mononuclear cells with that of cytokines through the use of an antiserum to interleukin 1. *Endocrinology*, 121:1164–1170, 1987.
69. Sato, K., Y. Fujii, K. Kasono, M. Saji, T. Tsushima, and K. Shizume, Stimulation of prostaglandin E_2 and bone resorption by recombinant human interleukin 1 alpha in fetal mouse bones. *Biochem. Biophys. Res. Commun.*, 138:618–624, 1986.
70. Tatakis, D.N., G. Schneeberger, and R. Dziak, Recombinant interleukin-1 stimulates prostaglandin E_2 production by osteoblastic cells: synergy with parathyroid hormone. *Calcif. Tissue Int.*, 42:358–362, 1988.
71. Tashjian, A.H., Jr., E.F. Voelkel, M. Lazzaro, D. Goad, T. Bosma, and L. Levine, Tumor necrosis factor-alpha (cachectin) stimulates bone resorption in mouse calvaria via a prostaglandin-mediated mechanism. *Endocrinology*, 120:2029–2036, 1987.
72. Tashjian, A.H., E.F. Voelkel, M. Lazzaro, J.R. Singer, A.B. Roberts, R. Derynck, M.E. Winkler, and L. Levine, Alpha and beta human transforming growth factors stimulate prostaglandin production and bone resorption in cultured mouse calvaria. *Proc. Natl. Acad. Sci. U.S.A.*, 82:4535–4538, 1985.
73. Shupnik, M.A., N.Y. Ip, and A.H., Tashjian, Jr., Characterization and regulation of receptors for epidermal growth factor in mouse calvaria. *Endocrinology*, 107:1738–1746, 1980.
74. Tashjian, A.H., Jr., E.L. Hohmann, H.N. Antoniades, and L. Levine, Platelet-derived growth factor stimulates bone resorption via a prostaglandin-mediated mechanism. *Endocrinology*, 111:118–124, 1982.
75. Chyun, Y.S., B.E. Kream, and L.G. Raisz, Cortisol decreases bone formation by inhibiting periosteal cell proliferation. *Endocrinology*, 114:477–480, 1984.
76. Gowen, M., D.D. Wood, E.J. Ihrie, M.K.B. McGuire, and R.G.R. Russell, An interleukin 1 like factor stimulates bone resorption in vitro. *Nature*, 306:378–380, 1983.
77. Heath, J.K., J. Saklatvala, M.C. Meikle, S.J. Atkinson, and J.J. Reynolds, Pig interleukin 1 (catabolin) is a potent stimulator of bone resorption in vitro. *Calcif. Tissue Int.*, 37:95–97, 1985.
78. Dewhirst, F.E., P.P. Stashenko, J.E. Mole, and T. Tsurumachi, Purification and partial sequence of human osteoclast-activating factor: identity with interleukin 1 beta. *J. Immunol.*, 135:2562–2568, 1985.
79. Stashenko, P., F.E. Dewhirst, W.J. Peros, R.L. Kent, and J.M. Ago, Synergistic interactions between interleukin 1, tumor necrosis factor, and lymphotoxin in bone resorption. *J. Immunol.*, 138:1464–1468, 1987.
80. Smith, D.D., M. Gowen, and G.R. Mundy, Effects of interferon-gamma and other cytokines on collagen synthesis in fetal rat bone cultures. *Endocrinology*, 120:2494–2499, 1987.

81. Canalis, E., Interleukin-1 has independent effects on deoxyribonucleic acid and collagen synthesis in cultures of rat calvariae. *Endocrinology,* 118:74–81, 1986.
82. Stashenko, P., F.E. Dewhirst, M.L. Rooney, L.A. Desjardins, and J.D. Heeley, Interleukin-1 beta is a potent inhibitor of bone formation in vitro. *J. Bone Miner. Res.,* 2:559–565, 1987.
83. Dinarello, C.A., Interleukin 1. *Rev. Infect. Dis.,* 6:51–95, 1984.
84. March, C.J., B. Mosley, A. Larsen, D.P. Cerretti, G. Braedt, V. Price, S. Gillis, C.S. Henney, S.R. Kronheim, K. Grabstein, P.J. Conlon, T.P. Hopp, and D. Cosman, Cloning, sequence and expression of two distinct human interleukin-1 complementary DNAs. *Nature,* 315:641–647, 1985.
85. Lomedico, P.T., U. Gubler, C.P. Hellmann, M. Dukovich, J.G. Giri, Y.E. Pan, K. Collier, R. Semionow, A.O. Chua, and S.B. Mizel, Cloning and expression of murine interleukin-1 cDNA in Escherichia coli. *Nature,* 312:458–462, 1984.
86. Mizel, S.B. and D. Mizel, Purification to apparent homogeneity of murine interleukin 1. *J. Immunol.,* 126:834–837, 1981.
87. Bird, T.A. and J. Saklatvala, Identification of a common class of high affinity receptors for both types of porcine interleukin-1 on connective tissue cells. *Nature,* 324:263–266, 1986.
88. Dower, S.K., S.R. Kronheim, T.P. Hopp, M. Cantrell, M. Deeley, S. Gillis, C.S. Henney, and D.L. Urdal, The cell surface receptors for interleukin-1 alpha and interleukin-1 beta are identical. *Nature,* 324:266–268, 1986.
89. Gray, P.W., D. Glaister, E. Chen, D.V. Goeddel, and D. Pennica, Two interleukin 1 genes in the mouse: cloning and expression of the cDNA for murine interleukin 1 beta. *J. Immunol.,* 137:3644–3648, 1986.
90. Huang, J.J., R.C. Newton, S.J. Rutledge, R. Horuk, J.B. Matthew, M. Covington, and Y. Lin, Characterization of murine IL-1 beta. Isolation, expression, and purification. *J. Immunol.,* 140:3838–3843, 1988.
91. Takacs, L., E.J. Kovacs, M.R. Smith, H.A. Young, and S.K. Durum, Detection of IL-1 alpha and IL-1 beta gene expression by in situ hybridization. Tissue localization of IL-1 mRNA in the normal C57BL/6 mouse. *J. Immunol.,* 141:3081-3095, 1988.
92. Thomson, B.M., J. Saklatvala, and T.J. Chambers, Osteoblasts mediate interleukin 1 stimulation of bone resorption by rat osteoclasts. *J. Exp. Med.,* 164:104-112, 1986.
93. Gowen, M., D.D. Wood, and G.G. Russell, Stimulation of the proliferation of human bone cells in vitro by human monocyte products with interleukin-1 activity. *J. Clin. Invest.,* 75:1223-1229, 1985.
94. Hanazawa, S., Y. Ohmori, S. Amano, K. Hirose, T. Miyoshi, M. Kumegawa, and S. Kitano, Human purified interleukin-1 inhibits DNA synthesis and cell growth of osteoblastic cell line (MC3T3-E1), but enhances alkaline phosphatase activity in the cells. *Calcif. Tissue Int.,* 203:279–284, 1986.
95. Ikeda, E., M. Kusaka, Y. Hakeda, K. Yokota, M. Kumegawa, and S. Yamamoto, Effects of interleukin 1 beta on osteoblastic clone MC3T3-E1 cells. *Calcif. Tissue Int.,* 43:162–166, 1988.
96. Harrison, J.R., S.J. Vargas, D.N. Petersen, A.T. Mador, J.A. Lorenzo, and B.E. Kream, Interleukin-1 alpha and phorbol ester inhibit procollagen gene expression in MC3T3-E1 cells by a transcriptional mechanism. *J. Bone Miner. Res.,* 4 (suppl. 1) (Abstr.), 1989.
97. Zhang, Y., J. Lin, Y.K. Yip, and J. Vilcek, Enhancement of cAMP levels and of protein kinase activity by tumor necrosis factor and interleukin 1 in human fibroblasts: role in the induction of interleukin 6. *Proc. Natl. Acad. Sci. U.S.A.,* 85:6802–6805, 1988.

98. Shirakawa, F., U. Yamashita, M. Chedid, and S.B. Mizel, Cyclic AMP-an intracellular second messenger for interleukin 1. *Proc. Natl. Acad. Sci. U.S.A.*, 85:8201-8205, 1988.
99. Harrison, J.R., H.A. Simmons, J.A. Lorenzo, B.E. Kream, and L.G. Raisz, Stimulation of prostaglandin E_2 production by interleukin-1 alpha and transforming growth factor-alpha in MC3T3-E1 cells is associated with increased prostaglandin H synthase mRNA levels. *J. Bone Miner. Res.*, 4 (suppl. 1), (Abstr.), 1989.
100. Sabatini, M., B. Boyce, T. Aufdemorte, L. Bonewald, and G.R. Mundy, Infustions of recombinant human interleukins 1 alpha and 1 beta cause hypercalcemia in normal mice. *Proc. Natl. Acad. Sci. U.S.A.*, 85:5235–5239, 1988.
101. Lotem, J. and L. Sachs, In vivo control of differentiation of myeloid leukemic cells by cyclosporine A and recombinant interleukin-1 alpha. *Blood,* 72:1595–1601, 1988.
102. Pfeilschifter, J., C. Chenu, A. Bird, G.R. Mundy, and G.D. Roodman, Interleukin-1 and tumor necrosis factor stimulate the formation of human osteoclastlike cells in vitro. *J. Bone Miner. Res.*, 4:113–118, 1989.
103. Amano, S., S. Hanazawa, K. Hirose, Y. Ohmori, and S. Kitano, Stimulatory effect on bone resorption of interleukin-1-like cytokine produced by an osteoblast-rich population of mouse calvarial cells. *Calcif. Tissue Int.*, 43:88–91, 1988.
104. Hanazawa, S., S. Amano, K. Nakada, Y. Ohmori, T. Miyoshi, K. Hirose, and S. Kitano, Biological characterization of interleukin-1-like cytokine produced by cultured bone cells from newborn mouse calvaria. *Calcif. Tissue Int.*, 41:31–37, 1987.
105. Hanazawa, S., Y. Ohmori, S. Amano, T. Miyoshi, M. Kumegawa, and S. Kitano, Spontaneous production of interleukin-1-like cytokine from a mouse osteoblastic cell line (MC3T3-E1). *Biochem. Biophys. Res. Commun.*, 131:774–779, 1985.
106. Bird, T.A., A.J.H. Gearing, and J. Saklatvala, Murine interleukin 1 receptor. *J. Biol. Chem.*, 263:12,063–12,069, 1988.
107. Sims, J.E., C.J. March, D. Cosman, M.B. Widmer, H.R. MacDonald, C.J. McMahan, C.E. Grubin, J.M. Wignall, J.L. Jackson, S.M. Call, D. Friend, A.R. Alpert, S. Gillis, D.L. Urdal, and S.K. Dower, cDNA expression cloning of the IL-1 receptor, a member of the immunoglobulin superfamily. *Science,* 241:585–589, 1988.
108. Pacifici, R., L. Rifas, S. Teitelbaum, E. Slatopolsky, R. McCracken, M. Bergfeld, W. Lee, L.V. Avioli, and W.A. Peck, Spontaneous release of interleukin 1 from human blood monocytes reflects bone formation in idiopathic osteoporosis. *Proc. Natl. Acad. Sci. U.S.A.*, 84:4616–4620, 1987.
109. Pacifici, R., L. Rifas, R. McCracken, I. Vered, C. McMurtry, L.V. Avioli, and W.A. Peck, Ovarian steroid treatment blocks a postmenopausal increase in blood monocyte interleukin 1 release. *Proc. Natl. Acad. Sci. U.S.A.*, 86:2398–2402, 1989.
110. Stock, J.L., J.A. Coderre, B. McDonald, and L.J. Rosenwasser, Effects of estrogen in vivo and in vitro on spontaneous interleukin-1 release by monocytes from postmenopausal women. *J. Clin. Endocrinol. Metab.*, 68:364–368, 1989.
111. Beutler, B. and A. Cerami, Cachectin and tumour necrosis factor as two sides of the same biological coin. *Nature,* 320:584–588, 1986.
112. Oliff, A., The role of tumor necrosis factor (cachectin) in cachexia. *Cell,* 54:141–142, 1988.
113. Beutler, B. and A. Cerami, Cachectin: more than a tumor necrosis factor. *N. Engl. J. Med.*, 316:379–385, 1987.
114. Old, L.J., Tumor necrosis factor (TNF). *Science,* 230:630–632, 1985.
115. Paul, N.L. and N.H. Ruddle, Lymphotoxin. *Annu. Rev. Immunol.*, 6:407–438, 1988.

116. Bertolini, D.R., G.E. Nedwin, T.S. Bringman, D.D. Smith, and G.R. Mundy, Stimulation of bone resorption and inhibition of bone formation in vitro by human tumour necrosis factors. *Nature,* 319:516–518, 1986.
117. Garrett, I.R., B.G.M. Durie, G.E. Nedwin, A. Gillespie, T. Bringman, M. Sabatini, D.R. Bertolini, and G.R. Mundy, Production of lymphotoxin, a bone-reborbing cytokine, by cultured human myeloma cells. *N. Engl. J. Med.,* 317:526–532, 1987.
118. Canalis, E., Effects of tumor necrosis factor on bone formation in vitro. *Endocrinology,* 121:1596–1604, 1987.
119. Centrella, M., T.L. McCarthy, and E. Canalis, Tumor necrosis factor-alpha inhibits collagen synthesis and alkaline phosphatase activity independently of its effect on deoxyribonucleic acid synthesis in osteoblast-enriched bone cell cultures. *Endocrinology,* 123:1442–1448, 1988.
120. Nanes, M.S., W.M. McKoy, and S.J. Marx, Inhibitory effects of tumor necrosis factor-alpha and interferon-gamma on deoxyribonucleic acid and collagen synthesis by rat osteosarcoma cells (ROS 17/2.8). *Endocrinology,* 124:339–345, 1989.
121. Johnson, R.A., B.F. Boyce, G.R. Mundy, and G.D. Roodman, Tumors producing human tumor necrosis factor induce hypercalcemia and osteoclastic bone resorption in nude mice. *Endocrinology,* 124:1424–1427, 1989.
122. Thomson, B.M., G.R. Mundy, and T.J. Chambers, Tumor necrosis factors alpha and beta induce osteoblastic cells to stimulate osteoclastic bone resorption. *J. Immunol.,* 138:775–779, 1987.
123. Gowen, M., B.R. MacDonald, and G.G. Russell, Actions of recombinant human gamma-interferon and tumor necrosis factor alpha on the proliferation and osteoblastic characteristics of human trabecular bone cells in vitro. *Arthr. Rheum.,* 31:1500–1507, 1988.
124. Suva, L.J., G.A. Winslow, R.E.H. Wettenhall, R.G. Hammonds, J.M. Moseley, H. Diefenbach-Jagger, C.P. Rodda, B.E. Kemp, H. Rodriguez, E.Y. Chen, P.J. Hudson, T.J. Martin, and W.I. Wood, A parathyroid hormone-related protein implicated in malignant hypercalcemia: cloning and expression. *Science,* 237:893–896, 1987.
125. Peterlik, M., O. Hoffmann, P. Swetly, K. Klaushofer, and K. Koller, Recombinant gamma-interferon inhibits prostaglandin-mediated and parathyroid hormone-induced bone resorption in cultured neonatal mouse calvaria. *FEBS Lett.,* 185:287–290, 1985.
126. Gowen, M. and G.R. Mundy, Actions of recombinant interleukin 1, interleukin 2, and interferon-gamma on bone resorption in vitro. *J. Immunol.,* 136:2478–2482, 1986.
127. Hoffmann, O., K. Klaushofer, H. Gleispach, H.J. Leis, T. Luger, K. Koller, and M. Peterlik, Gamma interferon inhibits basal and interleukin 1-induced prostaglandin production and bone resorption in neonatal mouse calvaria. *Biochem. Biophys. Res. Commun.,* 143:38–43, 1987.
128. Gowen, M., G.E. Nedwin, and G.R. Mundy, Preferential inhibition of cytokine-stimulated bone resorption by recombinant interferon gamma. *J. Bone Miner. Res.,* 1:469–474, 1986.
129. Jilka, R.L. and J.W. Hamilton, Inhibition of parathormone-stimulated bone resorption by type 1 interferon. *Biochem. Biophys. Res. Commun.,* 120:553–558, 1984.
130. Takahashi, N., G.R. Mundy, and G.D. Roodman, Recombinant human interferon-gamma inhibits formation of human osteoclast-like cells. *J. Immunol.,* 137:3544–3549, 1986.
131. Mamus, S.W., S. Beck-Schroeder, and E.D. Zanjani, Suppression of normal human erythropoiesis by gamma interferon in vitro. Role of monocytes and T lymphocytes. *J. Clin. Invest.,* 75:1496–1503, 1985.

132. Hattersley, G., E. Dorey, M.A. Horton, and T.J. Chambers, Human macrophage colony-stimulating factor inhibits bone resorption by osteoclasts disaggregated from rat bone. *J. Cell Physiol.*, 137:199–203, 1988.
133. Dexter, T.M., Blood cell development. The message in the medium. *Nature*, 309:746–747, 1984.
134. Metcalf, D., The granulocyte-macrophage colony-stimulating factors. *Science*, 229:16–22, 1985.
135. Schneider, G.B. and M. Relfson, The effects of transplantation of granulocyte-macrophage progenitors on bone resorption in osteopetrotic rats. *J. Bone Miner. Res.*, 2:225–232, 1988.
136. Schneider, G.B. and M. Relfson, A bone marrow fraction enriched for granulocyte-macrophage progenitors gives rise to osteoclasts in vitro. *Bone*, 9:303–308, 1988.
137. Sato, K., H. Mimura, D.C. Han, T. Kakiuchi, Y. Ueyama, H. Ohkawa, T. Okabe, Y. Kondo, N. Ohsawa, T. Tsushima, and K. Shizume, Production of bone-resorbing activity and colony-stimulating activity in vivo and in vitro by a human squamous cell carcinoma associated with hypercalcemia and leukocytosis. *J. Clin. Invest.*, 78:145–154, 1986.
138. Schneider, G.B. and M. Relfson, Pluripotent hemopoietic stem cells give rise to osteoclasts in vitro: effects of rGM-CSF. *Bone Miner.*, 5:129–138, 1989.
139. Van De Wijngaert, F.P., M.C. Tas, J.W.M. Van Der Meer, and E.H. Burger, Growth of osteoclast precursor-like cells from whole mouse bone marrow: inhibitory effect of CSF-1. *Bone Miner.*, 3:97–110, 1987.
140. MacDonald, B.R., G.R. Mundy, S. Clark, E.A. Wang, T.J. Kuehl, E.R. Stanley, and G.D. Roodman, Effects of human recombinant CSF-GM and highly purified CSF-1 on the formation of multinucleated cells with osteoclast characteristics in long-term bone marrow cultures. *J. Bone. Miner. Res.*, 1:227–233, 1986.
141. Shiina-Ishimi, Y., E. Abe, H. Tanaka, and T. Suda, Synthesis of colony-stimulating factor (CSF) and differentiation inducing factor (D-factor) by osteoblastic cells, clone MC3T3-E1. *Biochem. Biophys. Res. Commun.*, 134:400–406, 1986.
142. Sato, K., Y. Fujii, S. Asano, T. Ohtsuki, M. Kawakami, K. Kasono, T. Tsushima, and K. Shizume, Recombinant human interleukin 1 alpha and beta stimulate mouse osteoblast-like cells (MC3T3-E1) to produce macrophage-colony stimulating activity and prostaglandin E_2. *Biochem. Biophys. Res. Commun.*, 141:285–291, 1986.
143. Dedhar, S., L. Gaboury, P. Galloway, and C. Eaves, Human granulocyte-macrophage colony stimulating factor is a growth factor active on a variety of cell types of nonhemopoietic orgin. *Proc. Natl. Acad. Sci. U.S.A.*, 85:9253–9257, 1988.
144. Evans, D.B., R.A.D. Bunning, and R.G.G. Russell, The effects of recombinant human granulocyte-macrophage colony-stimulating factor (rhGM-CSF) on human osteoblast-like cells. *Biochem. Biophys. Res. Commun.*, 160:588–595, 1989.
145. Centrella, M. and E. Canalis, Local regulators of skeletal growth: a perspective. *Endocrinol. Rev.*, 6:544–551, 1985.
146. Centrella, M. and E. Canalis, Transforming and nontransforming growth factors are present in medium conditioned by fetal rat calvariae. *Proc. Natl. Acad. Sci. U.S.A.*, 82:7335–7339, 1985.
147. Robey, P.G., M.F. Young, K.C. Flanders, N.S. Roche, P. Kondaiah, A.H. Reddi, J.D. Termine, M.B. Sporn, and A.B. Roberts, Osteoblast synthesize and respond to transforming growth factor-type beta (TGF-beta) in vitro. *J. Cell Biol.*, 105:457–463, 1987.

148. Seyedin, S.M., T.C. Thomas, A.Y. Thompson, D.M. Rosen, and K.A. Piez, Purification and characterization of two cartilage-inducing factors from bovine demineralized bone. *Proc. Natl. Acad. Sci. U.S.A.,* 82:2267–2271, 1985.
149. Hauschka, P.V., A.E. Mavrakos, M.D. Iafrati, and S.E. Doleman, Growth factors in bone matrix. Isolation of multiple types by affinity chromatography on heparin-sepharose. *J. Biol. Chem.,* 261:12,665–12,674, 1986.
150. Centrella, M., T.L. McCarthy, and E. Canalis, Transforming growth factor beta is a bifunctional regulator of replication and collagen synthesis in osteoblast-enriched cell cultures from fetal rat bone. *J. Biol. Chem.,* 262:2869–2874, 1987.
151. Noda, M. and G.A. Rodan, Type beta transforming growth factor (TGF beta) regulation of alkaline phosphatase expression and other phenotype-related mRNAs in osteoblastic rat osteosarcoma cells. *J. Cell Physiol.,* 133:426–437, 1987.
152. Pfeilschifter, J., S.M. Seyedin, and G.R. Mundy, Transforming growth factor beta inhibits bone resorption in fetal rat long bone cultures. *J. Clin. Invest.,* 82:680–685, 1988.
153. Chenu, C., J. Pfeilschifter, G.R. Mundy, and G.D. Roodman, Transforming growth factor beta inhibits formation of osteoclast-like cells in long-term human marrow cultures. *Proc. Natl. Acad. Sci. U.S.A.,* 85:5683–5687, 1988.
154. Ibbotson, K.J., J. Harrod, M. Gowen, S. D'Souza, D.D. Smith, M.E. Winkler, R. Derynck, and G.R. Mundy, Human recombinant transforming growth factor alpha stimulates bone resorption and inhibits formation in vitro. *Proc. Natl. Acad. Sci. U.S.A.,* 83:2228–2232, 1986.
155. Stern, P.H., N.S. Krieger, R.A. Nissenson, R.D. Williams, M.E. Winkler, R. Derynck, and G.J. Strewler, Human transforming growth factor-alpha stimulates bone resorption in vitro. *J. Clin. Invest.,* 76:2016–2019, 1985.
156. Tashjian, A.H.Jr., E.F. Voelkel, W. Lloyd, R. Derynck, M.E. Winkler, and L. Levine, Actions of growth factors on plasma calcium. Epidermal growth factor and human transforming growth factor-alpha cause elevation of plasma calcium in mice. *J. Clin. Invest.,* 78:1405–1409, 1986.
157. Takahashi, N., B.R. MacDonald, J. Hon, M.E. Winkler, R. Derynck, G.R. Mundy, and G.D. Roodman, Recombinant human transforming growth factor-alpha stimulates the formation of osteoclast-like cells in long-term human marrow cultures. *J. Clin. Invest.,* 78:894–898, 1986.
158. Lorenzo, J.A., S.L. Sousa, and M. Centrella, Interleukin-1 in combination with transforming growth factor-alpha produces enhanced bone resorption in vitro. *Endocrinology,* 123:2194–2200, 1988.
159. DeLarco, J.E. and G.J. Todara, Growth factors from murine sarcoma virus-transformed cells. *Proc. Natl. Acad. Sci. U.S.A.,* 75:4001–4005, 1978.
160. Roberts, A.B., C.A. Frolik, M.A. Anzano, and M.B. Sporn, Transforming growth factors from neoplastic and nonneoplastic tissues. *Fed. Proc.,* 42:2621–2625, 1983.
161. Lorenzo, J.A., J. Quinton, S. Sousa, and L.G. Raisz, Effects of DNA and prostaglandin synthesis inhibitors on the stimulation of bone resorption by epidermal growth factor in fetal rat long-bone cultures. *J. Clin. Invest.,* 77:1897-1902, 1986.
162. Canalis, E. and L.G. Raisz, Effects of epidermal growth factor on bone formation in vitro. *Endocrinology,* 104:862-869, 1979.
163. Gutierrez, G.E., G.R. Mundy, R. Derynck, E.L. Hewlett, and M.S. Katz, Inhibition of parathyroid hormone-responsive adenylate cyclase in clonal osteoblast-like cells by transforming growth factor alpha and epidermal growth factor. *J. Biol. Chem.,* 262:15,845–15,850, 1987.

164. Madtes, D.K., E.W. Raines, K.S. Sakariassen, R.K. Assoian, M.B. Sporn, G.I. Bell, and R. Ross, Induction of transforming growth factor-alpha in activated human alveolar macrophages. *Cell,* 53:285–293, 1988.
165. Rappolee, D.A., D. Mark, M.J. Banda, and Z. Werb, Wound macrophages express TGF-alpha and other growth factors in vivo: analysis by mRNA phenotyping. *Science,* 241:708–712, 1988.
166. Canalis, E., Effect of platelet-derived growth factor on DNA and protein synthesis in cultured rat calvaria. *Metabolism,* 30:970–975, 1981.
167. Canalis, E., J. Lorenzo, W.H. Burgess, and T. Maciag, Effects of endothelial cell growth factor on bone remodeling in vitro. *J. Clin. Invest.,* 79:52–58, 1987.
168. Canalis, E., M. Centrella, and T. McCarthy, Effects of basic fibroblast growth factor on bone formation in vitro. *J. Clin. Invest.,* 81:1572–1577, 1988.
169. Abe, E., H. Tanaka, Y. Ishimi, C. Miyaura, T. Hayashi, H. Nagasawa, M. Tomida, Y. Yamaguchi, M. Hozumi, and T. Suda, Differentiation-inducing factor purified from conditioned medium of mitogen-treated spleen cell cultures stimulates bone resorption. *Proc. Natl. Acad. Sci. U.S.A.,* 83:5958–5962, 1986.
170. Abe, E., Y. Ishimi, N. Takahashi, T. Akatsu, H. Ozawa, H. Yamana, S. Yoshiki, and T. Suda, A differentiation-inducing factor produced by the osteoblastic cell line MC3T3-E1 stimulates bone resorption by promoting osteoclast formation. *J. Bone Miner. Res.,* 3:635–645, 1988.
171. Gouch, N.M., D.P. Gearing, J.A. King, T.A. Willson, D.J. Hilton, N.A. Nicola, and D. Metcalf, Molecular cloning and expression of the human homologue of the murine gene encoding myeloid leukemia-inhibitory factor. *Proc. Natl. Acad. Sci. U.S.A.,* 85:2623–2627, 1988.
172. de Vernejoul, M., M. Horowitz, J. Demignon, L. Neff, and R. Baron, Bone resorption by isolated chick osteoclasts in culture is stimulated by murine spleen cell supernatant fluids (osteoclast-activating factor) and inhibited by calcitonin and prostaglandin E_2. *J. Bone Miner. Res.,* 3:69–80, 1988.
173. Dewhirst, F.E., J.M. Ago, W.J. Peros, and P. Stashenko, Synergism between parathyroid hormone and interleukin 1 in stimulating bone resorption in organ culture. *J. Bone Miner. Res.,* 2:127–134, 1987.
174. Sato, K., Y. Fujil, K. Kasono, M. Ozawa, H. Imamura, Y. Kanaji, H. Kurosawa, T. Tsushima, and K. Shizume, Parathyroid hormone-related protein and interleukin-1 alpha synergistically stimulate bone resorption in vitro and increase the serum calcium concentration in mice in vivo. *Endocrinology,* 124:2172–2178, 1989.
175. Perry, H.M., III, Parathyroid hormone-lymphocyte interactions modulate bone resorption. *Endocrinology,* 119:2333–2339, 1986.

7. Involvement of Cytokines in Neurogenic Inflammation

EDWARD S. KIMBALL

INTRODUCTION

The inflammatory response of connective tissue and immune cells to neuropeptides, neurotransmitters, and vasoactive amines encompasses a phenomenon called neurogenic inflammation. Articular joints, the lungs, and various other tissues are innervated by nociceptive nerve fibers that are capable of secreting vasodilatory neuropeptides,[1] and the evidence for involvement of the nervous system and its products in inflammation is becoming more compelling.[2-10] For example, elevated levels of substance P (SP) have been detected in inflammatory exudates,[11] and intraarticular infusion of SP exacerbated adjuvant-induced arthritis in rats.[7] Besides SP, some other putative mediators include bradykinin (BK), neurokinin A (NK-A), neurokinin B (NK-B), and histamine. Fibroblasts, for example, respond to SP[2] and to BK[9] with increased DNA synthesis and by secreting prostaglandins.[3,12,13] Chondrocytes are stimulated to secrete collagenase by histamine,[14] and synovial fibroblasts respond to SP by proliferating and by secreting prostaglandins and collagenase.[3] Many of these responses duplicate those initiated by cytokines, especially interleukin-1 (IL-1) and tumor necrosis factor (TNF), secreted by cells of the immune system. In this chapter, evidence linking cytokines and neuroactive inflammatory substances in initiating and sustaining chronic inflammatory reactions will be presented.

NEUROPEPTIDES RESPONSIBLE FOR NEUROGENIC INFLAMMATION

Inflammatory responses consist of a multiplicity of cellular and tissue responses. Connective tissue activation results in hyperproliferation of fibroblasts and related cells, proteolytic enzyme secretion, release of mediators such as prostaglandins, shape changes of vascular endothelial cells that allow extravasation of inflammatory serum proteins, mast cell degranulation, and release of histamine. Lymphoid cell activation may result in synthesis and release of prostaglandins, platelet activating factor, IgE, rheumatoid factors, and cytokines. As will be discussed below, these

Table 1. Amino Acid Sequence of Neurokinins

	1	2	3	4	5	6	7	8	9	10	11
Substance P	Arg	Pro	Lys	Pro	Gln	Gln	Phe	Phe	Gly	Leu	Met - NH_2
SP_{4-11}				Pro	Gln	Gln	Phe	Phe	Gly	Leu	Met - NH_2
DPDT	Arg	-D-Pro-	Lys	Pro	Gln	Gln	-D-Trp-	Phe	-D-Trp-	Leu	Met - NH_2
NK-A		His	Lys	Thr	Asp	Ser	Phe	Val	Gly	Leu	Met - NH_2
NK-B		Asp	Met	His	Asp	Phe	Phe	Val	Gly	Leu	Met - NH_2

responses may engender a continuous feedback loop that can amplify and sustain a chronic inflammatory response.

Tachykinins

This family of neuropeptides contains three structurally related neuropeptides: substance P (SP), neurokinin-A (NK-A), and neurokinin-B (NK-B). As can be seen in Table 1, there is considerable amino acid sequence homology between SP, NK-A, and NK-B, particularly in their carboxy-terminal regions.

Receptors

The high degree of sequence homology suggests that there might be cross-reactivity in receptor-binding, and hence a certain degree of cross-reactivity in stimulating responses by nonneuronal tissues. Tachykinin receptors are widely distributed in tissues throughout the body.[15-18] Receptors for SP, NK-A, and NK-B are found on neuronal tissue. Cortical astrocytes[19] bind SP. Distinct binding sites in rat brain have been demonstrated for SP,[17] NK-A,[16] and NK-B.[18] NK-A binds to sites primarily in the cortex, especially in the hypothalamic region, whereas SP binds to sites primarily in the septum and striatum.[20] SP receptors also occur on vascular,[21] gastric,[15,22] and pulmonary tissues.[16]

SP receptors are known to exist on human T-cells,[23] particularly on T-helper cells, and on B-lymphocytes.[24] Most peripheral blood B-cells and B-cell lines do not possess extremely high numbers of SP receptors, although the IM-9 B-lymphoblastoid cell line was shown to possess 20,000 to 25,000 functional SP receptors.[25] In contrast, most lymphocytes contain fewer SP receptors (i.e., 7000 receptors per cell).[26,27] Cross-linking experiments revealed the presence of four molecular weight species with SP-binding properties[28] on IM-9 lymphoblastoid cells having molecular sizes of 33, 58, 78, and 116 kD. McGillis et al.[27] speculated that one possible configuration for the receptor is that of a heterodimeric or homodimeric complex of distinct subunits, where the various molecular forms observed comprise distinct subunits. This issue remains to be resolved, however. Three functional types of tachykinin receptor, SP-P, SP-E, and SP-N, recently redesignated NK-1, NK-2, and NK-3,[22] are defined on the basis of relative binding of SP, NK-A, and NK-B, and tachykinins, such as physoleimin, kassinin, etc., of marine origin. T-lymphocytes possess SP receptors of the SP-P (NK-1) type.[28]

SP receptors have also been identified on macrophages.[4] Receptor-binding studies on guinea pig macrophages with SP and marine homologues yielded data that suggested that the nature of the macrophage SP receptor is neither SP-P (NK-1) nor SP-E (NK-2).[4] This is consistent with data from our own laboratory regarding the relative activity of SP, NK-A, and NK-B on the mouse P388D1 macrophage cell line (NK-A ≥ NK-B > SP),[6] suggestive of either an NK-2 or NK-3-like receptor. The specific nature of neurokinin receptors on macrophages are still not totally clarified due to the high cross-reactivity of SP, NK-A, and NK-B for binding to these receptors.[4,16] The recent advent of more highly specific antagonists for NK-A and NK-B (i.e., Senktide) that would allow one to discriminate between the various receptor subclasses might help to clarify these receptor differences.

Inflammatory Mediator Release by Substance P

SP is the most extensively studied tachykinin in this regard. SP was discovered over 50 years ago by von Euler and Gaddum,[29] who characterized it by its ability to induce hypotension and smooth muscle contractions. SP is now known to stimulate synthesis and secretion of prostaglandins in a wide variety of somatic cells, that include synoviocytes,[3] macrophages,[4] and astrocytes,[30] and to promote kallikrein formation and plasma protein extravasation. The latter two processes would allow bradykinin to participate in any subsequent inflammatory processes.

SP is able to stimulate histamine release from mast cells. Intradermal injection of SP causes a wheal and flare response. Foreman et al.[31] and others[32,33] have shown that SP-induced flare is neurogenic, while the wheal response is not. The carboxy-terminal octapeptide of SP, SP_{4-11}, has been shown to be responsible for wheal formation,[31] but the first three amino-terminal residues are necessary for histamine release. NK-A is cosynthesized with SP[34] as part of a common polypeptide precursor and has high carboxy-terminal amino acid sequence homology to SP, but NK-A has no activity as a histamine inducer. Calcitonin gene-related peptide (CGRP) is made by the same sensory neurons as SP and NK-A.[35,36] CGRP is a potent vasodilator,[37] and is about one fourth as potent as SP as a histamine-inducer.[38] CGRP causes a wheal and flare response and, in addition, produces a long-lasting erythema that is accompanied by granulocyte infiltration to the site of inflammation. CGRP acts synergistically with other mediators, including bradykinin, platelet activating factor (PAF), and leukotriene B_4 (LTB_4), to promote vascular permeability, but has no such intrinsic activity.[39]

As discussed by Foreman,[8] Celander and Folkow[40] proposed an axon-reflex model for neurogenic inflammation. In this model, activation of polymodal nociceptors at one end of primary sensory neurons (i.e., C-fibers) by heat, pressure, or histamine, cause impulses to travel orthodromically (towards the central nervous system) and antidromically (back towards the periphery to a second efferent terminal) along the same neuron. Impulses of the latter type release neuronal SP and possibly NK-A and CGRP, which are present in the nerve termini. These neuropeptides act, in turn, on adjacent tissues to produce inflammation, release of PGE_2, and histamine, which subsequently activate nociceptors on neighboring sensory neurons. A second consequence of the presence of SP and histamine is increased blood flow to the site, introduction of plasma proteins, kallikrein activation, and chemoattraction of inflammatory cells, such as granulocytes, neutrophils, and monocytes, which subsequently exacerbate and sustain the inflammation.

Activation of Immune Cells by Substance P

In addition to mediator release, there are reports that suggest that SP may have a direct role in inducing activation of lymphoid cells and lymphoid cell precursors. A body of data exists[23-26,28,41] that shows that SP receptors occur on T-helper cells and in lower numbers on B-cells. In addition, the presence of SP correlates with increased IgA synthesis by Peyer's patch lymphocytes. Moore et al.[42] recently reported that SP acted as a costimulant for colony-stimulating factor 1 (CSF-1)-induced proliferation of marrow-derived mononuclear phagocyte precursors. These progenitor cells require two signals for initiation of colony formation, one of which can be supplied by CSF-1.[43] The second signal can be supplied by SP at 10^{-6} M to 10^{-8} M concentrations. This activity resides in the amino-terminal tetrapeptide (SP_{1-4}) (Arg-Pro-Lys-Thr), is absent from SP_{2-11}, and can be antagonized by a tuftsin analogue. Tuftsin is a tetrapeptide whose sequence is the reverse of that of SP_{1-4} (e.g., Thr-Lys-Pro-Arg).[44] Tuftsin also supplied a second signal in the same manner as SP and SP_{1-4}.[42] Other reports of SP as a regulator of phagocyte maturation and development derive from work by Bar-Shavit and colleagues.[45,46] Those reports are consistent with the one by Moore et al.[42] and show that SP and SP_{1-4} could stimulate phagocyte activation, as could tuftsin.[44,47-49]

A role for substance P in inflammatory processes is also suggested by observations that there are elevated levels of SP at sites of inflammation.[11,28] SP can influence chemotaxis of mononuclear and polymorphonuclear cells.[4,10,50] Besides promoting chemotaxis predominantly via binding to the f-Met-Leu-Phe binding site,[50] SP is capable of fully activating polymorphonuclear cells to release reactive oxygen species, mobilize calcium, activate phosphoinositol, and exocytose granule contents.[51] SP similarly promotes macrophage release of thromboxane and reactive oxygen species.[52] The participation of lymphocytes in inflammatory processes that are mediated by SP is further suggested by studies of Stanisz et al.[41] SP was shown to stimulate proliferation of rabbit T-cells obtained from the spleen, mesenteric lymph nodes, and from Peyer's patches. In addition, SP augmented *in vitro* IgA production by Peyer's patch-derived lymphocytes. This is consistent with other studies that demonstrated enhancement of human peripheral blood T-cell mitogenesis by substance P.[24] In those studies, SP promoted T-cell proliferation and augmented mitogen-stimulated T-cell mitosis.

Therefore, SP is a broad-spectrum mediator of inflammation that has effects on lymphocytes, monocytes, neutrophils, mast cells, and synoviocytes. It also increases vascular permeability, thereby allowing the influx of additional inflammatory cells and soluble mediators of inflammation. One such mediator may be bradykinin.[53]

Bradykinin

Bradykinin (BK) is a decapeptide cleavage product resulting from the action of kallikrein on kininogen.[54] The presence of thrombin in inflammatory sites will also promote kallikrein activation with subsequent BK formation.[55-57] BK is primarily known for its ability to induce hypotension through its vasodilatary action on vascular smooth muscle,[15,17] but it is also associated with nociception, peripheral transmission of pain, and induction of inflammatory reactions. BK is thought to be involved in the release of nonpeptide inflammatory mediators, such as histamine and serotonin (5-HT).[17]

Receptors

Two subtypes of BK receptor, B1 and B2, have been characterized.[17,54] B1 receptors are defined by their ability to bind desArg^9-BK and to be antagonized by desArg^9-(Leu^8)-BK. This compound is highly specific for B1 receptors and fails to cross-react with B2 receptor subtypes. B2 receptor antagonists are apparently less specific and are typified by antagonism of BK binding by D-Phe^7-Thi^{5,8}-BK. When these antagonists were tested on isolated vascular smooth muscle preparations from rabbits, B1 receptors were found to occur primarily on aorta and mesenteric vein and hence were held responsible for mediating contractions in those two tissues. B2 receptors have a different distribution and are thought to mediate contractile responses in carotid arteries and the jugular vein.

B1 receptor expression is stimulated by intravenous administration of LPS. Subsequently, various tissues respond to des-Arg^9-BK, a potent analogue specific for B1 receptors. Regoli[17] points out, though, that it is still not clear whether B1 receptors are generated during pathological conditions such as found in arthritic diseases and other chronic inflammatory diseases. Pathological reactions attributed to kinins involve activation of B2 receptors.[17] Thus, peripheral vasodilatation, increased vascular permeability, venoconstriction, and mobilization of inflammatory cells occur as a result of B2 activation in vascular endothelium of large and small arterial vessels.[58] The evidence supporting this comes from experiments that showed an absence of antagonism by B1 antagonists. Another B2 receptor-mediated phenomenon is the stimulation of afferent fibers, culminating in generation of pain, modulation of cardiovascular reflexes, and secondary afferent impulses. This too is thought to be a B2 receptor phenomenon. Supporting evidence, albeit inferential, derives from the finding that the specific B1 agonist desArg^9-BK is inactive in these systems, whereas BK is active.[54]

The presence of BK receptors has been inferred on other cell types by functional studies of BK activation of immune cells and embryonic fibroblast cell lines. BK receptors are imputed to occur on lymphoid cells, based on work by Duma et al.,[59] Burch and Axelrod[13] and Burch et al.[60] and work in our own laboratory.[61] In these studies the action of BK was followed on macrophage progenitor cell proliferation,[59] on monocyte/macrophage secretion of cytokines,[60,61] and on macrophage secretion of prostaglandins.[13] Work by Goldstein and Wall[9] demonstrated the presence of B1 receptors on fibroblasts. The presence of BK receptors on fibroblasts may also be inferred from work by Olsen et al.,[62] who recently demonstrated increases in intracellular Ca^{+2} mobilization in Swiss 3T3 fibroblasts as a result of BK treatment. Burch et al.[12] demonstrated enhanced phospholipase A_2 activation in BALB/3T3 fibroblasts as a result of interactions between IL-1 and BK, and Kimball and Fisher[5] demonstrated synergystic interactions between IL-1 and BK regarding promotion of DNA turnover and growth of BALB/3T3 fibroblasts.

Bradykinin as a Mediator of Inflammation

Activation of arachidonic acid metabolism in endothelium is mediated by kinins.[17,54,58] Thus, kinins promote release of prostacyclin from heart and kidney endothelial cells. This is consistent with the fact that smooth muscle contraction and production of pain in various organs are associated with release of prostaglandins (PG). BK can also stimulate PG production in aortic endothelial cells,[63] a fibrosarcoma cell line,[64] and mouse calvaria.[55,65]

One serious consequence of chronic inflammation in the joint is the destruction of cartilage and bone, as occurs in rheumatoid arthritis. A recent finding showed that BK can directly promote bone resorption *in vitro*[55] and that this effect may be due to the action of BK on both B1 and B2 subtypes on bone. BK-induced bone resorption, as determined by measuring ^{45}Ca release, could be inhibited by nonsteroidal anti-inflammatory drugs like indomethacin (1 μM) and naproxen (1 μM).[55] Goldstein and Wall[9] found that activation of B1 receptors on fibroblasts promotes the formation of collagen and also mediates cell multiplication. Regoli[17] speculated that these functions may be important to tissue repair in the lung, but may also be responsible for pathological tissue reactions in that organ. Similar reactions may occur in other sites of inflammation, such as arthritic joints. A role for BK in rheumatoid arthritis is suggested by a report by Keele and Eisen,[53] who examined kinin formation in rheumatoid arthritic patients.

CYTOKINES INVOLVED IN NEUROGENIC INFLAMMATION

Two cytokines frequently considered to be responsible for much of the pathophysiology accompanying acute and chronic inflammation are IL-1 and TNF. This is reviewed more extensively elsewhere in this volume, as well as in numerous other reviews.[66-70] TNF is exclusively a macrophage product, while IL-1 is predominantly a product of macrophages. IL-1 can be produced by a large variety of other cell types that include hepatomas,[71] endothelial cells,[72-74] and astroglial cells.[75,76] Additional sources of IL-1 include synovial fibroblasts, keratinocytes, and B-lymphocytes.[77] There is a rather striking parallelism, though, among the biological activities of IL-1 and TNF.[67] These are summarized in Table 2. The activities pertinent to inflammation are those that involve fibroblast and synoviocyte activation (prostaglandin production, protease release, proliferation, cytokine production) and endothelial cell activation (chemotactic factor release, cytokine production). These

Table 2. Inflammatory Activities of IL-1 and TNF[a]

Biological Activity	IL-1	TNF
Fibroblast proliferation	+	+
Fibroblast PGE$_2$	+	+
Fibroblast collagenase	+	+
Synoviocyte PGE$_2$	+	+
Synoviocyte collagenase	+	+
Chondrocyte PGE$_2$	+	−
Collagen synthesis	+	+
Bone resorption	+	+
Neutrophil superoxide release	+	+
Basophil degranulation	+	−
Plasma hypoferremia	+	+
Plasma zinc loss	+	+
Cytotoxicity to tumor cells	+	+
Acute phase protein synthesis	+	+
Decreased albumin synthesis	+	+
Decreased cytochrome P$_{450}$ synthesis	+	+
Fever	+	+
Slow wave sleep induction	+	+
Endothelial cell activation	+	+
Induction of IL-1 synthesis	+	+
Induction of other cytokines	+	+
T-cell activation	+	±
B-cell activation	+	+

[a] References 66, 67, 77.

Table 3. Common Fibroblast Activating Properties of IL-1 and SP

Biological Property	IL-1	SP
DNA synthesis	+	+
Prostaglandin synthesis	+	+
Collagenase secretion	+	+
De novo protein synthesis	+	+
Protein phosphorylation	+	+

processes are thought to make a major conribution to the immunopathology associated, for example, with rheumatoid arthritis or chronic lung inflammation.

INTERACTIONS BETWEEN NEUROTRANSMITTERS AND CYTOKINES

SP Potentiates IL-1

As summarized in Table 3, similar parallels exist between the fibroblast activating properties of IL-1 and for many of the nonneurological biological activities of

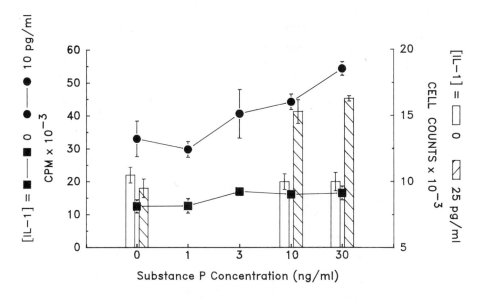

Figure 1. Augmentation of IL-1-induced DNA turnover and multiplication of BALB/3T3 fibroblasts by SP.[5]

SP. It was therefore of interest to us to investigate the potential interactive role for SP and other related neuropeptides and neurotransmitters with the inflammatory cytokine IL-1.

Interactions between SP and IL-1 were examined by monitoring proliferation of the BALB/3T3 mouse embryonic fibroblast cell line. These cells ordinarily exhibit density-dependent growth regulation unless specific stimuli are provided to drive the cells to overcome density-dependent growth regulation. Such signals are usually provided by platelet-derived growth factor (PDGF) as a competence factor and by epidermal growth factors (EGF) and/or fibroblast growth factors (FGF) as progression factors. In these studies, BALB/3T3 cells were used as an in *vitro* paradigm of synovial hyperplasia due to the action of IL-1 and inflammatory neuropeptides.

IL-1 alone has been shown to be a potent stimulant of DNA turnover in these and similar cells,[78-80] as has SP.[2] However, the combination of IL-1 and SP not only engendered an augmentation of DNA turnover, but also provided the necessary signals to initiate cell multiplication. This is illustrated in Figure 1. The potentiation of ^{3}H-thymidine incorporation was approximately twofold above the sum of the individual contributions by the IL-1 and the SP. This augmentation was observed with both IL-1_α and IL-1_β.[5]

IL-1 Potentiation by Brandykinin

BK shares the vasodilatory and bronchostrictive properties of SP and can gain access to the joint, for example, as a result of plasma extravasation via the action of SP. When tested on BALB/3T3 fibroblasts, BK was found to be a potent stimulant of radioactive thymidine incorporation. More importantly, BK consistently caused

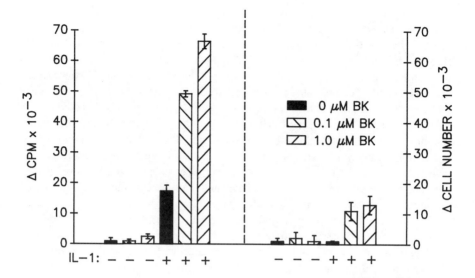

Figure 2. Synergism between IL-1_α and BK on BALB/3T3 fibroblast proliferation. ^{3}H-thymidine incorporation is indicated on the the left, and cell multiplication is indicated on the right.[5]

significant and dramatic increases in IL-1_α- and IL-1β-induced thymidine turnover, as shown in Figure 2 for IL-1_α and BK. The combination of IL-1 and BK also resulted in a net gain in cell numbers (approximately a 40 to 75% increase over cells stimulated with either IL-1 or BK separately).[5] A mechanism to explain synergy between BK and IL-1 may be derived from a report by Burch et al.[12] The authors of this study demonstrated that IL-1 treatment of BALB/3T3 fibroblasts resulted in an enhancement of BK-induced PGE$_2$ production. This might have come about due to the fact that both BK and IL-1 increase GTPase activity, which, as the authors suggest, is associated with G protein-coupled receptor reactions and cellular activation processes.[13] This mechanism may also be responsible for the enhanced mitotic rate observed for fibroblasts treated with IL-1 and BK. A recent report[59] revealed the interaction of BK with colony-stimulating factor 1 (CSF-1). In this study, however, instead of a potentiated response, BK caused a decrease in the responsiveness of adherent cells to proliferative signals supplied by CSF-1. The authors speculate that this might be due to the release of antiproliferative PGE$_2$. Prostaglandin release is strongly mediated by BK and may be part of the algesic principle of BK. This is also consistent with other findings that BK may either promote or potentiate the release of cytokine mediators of inflammation (i.e., IL-1, TNF) (see below).[60,61]

Interactions of Cytokines with Histamine and Serotonin

An indirect activity of BK is suggested to be the release of the inflammatory mediators histamine and 5-HT.[17] 5-HT is often released simultaneously with SP.[81]

SP is known to stimulate histamine release,[31-33,81] and synoviocytes[14] and chondrocytes[82] are known to possess functional histamine receptors. Therefore, two additional interactions we examined were those between IL-1 and 5-HT and between IL-1 and histamine. We found that 5-HT appeared to have no effect on fibroblast proliferation and also failed to potentiate IL-1-induced proliferation.[5] However, 5-HT was recently demonstrated to promote fibroblast growth factor (FGF)-induced DNA synthesis, adenylate cyclase production, and phospholipase C activation in quiescent Chinese hamster lung (CHL) fibroblasts.[84] This therefore suggests that 5-HT may not necessarily be involved in IL-1-related neurogenic proliferative processes, but it may yet be an important participant in a type of neurogenic inflammation culminating in other cellular activation processes, particularly those that can promote the activation of second messengers and the synthesis and release of soluble mediators, as shown to be promoted by FGF.

Histamine, when tested with IL-1, showed only weak (45% above an additive response) potentiation of IL-1-induced BALB/3T3 proliferation, and only when 10 μM histamine was used.[5] While this concentration of histamine may at times be achieved in nasal passages,[85] its physiological significance in other compartments, namely the joint or the lung, may be more problematical. Nevertheless, IL-1 and lower histamine concentrations demonstrated an additive response, suggesting that histamine may interact with cytokines in the inflammatory lesions.

PROMOTION OF CYTOKINE RELEASE BY NEUROPEPTIDES

Reports that SP was capable of extending adjuvant-induced synovial hyperplasia if administered intra-articularly during the induction phase[7] suggested that one of its roles might be to promote the release of inflammatory mediators. Subsequent *in vitro* studies revealed that SP could activate macrophages to release PGE_2[4] and synoviocytes to release PGE_2 and collagenase.[3] However, these mediators are not reputed to provide signals that will lead to a sustained inflammatory reaction. Therefore, it was possible that other inflammatory mediators, namely cytokines, were also being released by the intra-articularly injected SP and that cytokines were partly responsible for the sustained increases in paw volume observed in the treated animals.

IL-1 is currently thought to play a major role in the pathogenesis of rheumatoid arthritis. This is because it is secreted by macrophages and synoviocytes, is a potent inducer of fibroblast activation, and is elevated in the synovial fluids[86-88] and sera[88] of patients with rheumatoid arthritis. As a result of some of those observations, we sought to determine if generation of IL-1 or IL-1-like activity could be promoted by SP.

Using a well-established murine macrophage cell line noted for its ability to secrete IL-1, P388D1,[89,90] we found that SP was able to induce secretion of IL-1, or an IL-1-like activity,[79] if present in the cell cultures for as little as 4 hr at concentrations between 30 and 100 ng/ml. Similar results were observed if the C-terminal octapeptide of SP, SP_{4-11}, was substituted. SP_{4-11} is a more potent agonist

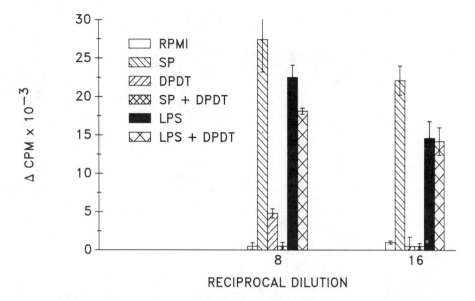

Figure 3. Stimulation of IL-1 production by SP, and specific inhibition by DPDT. Supernatants (1:8, 1:16) from P388D1 cells treated with SP (30 ng/ml), DPDT (100 ng/ml), LPS (300 ng/ml), or combinations as indicated.[6]

than SP,[31,81,91] and it appeared to be able to induce more IL-1 activity than did SP.[79] Similar orders of activity have been reported for SP and SP$_{4-11}$ on production of PGE$_2$ by guinea pig peritoneal macrophages.[4] These results are also consistent with earlier studies that demonstrated the ability of SP to augment phagocytosis by phagocytic cells[45,46] and to augment CSF-1-stimulated macrophage progenitor cell maturation.[42]

Although the nature of the SP receptors on macrophages and macrophage cell lines is still unknown, a synthetic analogue of SP, D-Pro^2-D-Trp^{7,9}-SP (DPDT), was examined for its ability to specifically inhibit SP-induced IL-1 secretion. DPDT, a specific antagonist of SP, is shown to be active *in vivo*.[92,93] When tested on P388D1 cells, it was found that DPDT had no significant activity as an IL-1 inducer, did not inhibit lipopolysaccharide (LPS)-induced IL-1 secretion, but did inhibit SP-induced IL-1 secretion. Therefore, one may presume that the IL-1 secreted by the SP-stimulated macrophages was probably due to specific interactions with a SP-like receptor.[79] These data are summarized in Figure 3.

Similar data were obtained from experiments with NK-A and NK-B. These two structurally related neurokinins were more potent IL-1 secretagogues than SP, but were slightly less potent than SP$_{4-11}$.[79] Table 4 shows comparative data for secreted IL-1 activity. The rank order for activity is not entirely consistent with relative receptor-binding activity of neurokinins to either SP-P (NK-1) or SP-E (NK-2) receptor subtypes, but is consistent with other studies pertaining to neurokinin receptor binding on guinea pig peritoneal macrophages.[4]

Table 4. IL-1 Production by Neurokinins

Inducer	Dose (ng/ml)	IL-1 Conc.[a] (U/ml ± SE)	n[b]
None		4.5 ± 5.4	8
SP	30	25.5 ± 11.7	8
SP	100	22.5 ± 9.5	3
SP_{4-11}	10	100 ± 1	3
NK-A	10	12.5	1
NK-A	30	28 ± 5	2
NK-A	100	53 ± 20	3
NK-B	10	12.5	1
NK-B	30	37.2 ± 24.8	2
NK-B	100	45.3 ± 33	2
LPS[c]	300	100	

[a] L-1 activity in P388D1 culture supernatants treated as indicated.
[b] Number of experiments.
[c] The activity in LPS treated cultures was arbitrarily set at 100 units per ml.

Those studies clearly demonstrated that the P388D1 macrophage cell line produces IL-1 or an IL-1-like substance in response to neurokinins, and by inference suggests that macrophages in the inflamed joint may respond in a similar way. Recent studies with human peripheral blood monocytes support those findings.[94] Thus, concentrations of SP ranging between 10 and 100 nM were able to optimally provoke IL-1 synthesis *in vitro*, and similar findings were reported for NK-A. In addition, human blood monocytes were shown to release TNF and IL-6.

TNF is a potent inflammatory mediator in the same way that IL-1 is, activates fibroblasts like IL-1 does, and synergizes with IL-1 to further promote fibroblast activation.[67] IL-6, a more recently described cytokine that is derived from monocytes and macrophages,[95,96] IL-1- and TNF-stimulated fibroblasts,[96] and T-cells,[97] is a potent acute phase promoter,[98] a B-cell activator,[99-102] and a promoter of T-cell growth and differentiation.[103] IL-6 is reported to have IL-1-like lymphocyte activating factor (LAF) activity on phytohemagglutinin (PHA) -stimulated thymocytes. However, there are significant differences between the activity of IL-1 and IL-6 in this system. IL-1 is active at concentrations less than 100 pg/ml. IL-6 concentrations greater than 1000-fold higher are required. IL-1 and IL-6 act synergistically, though, to promote thymocyte proliferation.[104] IL-6 is elevated in the synovial fluid and serum of patients with rheumatoid arthritis[105] and therefore may be equally as important as IL-1 in regulating and sustaining inflammatory processes.

A recent abstract[60] reported similar findings for bradykinin, whereby *in vitro* treatment of macrophages caused the induction of IL-1 and TNF secretion into the medium. Experiments in our laboratory are consistent with those findings and reveal that BK could potentiate LPS-stimulated IL-1 secretion (Figure 4) from elicited mouse peritoneal macrophages.

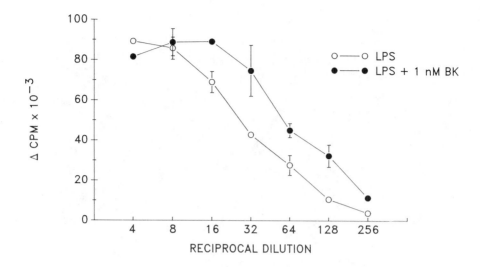

Figure 4. BK potentiation of IL-1 production by LPS-stimulated mouse adherent peritoneal exudate cells. 1×10^6 cells were cultured for 18 hr in medium containing 1 µg/ml LPS, with and without 1 nM BK. Cells were then washed and incubated for an additional 30 hr in serum-free medium prior to supernatant harvest for testing in a LAF assay.

CONCLUSION

The material in the foregoing chapter suggests that a cyclical mechanism of neuropeptide release and subsequent cytokine release occurs in arthritic disease and in possibly other chronic inflammatory diseases. The net result is an exacerbation of inflammation. A proposed model for the neurogenic inflammatory activity of SP and BK with IL-1 is shown in Figure 5. As shown, SP and BK may contribute to synovial cell activation directly and also to synovial hyperplasia as a result of their ability to amplify IL-1-induced fibroblast proliferation. The interactions between SP or BK with other cytokines, such as TNF, have yet to be tested. SP can also activate mast cells to secrete histamine, which may act on synovial fibroblasts, but in all probability is more likely to serve as a vasoactive amine promoting extravasation of serum proteins, such as complement, BK, and chemotactic factors. Circumstantial evidence showing the accumulation of mast cells adjacent to sites of joint erosion implicates mast cell products as chondrocyte and synoviocyte activators. Additional evidence suggests that IL-1 and histamine act synergistically to activate chondrocytes and that histamine can induce IL-1 production by synoviocytes.[106] Additional effects might occur through the action of PGE_2 secreted by SP-activated fibroblasts and macrophages. The latter cells can also secrete IL-1 and TNF, whose endothelial cell activating properties include chemotactic factor production. IL-6 produced either by macrophages or by fibroblasts can further activate rheumatoid factor-

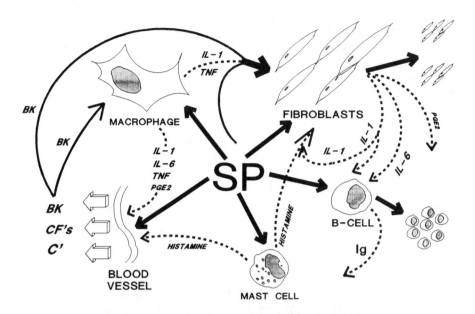

Figure 5. Model of cytokine involvement in neurogenic inflammation. BK, bradykinin; CFs, chemotactic factors; C', complement; Ig, immunoglobulin; PGE$_2$, prostaglandin E$_2$; SP, substance P. Broken arrows indicate secreted products.

producing B-cells. The latter can also be directly activated by SP. Finally, a recent abstract reports that intra-articular injection of either IL-1 or TNF results in the secretion of SP into the joint within 2 hr,[83] implying the existence of additional positive-feedback loops.

Synovial fluid contains a complex mixture of cytokines and growth factors whose individual effects have been studied. However, the properties of the unfractionated mixture still remain obscure. The report that FGF and 5-HT interact synergistically to activate second messenger pathways in fibroblasts[84] suggests that many more neuromodulator-immunomodulator interactions may exist. In addition, all three mammalian tachykinins[6] as well as bradykinin are capable of eliciting cytokine secretion (this report).[60,61] Therefore, neurogenic mechanisms may be responsible for sustaining the chronic inflammatory state via the release of IL-1 (and also TNF and IL-6) from infiltrating macrophages and from resident synoviocytes, as well as by amplifying the biologic activity of these cytokines thus secreted on cells already residing in the joint.

Neurogenic stimulation of immunologic mechanisms may play a significant role not only in the etiology of imflammatory disease, but also in autoimmune neuropathies, such as multiple sclerosis, or in cerebral scarring. Activation of astrocytes and microglial cells is thought to be associated with cerebral scarring, so it is important to note that glial cells produce an IL-1-like material.[75,76] SP has been shown to be able to activate astrocytes *in vitro*.[30] SP is also capable of activating immuno-

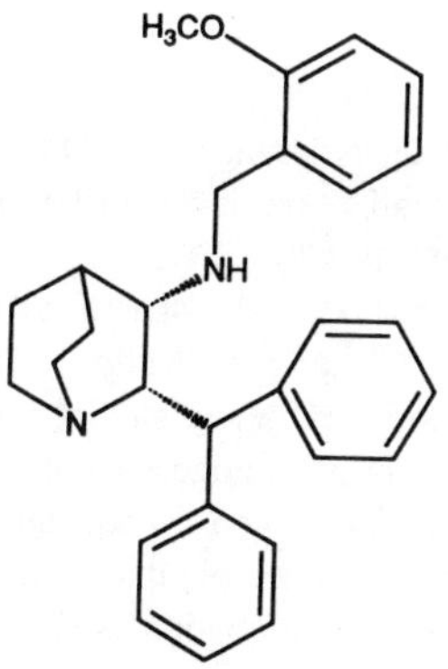

Figure 6. Structure of CP-96345, a nonpeptide substance P receptor antagonist.

globulin production by B-lymphocytes[41] and can provide stimulatory signals to T-helper cells.[24,41] Accordingly, the production of rheumatoid factor may be affected by SP or similarly acting neuropeptides. Thus, neurogenic stimulation and potentiation of immunologic mechanisms involved with autoimmune disease, inflammation, and immunologically mediated neuropathies could very well be a viable focus for therapeutic intervention.

ADDENDUM

Since the time when this chapter was initially submitted, efforts aimed at synthesizing highly potent nonpeptide antagonists of SP have been disclosed in the scientific literature. One compound, CP-96345,[107,108] (see Figure 6), is a stereoselective NK-1 receptor antagonist and has *in vitro* and *in vivo* pharmacological activity. It has a Kd of 0.22 ± 0.04 nM on guinea pig striatal membranes, and inhibits binding on labeled SP with an IC$_{50}$ of 3.4 ± 0.8 nM. In comparison, the peptide antagonist DPDT had an IC$_{50}$ of 2100 ± 480 nM and unlabeled SP had an IC$_{50}$ of 2.2 ± 0.3 nM. CP-96345 is also able to inhibit SP-mediated relaxation of contracted dog carotid artery *in vitro*, and administration of 3.4 mg/kg i.v. prevents or reduces SP-mediated salivation. The authors of those studies point out that "...CP-96345 may provide the means for determining the usefulness of an NK-1 receptor antagonist as therapy for disorders of the nervous system...".[108] As discussed above, this and compounds like this might also be applied to the treatment of nervous-system related inflammatory disorders.

REFERENCES

1. Hokfelt, T., J.O. Kellerth, G. Nilsson, and B. Pernow, Experimental immunohistochemical studies on the localization and distribution of substance P in cat primary sensory neurons. *Brain Res.*, 100:235, 1975.
2. Nilsson, J., A.M. von Euler, and C.J. Dalsgaard, Stimulation of connective tissue cell growth by substance P and substance K. *Nature*, 315:61, 1985.
3. Lotz, M., D.A. Carson, and J.H. Vaughn, Substance P activation of rheumatoid synoviocytyes: neural pathway in pathogenesis of arthritis. *Science*, 235:893, 1987.
4. Hartung, H.P., K. Wolters, and K. von Toyka, Substance P: binding properties and studies on cellular responses in guinea-pig macrophages. *J. Immunol.*, 136:3856, 1986.
5. Kimball, E.S. and M.C. Fisher, Potentiation of IL-1-induced Balb/3T3 fibroblast proliferation by neuropeptides. *J. Immunol.*, 141:4203, 1988.
6. Kimball, E.S., J. Vaught, and F.J. Persico, Substance P, neurokinin A, and neurokinin B induce generation of IL-1-like activity in P388D1 cells. Possible relevance to arthritic disease. *J. Immunol.*, 141:3564, 1988a.
7. Levine, J.D., R. Clark, M. Devor, C. Helms, M.A. Moskowitz, and A.I. Basbaum, Intraneuronal substance P contributes to the severity of experimental arthritis. *Science*, 226:547, 1984.
8. Foreman, J.C., Peptides and neurogenic inflammation. *Br. Med. Bull.*, 43:386, 1987.
9. Goldstein, R.H. and M. Wall, Activation of protein formation and cell division by bradykinin and des Arg^9 bradykinin. *J. Biol. Chem.*, 259:9263, 1984.
10. Goetzl, E.J., Ed., Neuromodulation of immunity and hypersensitivity. *J. Immunol.*, 135 (suppl.):739s, 1985.
11. Tissot, M., P. Pradelles, and J.P. Giroud, Substance-P-like levels in inflammatory exudates. *Inflammation*, 12:25, 1988.
12. Burch, R.M., J.R. Connor, and J. Axelrod Interleukin 1 amplifies receptor-mediated activation of phospholipase A_2 in 3T3 fibroblasts. *Proc. Natl. Acad. Sci. U.S.A.*, 85:6306, 1988a.
13. Burch, R.M. and J. Axelrod, Dissociation of bradykinin-induced prostaglandin formation from phosphatidylinositol turnover in Swiss 3T3 fibroblasts: evidence for G protein regulation of phospholipase A_2. *Proc. Natl. Acad. Sci. U.S.A.*, 84:6374, 1987.
14. Taylor, D.J. and D.E. Wooley, Evidence for both histamine H1 and H2 receptors on human articular chondrocytes. *Ann. Rheum. Dis.*, 46:431, 1987a.
15. Regoli, D., Neurohumoral regulation of precapillary vessels. the kallikrein-kinin system. *J. Cardiovasc. Pharmacol.*, 6 (suppl 2):401, 1984.
16. Regoli, D., G. Drapeau, S. Dion, and P. D'Orleans-Juste, Pharmacological receptors for substance P and neurokinins. *Life Sci.*, 40:109, 1987.
17. Regoli, D., Kinins. *Br. Med. Bull.*, 43:270, 1987.
18. Laufer, R., C. Gilon, M. Chorov and Z. Selinger, Characterization of a neurokinin B receptor site in rat brain using a highly selective radioligand. *J. Biol. Chem.*, 261:10257, 1986.
19. Torrens, Y., J.C. Beaujouan, M. Safroy, M.C. Daguet de Montety, L. Bergstrom, and J. Glowinski, Substance P receptors in primary cultures of cortical astrocytes from the mouse. *Proc. Natl. Acad. Sci. U.S.A.*, 83:9216, 1986.
20. Shults, C.W., S.H. Buck, E. Burdher, T.N. Chase, and T.L. O'Donohue, Dinstinct binding sites for substance P and neurokinin A (substance K): co-transmitters in rat brain. *Peptides*, 6:343, 1985.

21. Edvinsson, L. and R. Uddman, Immunohistochemical localization and dilatatory effect of substance P on human cerebral vessels. *Brain Res.*, 232:466, 1982.
22. Jacoby, H.I., Gastrointestinal tachykinin receptors. *Life Sci.*, 43:2203, 1988.
23. Payan, D.G., D.R. Brewster, A. Missirian-Bastian, and E.J. Goetzl, Substance P recognition by a subset of human T lymphocytes. *J. Clin. Invest.*, 74:1532, 1984a.
24. Payan, D.G., D.R. Brewster, and E.J. Goetzl, Specific stimulation of human T lymphocytes by substance P. *J. Immunol.*, 131:1613, 1983.
25. Payan, D.G., D.R. Brewster, and E.J. Goetzl, Stereospecific receptors for substance P on cultured human IM-9 lymphoblasts. *J. Immunol.*, 133:3260, 1984b.
26. Payan, D.G., J.P. McGillis, and E.J. Goetzl, Neuroimmunology. *Adv. Immunol.*, 39:299, 1986a.
27. McGillis, J.P., M.L. Organist, and D.G. Payan, Substance P and immunoregulation. *Fed. Proc.*, 46:196, 1987.
28. Payan, D.G., J.P. McGillis, and M.L. Organist, Binding characteristics and affininty labeling of protein consitituents of the human IM-9 lymphoblast receptor for substance P. *J. Biol. Chem.*, 261:14321, 1986b.
29. Von Euler, U.S. and J.H. Gaddum, An unidentified depressor substance in certain tissue extracts. *J. Physiol. (Lond.)*, 72:74, 1931.
30. Hartung, H.P., K. Heininger, B. Schafer, and K. von Toyka, Substance P and astrocytes: stimulation of the cyclooxygenase pathway of arachidonic acid metabolism. *FASEB J.*, 2:48, 1988.
31. Foreman, J.C., C.C. Jordan, P. Oehme, and H. Renner, Structure-activity relationships for some SP related peptides that cause wheal and flare reactions in human skin. *J. Physiol.*, 335:449, 1983.
32. Foreman, J.C. and C.C. Jordan, Histamine release and vascular changes induced by neuropeptides. *Agents Actions*, 13:233, 1983.
33. Hagermark, O., T. Hokfelt, and B. Pernow, Flare and itch induced by substance P in human skin. *J. Invest. Dermatol.*, 7:233, 1978.
34. Nawa, H., T. Hirose, H. Takashima, S. Inayama, and S. Nakanishi, Nucleotide sequences of cloned cDNAs for two types of bovine brain substance P precursors. *Nature*, 301:32, 1983.
35. Fisher, J., W.G. Forssman, T. Hokfelt, J.M. Lundberg, M. Reinecke, F.A. Tschopp, and Z. Wiesenfeld-Hallin, Immunoreactive calcitonin gene-related peptide and substance P: co-existence in sensory neurones and behavioral interaction after intrathecal administration in the rat. *J. Physiol.*, 362:92P, 1985.
36. Rosenfeld, M.G., J.J. Mermod, S.G. Amara, et al., Production of a novel neuropeptide encoded by the calcitonin gene via tissue-specific RNA processing. *Nature*, 304:129, 1983.
37. Brain, S.D., T.J. Williams, J.R. Tipins, H.R. Morris, and I. MacIntyre, Calcitonin gene-related peptide is a potent vasodilator. *Nature*, 313:54, 1985.
38. Piotrowski, W. and J.C. Foreman, Some effects of calcitonin gene-related peptide in human skin and on histamine release. *Br. J. Dermatol.*, 114:37, 1986.
39. Brain, S.D. and T.J. Williams, Inflammatory oedema induced by synergism between calcitonin gene-related peptide (CGRP) and mediators of increased vascular permeability. *Br. J. Pharmacol.*, 86:855, 1985.
40. Celander, O. and B. Folkow, The nature and the distribution of afferent fibres provided with the axon-reflex arrangement. *Acta Physiol. Scand.*, 29:359, 1953.

41. Stanisz, A.M., D. Befus, and J. Bienenstock, Differential effects of vasoactive intestinal peptide, substance P, and somatostatin on immunoglobulin synthesis and proliferation by lymphocytes from Peyer's patches, mesenteric lymph nodes and spleen. *J. Immunol.,* 136:152, 1986.
42. Moore, R.A., A.P. Osmand, J.A. Dunn, J.G. Joshi, and B.T. Rouse, Substance P augmentation of CSF-1-stimulated in vitro myelopoiesis. A two-signal progenitor-restricted, tuftsin-like effect. *J. Immunol.,* 141:2699, 1988.
43. Moore, R.N., J.G. Joshi, D.G. Deana, F.J. Pitruzzello, D.W. Horohov, and B.T. Rouse, Characterization of a two-signal-dependent, Ia$^+$ mononuclear phagocyte progenitor subpopulation that is sensitive to inhibition by ferritin. *J. Immunol.,* 136:1605, 1986.
44. Fridkin, M., Y. Stabinsky, V. Zakuth, and Z. Spirer, Tuftsin and some analogs. Synthesis and interaction with human polymorphonuclear leukocytes. *Biochim. Biophys. Acta,* 496:203, 1977.
45. Bar-Shavit, Z., R. Goldman, Y. Stabinsky, P. Gottlieb, M. Fridkin, V.I. Teichberg, and S. Blumberg, Enhancement of phagocytosis — a newly found activity of substance P residing in its N-terminal tetrapeptide sequence. *Biochem. Biophys. Res. Commun.,* 94:1445, 1980.
46. Goldman, R. and Z. Bar-Shavit, On the mechanism of the augmentation of the pagocytic capability of phagocytic cells by tuftsin, substance P, neurotensin, and kentsin and the interrelationship between their receptors. *Ann. N.Y. Acad. Sci.,* 491:143, 1983.
47. Najjar, V.A., Tuftsin, a natural activator of phagocyte cells: an overview. *Ann. N.Y. Acad. Sci.,* 419:1, 1983.
48. Fridkin, M. and P. Gottlieb, Tuftsin, Thr-Lys-Pro-Arg. Anatomy of an immunologically active peptide. *Mol. Cell. Biochem.,* 41:73, 1981.
49. Tzevohal, E., S. Segal., Y. Stabinsky, M. Fridkin, Z. Spirer, and M. Feldman, Tuftsin (an Ig-associated tetrapeptide) triggers the immunogenic function of macrophages: implications for activation of programmed cells. *Proc. Natl. Acad. Sci. U.S.A.,* 75:3400, 1978.
50. Marasco, W.A., H.J. Showell, and E.L. Becker, Substance P binds to the formylpeptide chemotaxis receptor on the rabbit neutrophil. *Biochem. Biophys. Res. Commun.,* 99:1065, 1981.
51. Serra, M.C., F. Bazzoni, V. Della Bianca, M. Greskowiak, and F. Rossi, Activation of human neutrophils by substance P. Effect on oxidative metabolism, exocytosis, cytosolic Ca^{2+} concentration and inositol phosphate formation. *J. Immunol.,* 141:2118, 1988.
52. Hartung, H.P. and K. von Toyka, Activation of macrophages by substance P: induction of oxidative burst and thromboxane release. *Eur. J. Pharmacol.,* 89:301, 1983.
53. Keele, C.A. and V. Eisen, Plasma kinin formation in rheumatoid arthritis. *Adv. Exp. Med. Biol.,* 8:471, 1970.
54. Regoli, D. and J. Barabe, Pharmacology of bradykinin and related kinins. *Pharmacol. Rev.,* 32:1, 1980.
55. Lerner, U.H., I.L. Jones, and G.T. Gustafson, Bradykinin, a new potential mediator of inflammation-induced bone resorption. *Arthr. Rheum.,* 30:530, 1987.
56. Meier, H.L., J.V. Pierce, R.W. Colman, and A.P. Kaplan, Activation and function of human Hageman factor: the role of high molecular weight kininogen and prekallikrein. *J. Clin Invest.,* 60:18, 1977.
57. Kaplan, A.P., The intrinsic coagulation, fibrinolytic, and kinin-forming pathways of man, in *Textbook of Rheumatology,* W.N. Kelley, E.D. Harris, Jr., S. Ruddy, and C.B. Sledge, Eds., W.B. Saunders, Philadelphia, 1981, 97.

58. Marceau, F., A. Lussier, D. Regoli, and P. Giroud, Pharmacology of kinins: their relevance to tissue injury and inflammation. *Gen. Pharmacol.*, 14:209, 1983.
59. Duma, E.M., J.S. Foster,and Moore, R.N, Bradykinin sensitization of CSF-1-responsive murine mononuclear phagocyte precursors to prostaglandin E. *J. Immunol.*, 141:3186, 1988.
60. Burch, R.M., J.M. Connor, and C. Tiffany, The kallikrein-kininogen-kinin system in chronic inflammation. Abstr. 79. 4th Int. Conf. Inflammation Research Association, 1988b.
61. Kimball, E.S. and C.R. Schneider, unpublished results.
62. Olsen, R., K. Santone, D. Melder, S.G. Oakes, R. Abraham, and G. Powls, An increase in intracellular free Ca^{+2} associated with serum-free growth stimulation of Swiss 3T3 fibroblasts by epidermal growth factor in the presence of bradykinin. *J. Biol. Chem.*, 263:18030, 1989.
63. Whorton, A.R., S.L. Young, J.L. Oata, A. Barchowsky, and R.S. Kent, Mechanism of bradykinin-stimulated prostaglandin synthesis in porcine aortic endothelial cells. *Biochim. Biophys. Acta,* 712:79, 1982.
64. Becherer, P.R., L.F. Mertz, and N.L. Baenziger, Regulation of prostaglandin synthesis mediated by thrombin and B2 bradykinin receptors in a fibrosarcoma cell line. *Cell,* 30:243, 1982.
65. Gustafson, G.T., O. Ljunggren, P. Boonekamp, and U. Lerner, Stimulation of bone resorption in cultured mouse calvaria by lys-bradykinin (kallidin), a potential mediator of bone resorption linking anaphylaxis processes to rarefying osteitis. *Bone Miner.*, 1:267, 1986.
66. Dinarello, C.A., Interleukin 1. *Rev. Infect. Dis.*, 6:51, 1984.
67. Dinarello, C.A., The biology of interleukin 1 and comparison to tumor necrosis factor. *Immunol. Lett.*, 16:227, 1988a.
68. Kimball, E.S. and S.C. Gilman, Relevance of monocyte/macrophage-derived cytokines to human health and disease, in *Immunopharmacology,* T. Rogers and S.C. Gilman, Eds., Telford Press, Caldwell, NJ, 1991, 253.
69. Di Giovione, F.S., J.A. Symons, J. Manson, and G.W. Duff, Soluble mediators of immunity: interleukins, in *Immunopathogenetic Mechanisms of Arthritis,* J.A. Goodacre and W.C. Dick, Eds., MTP Press, Lancaster, U.K., 1988, 101.
70. Allison, A.C., Immunopathogenetic mechanisms of arthritis and modes of action of antirheumatic therapies, in *Immunopathogenetic Mechanisms of Arthritis,* J.A. Goodacre and W.C. Dick, Eds., MTP Press, Lancaster, U.K., 1988, 211.
71. Doyle, M.V., L. Brindley, E. Kawasaki, and J. Larrick, High level human interleukin 1 production by a hepatoma cell line. *Biochem. Biophys. Res. Commun.*, 130:768, 1985.
72. Howells, G.L., D. Chantry, and M. Feldmann, Interleukin 1 (IL-1) and tumor necrosis factor synergise in the induction of IL-1 synthesis by human vascular endothelial cells. *Immunol. Lett.*, 19:169, 1988.
73. Libby, P., J.M. Ordovas, K.R. Auger, A.H. Robbins, L.K. Birinyi, and C.A. Dinarello, Endotoxin and tumor necrosis factor induce interleukin-1 beta gene expression in adult human vascular endothelial cells. *Am. J. Pathol.*, 124:179, 1986.
74. Warner, S.J.C., K.R. Auger, and P. Libby, Interleukin-1 induces interleukin-1. II. Recombinant human interleukin-1 induces interleukin-1 production by adult human vascular endothelial cells. *J. Immunol.*, 139:1911, 1987.
75. Fontana, A., F. Kristenson, R. Dubs, D. Gemsa, and E. Weber, Production of prostaglandin E and an interleukin-1-like factor by cultured astrocytes and C6 glioma cells. *J. Immunol.*, 129:2413, 1982.

76. Fontana, A. and P.J. Grob, Astrocyte-derived interleukin-1-like factors. *Lymphokine Res.*, 3:11, 1984.
77. Dinarello, C.A., Biology of interleukin 1. *FASEB J.*, 2:108, 1988b.
78. Schmidt, J.A., S.B. Mizel, D. Cohen and I. Green, Interleukin 1, a potential regulator of fibroblast proliferation. *J. Immunol.*, 128:2177, 1982.
79. Kimball, E.S., M.C. Fisher, and F.J. Persico, Potentiation of Balb/3t3 fibroblast proliferative response by IL-1 and EGF. *Cell. Immunol.*, 113:341, 1988b.
80. Dukovich, M., J.M. Severin, S.J. White, S. Yamazaki, and S.B. Mizel, Stimulation of fibroblast proliferation and prostaglandin production by purified recombinant interleukin 1. *Clin. Immunol. Immunopathol.*, 38:381, 1986.
81. Pernow, B., Substance P. *Pharmacol. Rev.*, 35:85, 1983.
82. Taylor, D.J. and D.E. Woolley, Histamine H1 receptors on adherent rheumatoid synovial cells in culture: demonstration by radioligand binding and inhibition of histamine-stimulated prostaglandin E production by histamine H1 antagonists. *Ann. Rheum. Dis.*, 46:425, 1987b.
83. O'Byrne, E.M., V. Blancussi, D. Wilson, J. Whalen, A. Rubin, M. Wong, M.D. Erion, J.P. Simke, J.P. Gilligan, and A.Y. Jeng, Increased levels of substance P (SP) in rabbit joints following intra-articular injection of interleukin-1 (IL-1) and tumor necrosis factor (TNF), Abstr. 336, Proc. 3rd Interscience World Conference on Inflammation, 1989.
84. Seuwen, K., I. Magnaldo, and J. Pouyssegur, Serotonin stimulates DNA synthesis in fibroblasts acting through 5-HT$_{1B}$ receptors coupled to a G$_i$-protein. *Nature*, 335:254, 1988.
85. Naclerio, R.M., The pathophysiology of allergic rhinitis: impact of therapeutic intervention. *J. Allergy Clin. Immunol.*, 82:927, 1988.
86. Nouri, A.M.E., G.S. Panayi, and S.M. Goodman, Cytokines and the chronic inflammation of rheumatic disease. I. The presence of interleukin 1 in synovial fluid. *Clin. Exp. Immunol.*, 55:295, 1984.
87. Wood, D.D., E.J. Ihrie, C.A. Dinarello, and P.L. Cohen, Isolation of an interleukin-1 like factor from human joint effusions. *Arthr. Rheum.*, 26:975, 1983.
88. Eastgate, J.A., N.C. Wood, F.S. Di Giovine, J.A. Symons, F.M. Grinlinton, and G.W. Duff, Correlation of plasma interleukin 1 levels with disease activity in rheumatoid arthritis. *Lancet*, Sept. 24:706, 1988.
89. Mizel, S.B., J.J. Oppenheim, and D.L. Rosenstreich, Characterization of lymphocyte activating factor (LAF) produced by the macrophage cell line, P388D1. I. Enhancement of LAF production by activated T lymphocytes. *J. Immunol.*, 120:1497, 1978a.
90. Mizel, S.B., J.J. Oppenheim, and D.L. Rosenstreich, Characterization of lymphocyte activating factor (LAF) produced by the macrophage cell line, P388D1. II. Biochemical characterization of LAF induced by activated T cells and LPS. *J. Immunol.*, 120:1504, 1978b.
91. Dutta, A.S., Agonists and antagonists of substance P. *Drugs Future*, 12:781, 1987.
92. Holmdahl, G., R. Hakanson, S. Leander, S. Rosell, K. Folkers, and F. Sundler, A SP antagonist, [D-Pro^2, D-Trp^{7,9}]-SP, inhibits inflammatory responses in the rabbit eye. *Science*, 214:1029, 1981.
93. Mandahl, A. and A. Bill, In the eye [D-Pro^2, D-Trp^{7,9}]-SP is a SP agonist, which modifies the response to SP, PGE$_1$ and antidromic trigeminal nerve stimulation. *Acta Physiol. Scand.*, 117:139, 1981.
94. Lotz, M., J.H. Vaughn, and D. Carson, Effect of neuropeptides on production of inflammatory cytokines by human monocytes. *Science*, 241:1218, 1988.

95. Nordan, R.P. and M. Potter, A macrophage-derived factor required by plasmacytomas for survival and proliferation in vitro. *Science,* 233:566, 1986.
96. May, L.T., J. Ghrayeb, U. Santhanam, S.B. Tatter, Z. Stohoeger, D.C. Helfgott, N. Chiorazzi, G. Grieninger, and P.B. Sehgal, Synthesis and secretion of multiple forms of beta$_2$-interferon/B-cell differentiation factor 2/ hepatocyte-stimulating factor by human fibroblasts and monocytes. *J. Biol. Chem.,* 263:7760, 1988.
97. Van Snick, J., S. Cayphas, A. Vink, C. Uyttenhove, P.G. Coulie, and R.J. Simpson, Purification and NH$_2$-terminal amino acid sequence of a new T-cell-derived lymphokine with growth factor activity for B cell hybridomas. *Proc. Natl. Acad. Sci. U.S.A.,* 83:9679, 1986.
98. Gauldie, J., C. Richards, D. Harnish, P. Lansdorp, and H. Baumann, Interferon beta$_2$/B cell stimulatory factor 2 shares identity with monocyte-derived hepatocyte-stimulating factor and regulates the major acute phase protein response in liver cells. *Proc. Natl. Acad. Sci. U.S.A.,* 84:7251, 1987.
99. Poupart, P., P. Vandenabeele, S. Cayphas, J. Van Snick, G. Haegeman, V. Kruys, W. Fiers, and J. Content, B cell growth modulating and differentiating activities of recombinant human 26-kD protein. *EMBO J.,* 6:1219, 1987.
100. Van Damme, J., G. Opdenakker, R.J. Simpson, M.R. Rubira, S. Cayphas, A. Vink, A. Billiau, and J. Van Snick, Identification of the human 26-kD protein, interferon beta$_2$ (IFN-beta$_2$), as a B cell hybridoma/plasmacytoma growth factor induced by interleukin 1 and tumor necrosis factor. *J. Exp. Med.,* 165:914, 1987.
101. Hirano, T., T. Taga, N. Nakano, K. Yasukawa, S. Kashiwamura, K. Shimizu, K. Nakajima, K.H. Pyun, and T. Kishimoto, Purification to homogeneity and characterization of human B-cell differentiation factor (BCDF or BSFp-2). *Proc. Natl. Acad. Sci. U.S.A.,* 82:5490, 1985.
102. Hirano, T., K. Yasukawa, H. Harada, T. Taga, Y. Watanabe, T. Matsuda, S. Kashiwamura, K. Nakajima, K. Koyama, A. Iwamatsu, S. Tsunasawa, F. Sakiyama, H. Matsui, Y. Takahara, T. Taniguchi, and T. Kishimoto, Complementary DNA for a novel human interleukin (BSF-2) that induces B lymphocytes to produce immunoglobulin. *Nature,* 324:73, 1986.
103. Van Snick, J.A., A. Vink, S. Cayphas,and C. Uyttenhove, Interleukin-HP1, a T cell-derived hybridoma growth factor that supports the in vitro growth of murine plasmacytomas. *J. Exp. Med.,* 165:641, 1987.
104. Elias, J.A., G. Trinchiere, J.M. Beck. P.L. Simon, P.B. Sehgal., L.T. May, and J.A. Kern, A synergistic interaction of IL-6 and IL-1 mediates the thymocyte-stimulating activity produced by recombinant IL-1-stimulated fibroblasts. *J. Immunol.,* 142:509, 1989.
105. Houssiau, F.A., J.-P. Devogelaer, J. Van Damme, C. Nagant de Deuxchaisnes, and J. Van Snick, Interleukin-6 in synovial fluid and serum of patients with rheumatoid arthritis and other inflammatory arthritides. *Arthr. Rheum.,* 31:784, 1988.
106. Wooley, D.E., J.S. Bartholomew, D.J. Taylor, and J.M. Evanson, Mast cells and rheumatoid arthritis, Abstr. 201. Proc. 3rd Interscience World Conf. Inflammation, 1989.
107. Snider, R.M., J.W. Constantine, J.A. Lowe III, et al., A potent nonpeptide antagonist of the substance P (NK1) receptor. *Science,* 251:435, 1991.
108. McLean, S., A.H. Ganong, T.F. Seeger, et al., Activity and distribution of binding sites in brain of a nonpeptide substance P (NK1) receptor antagonist. *Science,* 251:437, 1991.

8. The Role of Cytokines in Acute Pulmonary Vascular Endothelial Injury

SIMEON E. GOLDBLUM

INTRODUCTION

Dissimilar clinical settings, including sepsis, shock, and trauma, can provoke acute pulmonary vascular endothelial injury or the adult respiratory distress syndrome (ARDS).[1-3] A common thread among these seemingly disparate insults to the host is the acute phase response to stress.[4] Multiple components of the acute phase response are mediated by the two cytokines, interleukin-1 (IL-1)[4,5] and tumor necrosis factor/cachectin (TNFα).[6,7] Experimental models of acute lung injury have been developed to study the pathogenesis of ARDS and include endotoxin[8] and thrombin[9,10] in sheep, complement activation in rats,[11] as well as administration of platelet activating factor (PAF),[12-14] phorbol myristate acetate (PMA),[15-18] and arachidonate metabolites[19-21] to rabbits. Again, these same two endogenous mediators, IL-1 and TNFα, could be central to all these diverse processes that lead to pulmonary vascular endothelial injury (Table 1). Endotoxin,[5,22,23] thrombin,[24] C5 cleavage products,[25,26] and PMA[23] can all induce intracellular synthesis and release of IL-1. In fact, endotoxin[24,27-30] and thrombin[24] have been shown to induce IL-1 production at the endothelial surface. On the other hand, IL-1 has been shown to induce endothelial biosynthesis of both arachidonate metabolites[31-33] and PAF.[34] Similarly, endotoxin,[35-39] C5 cleavage products,[40] and PMA[41] can each induce TNFα production, which, in turn, can induce biosynthesis of both arachidonate metabolites[42-46] and PAF.[47,48] Therefore, established inducers of experimental lung injury together with known clinical antecedents to ARDS all incriminate IL-1 and/or TNFα in the pathogenesis of acute lung injury.

POSTULATED SCHEMA OF GRANULOCYTE-DEPENDENT PULMONARY VASCULAR ENDOTHELIAL INJURY

Postulated mechanisms of acute lung injury involve the generation of endogenous mediators that can sequester and activate circulating granulocytes within the pulmonary microvasculature. These mediators have included C5 cleavage products,[11] arachidonate metabolites,[19-21] and PAF.[12-14] Granulocyte sequestration within the lung strategically positions these cells in close proximity to the pulmonary microvascular endothelial surface.[49,50] Here the granulocytes can be stimulated to generate and release endothelium-damaging substances that compromise the integrity of the alveolar-capillary barrier, resulting in pulmonary edema.[11,16,17,51-55] That pulmonary leukostasis is essential to the subsequent development of acute lung injury has been suggested in experimental systems where granulocyte depletion has prevented increased pulmonary vascular permeability.[9,11,16,56,57] Some investigators have explained granulocyte-mediated pulmonary vascular endothelial damage through granulocyte-derived toxic oxygen intermediates.[11,16,17,51-53] Enzymes and/or scavengers directed against these toxic oxygen radicals protect against insults to the lung.[11,58-60] Granulocyte release of proteases can also induce endothelial injury.[54,55] However, granulocyte-depletion has not been consistently protective against all types of experimental acute lung injury.[61-65] In fact, ARDS has been reported in profoundly neutropenic patients.[66-69] Therefore, some aspects or subgroups of acute lung injury and ARDS may not be granulocyte dependent. In the context of this schema for granulocyte-dependent pulmonary vascular endothelial injury, IL-1 and TNFα exert biological effects upon two highly relevant target tissues, circulating granulocytes and the endothelium, as well as modulate granulocyte-to-endothelial surface interactions. These two cytokines share biological properties that implicate them as candidate mediators of acute lung injury. These bioactivities include their abilities to induce host tissue generation of arachidonate metabolites,[31-33,42-46,70-80] PAF,[34,47] and toxic oxygen radicals,[81-85] all relevant to the pathogenesis of acute lung injury (Figure 1).

CYTOKINE-INDUCED PULMONARY LEUKOSTASIS

The cytokines IL-1 and TNFα exert direct effects on granulocytes[81,83,86-88] (Table 2) and endothelial cells[88-93] (Table 3) *in vitro* that may explain granulocyte-to-endothelial surface interaction *in vivo*. Preexposure of granulocytes to IL-1 *in vitro* does not augment their adherence to untreated endothelial monolayers,[89] whereas rTNFα exposure rapidly increases granulocyte adherence to endothelial monolayers[88] as well as to nylon fiber columns.[83] The granulocyte adherence to endothelial monolayers is maximal after a 5 min TNFα exposure; it is not blocked by either protein synthesis inhibition by cycloheximide or mRNA synthesis inhibition by actinomycin D. This increased adherence can be blocked with the monoclonal antibody (MoAb) 60.3,[88,92] which recognizes an epitope of the neutrophil surface glycoprotein complex, CDw18, essential to optimal adherence. In fact, rTNFα has been shown to rapidly induce granulocyte surface expression of adhesion molecules CDw18, which is the receptor for iC3b (CR3).[79,86-88] These direct effects of cytokines on granulocyte adherence to endothelial surfaces may explain the rapidly appearing margination and granulocytopenia that follows intravenous infusion of cytokine.[94-97]

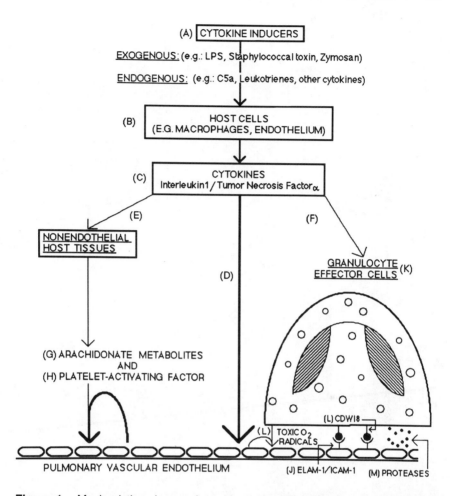

Figure 1. Mechanistic schema of cytokine-induced pulmonary vascular endothelial injury. Numerous exogenous and endogenous stimuli (A) induce host cell (B) production of cytokines (C), including IL-I and TNFα. These two cytokines can directly (D) or indirectly (E or F) provoke pulmonary vascular endothelial injury. IL-I and TNFα can directly (D) bind to receptors on the endothelial cell surface stimulating arachidonate metabolism (G), PAF biosynthesis (H), and generation of toxic oxygen intermediates (I). IL-I and TNFα also induce surface expression of endothelial leukocyte adhesion molecules (J) (ELAM-I and ICAM-I), rendering the endothelial surface hyperadhesive for granulocytes. These two cytokines can also indirectly (E) provide increased arachidonate metabolites (G) and PAF (H) to the lung by stimulating their biosynthesis in numerous extrapulmonary and/or pulmonary nonendothelial tissues. The cytokines can also indirectly attack the endothelial barrier through granulocyte effector cells (K). IL-I and TNFα are both neutrophil secretagogues, and TNFα augments granulocyte adherence via the CDw18 complex (L) and increases granulocyte generation of toxic O_2^- (I). The hyperadherent granulocytes release toxic oxygen radicals (I) and proteases (M), which may injure the pulmonary vascular endothelium.

Table 1. Established Inducers of Both Lung Injury and Cytokines

Inducer	Animal Model	Reference	Target Cell/Species	Cytokine	Reference
Endotoxin	Rats	193	Monocyte/macrophage	IL-1	22,23,127
	Sheep	56	Endothelium	IL-1	24,27-30
			Mice	TNF	35,37
			Rabbit	TNF	36,39
			Dog (alveolar macrophages)	IL-1, TNF	220, 220
			Human	IL-1	216
				TNF	38,216,219
Phorbyl Myristate acetate (PMA)	Rabbits	15-18	Monocyte/macrophage	IL-1	23
				TNF	41
Thrombin	Sheep	9,10	Endothelium	IL-1	24
C5 cleavage products	Mice	60	Macrophage	IL-1	25,26
	Rats	11,58		TNF	40
	Rabbits	191,192			

Table 2. Effects of IL-1 and TNF on Granulocytes

Effects	IL-1	Reference	TNF	Reference
Myelopoiesis (G-CSF/GM-CSF)	+	132-140	+	130,131,140
Early neutropenia	+	94-96	+	116,144,196,255
Neutrophilia	+	94,95,144	+	141-144,255
Adherence *in vitro*	+	83	+	83,88
	−	89		
Pulmonary leukostasis	+	94-97,116	+	97,116,117,196
Chemokinesis	−	152	NR	
Chemotaxis *in vitro*	+	*	+	157
	−	83,152	−	83,149
Recruitment *in vivo*	+	109-115	+	110,114,115
Surface expression of receptors	+	82†	+	81,86,87,148‡
Phagocytosis	+	NR	+	81,86,150,155
	−	152	−	NR
Degranulation	+	145,146,151	+	81,83
	−	83,152	NR	
PAF synthesis	NR		+	48
Oxygen-dependent metabolism	+	83,85	+	81-83,86,149,150
	−	152		154
Cytotoxicty	NR		+	86,149,150,155 156

NR = not reported.
* = early reports of *in vitro* chemotactic activity of IL-1 for granulocytes not included.
† = receptors for fMLP.
‡ = receptors for C3b, iC3b, and fMLP.

In addition to exerting direct biological effects on circulating myeloid cells, rIL-1 or rTNFα pretreatments can render endothelial monolayers hyperadhesive for untreated granulocytes in a dose- and time-dependent manner[88-93] (Table 3). A cytokine exposure time as short as 5 min[88] followed by a 1- to 2-hr lag time[89,90] is required before increased adhesiveness is evident. The endothelial hyperadhesiveness is temperature-dependent, clearly demonstrable at 37°C, but not at 22°C or 4°C.[91] IL-1-treated endothelial monolayers maintain adhesiveness after fixation with paraformaldehyde.[89] The process can be blocked by either protein or mRNA synthesis inhibition, but not by cyclooxygenase inhibition.[88,89,91,98] The supernatants from IL-1-treated endothelium do not augment adhesiveness of fresh endothelium,[90] and subsequent mild trypsin treatment of IL-1-treated endothelium abolishes its hyperadhesive state.[89] Preincubation of granulocytes with MoAb 60.3 blocks granulocyte adherence to cytokine-treated endothelium,[88,92] whereas pretreatment of stimulated endothelium with the same antibody has no effect. Cytokines induce endothelial surface expression of an antigen, the endothelial-leukocyte adhesion molecule 1 (E-LAM 1), that can be recognized by MoAbs H4/18 and H18/7.[98-103] In preliminary studies, intact MoAbs H4/18 and H18/7 and F(ab')$_2$

Table 3. Interactions of IL-I and TNF with Endothelium

Effects	IL-I	References	TNF	References
Receptors	+	158	+	159
Morphological changes	+	160,163	+	161,162,164
Cytotoxicity	-	89,91	+	162,164
			-	160-163
Superoxide anion production	+	84	NR	
Arachidonate metabolism	+	31-33	+	46
PAF synthesis	+	34	+	48
Procoagulant	+	104-107,166	+	107,108
Protein C	-	106	-	108
Plasminogen activator inhibitor	+	166-169,171	+	171
Hyperadhesiveness for granulocytes	+	89-92	+	88,92,93
Endothelial-leukocyte adhesion molecule 1	+	98,100-103	+	98,100-103
ICAM-I	+	100	+	100-102
IL-I	NR		+	27-29
GM-CSF	+	132,134,135,137	+	131
PDGF	NR		+	178

NR = not reported.

fragments partially inhibited cytokine-induced endothelial hyperadhesiveness for HL-60 promyelocytic cells and neutrophils.[98,103] Although cytokines also augment endothelial surface procoagulant activity,[104-108] hiruden, a potent thrombin antagonist, did not inhibit cytokine-induced adhesiveness for granulocytes.[89] In summary, these studies suggest that IL-l and TNFα interact with endothelial receptors to stimulate endothelial synthesis and expression of surface protein(s) that recognize granulocyte surface adhesion molecules (CDw18 complex) and enhance endothelial adhesiveness for granulocytes. The initial cytokine stimulation is energy dependent but once the antigen is expressed on the endothelial cell surface, endothelial cell metabolism is not required. Neither thrombin nor cyclooxygenase metabolites were essential to the hyperadhesive state. These cytokine effects on either circulating granulocytes and/or the microvascular endothelial surface may promote tissue leukostasis *in vivo*.

Several studies have demonstrated that cytokine administration can recruit granulocytes *in vivo*[109-113] (Table 2). Granstein et al. found that injection of murine rIL-l into the footpads of mice caused an influx of neutrophils with intravascular margination within 1 hr, with peak accumulation at 4 hr.[111] Sayers et al. showed that intraperitoneal administration of either human rIL-l or rTNFα to mice resulted in a rapid neutrophil influx into the peritoneal cavity.[114] Several investigators have shown that intradermal injection of IL-l into rabbits induced accumulation of ^{51}Cr-labeled neutrophils at the injection site.[109,110,112,113,115] They also found that IL-l could serve as either the preparatory or provocative stimulus for a local Schwartzman reaction. Using a similar rabbit model, Movat demonstrated that rTNFα can also mobilize granulocytes to the site of administration in rabbits.[110,115] He also found that IL-l-induced neutrophil accumulation could be blocked by coadministration of either cycloheximide or actinomycin D, and that IL-1 and TNFα could synergistically induce local tissue leukostasis.

More pertinent to the pathogenesis of granulocyte-dependent pulmonary vascular endothelial injury, cytokines have been shown to generate pulmonary leukostasis[94-97,116,117] (Table 2). IL-1 and TNFα might contribute to the sequestration of granulocytes within the pulmonary microvasculature either through (1) a local intrapulmonary production or (2) systemically increased circulating levels of these same mediators. Bochner et al. have demonstrated that cultured human lung fragments release a factor(s) that augments granulocyte adherence to cultured endothelial monolayers that can be blocked by anti-IL-1 antibody.[118] Alveolar macrophages from mice,[119] rats,[120,121] rabbits,[122-124] and humans[125-127] have been shown to produce and release IL-1. In a rat model of monocrotaline-induced lung injury, Gillespie et al. have demonstrated that bronchoalveolar lavage fluid (BALF) IL-1 bioactivity serially increased coincident with increasing lung granulocyte burden as measured by myeloperoxidase (MPO) activity per gram lung tissue and BALF leukocyte counts.[128] In response to a pulmonary insult, alveolar macrophages and/or other resident lung cells may elaborate cytokines that can induce a hyperadhesive pulmonary vascular endothelial surface that could then trap granulocyte traffic within the lung. These cytokines could also indirectly generate pulmonary leukostasis by stimulating alveolar macrophage production of the recently described monocyte-derived neutrophil chemotactic factor (MDNCF).[129]

Any extrapulmonary process that introduces cytokines into the venous bloodstream also may provide conditions for pulmonary leukostasis. The first microvascular bed to be traversed by the systemic venous circulation is the lung. Here, cytokines could recognize and bind to endothelial receptors, inducing an endothelial hyperadhesive state, again trapping circulating granulocytes. To determine whether cytokines could induce granulocyte sequestration within the pulmonary microvasculature, we intravenously infused into rabbits a single bolus of a murine P388D$_1$ macrophage-derived monokine preparation and studied the animals for pulmonary leukostasis.[94] The monokine preparation had a 10,000 to 30,000 molecular weight range and was standardized on the basis of IL-1 bioactivity as measured by the standard murine thymocyte comitogen assay and contained neither TNFα nor C5 bioactivities. A single 50- or 100-unit bolus infusion into rabbits induced profound, but transient, circulating granulocytopenia and marked pulmonary leukostasis.[94] The monokine preparation produced significant increases in both histological (mean alveolar septal wall granulocytes per high power field) and biochemical (mean myeloperoxidase activity per gram lung tissue) markers of tissue leukostasis, which were sustained for at least 24 hr. Because the active fraction of this monokine preparation was compatible with IL-1, we studied the effects of purified human rIL-1β on pulmonary leukostasis in the same rabbit model.[95] The rIL-1β (200 or 500 units) had a dose-dependent effect on both circulating granulocytopenia and pulmonary tissue leukostasis.

TNFα can also induce pulmonary leukostasis[97,116,117] (Table 2). Tracey et al. reported pulmonary intravascular granulocyte thrombi in rTNFα-challenged rats (1.8 mg/kg BW).[117] Stephens et al. found that rTNFα infusion into guinea pigs (0.14 mg/kg BW) induced profound neutropenia and increased neutrophils per alveolus on lung histology.[97] They failed to demonstrate increased recovery of neutrophils in BALF. Okusawa et al. have demonstrated that coadministration of rIL-1 (1 μg/kg) and rTNFα (1 μg/kg) followed by constant infusion of each cytokine (at 5 ng/kg/min) synergistically induces pulmonary leukostasis in rabbits.[116] Although these studies demonstrate that cytokines can induce pulmonary leukostasis at the expense of the circulating granulocyte pool, it is not clear whether pulmonary leukostasis is a prerequisite lesion to cytokine-induced lung injury.

EFFECT OF CYTOKINES ON GRANULOCYTES

IL-1 and TNFα exert both direct and indirect biological effects on the myeloid cell lineage from early progenitor cells to mature neutrophils (Table 2). IL-1 and TNFα have been shown to stimulate production of colony-stimulating activities, including G-CSF and GM-CSF, in fibroblasts, endothelium and stromal cells *in vitro*[130-139] and in mice *in vivo*.[140] Administration of either cytokine, IL-1 or TNFα, to experimental animals produces a rapid bone marrow release of mature neutrophils with resulting neutrophilia.[94,95,141-143] We have demonstrated up to 500% increases in circulating granulocytes after infusion of either murine P388D$_1$ cell-derived IL-1,[94] or human rIL-1β[95] into rabbits. We also have demonstrated ≥ 300% increase in blood neutrophils in rTNFα-challenged rabbits.[143] Investigators have confirmed this rTNFα-induced neutrophilia in mice[141,142] as well as in rats.[144]

For direct interaction with myeloid cells, several lines of experimental evidence suggest the presence of binding sites for these two cytokines on neutrophils. Smith et al. found that neutrophils pretreated with purified human monocyte-derived IL-1 became unresponsive to a subsequent IL-1 stimulus without cross-desensitization to *N*-formylmethionyleucylphenylalanine (fMLP), leukotriene B4 (LTB4), or 1-O-hexadecyl/octadecyl-2-O-acetyl-sn-glyceryl-3-phosphorylcholine (AGEPC).[145,146] These data suggested a specific neutrophil receptor for IL-1. Kahaleh et al. showed that neutrophils possess high affinity receptors for ^{125}I-labeled rIL-1α.[147] TNFα binding sites have also been demonstrated on myeloid cell lines,[148,149] and TNFα can induce monocyte differentiation of the human HL-60 promyelocytic cell line.[150]

IL-1 is reportedly a neutrophil secretagogue.[145,146,151.] Monocyte-derived IL-1 induces exocytosis of both azurophilic and specific granules. More recently, studies using pure rIL-1β could not demonstrate neutrophil degranulation of lysosomal enzymes.[152] These same studies also showed that rIL-1β failed to alter neutrophil intracellular free Ca^{++}, superoxide anion production, phagocytosis, chemokinesis, and chemotaxis. Sullivan et al. have confirmed the inability of IL-1 to stimulate either granulocyte release of lysozyme or chemotaxis.[83] However, they did find that rIL-1 stimulates neutrophil superoxide anion production. Furthermore, IL-1 primed neutrophils for increased superoxide generation in response to an fMLP stimulus. In other studies, both monocyte-derived and recombinant IL-1 stimulate histamine release from basophils and mast cells.[153] Thus, the effects of IL-1 upon neutrophil generation of superoxide anion and degranulation might require further clarification.

rTNFα binds to granulocyte receptors[154] and vigorously up regulates granulocyte metabolism and function *in vitro*[81-83,86,87,148,150,154-156] (Table 2). It rapidly induces neutrophil surface expression of receptors for C3b (CRl) and iC3b (CR3).[81,86,87] These cells display increased *in vitro* adherence[83,88] as well as increased phagocytosis of zymosan[81,86,150] and latex particles.[155] TNFα is reportedly chemotactic for monocytes and neutrophils *in vitro*.[157] TNFα is a neutrophil secretagogue affecting the specific granules.[81,87] TNFα also primes neutrophils for enhanced lysozyme release in response to fMLP.[83] This cytokine stimulates neutrophil synthesis and release of PAF.[48] TNFα enhances granulocyte-mediated inhibition of fungal growth[156] as well as antibody-dependent cell-mediated cytotoxicity (ADCC) for chicken erythrocytes[155] and P815Y murine mastocytoma tumor cells.[86,150] TNFα stimulates granulocyte production of superoxide anion[149,154] and, further, primes granulocytes for increased PMA-induced chemiluminescence[86,150] and fMLP-stimulated superoxide anion production.[82,83,86] Furthermore, rTNFα-activated granulocytes compromise the morphological integrity of human umbilical vein-derived endothelial cell monolayers.[149] Therefore, IL-1 and TNFα can increase bone marrow production and release of hyperadherent and metabolically activated granulocytes that may serve as effector cells for host target tissues, including the pulmonary microvascular endothelium.

EFFECTS OF CYTOKINES ON ENDOTHELIUM

IL-1 and TNFα also act directly on the endothelium (Table 3). Both cytokines recognize and bind to specific endothelial cell receptors[158,159] and induce morphological changes in cultured endothelial cells that include cellular elongation and overlapping.[160-164] After TNFα exposure, actin filaments become reorganized and the fibronectin matrix altered. TNFα has also been shown to exert a cytostatic influence on endothelial cells *in vitro*.[164,165] Cells exposed at subconfluence undergo growth arrest with decreased ^{3}H-thymidine incorporation. On the other hand, TNFα has been shown to stimulate angiogenesis with neovascularization in the rabbit cornea *in vivo*.[165] Multiple studies have failed to demonstrate either IL-1-[89,91] or TNFα[161,162,164,165]-induced human endothelial cytotoxicity. In one report, TNFα was shown to induce cytotoxicity in normal bovine capillary endothelium *in vitro*.[162] Matsubara and Ziff have demonstrated a possible mechanism for direct cytokine-induced endothelial injury;[84] they found that IL-1 enhanced superoxide anion release from cultured endothelial cells. The relative contributions of endothelial cell-generated toxic oxygen radicals and granulocyte-derived oxygen intermediates to cumulative endothelial injury have not been delineated.

IL-1 and TNFα can both stimulate endothelial cell arachidonate metabolism.[31-33]

IL-l stimulates human umbilical vein biosynthesis of the two vasoactive arachidonate metabolites, PGI_2[31,33] and PGE_2.[32] TNFα has been shown to induce phospholipase A_2, release of ^3H-arachidonate metabolites, and increased permeability to radiolabeled tracer in cultured bovine endothelial cells that could be inhibited by BW755C (combined cyclooxygenase and lipoxygenase inhibition).[46] In addition, IL-l and TNFα induce endothelial synthesis of another phospholipase A_2-driven product, PAF.[34,48] As mentioned above, PAF is an established mediator of acute lung injury.[12-14]

IL-l and TNFα also modulate the hemostatic state at the endothelial surface.[24,104-108,166-171] These two cytokines each can stimulate synthesis and cell surface expression of procoagulant activity in human endothelial cell monolayers.[104-108] The tissue factor-like procoagulant activity in the presence of factor VII and Ca^{++} can activate factor X with subsequent conversion of prothrombin to thrombin.[172] In an intact sheep model, α thrombin administration has been shown to augment pulmonary lymph flow, indicative of increased pulmonary vascular permeability.[9,10,173-175] In addition to generating procoagulant activity, IL-l suppresses endothelial thrombin-mediated protein C activation and assembly of functional protein C-protein S complex in the aorta of IL-l-infused rabbits.[106] IL-l and TNFα also induce human endothelial cell synthesis of plasminogen activator inhibitor.[166-169,171] Similarly, TNFα also suppresses the antithrombotic protein C pathway.[108] The increased expression of PAF and procoagulant activity combined with inhibition of protein C activation and increased plasminogen activator inhibitor shift the balance at the endothelial surface towards a prothrombotic state. In fact, TNFα infusion into tumor-bearing mice produces fibrin deposition with formation of occlusive intravascular thrombi within the tumor vascular bed.[170] The vasculo-occlusive process was limited to the tumor microvasculature and was associated with reduction of tumor blood flow.

In addition to their effects on endothelial arachidonate metabolism, PAF biosynthesis, and endothelial surface hemostasis, IL-l and TNFα also induce endothelial cell production of other cytokines.[27,29,130-132,134,135,137,176] IL-l and TNFα each can induce endothelial synthesis of GM-CSF.[130-132,134,135,137] TNFα also stimulates endothelial biosynthesis of platelet-derived growth factor (PDGF)[176] as well as IL-l,[27-29] itself a mediator of lung injury.[95,96] The relevance of endothelial expression of GM-CSF and PDGF to acute tissue injury warrants further investigation.

More recently IL-l and TNFα have been shown to induce expression of a new endothelial cell surface protein, ELAM-l, recognized by MoAbs H4/18 and H18/7.[98-103] These MoAbs were raised by immunization with IL-l-treated endothelial cells. The expression of this cytokine-induced endothelial surface antigen, ELAM-l, is dose dependent, reversible, and also blocked by either protein- or mRNA-synthesis inhibition.[98] MoAbs H4/18 and H18/7 immunoprecipitate two biosynthetically labeled polypeptides (Mr 115,000; Mr 95,000) from cytokine-stimulated endothelium, but not from control endothelium. Expression of these antigens peaks at 4 to 6 hr and declines to near basal levels by 24 hr. These antigens are not found in normal human tissues, but are rapidly expressed with inflammation and delayed hypersensitivity reactions.[99] The H4/18 and H18/17 staining preferentially occurs in postcapillary venular endothelium in cultured human foreskin.[101] These two MoAbs recognize two distinct epitopes on ELAM-l. IL-l and TNFα can also induce endothelial intercellular adhesion molecule (ICAM-l).[100,102] In addition, IL-l and TNFα have been shown to induce endothelial cell target antigens that increase endothelial susceptibility to lysis by acute Kawasaki's disease sera.[177] Thus, cytokine-primed endothelial surfaces express antigens that render the endothelium more vulnerable to serum- and/or granulocyte-dependent injury.

EFFECT OF CYTOKINES ON ARACHIDONATE METABOLISM

In addition to the above-mentioned cytokine-induced changes in endothelial cell arachidonate metabolism,[31-33,46] IL-l and TNFα can also generate arachidonate metabolites in nonendothelial tissues[42-45,70-80,178-186] (Table 4). IL-l can induce phospholipase A_2 in rabbit chondrocytes[78] and cultured 3T3 fibroblasts,[186] as well as cyclooxygenase in human skin fibroblasts.[80] IL-l has been shown to stimulate synthesis of the cyclooxygenase product, PGE_2, in peripheral blood monocytes,[178] peritoneal macrophages,[179,180] fibroblasts,[42,44,74,76,181] muscle cells,[71,79] chondrocytes,[72] synovium,[42,45,70,73,75,77] and brain tissue.[182-184] IL-l has also been shown to stimulate synthesis of other cyclooxygenase metabolites, including $PGF_{2\alpha}$ and 6-keto $PGF_{1\alpha}$ (the stable metabolite of PGI_2),[31,32,79] as well as TxB_2.[184] In addition, IL-l has been shown to generate the lipoxygenase pathway products 5- and 15-hydroxyeicosatetraenoic acids (HETE).[185] One lipoxygenase product, leukotriene B_4, is a postulated mediator of increased pulmonary vascular permeability in sheep.[187,188] TNFα also stimulates PGE_2 production in nonendothelial tissues,[42-45] including lung fibroblasts.[44]

Systemically elevated circulating levels or increased endothelial cell production of arachidonate metabolites may be pertinent to vascular permeability changes, in general,[189,190] and, in particular, to the alveolar-capillary leak of acute lung injury.[187,188] Increased cutaneous vascular permeability that occurs in response to intradermal injections of various stimuli in rabbits can be greatly enhanced by the addition of either PGE_2 or PGI_2.[189,190] Intradermal injections of either metabolite alone does not affect plasma exudation. Their enhancement of vascular permeability can be closely correlated with increments in local blood flow. More pertinent to acute lung injury, Henson and co-workers have found that intrabronchial instillation of C5 fragments, only in concert with PGE_2 or hypoxia, can induce lung injury[191] and that lung injury induced by intravascular complement activation together with hypoxia can be blocked by cyclooxygenase inhibition.[192] In an intact rabbit model, Okusawa et al. have recently demonstrated that administration of either rIL-l α or rTNFα induces systemic hypotension with pulmonary hemorrhage and edema.[116] Coadministration of the two cytokines synergistically induced circulatory shock and lung injury that could be blocked by cyclooxygenase inhibition. Therefore, cytokines

Table 4. Effects of Cytokines on Arachidonate Metabolism

Cytokine	Target Cell/Organ	Species	Effect	Reference
IL-l	Endothelium	Human	PGE_2	32
			PGI_2	31,33
	Monocyte/macrophage	Human	PGE_2	178
		Murine	PGE_2	179,180
	Lymphocytes	Human	5-HETE,15-HETE	185
	Fibroblasts	Human	Cyclooxygenase	80
			PGE_2	44,74,76,181
	Chondrocytes	Rabbit	Phospholipase A_2	78
			PGE_2	72
		Murine	PGE_2	180
	Synovium	Human	Phospholipase	45
			^{3}H-arachidonate	45
			PGE_2	70,75,77
		Human	PGE_2	198
	Smooth muscle	Monkey	PGE_2, $PGF_2\alpha$, PGI_2	79
		Rat	PGI_2	31,33
	Skeletal muscle	Rat	PGE_2	71
	Hepatocytes	Rat	PGE_2	79
	Brain	Rabbit	PGE_2	182,183
		Cat	PGE_2, TxB_2	184
TNF	Endothelium	Bovine	Phospholipase A_2	46
			^{3}H-arachidonate release	46
	Macrophage	Murine	PGE_2	43
	Fibroblasts	Human	PGE_2	42,44
	Synovium	Human	Phospholipase	45
			^{3}H-arachidonate release	45
			PGE_2	42
	Lung	Sheep	PGE_2, PGI_2, TxB_2	196

may generate systemically increased circulating levels and/or locally increase arachidonate metabolites at the pulmonary microvascular endothelial surface. Here, these metabolites may modulate microvascular blood flow or intraluminal hydrostatic pressures that could influence edema formation associated with lung injury.

CYTOKINE-INDUCED PAF SYNTHESIS

Another endogenous mediator through which cytokines may act upon host tissues is PAF. As mentioned in the previous section, both IL-1[78] and TNFα[46] can induce phospholipase A_2 activity, a prerequisite to PAF generation. Both IL-l and TNFα have been shown to stimulate biosynthesis of PAF in cultured human endothelial monolayers.[34,48] TNFα also induces PAF in rat peritoneal macrophages and neutrophils.[48] TNFα has been shown to induce PAF production in rat intestinal tissue, and TNFα-induced systemic hypotension and bowel necrosis could be prevented by PAF receptor antagonism.[47] Although a specific role for PAF in TNFα-induced lung injury has not been reported, PAF receptor antagonism attenuates pulmonary extravasation of ^{125}I-albumin in endotoxin-challenged rats.[193]

EFFECTS OF CYTOKINES ON HEMODYNAMIC PARAMETERS

Hemodynamic alterations is another mechanism through which cytokines may provoke acute lung injury or "shock lung".[1-3] rIL-l has been shown to induce systemic hypotension in rabbits (0.75 μg/kg).[194] In a preliminary report, Schnizlein-Bick et al. showed that human IL-lα could increase lung weight without altering pulmonary microvascular pressures in an isolated perfused rat lung.[230] This strongly suggested that IL-l-induced pulmonary edema is caused by increased permeability rather than by increased hydrostatic pressure. rTNFα can also induce hypotension in rats (3.6 mg/kg),[117] rabbits (5 μg/kg or 50 μg/kg),[116,194] and guinea pigs (0.14 mg/kg).[97] Antihuman rTNFα murine monoclonal antibody has been shown to protect against an LD_{100} dose of *Escherichia coli*-induced systemic hypotension and decreased cardiac output in anesthetized baboons.[195] Goat antihuman rTNFα polyclonal antibody prevents the profound decrease in mean arterial pressure in endotoxin-challenged rabbits.[39] In rabbits, rTNFα infusion at 5 mg/kg BW not only induced sustained hypotension, but decreased systemic vascular resistance and central venous pressure and increased heart rate and cardiac output.[116] Stephens et al. have demonstrated that rTNFα-induced pulmonary extravasation of ^{125}I-albumin in guinea pigs was not associated with significant increments of mean left atrial and pulmonary artery pressures.[97] Although they did not measure cardiac output to determine pulmonary vascular resistance, they suggested that the increase in transvascular albumin flux was independent of increased hydrostatic pressures. More recently, Horvath et al.[255] as well as Johnson et al.[196] have both shown that rTNFα infusion (12 μg/kg BW and 0.01 mg/kg BW, respectively) into awake sheep produces rapid-onset, transient increases in pulmonary artery pressure and pulmonary vascular resistance; cardiac output does not change. Wheeler et al. also studied pulmonary function following rTNFα infusion and found increased resistance to air flow, decreased dynamic compliance, and an increased alveolar-arterial O_2 gradient.[197] The contribution of these hemodynamic alterations to TNFα-induced lung injury has not been defined.

It is also not clear whether cytokine-induced shock is a direct and/or indirect effect. Anti-TNFα antibodies have been shown to protect against shock associated with either *E. coli*[195] or endotoxin[39] challenge. This immunoblockade could abort direct TNFα-receptor interaction and/or the generation of secondary endogenous mediators also deleterious to host tissues. IL-l exerts biological effects on human vascular smooth muscle cells *in vitro*.[198-200] Endotoxin[198] and IL-l itself[199] can each induce IL-l gene expression in these cells. IL-l also stimulates vascular smooth muscle PGE_2 production and exerts mitogenic activity on vascular smooth muscle cells that can be inhibited by prostanoids.[200] TNFα produces a decrease in resting transmembrane potential difference in rat skeletal muscle fibers.[201] TNFα effects on smooth muscle have not been reported. Direct cytokine interactions with vascular smooth muscle may contribute to changes in pulmonary and systemic vasomotor tone. Whether direct cytokine actions on the vascular smooth musculature play a role in pulmonary hypertension and edema formation or systemic hypotension and circulatory shock is not known. Cytokine-induced hemodynamic alterations may also involve secondary endogenous mediators. Okusawa et al. have blocked the synergistic hypotension induced by IL-l and TNFα coadministration in rabbits with prior cyclooxygenase inhibition.[116] Sun and Hsueh prevented TNFα-induced bowel necrosis in rats, using PAF receptor antagonism.[47] Both cyclooxygenase products and PAF are established mediators of shock[116,193] and acute lung injury.[187,188,193] Therefore, cytokines may provoke shock and tissue injury directly and/or through arachidonate metabolites, PAF, and/or other endogenous mediators.

EVIDENCE FOR ROLE FOR CYTOKINES IN EXPERIMENTAL AND CLINICAL LUNG INJURY

Evidence for a definite and specific role for cytokines in the pathogenesis of acute pulmonary vascular endothelial injury and ARDS continues to unfold. In support of the hypothesis that IL-l and TNFα participate in or are central to this pathological process, the following arguments are presented.

Appropriate to the Diversity of Stress and Inducers of Lung Injury, In Vitro Monocyte Production of IL-l and TNFα Can Be Triggered by a Wide Range of Stimuli

Endogenous mediators, including thrombin,[24] C5 cleavage products,[25,26] leukotrienes B_4 and D_4,[202] colony-stimulating factors,[203-205] γ-interferon,[127,206,207] and IL-3,[205] have been shown to induce cytokine production. Exogenous factors, including bacterial,[22-24,27,35-39,41,208] fungal,[23,209] and viral agents,[210-212] or their constituents, are also potent inducers of cytokine synthesis. Many of these stimulants are relevant to experimental systems and clinical settings that can lead to acute lung injury and ARDS.

Increased Circulating or Local Levels of Cytokines Have Been Demonstrated in Both Experimental Models of Lung Injury and/or Clinical Antecedents of ARDS

Circulating levels of IL-l bioactivity are increased in patients with septic processes.[213-216] Thermal injury patients have increased IL-l in burn blister fluid,[217] and head trauma patients have increased ventricular fluid IL-l.[218] Elevated circulating levels of TNFα have been reported in rabbits[39] and humans[38] following endotoxin infusion, as well as in patients with meningococcemia.[216,219] All of these conditions can be complicated by acute lung injury and ARDS.

Increased Levels of Cytokines Have Been Demonstrated in Relevant Body Compartments (Lung Parenchyma, Bronchoalveolar Lavage Fluid) During Experimental Lung Injury and in ARDS Patients

Bochner et al. found that cultured human lung fragments spontaneously release IL-1α and IL-1β, but not TNFα.[118] Tabor et al. showed that alveolar macrophages harvested from endotoxin-challenged dogs produced increased levels of both IL-1 and TNFα.[220] Kobayashi et al. demonstrated that intratracheal administration of BCG to immunized mice induces pulmonary granuloma formation bioactivity;[221] extracts of the lung granulomata contained increased IL-1 bioactivity. Although this was a different model of lung injury, tissue IL-1 activity increased acutely, peaking on day 1. Although the capacity of pulmonary alveolar macrophages to synthesize and release IL-1 is controversial,[120,121,125-127] increased alveolar macrophage release of IL-1 activity has been demonstrated after several injurious stimuli to the lung.[119,128,222,223] Lamontagne et al. showed that during murine infection with the nematode *Nippostrongylus brasiliensis*, which involves larvae migration through lung parenchyma, alveolar macrophages spontaneously released increased IL-1 *in vitro*.[119] Schwartz et al. demonstrated increased IL-1 bioactivity released from alveolar macrophages *in vitro* and in BALF from rabbits with concanavalin A-induced interstitial edema.[222] In an established rat model of monocrotaline-induced lung injury, Gillespie et al. demonstrated serially increased IL-1 bioactivity in BALF.[128] This lung injury model is subacute, but involves rearrangement of pulmonary perivascular architecture, resulting in pulmonary hypertension. Blanchard et al. intratracheally inoculated mice with *Legionella pneumophilia*, which resulted in increased TNFα release by adherent lung leukocytes, as well as increased TNFα in BALF.[223] In the face of this increased local TNFα bioactivity, no circulating TNFα could be demonstrated. These findings are compatible with the study of Martinet et al.,[224] which demonstrated that LPS-activated alveolar macrophages express increased TNFα mRNA and release increased protein product compared to blood monocytes obtained from the same individual. More recently, increased immunoreactive IL-1 has been recovered in the BALF of ARDS patients.[225] Therefore, increased cytokine bioactivity within lung tissue and/or the bronchoalveolar compartment has been shown to be temporally coincident with experimental lung injury and ARDS.

Cytokines Have Been Shown to Induce Vascular Endothelial Injury in Extrapulmonary Tissues

Beck et al. have demonstrated that murine IL-1 (3 to 29 units) can serve as either the preparatory or provocative mediator for a local Schwartzman reaction in rabbits.[109] In their model, the reaction involved cutaneous intravascular fibrin platelet thrombi, perivascular hemorrhage, and edema. Martin et al. showed that rIL-1β augments cutaneous extravasation of Evans blue dye in rats.[226] Movat has demonstrated that intradermal coadministration of IL-1β (0.5 μg) and TNFα (1 μg) to rabbits induces granulocyte-mediated vascular endothelial injury that included leukocyte thrombi, endothelial lysis, and neutrophil granules abutting against residual endothelial basement membrane.[110] Remick et al. have demonstrated that TNFα (1 and 10 μg) produces intestinal vascular endothelial cell damage in mice.[142] At the higher dose (10 μg per mouse), all animals had watery diarrhea. Ultrastructural studies revealed endothelial cell blebbing, gap formation, and exposed subendothelial basement membrane. Mice preinfused with ^{125}I-albumin displayed increased small bowel vascular permeability to the radiolabeled tracer. Therefore, four investigators have clearly shown that cytokines can induce vascular endothelial injury in three different animal models.

IL-1-INDUCED PULMONARY VASCULAR ENDOTHELIAL INJURY

Increased systemic and/or local levels of cytokines that were temporally coincident with evolving lung injury have been demonstrated. These findings have revealed important associations between cytokines and lung injury, but have not proven causality. As a first test of the direct involvement of cytokines in acute lung injury, we determined whether a single bolus infusion of an IL-1-containing murine monokine preparation (described above) could induce acute lung injury in rabbits.[94] This same monokine preparation displayed *in vivo* bioactivity in rabbits, including neutrophilia,[94] hypozincemia, and hypoferremia.[227] A single bolus infusion containing 100 units of IL-1 bioactivity failed to induce demonstrable lung injury. Because IL-1 reportedly has an extremely rapid rate of clearance (t 1/2 < 2 min),[228] we then administered the monokine preparation (500 units of IL-1 bioactivity) in divided-dose bolus infusions.[96] Animals preinfused with ^{125}I-albumin were studied for pulmonary edema formation and pulmonary extravasation of ^{125}I-albumin. After 5 hr, the divided-dose infusions increased pulmonary edema formation (Figure 2A) as well as pulmonary extravasation of ^{125}I-albumin across the alveolar-capillary barrier into the intra-alveolar compartment (Figure 2B). Whether the radiolabeled albumin was administered prior to monokine challenge or 0.5 hr prior to sacrifice (i.e., 2.5 hr following the last monokine bolus infusion), increased pulmonary extravasation of the tracer was evident. Electron microscopic studies of lung sections obtained from monokine-challenged animals demonstrated endothelial cell injury with blebbing and gap formation, perivascular edema, and extravasation of preinfused electron-dense ultrastructural permeability tracer (Figure 3).[96] To exclude endotoxin-mediated bioactivity, equivalent doses of the murine monokine preparation, passed through columns constructed with agarose beads coated with immobilized polymyxin B, were administered to additional animals. Prior polymyxin B treatment did not diminish the ability of the monokine preparation to increase either pulmonary edema formation (Figure 2A) or pulmonary extravasation of ^{125}I-albumin (Figure 2B).[229] To further control for endotoxin, animals were administered divided-dose bolus infusions of *Escherichia coli* O111:B4 lipopolysaccharide for a total endotoxin challenge that was 1000-fold the LPS concentration in the monokine preparation. The endotoxin-challenged animals did not develop increased pulmonary edema formation (Figure 2A) or pulmonary extravasation of ^{125}I-albumin (Figure

2B).[229] Since the active fraction of our monokine fraction was compatible with IL-1, we studied whether human rIL-1α and rIL-1β could simulate the monokine-induced pulmonary vascular endothelial injury in rabbits.[95] In the same intact animal model, rIL-1α induced dose-dependent transient granulocytopenia followed by marked neutrophilia and significant pulmonary leukostasis (1.25 to 125 µg/kg). Both human rIL-1 species (125 µg/kg) induced pulmonary edema formation. rIL-1β significantly increased pulmonary extravasation of ^{125}I-albumin, whereas rIL-1α-infused animals only displayed a trend toward significance (p <0.07). There were no significant differences in either pulmonary edema formation or pulmonary vascular leak of ^{125}I-albumin between rIL-1α- and rIL-β-infused animals. Electron microscopy of lung sections obtained from rIL-1-infused animals revealed the same endothelial injury, perivascular edema and extravasation of an ultrastructural permeability tracer (as in Figure 3). Therefore, an LPS-induced macrophage product with a molecular weight within the 10,000 to 30,000 range that displayed both *in vitro* and *in vivo* IL-1 bioactivities could induce pulmonary vascular endothelial injury in rabbits. Prior polymyxin B treatment of the monokine preparation failed to abort these changes, and 1000-fold concentrations of endotoxin could not simulate them. In addition, pure human rIL-1 could induce this same pulmonary vascular endothelial injury. Although these studies demonstrated that under certain experimental conditions a macrophage product compatible with IL-1, and IL-1 itself, can induce lung injury, they do not exclude the potential importance of other macrophage-derived factors. Similarly, although IL-1-induced endothelial injury cannot be ascribed solely to endotoxin contamination, the data do not exclude the possibility of a synergistic action between IL-1 and endotoxin.

To facilitate the study of the direct effects of our murine monokine preparation and human rIL-1 on the pulmonary vascular endothelial surface, we utilized a cultured porcine pulmonary artery endothelial monolayer system.[229] Pulmonary artery endothelial cells were grown to confluence on membrane filters mounted in modified Boyden chambers. The monokine preparation induced a dose- (Figure 4) and time- (Figure 5A) dependent increase in transendothelial albumin flux. Endothelial cell release of lactate dehydrogenase as a marker of cell death was not significantly increased for at least 6 hr (Figure 5B), but increased albumin transfer was evident. As in the *in vivo* studies described above, prior polymyxin B treatment of the monokine preparation failed to decrease its activity, and a comparable concentration of endotoxin failed to mimic it (Figure 6). When we attempted to confirm the murine monokine-induced *in vitro* endothelial permeability changes using human rIL-1, greater concentrations were required (Figure 4). These findings suggest the involvement of non-IL-1 macrophage-derived factor(s) or cofactor(s). More recently, Okusawa et al. have reported no remarkable changes in lungs of rabbits after single low-dose bolus infusions of rIL-1β (5 µg/kg) followed by a 2-hr infusion at 10 µg/kg/min.[116] Perhaps the lower dosage and shorter study period explain their negative findings. Although it is clear that IL-1 can induce pulmonary vascular endothelial injury in rabbits, its role in ARDS in human patients remains undefined. To date no studies have utilized immunological intervention in either experimental models of lung injury or human ARDS, with anti-IL-1 antibodies. In another form of lung parenchy-

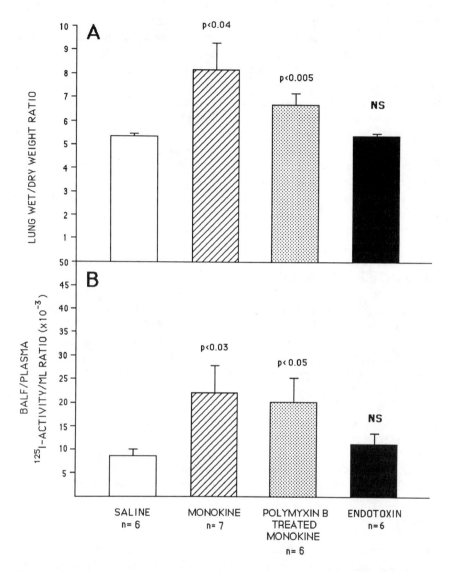

Figure 2. Effect of monokine preparation on (A) pulmonary edema formation and (B) pulmonary extravasation of ^{125}I-albumin into the intra-alveolar compartment in rabbits. (A) Mean (± SEM) lung wet-to-dry weight ratios in six saline-, seven monokine-, and six endotoxin-infused animals. An additional six animals infused with the monokine preparation after passage through immobilized polymyxin B were similarly studied. Vertical bars represent mean wet-to-dry lung weight ratios and brackets represent SEM. (B) mean (± SEM) BALF/plasma ^{125}I-activity ratios in six saline-, seven monokine-, and six endotoxin-infused animals. An additional six animals infused with the monokine preparation after passage through immobilized polymyxin B were similarly studied. Vertical bars represent mean BALF/plasma ^{125}I-activity ratios and brackets represent SEM.

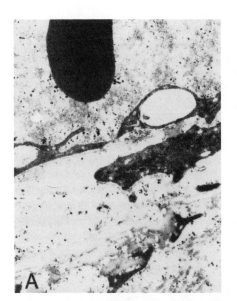 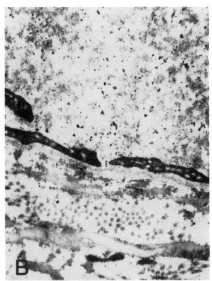

Figure 3. High-magnification electron micrograph of pulmonary venule endothelium from monokine-infused rabbits. (A) Note endothelial blebs, electron lucent matrix of connective tissue in venular wall indicating interstitial edema, and extravasation of glycogen granules into the interstitium. (Magnification × 20,000.) (B) Note endothelial "gap" with glycogen particles within vascular lumen, the interstitium, as well as within the gap itself (Magnification × 26,000.) (From Goldblum, S. E., *J. Appl. Physiol.*, 63:2093–2100, 1987. With permission.)

mal injury, Kasahara et al. have demonstrated that intratracheal installation of IL-1-coated sepharose beads induces granuloma formation in mice, whereas administration of sepharose beads without the IL-1 coating does not.[231] This form of IL-1-induced alterations in lung architecture are covered elsewhere in this volume.

TNFα-INDUCED PULMONARY VASCULAR ENDOTHELIAL INJURY

Tracey et al. have demonstrated that human rTNFα (1.8 mg/kg BW) induces shock and tissue injury in rats.[117] They found that lungs obtained from these rTNFα-challenged rats were grossly hemorrhagic, and light microscopy revealed alveolar membrane thickening, interstitial and peribroncheolar inflammation, and intravascular granulocyte thrombi. Talmadge et al. described pulmonary venous thromboses in rTNFα-challenged (10 µg per animal) 4-week-old mice.[232] Stephens et al. found that rTNFα infusion (0.14 mg/kg BW) into guinea pigs increased pulmonary edema formation and pulmonary vascular permeability to ^{125}I-albumin.[97] These animals displayed decreasing alveolar-arterial oxygen differences without any significant changes in arterial pH. As discussed above, the TNFα-challenged animals did not have increased mean left atrial or pulmonary artery pressures. Although cardiac

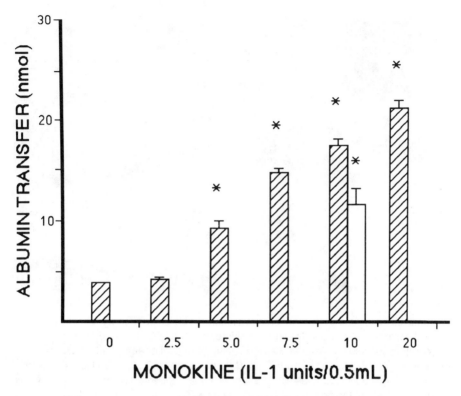

Figure 4. Effect of IL-I-containing monokine preparation and human recombinant IL-1β on subsequent transfer of albumin across cultured porcine pulmonary artery endothelial monolayers. Cells were washed with M-199 and then incubated for 24 hr with M-199 enriched with 5% fetal bovine serum and various doses of the IL-I-containing monokine preparation or 20 units of human rIL-Iβ (open bar) per ml. After washing cells again in M-199, transendothelial albumin transfer was measured over a 1 hr period. The concentration of albumin added to the upper chamber during the subsequent 1 hr incubation was 200 μM. Values are mean ± SE, n = 6. * = significantly increased at $p < 0.05$ compared to media control.

outputs were not measured to calculate pulmonary vascular resistance, they suggested that the increased transvascular albumin flux could not be explained by increased pulmonary vascular pressures. These same investigators have shown that prior granulocyte depletion aborts the rTNFα-induced lung injury.[233] In a recent report, Horvath et al. demonstrated that rTNFα infusion (12 μg/kg BW) into sheep augmented pulmonary artery pressures, pulmonary vascular resistance, and pulmonary lymph flow and transvascular protein clearance rates.[255] In this study, prior granulocyte depletion did not abate the increased pulmonary vascular permeability. These same investigators also demonstrated rTNFα-induced increments in ^{125}I-albumin transfer across ovine endothelial monolayers. In another recent report, Johnson et al. confirmed that rTNFα challenge (0.01 mg/kg BW) to sheep induces pulmonary

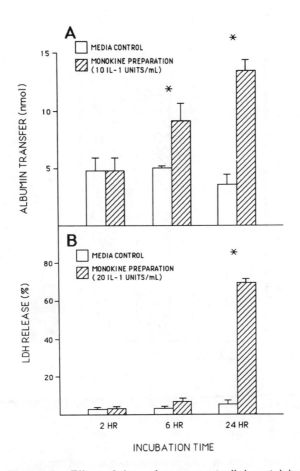

Figure 5. Effect of time of exposure to IL-I-containing monokine preparation on (A) subsequent albumin transfer across cultured porcine pulmonary artery endothelial monolayers and (B) porcine pulmonary artery endothelial cell release of LDH. (A) Cells were washed with M-199 and incubated for either 2, 6, or 24 hr with M-199 enriched with 5% fetal bovine serum with or without the monokine preparation containing 10 units IL-I per ml of culture medium. After the cells were washed again with M-199, albumin transfer was measured over a 1 hr period with M-199 containing 200 μM albumin. Values are mean ± SE, n = 6. * = significantly increased at $p < 0.05$ compared to simultaneous media control. (B) Porcine endothelial cells were washed with M-199 and incubated for 2, 6, or 24 hr with M-199 enriched with 5% fetal bovine serum with or without the monokine preparation containing 20 u IL-I per ml of culture media. Values are mean ± SE, n = 6. * = significantly increased at $p < 0.001$ compared to simultaneous media control.

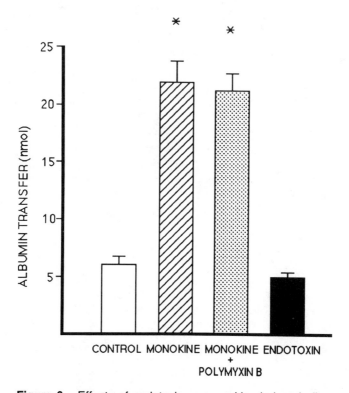

Figure 6. Effects of endotoxin on monokine-induced albumin transfer across cultured porcine pulmonary artery endothelial monolayers. Cells were washed with M-199 and incubated for 24 hr with M-199 enriched with 5% fetal bovine serum with or without the monokine preparation containing 10 units IL-1 activity per ml of culture medium. In selected experiments the cells were treated with polymyxin B-adsorbed monokine or endotoxin alone. After repeat washing with M-199, albumin transfer was measured over a 1 hr period with M-199 containing 200 µM albumin. Values are mean ± SE, n = 6. * = significantly increased at $p < 0.001$ compared to simultaneous media control.

artery hypertension, hypoxemia, and increases in lung lymph flow.[196] All of these data suggest that TNFα can promote sequestration of circulating granulocytes within the pulmonary microvasculature, augment pulmonary vascular permeability, and increase pulmonary edema formation. TNFα administration may also generate pulmonary hypertension. Whether and to what extent these pressure changes contribute to transpulmonary vascular fluid and/or protein flux is not yet clear. Recently, we have shown that infusion of doses of human rTNFα to rabbits that do not induce neutropenia or pulmonary leukostasis (5 µg/kg) can cause pulmonary vascular endothelial injury and increased lung water.[143] Animals were administered five equivalent bolus infusions every 0.5 hr over the first 2 hr of a 5-hr study period. Not only did these animals not display neutropenia, they had marked neutrophilia at 1, 2, and 3 hr. In our system, the rTNFα augmented pulmonary edema formation (151%) and pulmonary extravasation of ^{125}I-albumin (376%) relative to saline-injected controls. Electron microscopy revealed endothelial injury with subendothelial blebbing and gap formation, perivascular edema, and extravasation of an ultrastructural permeability tracer. Okusawa et al. have now demonstrated that coadministration of IL-1 and rTNFα synergistically induce pulmonary leukostasis, edema and hemorrhage in rabbits.[116] The TNFα or synergistic lung injury (TNFα + IL-l) could be blocked by ibuprophen, a cyclooxygenase inhibitor. Using the cultured porcine pulmonary artery endothelial monolayer system described above, we have now found that rTNFα directly increases transendothelial albumin flux in both a dose- and time-dependent manner.[143] rTNFα-induced increase in transendothelial albumin flux could not be demonstrated at 2 hr, but was evident at 6 and 24 hr. In this same in vitro system, there was no increased lactate dehydrogenase (LDH) release at 2 or 6 hr. At 24 hr there was ≤ 8% increased LDH release. Therefore, rTNFα could induce endothelial permeability changes without endothelial cytotoxicity in vitro in the absence of granulocyte effector cells. This in vitro system also obviates the pressure changes that come into play in vivo. In a similar in vitro system using a bovine pulmonary artery endothelial cell line and ^{125}I-protein A as the tracer, Clark et al. also have studied the effect of rTNFα on endothelial permeability.[46] They reported increased permeability to ^{125}I-protein A within minutes. They also demonstrated rTNFα-induced endothelial cell retraction on phase contrast microscopy. These investigators showed that rTNFα exposure increased phospholipase A_2 activity and ^{3}H-arachidonate release in the endothelial monolayers. There was no increase in phospholipase C activity as measured by phosphatidylinositol turnover. The permeability change could be blocked with BW755C (combined cyclooxygenase and lipoxygenase inhibition) as well as by *Bordetella pertussis* toxin. Their findings suggest that arachidonate metabolism contributes to rTNFα-induced changes in endothelial barrier function and that rTNFα receptor-ligand interaction is coupled to a G-protein that can transduce a not-yet-defined signal for endothelial retraction and increased permeability.

Although several investigators have demonstrated rIL-1-α and rTNFα-induced lung injury, White et al. have shown that pretreatment with both rIL-1 (10 μg) and rTNFα (10 μg) protects rats from lung injury after hyperoxia.[234] Coadministration of rIL-1 and rTNFα increased ratios of reduced to oxidized glutathione in lung tissue of hyperoxia-exposed rats compared to saline-infused controls. No differences were found for superoxide dismutase, glutathione peroxidase, glutathione reductase, glucose-6-phosphate dehydrogenase, and catalase activities in lungs of treated vs. control rats. These studies suggest that combined IL-1/TNFα pretreatment favorably alters lung glutathione redox status (i.e., decreases oxidized glutathione), which protects against hyperoxia.

INTERLEUKIN-2 LYMPHOKINE-ACTIVATED KILLER CELL-MEDIATED PULMONARY VASCULAR ENDOTHELIAL INJURY

Human rIL-2 has been shown to generate lymphokine-activated killer (LAK) cells[235-237] and to mediate regression of malignant tumors in experimental animals[238,239] and humans.[240-242] A major adverse effect of rIL-2, alone or in combination with LAK cell administration, is a generalized vascular leak syndrome that involves the pulmonary microvasculature.[240-244] The vascular leak syndrome has been demonstrated in mice,[243] rats,[245] sheep,[244,246] and human subjects.[240-242] Rosenberg and co-workers studied the effects of rIL-2 administration to patients with advanced cancer.[240-242] These patients displayed significant weight gain and systemic hypotension, which was ascribed to the vascular leak syndrome and intravascular depletion. Rosenstein et al. studied the rIL-2-induced vascular leak syndrome in mice preinfused with ^{125}I-albumin.[243] They found increased pulmonary vascular permeability that could be prevented by prior irradiation, cyclophosphamide, or cortisone acetate. Further, the vascular leak was not observed in rIL-2-challenged athymic nude mice. Ettinghausen et al. have shown that LAK cells and rIL-2-induced vascular leak in irradiated mice, whereas rIL-2, with either splenocytes cultured without IL-2 or IL-2 with irradiated LAK cells did not.[247] Gately et al. found that rIL-2-challenged mice displayed marked tissue lymphoid cell infiltration and that 55 to 83% of these cells were asialo-GM$_1$ positive.[248] They then demonstrated that pretreatment of mice with anti-asialo-GM$_1$-γ-globulin abolished or greatly reduced the severity of the rIL-2-induced vascular leak syndrome. These studies strongly suggested that host lymphoid elements bearing asialo-GM$_1$ antigen were essential to rIL-2-induced changes in vascular permeability. rIL-2 administration also has been shown to expand the peripheral lymphocyte pool,[235,236] to induce pulmonary lymphocytic infiltration,[244] and to augment LAK cell proliferation in lung tissue.[237,248] Therefore, rIL-2 mobilizes lymphocytes to the relevant body compartment temporally coincident with pulmonary edema formation. On a cellular level, rIL-2 has been shown to augment lymphocyte adhesiveness to endothelial surfaces and to induce endothelial cell lysis *in vitro*.[249,250] Large granular lymphocytes have cytoplasmic granules that contain cytolytic activity.[251] These granule-derived cytolysins induce ring-shaped membrane lesions in target cells. Therefore, rIL-2-activated lymphocyte-mediated endothelial cytoxicity may contribute to the rIL-2-induced vascular leak syndrome.

IL-2 may also promote pulmonary vascular endothelial injury by generating other endogenous mediators.[252,253] Nedwin et al. have shown that human peripheral blood mononuclear cells can be induced by rIL-2 to secrete both TNFα and lymphotoxin (TNFβ).[252] Cotran et al. studied skin biopsies, obtained from rIL-2-treated subjects, for endothelial surface expression of three activation antigens.[253] After 5 d of rIL-2 administration, all patients demonstrated marked expression of all three antigens. rIL-2 failed to induce these same antigens in cultured endothelial cells, whereas TNFα, TNFβ, and IL-1 were potent inducers.[89-93,98-102] These findings suggest that rIL-2 administration induces *in vivo* generation of one or more of these endogenous mediators that are known to exert biological effects on the endothelium. TNFα and IL-1 are both candidate co-mediators for the rIL-2-induced vascular leak syndrome. rIL-2-treated subjects become febrile;[240,242] TNFα and IL-1 have both been established as endogenous pyrogens.[4-6] Murine IL-1,[96] human rIL-1,[95] and rTNFα[97,117,143] have each been shown to induce pulmonary vascular endothelial injury. The monokine (IL-1/TNFα)-induced lung injury differed from the rIL-2-induced changes. The former involved endothelial bleb and gap formation[95,96,143] and pulmonary sequestration of granulocytes.[94,95,97,116,117] rIL-2-induced endogenous production of lower plasma and/or BALF concentrations of either IL-1 and/or TNFα could result in more subtle pulmonary vascular endothelial changes that could contribute to rIL-2-induced lung injury. IL-2 also can activate the lipoxygense pathway with production of lipoxygenase metabolites which reportedly play a role in pulmonary vascular permeability.[187,188]

Electron microscopy of lungs obtained from rIL-2-treated rabbits have revealed extensive pulmonary arteriolar endothelial injury with complete sparing of venules.[256] This pattern of pulmonary vascular endothelial injury implicates pressure-mediated changes on the arterial side of the pulmonary circulation, rather than an inflammatory process, which usually predominates in the postcapillary venular endothelium. Hemodynamic alterations have been noted in human subjects[241,242,254] as well as in sheep.[244] In humans, the hemodynamic alterations can be iatrogenically altered by the intravenous administration of fluids. In a clinical study,[241] 35% of IL-2-treated patients experienced hypotension, while in another study[242] 64% of patients receiving constant-infusion of rIL-2 required pressor therapy. The hypotension has been ascribed to the vascular leak syndrome with resultant intravascular depletion. Glauser et al.[244] have conducted hemodynamic studies in sheep during a 72-hr continuous intravenous infusion of rIL-2. They demonstrated dose-dependent systemic hypotension and pulmonary artery hypertension in the face of a normal pulmonary wedge pressure. Lung water and lung lymph flow were both increased. At the lower rIL-2 dose, lymph flow was increased in the absence of pulmonary hypertension. These studies suggest a pressure-independent increase in pulmonary microvascular permeability as well as increased pulmonary hypertension at higher rIL-2 doses.

The rIL-2-induced pulmonary vascular leak syndrome may involve complex mechanisms involving both immunological and hemodynamic processes. Unlike other forms of noncardiogenic pulmonary edema, rIL-2-induced pulmonary vascular endothelial injury involves a T-lymphocyte, not a granulocyte effector cell. It is questionable whether the extremely large quantities of rIL-2 that are prerequisite to this iatrogenic vascular leak syndrome would ever be spontaneously operative under physiological or pathological conditions. Still, a schema for pulmonary vascular endothelial injury in which IL-2 is a central mediator and the LAK cell is the effector cell for endothelial target tissue is novel and may increase our understanding of acute and/or chronic lung injury in general.

SUMMARY OF ROLE OF CYTOKINES IN THE PATHOGENESIS OF ACUTE PULMONARY VASCULAR ENDOTHELIAL INJURY

Numerous endogenous and exogenous mediators can induce host cell production of the cytokines IL-1 and TNFα (Figure 1). Experimental models of pulmonary vascular endothelial injury and clinical antecedents for ARDS both can be associated with increased circulating levels of these two cytokines. Under specific experimental conditions, local increases of the cytokines have been demonstrated in anatomical sites relevant to lung injury. These two cytokines can directly or indirectly contribute to pulmonary vascular endothelial injury. They recognize and bind to receptors on host tissues, including the endothelium, where they can induce microvascular endothelial synthesis of arachidonate metabolites and PAF, both postulated mediators of lung injury. The cytokines can also stimulate nonendothelial tissue synthesis and systemic release of these same mediators into the venous blood for presentation to the pulmonary microvasculature.

IL-1 and TNFα also can induce endothelial cell surface expression of leukocyte adhesion molecules, which render endothelial surfaces hyperadhesive for circulating granulocytes. TNFα can also induce granulocyte surface expression of glycoproteins essential to optimal granulocyte adherence. Cytokines augment granulocyte-to-endothelial interaction and pulmonary leukostasis. Granulocyte sequestration within the pulmonary microvasculature strategically positions these cells in close proximity to the endothelial surface, where they can release toxic oxygen intermediates and proteases injurious to the endothelial surface. Cytokines might prime either granulocytes or endothelium for enhanced responsiveness to other stimuli (C5 cleavage products, endotoxin, etc.), or themselves provide the definitive stimulus. Although granulocytes might play a role in cytokine-induced pulmonary vascular endothelial injury *in vivo*, cytokines directly induce endothelial injury *in vitro* in the absence of granulocyte effector cells. Therefore, cytokines can directly compromise endothelial surface integrity and might contribute to the pathogenesis of ARDS in profoundly neutropenic patients. Furthermore IL-1 and TNFα, as well as other cytokines (γ-interferon), synergistically exert biological effects on target tissues, including the lung.

Another cytokine that induces pulmonary vascular leak syndrome is IL-2. IL-2 generates and recruits lymphoid elements to lung tissue, where they adhere to the microvascular endothelial surface. Here the IL-2 and lymphokine-activated killer cells mediate pulmonary vascular endothelial injury. Whether this IL-2-induced, T-lymphocyte-mediated process is relevant to acute and/or chronic lung injury in general is not known.

ADDENDUM

Since the writing of this chapter, additional information has appeared regarding the role of cytokines in acute pulmonary vascular endothelial injury and ARDS.

Human rTNFα not only induces experimental acute lung injury in numerous species (e.g., rats, guinea pigs, rabbits, dogs, and sheep), but causes tachypnea with fluid retention in cancer patients.[257] Furthermore, septic shock patients with detectable plasma TNFα have a higher incidence and severity of ARDS,[258] and increased TNFα bioactivity has been recovered in the BALF of ARDS patients.[259]

Additional data has also been presented regarding the cellular mechanism(s) of TNFα-induced changes in pulmonary vascular endothelial barrier function. There are reports that claim[260] or discount[261] PAF as a secondary or more distal mediator for TNFα-induced changes in permeability *in vivo*. As already noted above, TNFα has been shown to augment movement of macromolecules across the endothelial barrier in an *in vitro* system that precludes both hemodynamic changes and granulocyte effector cells.[143] More recently, we have demonstrated a dose-, time-, and temperature-dependent rTNFα-induced increment in ^{14}C-albumin flux across bovine pulmonary artery endothelial cell monolayers.[262] rTNFα exposure times as brief as 5 min increased transendothelial albumin flux, detectable only after ≥4 hr stimulus-to-response lag time. The changes were reversible and occurred in the absence of cell injury or death. The rTNFα-induced changes were not protein synthesis-dependent. In fact, prior cycloheximide or actinomycin D treatment greatly enhanced the rTNFα-induced increments in ^{14}C-albumin flux.[262,263] The influence of rTNFα on endothelial barrier function wan not albumin specific; other macromolecules were similarly affected.[263] Although granulocyte effector cells are not a prerequisite for the rTNFα-induced changes in endothelial permeability, TNFα has been shown to enhance susceptibility of vascular endothelial cells to granulocyte-mediated cytotoxicity.[264] TNFα has been shown to induce F-actin reorganization, cell retraction, and intercellular gap formation in human umbilical vein[161] and bovine aortic[265] endothelial cells. Using a fluorescein-labeled actin probe, we have confirmed these findings in bovine pulmonary artery endothelial cells.[266] We found isolated disruptions within the F-actin lattice exclusively at the endothelial cell-to-cell interface and significantly increased monomeric G-actin, both temporarily coincident with changes in endothelial barrier function. Furthermore, preloading of nonpermeabilized living endothelial cells with the F-actin stabilizing agent, phallicidin, protected against rTNFα-induced increments in G-actin and transendothelial ^{14}C-albumin flux. These findings suggest that rTNFα influences endothelial cell barrier function through endothelial cell actin rearrangement and intercellular gap formation, thereby generating a paracellular leak.

REFERENCES

1. Connors, A.F., Jr., D.R. McCaffree, and R.M. Rogers, The adult respiratory distress syndrome. *Dis. Mon.*, 27:10–75, 1981.
2. Fowler, A.A., R.F. Hamman, J.T. Good, K.N. Benson, M. Baird, D.J. Eberle, T.L. Petty, and T.M. Hyers, Adult respiratory distress syndrome: risk with common predispositions. *Ann. Intern. Med.*, 98:593–597, 1983.

3. Hyers, T.M. and A.A. Fowler, Adult respiratory distress syndrome: causes, morbidity, and mortality. *Fed. Proc.*, 45:25–29, 1986.
4. Dinarello, C.A., Interleukin-1 and the pathogenesis of the acute phase response. *N. Engl. J. Med.*, 311:1413–1418, 1984.
5. Dinarello, C.A., Interleukin-1. *Rev. Infect. Dis.*, 6:51–95, 1984.
6. Dinarello, C.A., J.G. Cannon, S.M. Wolff, H.A. Bernheim, B. Beutler, A. Cerami, I.S. Figari, M.A. Palladino, Jr., and J.V. O'Connor, Tumor necrosis factor (cachectin) is an endogenous pyrogen and induces production of interleukin-1. *J. Exp. Med.*, 163:1433–1450, 1986.
7. Perlmutter, D.H., C.A. Dinarello, P.I. Punsal, and H.R. Colten, Cachectin/tumor necrosis factor regulates hepatic acute-phase gene expression. *J. Clin. Invest.*, 78:1349–1354, 1986.
8. Brigham, K.L. and B. Meyrick, Endotoxin and lung injury. *Am. Rev. Respir. Dis.*, 133:913–927, 1986.
9. Tahamont, M.V. and A.B. Malik, Granulocytes mediate the increase in pulmonary vascular permeability after thrombin embolism. *J. Appl. Physiol.*, 54:1489–1495, 1983.
10. Johnson, A. and A.B. Malik, Pulmonary transvascular fluid and protein exchange after thrombin-induced microembolism. Differential effects of cyclooxygenase inhibitors. *Am. Rev. Respir. Dis.*, 132:70–76, 1985.
11. Till, G.O., K.J. Johnson, R. Kunkel, and P.A. Ward, Intravascular activation of complement and acute lung injury. Dependency on neutrophils and toxic oxygen metabolites. *J. Clin. Invest.*, 69:1126–1135, 1982.
12. McManus, L.M., D.J. Hanahan, C.A. Demopoulos, and R.N. Pinckard, Pathobiology of the intravenous infusion of acetyl glyceryl ether phosphorylcholine (AGEPC), a synthetic platelet-activating factor (PAF), in the rabbit. *J. Immunol.*, 124:2919–2924, 1980.
13. McManus, L.M. and R.N. Pinckard, Kinetics of acetyl glyceryl ether phosphorylcholine (AGEPC)-induced acute lung alterations in the rabbit. *Am. J. Pathol.*, 121:55–68, 1985.
14. Worthen, G.S., A.J. Goins, B.C. Mitchel, G.L. Larsen, J.R. Reeves, and P.M. Henson, Platelet-activating factor causes neutrophil accumulation and edema in rabbit lungs. *Chest*, 83:13S–15S, 1983.
15. O'Flaherty, J.T., S. Cousart, A.S. Lineberger, E. Bond, D.A. Bass, L.R. DeChatelet, E.S. Leake, and C.E. McCall, Phorbol myristate acetate. *In vivo* effects upon neutrophils, platelets, and lung. *Am. J. Pathol.*, 101:79–92, 1980.
16. Shasby, D.M., K.M. Vanbenthuysen, R.M. Tate, S.S. Shasby, I. McMurtry, and J.E. Repine, Granulocytes mediate acute edematous lung injury in rabbits and in isolated rabbit lungs perfused with phorbol myristate acetate: role of oxygen radicals. *Am. Rev. Respir. Dis.*, 125:443–447, 1982.
17. Shasby, D.M., S.S. Shasby, and M.J. Peach, Granulocytes and phorbol myristate acetate increase permeability to albumin of cultured endothelial monolayers and isolated perfused lungs. Role of oxygen radicals and granulocyte adherence. *Am. Rev. Respir. Dis.*, 127:72–76, 1983.
18. McCall, C.E., R.G. Taylor, S.L. Cousart, R.D. Woodruff, J.C. Lewis, and J.T. O'Flaherty, Pulmonary injury induced by phorbol myristate acetate following intravenous administration in rabbits. Acute respiratory distress followed by pulmonary interstitial pneumonitis and pulmonary fibrosis. *Am. J. Pathol.*, 111:258–262, 1983.
19. Silver, M.J., W. Hoch, J.J. Kocsis, C.M. Ingerman, and J.B. Smith, Arachidonic acid causes sudden death in rabbits. *Science*, 183:1085–1087, 1974.
20. Shasby, D.M., S.S. Shasby, and M.J. Peach, Polymorphonuclear leukocyte: arachidonate edema. *J. Appl. Physiol.*, 59:47–55, 1985.

21. Seeger, W., D. Walmrath, M. Menger, and H. Neuhof, Increased lung vascular permeability after arachidonic acid and hydrostatic challenge. *J. Appl. Physiol.*, 61:1781–1789, 1986.
22. Arend, W.P., S. D'Angelo, R.J. Massoni, and F.G. Joslin, *The Physiologic, Metabolic, and Immunologic Actions of Interleukin-1*, Alan R. Liss, New York, 1985, 399–407.
23. Lepe-Zuniga, J.L. and I. Gery, Production of intra- and extracellular interleukin-1 (IL-1) by human monocytes. *Clin. Immunol. Immunopathol.*, 31:222–230, 1984.
24. Stern, D.M., I. Bank, P.P. Nawroth, J. Cassimeris, W. Kisiel, J.W. Fenton, II, C. Dinarello, L. Chess, and E.A. Jaffe, Self-regulation of procoagulant events on the endothelial cell surface. *J. Exp. Med.*, 162:1223–1235, 1985.
25. Goodman, M.G., D.E. Chenoweth, and W.O. Weigle, Induction of interleukin 1 secretion and enhancement of humoral immunity by binding of human C5a to macrophage surface C5a receptors. *J. Exp. Med.*, 1156:912–917, 1982.
26. Okusawa, S., C.A. Dinarello, K.B. Yancey, S. Endres, T.J. Lawley, M.M. Frank, J.F. Burke, and J.A. Gelfand, C5a induction of human interleukin 1. Synergistic effect with endotoxin or interferon-γ. *J. Immunol.*, 139:2635–2640, 1987.
27. Libby, P., J.M. Ordovas, K.R. Auger, A.H. Robbins, L.K. Birinyi, and C.A. Dinarello, Endotoxin and tumor necrosis factor induce interleukin-1 gene expression in adult human vascular endothelial cells. *Am. J. Pathol.*, 124:179–185, 1986.
28. Miossec, P., D. Cavender, and M. Ziff, Production of interleukin 1 by human endothelial cells. *J. Immunol.*, 136:2486–2491, 1986.
29. Kurt-Jones, E.A., W. Fiers, and J.S. Pober, Membrane interleukin 1 induction on human endothelial cells and dermal fibroblasts. *J. Immunol.*, 139:2317–2324, 1987.
30. Wagner, C.R., R.M. Vetto, and D.R. Burger, Expression of I-region-associated antigen (Ia) and interleukin 1 by subcultured human endothelial cells. *Cell. Immunol.*, 93:91–104, 1985.
31. Rossi, V., F. Breviario, P. Ghezzi, E. Dejana, A. Mantovani, Prostacyclin synthesis induced in vascular cells by interleukin-1. *Science*, 229:174–176, 1985.
32. Albrightson, C.R., N.L. Baenziger, and P. Needleman, Exaggerated human vascular cell prostaglandin biosynthesis mediated by monocytes: role of monokines and interleukin 1. *J. Immunol.*, 135:1872–1877, 1985.
33. Dejana, E., F. Breviario, V. Rossi, P. Ghezzi, and A. Mantovani, *The Physiologic, Metabolic, and Immunologic Actions of Interleukin-1*, Alan R. Liss, New York, 1985, 55–61.
34. Bussolino, F., F. Breviario, C. Tetta, M. Aglietta, A. Mantovani, and E. Dejana, Interleukin 1 stimulates platelet-activating factor production in cultured human endothelial cells. *J. Clin. Invest.*, 77:2027–2033, 1986.
35. Carswell, E.A., L.J. Old, R.L. Kassel, S. Green, N. Fiore, and B. Williamson, An endotoxin-induced serum factor that causes necrosis of tumors. *Proc. Natl. Acad. Sci. U.S.A.*, 72:3666–3670, 1975.
36. Beutler, B.A., I.W. Milsark, and A. Cerami, Cachectin/tumor necrosis factor: production, distribution, and metabolic fate in vivo. *J. Immunol.*, 135:3972–3977, 1985.
37. Kiener, P.A., F. Marek, G. Rodgers, P-F. Lin, G. Warr, and J. Desiderio, Induction of tumor necrosis factor, IFN-γ, and acute lethality in mice by toxic and non-toxic forms of lipid A. *J. Immunol.*, 141:870–874, 1988.
38. Michie, H.R., K.R. Manogue, D.R. Spriggs, A. Revhaug, S. O'Dwyer, C.A. Dinarello, A. Cerami, S.M. Wolff, and D.W. Wilmore, Detection of circulating tumor necrosis factor after endotoxin administration. *N. Engl. J. Med.*, 318:1481–1486, 1988.

39. Mathison, J.C., E. Wolfson, and R.J. Ulevitch, Participation of tumor necrosis factor in the mediation of Gram negative bacterial lipopolysaccharide-induced injury in rabbits. *J. Clin. Invest.*, 81:1925–1937, 1988.
40. Okusawa, S., K.B. Yancey, J.W.M. Van Der Meer, S. Endres, G. Lonnemann, K. Hefter, M.M. Frank, J.F. Burke, C.A. Dinarello, and J.A. Gelfand, C5a stimulates secretion of tumor necrosis factor from human mononuclear cells *in vitro*. Comparison with secretion of interleukin 1β and interleukin 1α. *J. Exp. Med.*, 168:443–448, 1988.
41. Aggarwal, B.B., W.J. Kohr, P.E. Hass, B. Moffat, S.A. Spencer, W.J. Henzel, T.S. Bringman, G.E. Nedwin, D.V. Goeddel, and R.N. Harkins, Human tumor necrosis factor. Production, purification, and characterization. *J. Biol. Chem.*, 260:2345–2354, 1985.
42. Dayer, J.-M., B. Beutler, and A. Cerami, Cachectin/tumor necrosis factor stimulates collagenase and prostaglandin E_2 production by human synovial cells and dermal fibroblasts. *J. Exp. Med.*, 162:2163–2168, 1985.
43. Bachwich, P.R., S.W. Chensue, J.W. Larrick, and S.L. Kunkel, Tumor necrosis factor stimulates interleukin-1 and prostaglandin E_2 production in resting macrophages. *Biochem. Biophys. Res. Commun.*, 136:94–101, 1986.
44. Elias, J.A., K. Gustilo, W. Baeder, and B. Freundlich, Synergistic stimulation of fibroblast prostaglandin production by recombinant interleukin 1 and tumor necrosis factor. *J. Immunol.*, 138:3812–3816, 1987.
45. Godfrey, R.W., W.J. Johnson, and S.T. Hoffstein, Recombinant tumor necrosis factor and interleukin-1 both stimulate human synovial cell arachidonic acid release and phospholipid metabolism. *Biochem. Biophys. Res. Commun.*, 142:235–241, 1987.
46. Clark, M.A., M.-J. Chen, S.T. Crooke, and J.S. Bomalaski, Tumor necrosis factor (cachectin) induces phospholipase A_2 activity and synthesis of a phospholipase A_2-activating protein in endothelial cells. *Biochem. J.*, 250:125–132, 1988.
47. Sun, X.-M. and W. Hsueh, Bowel necrosis induced by tumor necrosis factor in rats is mediated by platelet-activating factor. *J. Clin. Invest.*, 81:1328–1331, 1988.
48. Camussi, G., F. Bussolino, G. Salvidio, and C. Baglioni, Tumor necrosis factor/cachectin stimulates peritoneal macrophages, polymorphonuclear neutrophils, and vascular endothelial cells to synthesize and release platelet activating factor. *J. Exp. Med.*, 166:1390–1404, 1987.
49. Brigham, K.L. and B. Meyrick, Interactions of granulocytes with the lungs. *Circ. Res.*, 54:623–635, 1984.
50. Brigham, K.L. and B. Meyrick, Granulocyte-dependent injury of pulmonary endothelium: a case of miscommunication? *Tissue Cell*, 16:137–155, 1984.
51. Sacks, T., C.F. Moldow, P.R. Craddock, T.K. Bowers. and H.S. Jacob, Oxygen radicals mediate endothelial cell damage by complement-stimulated granulocytes. An in vitro model of immune vascular damage. *J. Clin. Invest.*, 61:1161–1167, 1978.
52. Weiss, S.J., J. Young, A.F. LoBuglio, A. Slivka, Role of hydrogen peroxide in neutrophil-mediated destruction of cultured endothelial cells. *J. Clin. Invest.*, 68:714–721, 1981.
53. Martin, W.J., II, Neutrophils kill pulmonary endothelial cells by a hydrogen peroxide-dependent pathway. An in vitro model of neutrophil-mediated lung injury. *Am. Rev. Respir. Dis.*, 130:209–213, 1984.
54. Weiss, S.J. and S. Regiani, Neutrophils degrade subendothelial matrices in the presence of alpha-l-proteinase inhibitor. Cooperative use of lysosomal proteinases and oxygen metabolites. *J. Clin. Invest.*, 73:1297–1303, 1984.

55. Weiss, S.J., J.T. Curnutte, and S. Regiani, Neutrophil-mediated solubilization of the subendothelial matrix: oxidative and nonoxidative mechanisms of proteolysis used by normal and chronic granulomatous disease phagocytes. *J. Immunol.*, 136:636–641, 1986.
56. Heflin, A.C., Jr. and K.L. Brigham, Prevention by granulocyte depletion of increased vascular permeability of sheep lung following endotoxemia. *J. Clin. Invest.*, 68:1253–1260, 1981.
57. Johnson, A. and A.B. Malik, Pulmonary edema after glass bead microembolization: protective effect of granulocytopenia. *J. Appl. Physiol.*, 52:155–161, 1982.
58. Ward, P.A., G.O. Till, R. Kunkel, and C. Beauchamp, Evidence for role of hydroxyl radical in complement and neutrophil-dependent tissue injury. *J. Clin. Invest.*, 72:789–801, 1983.
59. Flick, M.R., J.M. Hoeffel, and N.C. Staub, Superoxide dismutase with heparin prevents increased lung vascular permeability during air emboli in sheep. *J. Appl. Physiol.*, 55:1284–1291, 1983.
60. Tvedten, H.W., G.O. Till, and P.A. Ward, Mediators of lung injury in mice following systemic activation of complement. *Am. J. Pathol.*, 119:92–100, 1985.
61. Martin, W.J., II, J.E. Gadek, G.W. Hunninghake, and R.G. Crystal, Oxidant injury of lung parenchymal cells. *J. Clin. Invest.*, 68:1277–1288, 1981.
62. Fairman, R.P., F.L. Glauser, and R. Falls, Increases in lung lymph and albumin clearance with ethchlorvynol. *J. Appl. Physiol.*, 50:1151–1155, 1981.
63. Raj, J.U. and R.D. Bland, Neutrophil depletion does not prevent oxygen-induced lung injury in rabbits. *Chest*, 83:20S–21S, 1983.
64. Hofman, W.F. and I.C. Ehrhart, Permeability edema in dog lung depleted of blood components. *J. Appl. Physiol.*, 57:147–153, 1984.
65. Julien, M., J.M. Hoeffel, and M.R. Flick, Oleic acid lung injury in sheep. *J. Appl. Physiol.*, 60:433–440, 1986.
66. Braude, S., J. Apperley, T. Krausz, J.M. Goldman, and D. Royston, Adult respiratory distress syndrome after allogeneic bone-marrow transplantation: evidence for a neutrophil-independent mechanism. *Lancet*, 1:1239–1242, 1985.
67. Maunder, R.J., R.C. Hackman, E. Riff, R.K. Albert, and S.C. Springmeyer, Occurrence of the adult respiratory distress syndrome in neutropenic patients. *Am. Rev. Respir. Dis.*, 133:313–316, 1986.
68. Ognibene, F.P., S.E. Martin, M.M. Parker, T. Schlesinger, P. Roach, C. Burch, J.H. Shelhamer, and J.E. Parrillo. Adult respiratory distress syndrome in patients with severe neutropenia. *N. Engl. J. Med.*, 315:547–551, 1986.
69. Laufe, M.D., R.H. Simon, A. Flint, and J.B. Keller, Adult respiratory distress syndrome in neutropenic patients. *Am. J. Med.*, 80:1022–1026, 1986.
70. Mizel, S.B., J.-M. Dayer, S.M. Krane, and S.E. Mergenhagen, Stimulation of rheumatoid synovial cell collagenase and prostaglandin production by partially purified lymphocyte-activating factor (interleukin 1). *Proc. Natl. Acad. Sci. U.S.A.*, 78:2474–2477, 1981.
71. Baracos, V., H.P. Rodemann, C.A. Dinarello, and A.L. Goldberg, Stimulation of muscle protein degradation and prostaglandin E_2 release by leukocytic pyrogen (interleukin-l). A mechanism for the increased degradation of muscle proteins during fever. *N. Engl. J. Med.*, 308:553–558, 1983.
72. Evequoz, V., F. Bettens, F. Kristensen, U. Trechsel, B.M. Stadler, J.-M. Dayer, A.L. DeWeck, and H. Fleisch, Interleukin 2-independent stimulation of rabbit chondrocyte collagenase and prostaglandin E_2 production by an interleukin 1-like factor. *Eur. J. Immunol.*, 14:490–495, 1984.

73. Dayer, J.-M., C. Zavadil-Grob, C. Ucla, and B. Mach, Induction of human interleukin 1 mRNA measured by collagenase- and prostaglandin E_2-stimulating activity in rheumatoid synovial cells. *Eur. J. Immunol.*, 14:898–901, 1984.
74. Bernheim, H.A. and C.A. Dinarello, Effects of purified human interleukin-1 on the release of prostaglandin E_2 from fibroblasts. *Br. J. Rheumatol.*, 24 (suppl.):122–127, 1985.
75. Dayer, J-M., B. de Rochemonteix, B. Burrus, S. Demczuk, and C.A. Dinarello, Human recombinant interleukin 1 stimulates collagenase and prostaglandin E_2 production by human synovial cells. *J. Clin. Invest.*, 77:645–648, 1986.
76. Zucali, J.R., C.A. Dinarello, D.J. Oblon, M.A. Gross, L. Anderson, and R.S. Weiner, Interleukin 1 stimulates fibroblasts to produce granulocyte-macrophage colony-stimulating activity and prostaglandin E_2. *J. Clin. Invest.*, 77:1857–1863, 1986.
77. Rupp, E.A., P.M. Cameron, C.S. Ranawat, J.A. Schmidt, and E.K. Bayne, Specific bioactivities of monocyte-derived interleukin 1α and interleukin 1β are similar to each other on cultured murine thymocytes and on cultured human connective tissue cells. *J. Clin. Invest.*, 78:836–839, 1986.
78. Chang, J., S.C. Gilman, and A.J. Lewis, Interleukin 1 activates phospholipase A_2 in rabbit chondrocytes: a possible signal for IL 1 action. *J. Immunol.*, 136:1283–1287, 1986.
79. Xiao, D.-M. and L. Levine, Stimulation of arachidonic acid metabolism: differences in potencies of recombinant human interleukin-1α and interleukin-1β on two cell lines. *Prostaglandins*, 32:709–718, 1986.
80. Raz, A., A. Wyche, N. Siegel, and P. Needleman, Regulation of fibroblast cyclooxygenase synthesis by interleukin-1. *J. Biol. Chem.*, 263:3022–3028 1988.
81. Klebanoff, S.J., M.A. Vadas, J.M. Harlan, L.H. Sparks, J.R. Gamble, J.M. Agosti, and A.M. Waltersdorph, Stimulation of neutrophils by tumor necrosis factor. *J. Immunol.*, 136:4220–4225, 1986.
82. Atkinson, Y.H., W.A. Marasco, A.F. Lopez, and M.A. Vadas, Recombinant human tumor necrosis factor-α regulation of *N*-formylmethionylleucylphenylalanine receptor affinity and function on human neutrophils. *J. Clin. Invest.*, 81:759–765, 1988.
83. Sullivan, G.W., H.T. Carper, W.J. Novick, Jr., and G.L. Mandell, Inhibition of the inflammatory action of interleukin-1 and tumor necrosis factor (alpha) on neutrophil function by pentoxifylline. *Infect. Immun.*, 56:1722–1729, 1988.
84. Matsubara, T. and M. Ziff, Increased superoxide anion release from human endothelial cells in response to cytokines. *J. Immunol.*, 137:3295–3298, 1986.
85. Klempner, M.S., C.A. Dinarello, W.R. Henderson, and J.I. Gallin, Stimulation of neutrophil oxygen-dependent metabolism by human leukocytic pyrogen. *J. Clin. Invest.*, 64:996–1002, 1979.
86. Perussia, B., M. Kobayashi, M.E. Rossi, I. Anegon, and G. Trinchieri, Immune interferon enhances functional properties of human granulocytes: role of Fc receptors and effect of lymphotoxin, tumor necrosis factor, and granulocyte-macrophage colony-stimulating factor. *J. Immunol.*, 138:765–774, 1987.
87. Berger, M., E.M. Wetzler, and R.S. Wallis, Tumor necrosis factor is the major monocyte product that increases complement receptor expression on mature human neutrophils. *Blood*, 71:151–158, 1988.
88. Gamble, J.R., J.M. Harlan, S.J. Klebanoff, and M.A. Vadas, Stimulation of the adherence of neutrophils to umbilical vein endothelium by human recombinant tumor necrosis factor. *Proc. Natl. Acad. Sci. U.S.A.*, 82:8667–8671, 1985.

89. Bevilacqua, M.P., J.S. Pober, M.E. Wheeler, R.S. Cotran, and M.A. Gimbrone, Jr., Interleukin 1 acts on cultured human vascular endothelium to increase the adhesion of polymorphonuclear leukocytes, monocytes, and related leukocyte cell lines. *J. Clin. Invest.*, 76:2003–2011, 1985.
90. Dunn, C.J. and W.E. Fleming, The role of interleukin-1 in the inflammatory response with particular reference to endothelial cell-leukocyte adhesion, in *The Physiologic, Metabolic, and Immunologic Actions of Interleukin-1*, Alan R. Liss, New York, 1985, 45–54.
91. Schleimer, R.P. and B.K. Rutledge, Cultured human vascular endothelial cells acquire adhesiveness for neutrophils after stimulation with interleukin 1, endotoxin, and tumor-promoting phorbol diesters. *J. Immunol.*, 136:649–654, 1986.
92. Pohlman, T.H., K.A. Stanness, P.G. Beatty, H.D. Ochs, and J.M. Harlan, An endothelial cell surface factor(s) induced in vitro by lipopolysaccharide, interleukin 1, and tumor necrosis factor-increases neutrophil adherence by a CDw18-dependent mechanism. *J. Immunol.*, 136:4548–4553, 1986.
93. Broudy, V.C., J.M. Harlan, and J.W. Adamson, Disparate effects of tumor necrosis factor-α/cachectin and tumor necrosis factor-β/lymphotoxin on hematopoietic growth factor production and neutrophil adhesion molecule expression by cultured human endothelial cells. *J. Immunol.*, 138:4298–4302, 1987.
94. Goldblum, S.E., D.A. Cohen, M.N. Gillespie, and C.J. McClain, Interleukin-1-induced granulocytopenia and pulmonary leukostasis in rabbits. *J. Appl. Physiol.*, 62:122–128, 1987.
95. Goldblum, S.E., K. Yoneda, D.A. Cohen, and C.J. McClain, Provocation of pulmonary vascular endothelial injury in rabbits by human recombinant interleukin-1β. *Infect. Immun.*, 56:2255–2263, 1988.
96. Goldblum, S.E., M. Jay, K. Yoneda, D.A. Cohen, C.J. McClain, and M.N. Gillespie, Monokine-induced acute lung injury in rabbits. *J. Appl. Physiol.*, 63:2093–2100, 1987.
97. Stephens, K.E., A. Ishizaka, J.W. Larrick, and T.A. Raffin, Tumor necrosis factor causes increased pulmonary permeability and edema. Comparison to septic acute lung injury. *Am. Rev. Respir. Dis.*, 137:1364–1370, 1988.
98. Pober, J.S., M.P. Bevilacqua, D.L. Mendrick, L.A. Lapierre, W. Fiers, and M.A. Gimbrone, Jr., Two distinct monokines, interleukin 1 and tumor necrosis factor, each independently induce biosynthesis and transient expression of the same antigen on the surface of cultured human vascular endothelial cells. *J. Immunol.*, 136:1680–1687, 1986.
99. Cotran, R.S., M.A. Gimbrone, Jr., M.P. Bevilacqua, D.L. Mendrick, and J.S. Pober, Induction and detection of a human endothelial activation antigen in vivo. *J. Exp. Med.*, 164:661–666, 1986.
100. Pober, J.S., M.A. Gimbrone, Jr., L.A. Lapierre, D.L. Mendrick, W. Fiers, R. Rothlein, and T.A. Springer, Overlapping patterns of activation of human endothelial cells by interleukin 1, tumor necrosis factor, and immune interferon. *J. Immunol.*, 137:1893–1896, 1986.
101. Messadi, D.V., J.S. Pober, W. Fiers, M.A. Gimbrone, Jr., and G.F. Murphy, Induction of an activation antigen on postcapillary venular endothelium in human skin organ culture. *J. Immunol.*, 139:1557–1562, 1987.
102. Pober, J.S., L.A. Lapierre, A.H. Stolpen, T.A. Brock, T.A. Springer, W. Fiers, M.P. Bevilacqua, D.L. Mendrick, and M.A. Gimbrone, Jr., Activation of cultured human endothelial cells by recombinant lymphotoxin: comparison with tumor necrosis factor and interleukin 1 species. *J. Immunol.*, 138:3319–3324, 1987.

103. Bevilacqua, M.P. and M.A. Gimbrone, Jr., Inducible endothelial functions in inflammation and coagulation. *Semin. Thromb. Hemost.*, 13:425–433, 1987.
104. Bevilacqua, M.P., J.S. Pober, G.R. Majeau, R.S. Cotran, and M.A. Gimbrone, Jr., Interleukin 1 (IL-l) induces biosynthesis and cell surface expression of procoagulant activity in human vascular endothelial cells. *J. Exp. Med.*, 160:618–623, 1984.
105. Bevilacqua, M.P., J.S. Pober, M.E. Wheeler, R.S. Cotran, and M.A. Gimbrone, Jr., Interleukin-1 activation of vascular endothelium. Effects on procoagulant activity and leukocyte adhesion. *Am. J. Pathol.*, 121:393–403, 1985.
106. Nawroth, P.P., D.A. Handley, C.T. Esmon, and D.M. Stern, Interleukin 1 induces endothelial cell procoagulant while suppressing cell-surface anticoagulant activity. *Proc. Natl. Acad. Sci. U.S.A.*, 83:3460–3464, 1986.
107. Bevilacqua, M.P., J.S. Pober, G.R. Majeau, W. Fiers, R.S. Cotran, and M.A. Gimbrone, Jr., Recombinant tumor necrosis factor induces procoagulant activity in cultured human vascular endothelium: characterization and comparison with the actions of interleukin 1. *Proc. Natl. Acad. Sci. U.S.A.*, 83:4533–4537, 1986.
108. Nawroth, P.P. and D.M. Stern, Modulation of endothelial cell hemostatic properties by tumor necrosis factor. *J. Exp. Med.*, 163:740–745, 1986.
109. Beck, G., G.S. Habicht, J.L. Benach, and F. Miller, Interleukin 1: a common endogenous mediator of inflammation and the local Schwartzman reaction. *J. Immunol.*, 136:3025–3031, 1986.
110. Movat, H.Z., Tumor necrosis factor and interleukin-l, role in acute inflammation and microvascular injury. *J. Lab. Clin. Med.*, 110:668–681, 1987.
111. Granstein, R.D., R. Margolis, S.B. Mizel, and D.N. Sauder, In vivo inflammatory activity of epidermal cell-derived thymocyte activating factor and recombinant interleukin 1 in the mouse. *J. Clin. Invest.*, 77:1020–1027, 1986.
112. Cybulsky, M.I., I.G. Colditz, and H.Z. Movat, The role of interleukin-l in neutrophil leukocyte emigration induced by endotoxin. *Am. J. Pathol.*, 124:367–372, 1986.
113. Habicht, G.S. and G. Beck, *The Physiologic, Metabolic, and Immunologic Actions of Interleukin-1*, Alan R. Liss, New York, 1985, 13–24.
114. Sayers, T.J., T.A. Wiltrout, C.A. Bull, A.C. Denn, III, A.M. Pilaro, and B. Lokesh, Effect of cytokines on polymorphonuclear neutrophil infiltration in the mouse. Prostaglandin- and leukotriene-independent induction of infiltration by IL-l and tumor necrosis factor. *J. Immunol.*, 141:1670–1677, 1988.
115. Cybulsky, M.I., D.J. McComb, and H.Z. Movat. Neutrophil leukocyte emigration induced by endotoxin. Mediator roles of interleukin 1 and tumor necrosis factor α. *J. Immunol.*, 140:3144–3149, 1988.
116. Okusawa, S., J.A. Gelfand, T. Ikejima, R.J. Connolly, and C.A. Dinarello, Interleukin 1 induces a shock-like state in rabbits. Synergism with tumor necrosis factor and the effect of cyclooxygenase inhibition. *J. Clin. Invest.*, 81:1162–1172, 1988.
117. Tracey, K.J., B. Beutler, S.F. Lowry, J. Merryweather, S. Wolpe, I.W. Milsark, R.J. Hariri, T.J. Fahey, III, A. Zentella, J.D. Albert, G.T. Shires, and A. Cerami, Shock and tissue injury induced by recombinant human cachectin. *Science*, 234:470–474, 1986.
118. Bochner, B.S., S.D. Landy, M. Plaut, C.A. Dinarello, and R.P. Schleimer, Interleukin 1 production by human lung tissue. 1. Identification and characterization. *J. Immunol.*, 139:2297–2302, 1987.
119. Lamontagne, L., J. Gauldie, A. Stadnyk, C. Richards, and E. Jenkins, In vivo initiation of unstimulated in vitro interleukin-l release by alveolar macrophages. *Am. Rev. Respir. Dis.*, 131:326–330, 1985.

120. Shellito, J. and H.B. Kaltreider, Heterogeneity of immunologic function among subfractions of normal rat alveolar macrophages. *Am. Rev. Respir. Dis.*, 129:747–753, 1984)
121. Oghiso, Y., Morphologic and functional heterogeneity among rat alveolar macrophage fractions isolated by centrifugation on density gradients. *J. Leukoc. Biol.*, 42:188–196, 1987.
122. Murphy, P.A., P.L. Simon, and W.F. Willoughby, Endogenous pyrogens made by rabbit peritoneal exudate cells are identical with lymphocyte-activating factors made by rabbit alveolar macrophages. *J. Immunol.*, 124:2498–2501, 1980.
123. Simon, P.L. and W.F. Willoughby, The role of subcellular factors in pulmonary immune function: physicochemical characterization of two distinct species of lymphocyte-activating factor produced by rabbit alveolar macrophages. *J. Immunol.*, 126;1534–1541, 1981.
124. Furutani, Y., M. Notake, M. Yamayoshi, J.-I. Yamagishi, H. Nomura, M. Ohue, R. Furuta, T. Fukui, M. Yamada, and S. Nakamura, Cloning and characterization of the cDNAs for human and rabbit interleukin-1 precursor. *Nucleic Acids Res.*, 13:5869–5882, 1985.
125. Wewers, M.D., S.I. Rennard, A.J. Hance, P.B. Bitterman, and R.G. Crystal, Normal human alveolar macrophages obtained by bronchoalveolar lavage have a limited capacity to release interleukin-1. *J. Clin. Invest.*, 74:2208–2218, 1984.
126. Elias, J.A., A.D. Schreiber, K. Gustilo, P. Chien, M.D. Rossman, P.J. Lammie, and R.P. Daniele, Differential interleukin 1 elaboration by unfractionated and density fractionated human alveolar macrophages and blood monocytes: relationship to cell maturity. *J. Immunol.*, 135:3198–3204, 1985.
127. Eden, E. and G.M. Turino, Interleukin-1 secretion by human alveolar macrophages stimulated with endotoxin is augmented by recombinant immune (gamma) interferon. *Am. Rev. Respir. Dis.*, 133:455–460, 1986.
128. Gillespie, M.N., S.E. Goldblum, D.A. Cohen, and C.J. McClain, Interleukin 1 bioactivity in the lungs of rats with monocrotaline-induced pulmonary hypertension. *Proc. Soc. Exp. Biol. Med.*, 187:26–32, 1988.
129. Matsushima, K., K. Morishita, T. Yoshimura, S. Lavu, Y. Kobayashi, W. Lew, E. Appella, H.F. Kung, E.J. Leonard, and J.J. Oppenheim, Molecular cloning of a human monocyte-derived neutrophil chemotactic factor (MDNCF) and the induction of MDNCF mRNA by interleukin 1 and tumor necrosis factor. *J. Exp. Med.*, 167:1883–1893, 1988.
130. Munker, R., J. Gasson, M. Ogawa, and H.P. Koeffler, Recombinant human TNF induces production of granulocyte-monocyte colony-stimulating factor. *Nature*, 323:79–82, 1986.
131. Broudy, V.C., K. Kaushansky, G.M. Segal, J.M. Harlan, and J.W. Adamson, Tumor necrosis factor type α stimulates human endothelial cells to produce granulocyte/macrophage colony-stimulating factor. *Proc. Natl. Acad. Sci. U.S.A.*, 83:7467–7471, 1986.
132. Bagby, G.C., Jr., C.A. Dinarello, P. Wallace, C. Wagner, S. Hefeneider, and E. McCall, Interleukin 1 stimulates granulocyte macrophage colony-stimulating activity release by vascular endothelial cells. *J. Clin. Invest.*, 78:1316–1323, 1986.
133. Mochizuki, D.Y., J.R. Eisenman, P.J. Conlon, A.D. Larsen, and R.J. Tushinski, Interleukin 1 regulates hematopoietic activity, a role previously ascribed to hemopoietin 1. *Proc. Natl. Acad. Sci. U.S.A.*, 84:5267–5271, 1987.
134. Sieff, C.A., S. Tsai, and D.V. Faller, Interleukin 1 induces cultured human endothelial cell production of granulocyte-macrophage colony-stimulating factor. *J. Clin. Invest.*, 79:48–51, 1987.

135. Segal, G.M., E. McCall, T. Stueve, and G.C. Bagby, Jr., Interleukin 1 stimulates endothelial cells to release multilineage human colony-stimulating activity. *J. Immunol.*, 138:1772–1778, 1987.
136. Yang, Y.-C., S. Tsai, G.G. Wong, and S.C. Clark, Interleukin-1 regulation of hematopoietic growth factor production by human stromal fibroblasts. *J. Cell. Physiol.*, 134:292–296, 1988.
137. Zsebo, K.M., V.N. Yuschenkoff, S. Schiffer, D. Chang, E. McCall, C.A. Dinarello, M.A. Brown, B. Altrock, and G.C. Bagby, Jr., Vascular endothelial cells and granulopoiesis: interleukin-1 stimulates release of G-CSF and GM-CSF. *Blood*, 71:99–103, 1988.
138. Fibbe, W.E., J. van Damme, A. Billiau, H.M. Goselink, P.J. Voogt, G. van Eeden, P. Ralph, B.W. Altrock and J.H.F. Falkenburg, Interleukin 1 induces human marrow stromal cells in long-term culture to produce granulocyte colony-stimulating factor and macrophage colony-stimulating factor. *Blood*, 71:430–435, 1988.
139. Fibbe, W.E., J. van Damme, A. Billiau, N. Duinkerken, E. Lurvink, P. Ralph, B.W. Altrock, K. Kaushansky, R. Willemze, and J.H.F. Falkenburg, Human fibroblasts produce granulocyte-CSF, macrophage-CSF, and granulocyte-macrophage-CSF following stimulation by interleukin-l and poly (rI). Poly (rc). *Blood*, 72:860–866, 1988.
140. Vogel, S.N., S.D. Douches, E.N. Kaufman, and R. Neta, Induction of colony stimulating factor in vivo by recombinant interleukin 1 and recombinant tumor necrosis factor. *J. Immunol.*, 138:2143–2148, 1987.
141. Remick, D.G., J. Larrick, and S.L. Kunkel, Tumor necrosis factor-induced alterations in circulating leukocyte populations. *Biochem. Biophys. Res. Commun.*, 141:818–824, 1986.
142. Remick, D.G., R.G. Kunkel, J.W. Larrick, and S.L. Kunkel, Acute in vivo effects of human recombinant tumor necrosis factor. *Lab. Invest.*, 56:583–590, 1987.
143. Goldblum, S.E., B. Hennig, M. Jay, K. Yoneda, and C.J. McClain, Tumor necrosis factor α-induced pulmonary vascular endothelial injury. *Infect. Immun.*, 57:1218–1226, 1989.
144. Ulich, T.R., J. Del Castillo, M. Keys, G.A. Granger, and R.-X. Ni, Kinetics and mechanisms of recombinant human interleukin 1 and tumor necrosis factor-α-induced changes in circulating numbers of neutrophils and lymphocytes. *J. Immunol.*, 139:3406–3415, 1987.
145. Smith, R.J., S.C. Speziale, and B.J. Bowman, Properties of interleukin-l as a complete secretagogue for human neutrophils. *Biochem. Biophys. Res. Commun.*, 130:1233–1240, 1985.
146. Smith, R.J., B.J. Bowman, and S.C. Speziale, *The Physiologic, Metabolic, and Immunologic Actions of Interleukin-1*, Alan R. Liss, New York, 1985, 31–43.
147. Rhyne, J.A., S.B. Mizel, R.G. Taylor, M. Chedid, and C.E. McCall, Characterization of the human interleukin 1 receptor on human polymorphonuclear leukocytes. *Clin. Immunol. Immunopathol.*, 48:354–361, 1988.
148. Schutze, S., P. Scheurich, C. Schluter, U. Ucer, K. Pfizenmaier, and M. Kronke, Tumor necrosis factor-induced changes of gene expression in U937 cells. Differentiation-dependent plasticity of the responsive state. *J. Immunol.*, 140:3000–3005, 1988.
149. Shalaby, M.R., M.A. Palladino, Jr., S.E. Hirabayashi, T.E. Eessalu, G.D. Lewis, H.M. Shepard, and B.B. Aggarwal, Receptor binding and activation of polymorphonuclear neutrophils by tumor necrosis factor-alpha. *J. Leukoc. Biol.*, 41:196–204, 1987.

150. Trinchieri, G., M. Kobayashi, M. Rosen, R. Loudon, M. Murphy, and B. Perussia, Tumor necrosis factor and lymphotoxin induce differentiation of human myeloid cell lines in synergy with immune interferon. *J. Exp. Med.*, 164:1206–1225, 1986.
151. Klempner, M.S., C.A. Dinarello, and J.I. Gallin, Human leukocytic pyrogen induces release of specific granule contents from human neutrophils. *J. Clin. Invest.*, 61:1330–1336, 1978.
152. Georgilis, K., C. Schaefer, C.A. Dinarello, and M.S. Klempner, Human recombinant interleukin 1β has no effect on intracellular calcium or on functional responses of human neutrophils. *J. Immunol.*, 138:3403–3407, 1987.
153. Subramanian, N. and M.A. Bray, Interleukin 1 releases histamine from human basophils and mast cells in vitro. *J. Immunol.*, 138:271–275, 1987.
154. Larrick, J.W., D. Graham, K. Toy, L.S. Lin, G. Senyk, and B.M. Fendly, Recombinant tumor necrosis factor causes activation of human granulocytes. *Blood*, 69:640–644, 1987.
155. Shalaby, M.R., B.B. Aggarwal, E. Rinderknecht, L.P. Svedersky, B.S. Finkle, and M.A. Palladino, Jr., Activation of human polymorphonuclear neutrophil functions by interferon-γ and tumor necrosis factors. *J. Immunol.*, 135:2069–2073, 1985.
156. Djeu, J.Y., D.K. Blanchard, D. Halkias, and H. Friedman, Growth inhibition of candida albicans by human polymorphonuclear neutrophils: activation by interferon-γ and tumor necrosis factor. *J. Immunol.*, 137:2980–2984, 1986.
157. Ming, W.J., L. Bersani, and A. Mantovani, Tumor necrosis factor is chemotactic for monocytes and polymorphonuclear leukocytes. *J. Immunol.*, 138:1469–1474, 1987.
158. Dower, S.K., S.R. Kronheim, C.J. March, P.J. Conlon, T.P. Hopp, S. Gillis, and D.L. Urdal, Detection and characterization of high affinity plasma membrane receptors for human interleukin-1. *J. Exp. Med.*, 162:501–515, 1985.
159. Nawroth, P.P., I. Bank, D. Handley, J. Cassimeris, L. Chess, and D. Stern, Tumor necrosis factor/cachectin interacts with endothelial cell receptors to induce release of interleukin 1. *J. Exp. Med.*, 163:1363–1375, 1986.
160. Montesano, R., L. Orci, and P. Vassalli, Human endothelial cell cultures: phenotypic modulation by leukocyte interleukins. *J. Cell. Physiol.*, 122:424–434, 1985.
161. Stolpen, A.H., E.C. Guinan, W. Fiers, and J.S. Pober, Recombinant tumor necrosis factor and immune interferon act singly and in combination to reorganize human vascular endothelial cell monolayers. *Am. J. Pathol.*, 123:16–24, 1986.
162. Sato, N., T. Goto, K. Haranaka, N. Satomi, H. Nariuchi, T. Mano-Hirano, and Y. Sawasaki, Actions of tumor necrosis factor on cultured vascular endothelial cells: morphologic modulation, growth inhibition, and cytotoxicity. *JNCL*, 76:1113–1121, 1986.
163. Ooi, B.S., E.P. MacCarthy, A. Hsu, and Y.M. Ooi, Human mononuclear cell modulation of endothelial cell proliferation. *J. Lab. Clin. Med.*, 102:428–433, 1983.
164. Kahaleh, M.B., E.A. Smith, Y. Soma, and E.C. LeRoy, Effect of lymphotoxin and tumor necrosis factor on endothelial and connective tissue cell growth and function. *Clin. Immunol. Immunopathol.*, 49:261–272, 1988.
165. Frater-Schroder, M., W. Risau, R. Hallmann, P. Gautschi, and P. Bohlen, Tumor necrosis factor type α, a potent inhibitor of endothelial cell growth in vitro, is angiogenic in vivo. *Proc. Natl. Acad. Sci. U.S.A.*, 84:5277–5281, 1987.
166. Bevilacqua, M.P., R.R. Schleef, M.A. Gimbrone, Jr., and D.J. Loskutoff, Regulation of the fibrinolytic system of cultured human vascular endothelium by interleukin 1. *J. Clin. Invest.*, 78:587–591, 1986.

167. Emeis, J.J. and T. Kooistra, Interleukin 1 and lipopolysaccharide induce an inhibitor of tissue-type plasminogen activator in vivo and in cultured endothelial cells. *J. Exp. Med.,* 163:1260–1266, 1986.
168. Nachman, R.L., K.A. Hajjar, R.L. Silverstein, and C.A. Dinarello, Interleukin 1 induces endothelial cell synthesis of plasminogen activator inhibitor. *J. Exp. Med.,* 163:1595–1600, 1986.
169. Gramse, M., F. Breviario, G. Pintucci, I. Millet, E. Dejana, J. van Damme, M.B. Donati, and L. Mussoni, Enhancement by interleukin-l (IL-1) of plasminogen activator inhibitor (PA-I) activity in cultured human endothelial cells. *Biochem. Biophys. Res. Commun.,* 139:720–727, 1986.
170. Nawroth, P., D. Handley, G. Matsueda, R. De Waal, H. Gerlach, D. Blohm, and D. Stern, Tumor necrosis factor/cachectin-induced intravascular fibrin formation in meth A fibrosarcomas. *J. Exp. Med.,* 168:637–647, 1988.
171. Schleef, R.R., M.P. Bevilacqua, M. Sawdey, M.A. Gimbrone, Jr., and D.J. Loskutoff, Cytokine activation of vascular endothelium. Effects on tissue-type plasminogen activator and type 1 plasminogen activator inhibitor. *J. Biol. Chem.,* 263:5797–5803, 1988.
172. Bloom, A.L. and D.P. Thomas Ed., *Haemostasis and Thrombosis.* Churchill Livingstone, New York, 1987.
173. Malik, A.B., A. Johnson, M.V. Tahamont, H. van der Zee, and F.A. Blumenstock, Role of blood components in mediating lung vascular injury after pulmonary vascular thrombosis. *Chest,* 83:21S–24S, 1983.
174. Bizios, R., F.L. Minnear, H. van der Zee, and A.B. Malik, Effects of cyclooxygenase and lipoxygenase on lung fluid balance after thrombin. *J. Appl. Physiol.,* 55:462–471, 1983.
175. Cooper, J.A., S.J. Solano, R. Bizios, J.E. Kaplan, and A.B. Malik, Pulmonary neutrophil kinetics after thrombin-induced intravascular coagulation. *J. Appl. Physiol.,* 57:826–832, 1984.
176. Hajjar, K.A., D.P. Hajjar, R.L. Silverstein, and R.L. Nachman, Tumor necrosis factor-mediated release of platelet-derived growth factor from cultured endothelial cells. *J. Exp. Med.,* 166:235–245, 1987.
177. Leung, D.Y., R.S. Geha, J.W. Newburger, J.C. Burns, W. Fiers, L.A. Lapierre, and J.S. Pober, Two monokines, interleukin 1 and tumor necrosis factor, render cultured vascular endothelial cells susceptible to lysis by antibodies circulating during Kawasaki syndrome. *J. Exp. Med.,* 164:1958–1972, 1986.
178. Dinarello, C.A., S.O. Marnoy, and L.J. Rosenwasser, Role of arachidonate metabolism in the immunoregulatory function of human leukocytic pyrogen/lymphocyte-activating factor/interleukin 1. *J. Immunol.,* 130:890–895, 1983.
179. Kunkel, S.L., S.W. Chensue, and S.H. Phan, Prostaglandins as endogenous mediators of interleukin 1 production. *J. Immunol.,* 136:186–192, 1986.
180. Phadke, K., D.G. Carlson, B.D. Gitter, and L.D. Butler, Role of interleukin 1 and interleukin 2 in rat and mouse arthritis models. *J. Immunol.,* 136:4085–4091, 1986.
181. Dinarello, C.A., J.G. Cannon, J.W. Mier, H.A. Bernheim, G. LoPreste, D.L. Lynn, R.N. Love, A.C. Webb, P.E. Auron, R.C. Reuben, A. Rich, S.M. Wolff, and S.D. Putney, Multiple biological activities of human recombinant interleukin 1. *J. Clin. Invest.,* 77:1734–1739, 1986.
182. Bernheim, H.A., T.M. Gilbert, and J.T. Stitt, Prostaglandin E levels in third ventricular cerebrospinal fluid of rabbits during fever and changes in body temperature. *J. Physiol. (Lond.),* 301:69–78, 1980.

183. Dinarello, C.A. and H.A. Bernheim, Ability of human leukocytic pyrogen to stimulate brain prostaglandin synthesis in vitro. *J. Neurochem.*, 37:702–708, 1981.
184. Coceani, F., I. Bishai, C.A. Dinarello, and F.A. Fitzpatrick, Prostaglandin E_2 and thromboxane B_2 in cerebrospinal fluid of afebrile and febrile cat. *Am. J. Physiol.*, 244:R785-R793, 1983.
185. Farrar, W.L. and J.L. Humes, The role of arachidonic acid metabolism in the activities of interleukin 1 and 2. *J. Immunol.*, 135:1153–1159, 1985.
186. Burch, R.M., J.R. Connor, and J. Axelrod, Interleukin 1 amplifies receptor-mediated activation of phospholipase A_2 in 3T3 fibroblasts. *Proc. Natl. Acad. Sci. U.S.A.*, 85:6306–6309, 1988.
187. Malik, A.B., M.B. Perlman, J.A. Cooper, T. Noonan, and R. Bizios, Pulmonary microvascular effects of arachidonic acid metabolites and their role in lung vascular injury. *Fed. Proc.*, 44:36–42, 1985.
188. Brigham, K.L., Metabolites of arachidonic acid in experimental lung vascular injury. *Fed. Proc.*, 44:43–45, 1985.
189. Williams, T.J., Prostaglandin E_2, prostaglandin I_2, and the vascular changes of inflammation. *Br. J. Pharmacol.*, 65:517–524, 1979.
190. Wedmore, C.V. and T.J. Williams, Control of vascular permeability by polymorphonuclear leukocytes in inflammation. *Nature*, 289:646–650, 1981.
191. Henson, P.M., G.L. Larsen, R.O. Webster, B.C. Mitchell, A.J. Goins, and J.E. Henson, Pulmonary microvascular alterations and injury induced by complement fragments: synergistic effect of complement activation, neutrophil sequestration, and prostaglandins. *Ann. N.Y. Acad. Sci.*, 384:287–300, 1982.
192. Larsen, G.L., R.O. Webster, G.S. Worthen, R.S. Gumbay, and P.M. Henson, Additive effect of intravascular complement activation and brief episodes of hypoxia in producing increased permeability in the rabbit lung. *J. Clin. Invest.*, 5:902–910, 1985.
193. Chang, S.-W., C.O. Feddersen, P.M. Henson, and N.F. Voelkel, Platelet-activating factor mediates hemodynamic changes and lung injury in endotoxin-treated rats. *J. Clin. Invest.*, 79:1498–1509, 1987.
194. Weinberg, J.R., D.J.M. Wright, and A. Guz, Interleukin-1 and tumor necrosis factor cause hypotension in the conscious rabbit. *Clin. Sci.*, 75:251–255, 1988.
195. Tracey, K.J., Y. Fong, D.G. Hesse, K.R. Manogue, A.T. Lee, G.C. Kuo, S.F. Lowry, and A. Cerami, Anti-cachectin/TNF monoclonal antibodies prevent septic shock during lethal bacteremia. *Nature*, 330:662–664, 1987.
196. Johnson, J.E., B. Meyrick, G. Jesmok, and K.L. Brigham, Human recombinant tumor necrosis factor alpha infusion mimics endotoxemia in awake sheep. *J. Appl. Physiol.*, 66:1448–1454, 1989.
197. Wheeler, A.P., G. Jesmok, and K.L. Brigham, Human recombinant tumor necrosis factor alpha (rTNF) causes endotoxin-like changes in sheep hemodynamics and lung mechanics. *Am. Rev. Respir. Dis.*, 137:A243, 1988.
198. Libby, P., J.M. Ordovas, L.K. Birinyi, K.R. Auger, and C.A. Dinarello, Inducible interleukin-1 gene expression in human vascular smooth muscle cells. *J. Clin. Invest.*, 78:1432–1438, 1986.
199. Warner, S.J.C., K.R. Auger, and P. Libby, Human interleukin 1 induces interleukin 1 gene expression in human vascular smooth muscle cells. *J. Exp. Med.*, 165:1316–1331, 1987.
200. Libby, P., S.J.C. Warner, and G.B. Friedman, Interleukin 1: a mitogen for human vascular smooth muscle cells that induces the release of growth inhibitory prostanoids. *J. Clin. Invest.*, 81:487–498, 1988.

201. Tracey, K.J., S.F. Lowry, B. Beutler, A. Cerami, J.D. Albert, and G.T. Shires, Cachectin/ tumor necrosis factor mediates changes of skeletal muscle plasma membrane potential. *J. Exp. Med.*, 164:1368–1373, 1986.
202. Rola-Pleszczynski, M. and I. Lemaire, Leukotrienes augment interleukin 1 production by human monocytes. *J. Immunol.*, 135:3958–3961, 1985.
203. Warren, M.K. and P. Ralph, Macrophage growth factor CSF-1 stimulates human monocyte production of interferon, tumor necrosis factor, and colony stimulating activity. *J. Immunol.*, 137:2281–2285, 1986.
204. Cannistra, S.A., A. Rambaldi, D.R. Spriggs, F. Herrmann, D. Kufe, and J.D. Griffin, Human granulocyte-macrophage colony-stimulating factor induces expression of the tumor necrosis factor gene by the U937 cell line and by normal human monocytes. *J. Clin. Invest.*, 79:1720–1728, 1987.
205. Cannistra, S.A., E. Vellenga, P. Groshek, A. Rambaldi, and J.D. Griffin, Human granulocyte-monocyte colony-stimulating factor and interleukin 3 stimulate monocyte cytotoxicty through a tumor necrosis factor-dependent mechanism. *Blood*, 71:672–676, 1988.
206. Collart, M.A., D. Belin, J.-D. Vassalli, S. De Kossodo, and P. Vassalli, γ-Interferon enhances macrophage transcription of the tumor necrosis factor/cachectin, interleukin 1, and urokinase genes, which are controlled by short-lived repressors. *J. Exp. Med.*, 164:2113–2118, 1986.
207. Gerrard, T.L., J.P. Siegel, D.R. Dyer, and K.C. Zoon, Differential effects of interferon-α and interferon-γ on interleukin 1 secretion by monocytes. *J. Immunol.*, 138:2535–2540, 1987.
208. Chensue, S.W., M.P. Davey, D.G. Remick, and S.L. Kunkel, Release of interleukin-1 by peripheral blood mononuclear cells in patients with tuberculosis and active inflammation. *Infect. Immun.*, 52:341–343, 1986.
209. Vismara, D., G. Lombardi, E. Piccolella, and V. Colizzi, Dissociation between interleukin-1 and interleukin-2 production in proliferative response to microbial antigens: restorative effect of exogenous interleukin-2. *Infect. Immun.*, 49:298–304, 1985.
210. Aderka, D., H. Holtmann, L. Toker, T. Hahn, and D. Wallach, Tumor necrosis factor induction by Sendai virus. *J. Immunol.*, 136:2938–2942, 1986.
211. Lotz, M., C.D. Tsoukas, S. Fong, C.A. Dinarello, D.A. Carson, and J.H. Vaughan, Release of lymphokines after Epstein Barr virus infection in vitro. I. Sources of and kinetics of production of interferons and interleukins in normal humans. *J. Immunol.*, 136:3636–3642, 1986.
212. Wright, S.C., A. Jewett, R. Mitsuyasu, and B. Bonavida, Spontaneous cytotoxicity and tumor necrosis factor production by peripheral blood monocytes from AIDS patients. *J. Immunol.*, 141:99–104, 1988.
213. Wannemacher, R.W., Jr., H.L. DuPont, R.S. Pekarek, M.C. Powanda, A. Schwartz, R.B. Hornick, and W.R. Beisel, An endogenous mediator of depression of amino acids and trace metals in serum during typhoid fever. *J. Infect. Dis.*, 126:77–86, 1972.
214. Wannemacher, R.W., Jr., R.S. Pekarek, A.S. Klainer, P.J. Bartelloni, H.L. Dupont, R.B. Hornick, and W.R. Beisel, Detection of a leukocytic endogenous mediator-like mediator of serum amino acid and zinc depression during various infectious illnesses. *Infect. Immun.*, 11:873–875, 1975.
215. Ridel, P.-R., P. Jamet, Y. Robin, and M-A. Bach, Interleukin-1 released by blood-monocyte-derived macrophages from patients with leprosy. *Infect. Immun.*, 52:303–308, 1986.

216. Girardin, E., G.E. Grau, J.-M. Dayer, P. Roux-Lombard, the J5 Study Group, and P.-H. Lambert, Tumor necrosis factor and interleukin-1 in the serum of children with severe infectious purpura. *N. Engl. J. Med.,* 319:397–400, 1988.
217. Kupper, T.S., E.A. Deitch, C.C. Baker, and W. Wong, The human burn wound as a primary source of interleukin-1 activity. *Surgery,* 100:409–414, 1986.
218. McClain, C.M., D. Cohen, L. Ott, C.A. Dinarello, and B. Young, Ventricular fluid interleukin-1 activity in patients with head injury. *J. Lab. Clin. Med.,* 110:48–54, 1987.
219. Waage, A., A. Halstensen, and T. Espevik, Association between tumor necrosis factor in serum and fatal outcome in patients with meningococcal disease. *Lancet,* 1:355–357, 1987.
220. Tabor, D.R., S.K. Burchett, and R.F. Jacobs, Enhanced production of monokines by canine alveolar macrophages in response to endotoxin-induced shock. *Proc. Soc. Exp. Biol. Med.,* 187:408–415, 1988.
221. Kobayashi, K., C. Allred, R. Castriotta, and T. Yoshida, Strain variation of bacillus calmette-guerin-induced pulmonary granuloma formation is correlated with anergy and the local production of migration inhibition factor and interleukin 1. *Am. J. Pathol.,* 119:223–235, 1985.
222. Schwartz, L.W., P.L. Simon, M.G. Boy, and C.E. Card, Interleukin-1 production associated with concanavalin A-induced interstitial pneumonia in the rabbit. *Fed. Proc.,* 44:1260, 1985.
223. Blanchard, D.K., J.Y. Djeu, T.W. Klein, H. Friedman, and W.E. Stewart, II, Induction of tumor necrosis factor by *Legionella pneumophila*. *Infect. Immun.,* 55:433–437, 1987.
224. Martinet, Y., K. Yamauchi, and R.G. Crystal, Differential expression of the tumor necrosis factor/cachectin gene by blood and lung mononuclear phagocytes. *Am. Rev. Respir. Dis.,* 138:659–665, 1988.
225. Siler, T.M., J.E. Swierkosz, and R.O. Webster, Immunoreactive IL-1 is increased in patients at high risk and with established adult respiratory distress syndrome. *Exp. Lung. Res.,* 15:881–894, 1989.
226. Martin, S., K. Maruta, V. Burkart, S. Gillis, and H. Kobl, IL-1 and IFNγ increase vascular permeability. *Immunology,* 64:301–305, 1988.
227. Goldblum, S.E., D.A. Cohen, M. Jay, and C.J. McClain, Interleukin I-induced depression of iron and zinc: role of granulocytes and lactoferrin. *Am. J. Physiol.,* 252:E27-E32, 1987.
228. Kampschmidt, R.F. and T. Jones, Rate of clearance of interleukin-1 from the blood of normal and nephrectomized rats. *Proc. Soc. Exp. Biol. Med.,* 180:170–173, 1985.
229. Hennig, B., S.E. Goldblum, and C.J. McClain, Interleukin-1 (IL–1) and tumor necrosis factor/cachectin (TNF) increase endothelial permeability *in vitro*. *J. Leukoc. Biol.,* 42:551–552, 1987.
230. Schnizlein-Bick, C.T., J.W. Barnard, J.E. Roepke, and R.A. Rhoades, Interleukin-1 increases lung weight without altering vascular pressure in the isolated perfused rat lung. *Am. Rev. Respir. Dis.,* 135:A258 1987.
231. Kasahara, K., K. Kobayashi, Y. Shikama, I. Yoneya, K Soezima, H. Ide, and T. Takahashi, Direct evidence for granuloma-inducing activity of interleukin-1. Induction of experimental pulmonary granuloma formation in mice by interleukin-1-coupled beads. *Am. J. Pathol.,* 130:629–638, 1988.
232. Talmadge, J.E., O. Bowersox, H. Tribble, S.H Lee, H.M. Shepard, and D. Liggitt, Toxicity of tumor necrosis factor is synergistic with γ-interferon and can be reduced with cyclooxygenase inhibitors. *Am. J. Pathol.,* 128:410–425, 1987.

233. Stephens, K.E., A. Ishizaka, Z. Wu, J.W. Larrick, and T.A. Raffin, Granulocyte depletion prevents tumor necrosis factor-mediated acute lung injury in guinea pigs. *Am. Rev. Respir. Dis.*, 138:1300–1307, 1988.
234. White, C.W., P. Ghezzi, C.A. Dinarello, S.A. Caldwell, I.F. McMurtry, and J.E. Repine, Recombinant tumor necrosis factor/cachectin and interleukin 1 pretreatment decreases lung oxidized glutathione accumulation, lung injury, and mortality in rats exposed to hyperoxia. *J. Clin. Invest.*, 79:1868–1873, 1987.
235. Ettinghausen, S.E., E.H. Lipford, III, J.J. Mule, and S.A. Rosenberg, Systemic administration of recombinant interleukin 2 stimulates *in vivo* lymphoid cell proliferation in tissues. *J. Immunol.*, 135:1488–1497, 1985.
236. Lotze, M.T., Y.L. Matory, S.E. Ettinghausen, A.A. Rayner, S.O. Sharrow, C.A.Y. Seipp, M.C. Custer, and S.A. Rosenberg, In vivo administration of purified human interleukin 2. II. Half-life, immunologic effects, and expansion of peripheral lymphoid cells in vivo with recombinant IL-2. *J. Immunol.*, 135:2865–2875, 1985.
237. Ettinghausen, S.E., E.H. Lipford, III, J.J. Mule, and S.A. Rosenberg, Recombinant interleukin 2 stimulates *in vivo* proliferation of adoptively transferred lymphokine-activated killer (LAK) cells. *J. Immunol.*, 135:3623–3635, 1985.
238. Mule, J.J., S. Shu, S.L. Schwarz, and S.A. Rosenberg, Adoptive immunotherapy of established pulmonary metastases with LAK cells and recombinant interleukin-2. *Science*, 225:1487–1489, 1984.
239. Donohue, J.H., M. Rosenstein, A.E. Chang, M.T. Lotze, R.J. Robb, and S.A. Rosenberg, The systemic administration of purified interleukin 2 enhances the ability of sensitized murine lymphocytes to cure a disseminated syngeneic lymphoma. *J. Immunol.*, 132:2123–2128, 1984.
240. Rosenberg, S.A., M.T. Lotze, L.M. Muul, S. Leitman, A.E. Chang, S.E. Ettinghausen, Y.L. Matory, J.M. Skibber, E. Shiloni, J.T. Vetto, C.A. Seipp, C. Simpson, and C.M. Reichert, Observations on the systemic administration of autologous lymphokine-activated killer cells and recombinant interleukin-2 to patients with metastatic cancer. *N. Engl. J. Med.*, 313:1485–1492, 1985.
241. Rosenberg, S.A., M.T. Lotze, L.M. Muul, A.E. Chang, F.P. Avis, S. Leitman, W.M. Linehan, C.N. Robertson, R.E. Lee, J.T. Rubin, C.A. Seipp, C.G. Simpson, and D.E. White, A progress report on the treatment of 157 patients with advanced cancer using lymphokine-activated killer cells and interleukin-2 or high-dose interleukin-2 alone. *N. Engl. J. Med.*, 316:889–897, 1987.
242. West, W.H., K.W. Tauer, J.R. Yannelli, G.D. Marshall, D.W. Orr, G.B. Thurman, and R.K. Oldham, Constant-infusion recombinant interleukin-2 in adoptive immunotherapy of advanced cancer. *N. Engl. J. Med.*, 316:898–905, 1987.
243. Rosenstein, M., S.E. Ettinghausen, and S.A. Rosenberg, Extravasation of intravascular fluid mediated by the systemic administration of recombinant interleukin 2. *J. Immunol.*, 137:1735–1742, 1986.
244. Glauser, F.L., G.G. DeBlois, D.E. Bechard, R.E Merchant, A.J. Grant, A.A. Fowler, and R.P. Fairman, Cardiopulmonary effects of recombinant interleukin-2 infusion in sheep. *J. Appl. Physiol.*, 64:1030–1037, 1988.
245. Fairman, R.P., F.L. Glauser, R.E. Merchant, D. Bechard, A.A. Fowler, Increase of rat pulmonary microvascular permeability to albumin by recombinant interleukin-2. *Cancer Res.*, 47:3528–3532, 1987.
246. Glauser, F.L., G.G. DeBlois, D.E. Bechard, R.E. Merchant, A.J. Grant, A.A. Fowler, and R.P. Fairman, A comparison of the cardiopulmonary effects of continuous versus bolus infusion of recombinant interleukin-2 in sheep. *Cancer Res.*, 48:2221–2225, 1988.

247. Ettinghausen, S.E., R.K. Puri, and S.A. Rosenberg, Increased vascular permeability in organs mediated by the systemic administration of lymphokine-activated killer cells and recombinant interleukin-2 in mice. *J. Natl. Cancer Inst.*, 80:177–188, 1988.
248. Gately, M.K., T.D. Anderson, and T.J. Hayes, Role of asialo-GM_1-positive lymphoid cells in mediating the toxic effects of recombinant IL-2 in mice. *J. Immunol.*, 141:189–200, 1988.
249. Damle, N.K., L.V. Doyle, J.R. Bender, and E.C. Bradley, Interleukin 2-activated human lymphocytes exhibit enhanced adhesion to normal vascular endothelial cells and cause their lysis. *J. Immunol.*, 138:1779–1785, 1987.
250. Aronson, F.R., P. Libby, E.P. Brandon, M.W. Janicka, and J.W. Mier, IL-2 rapidly induces natural killer cell adhesion to human endothelial cells. A potential mechanism for endothelial injury. *J. Immunol.*, 141:158–163, 1988.
251. Henkart, P.A., P.J. Millard, C.W. Reynolds, and M.P. Henkart, Cytolytic activity of purified cytoplasmic granules from cytotoxic rat large granular lymphocyte tumors. *J. Exp. Med.*, 160:75–93, 1984.
252. Nedwin, G.E., L.P. Svedersky, T.S. Bringman, M.A. Palladino, Jr., and D.V. Goeddel, Effect of interleukin 2, interferon-γ, and mitogens on the production of tumor necrosis factors α and β. *J. Immunol.*, 135:2492–2497, 1985.
253. Cotran, R.S., J.S. Pober, M.A. Gimbrone, Jr., T.A. Springer, E.A. Wiebke, A.A. Gaspari, S.A. Rosenberg, and M.T. Lotze, Endothelial activation during interleukin 2 immunotherapy. A possible mechanism for the vascular leak syndrome. *J. Immunol.*, 139:1883–1888, 1987.
254. Gaynor, E.R., L. Vitek, L. Sticklin, S.P. Creekmore, M.E. Ferraro, J.X. Thomas, Jr., S.G. Fisher, and R.I. Fisher, The hemodynamic effects of treatment with interleukin-2 and lymphokine-activated killer cells. *Ann. Int. Med.*, 109:953–958, 1988.
255. Horvath, C.J., T.J. Ferro, G. Jesmok, and A.B. Malik, Recombinant tumor necrosis factor increases pulmonary vascular permeability independent of neutrophils. *Proc. Natl. Acad. Sci. U.S.A.*, 85:9219–9223, 1988.
256. Goldblum, S.E., K. Yoneda, M. Cibull, T. Pearson, C. Hall, and M.E. Marshall, Human recombinant interleukin-2 provokes acute pulmonary vascular endothelial injury in rabbits. *J. Biol. Resp. Mod.*, 9:127–139, 1990.
257. Steinmetz, T., M. Schaadt, R. Gahl, V. Schenk, V. Diehl, and M. Pfreundschuh, Phase I study of 24-hour continuous intravenous infusion of recombinant human tumor necrosis factor. *J. Biol. Resp. Mod.*, 7:417–423, 1988.
258. Marks, J.D., C.B. Marks, J.M. Luce, A.B. Montgomery, J. Turner, C.A. Metz, and J. F. Murray, Plasma tumor necrosis factor in patients with septic shock. Mortality rate, incidence of adult respiratory distress syndrome, and effects of methylprednisolone administration. *Am. Rev. Resp. Dis.* 141:94–97, 1990.
259. Millar, A.B., N.M. Foley, M. Singer, N.McI. Johnson, A. Meager, and G.A.W. Rook, Tumor necrosis factor in bronchopulmonary secretions of patients with adult respiratory distress syndrome. *Lancet*, ii:712–714, 1989.
260. Hocking, D.C., P.G. Philips, J.T. Ferro, and A. Johnson, Mechanisms of pulmonary edema induced by tumor necrosis factor-α. *Circ. Res.*, 67:68–77, 1990.
261. Chang, S.W., N. Ohara, G. Kuo, and N.F. Voelkel, Tumor necrosis factor-induced lung injury is not mediated by platelet-activating factor. *Am. J. Physiol. (Lung Cell Mol. Physiol.)*, 257:L232–L239, 1989.
262. Goldblum, S.E. and W.L. Sun, Human recombinant tumor necrosis factor α augments pulmonary artery transendothelial albumin flux *in vitro*. *Am. J. Physiol. (Lung Cell Mol. Physiol.)*, 258:L57–L67, 1990.

263. Goldblum, S.E. and W.L. Sun, Enhancement of tumor necrosis factor α (TNFα) -induced changes in endothelial barrier function by protein synthesis inhibition. *Progr. Leukoc. Biol.*, 10B:217–223, 1990.
264. Varani, J., M.J. Bendelow, D.E. Sealey, S.L. Kunkel, D.E. Gannon, U.S. Ryan, and P.A. Ward, Tumor necrosis factor enhances susceptibility of vascular endothelial cells to neutrophil-mediated killing. *Lab. Invest.*, 59:292–295, 1988.
265. Brett, J., H. Gerlach, P. Nawroth, S. Steinberg, G. Godman, and D. Stern, Tumor necrosis factor/cachectin increases permeability of endothelial cell monolayers by mechanism involving regulatory G proteins. *J. Exp. Med.*, 169:1977–1991, 1989.
266. Goldblum, S.E., J.N. Thupari, and P.C. Phelps, Tumor necrosis factor α-induced actin reorganization and intercellular gap formation in pulmonary vascular endothelium: a mechanism for paracellular leak in adult respiratory distress syndrome associated with sepsis. 30th Interscience Conf. Antimicrobial Agents and Chemotherapy, Abstr. 296, Atlanta, GA, October 1990.

9. Hepatic Dysfunction Due to Cytokines

STEVEN I. SHEDLOFSKY AND CRAIG J. McCLAIN

INTRODUCTION

The liver plays a key role in defense against external microbial and toxic agents and in maintaining homeostasis after a variety of stresses. The cytokines interleukin-1 (IL-1), tumor necrosis factor (TNF)/cachectin, and IL-6 have all been reported to affect hepatic metabolism in various ways, the best known and most studied effects being the synthesis and secretion of a variety of glycoproteins during the acute phase response. Other effects of cytokines include changes in hepatic glucose, glycogen, and triglyceride metabolism (i.e., energy generating oxidative metabolism), lipoprotein metabolism, mineral metabolism (e.g., Fe, Zn, and Cu), and xenobiotic/steroid metabolism. All of these hepatic responses to cytokines probably contribute to host defense mechanisms.

However, significant hepatotoxicity has been noted in several well-described clinical states and experimental models whose pathobiology is thought to be mediated primarily by cytokines. In particular, studies of clinical and experimental sepsis and endotoxemia strongly suggest that cytokines can in fact play a role in the hepatic damage. Whether the hepatotoxicity represents an exaggeration or alteration of a normally beneficial homeostatic response or whether hepatotoxicity is caused by too high an exposure to one or more cytokines is unknown. Further suggesting possible detrimental effects of cytokines is that the clinical manifestations of several acute and chronic hepatic diseases might be cytokine mediated. In this chapter, we will review the currently known data regarding how cytokines, especially TNF, might cause hepatic dysfunction, as well as how cytokines participate in the pathobiology of various clinical hepatic diseases.

ROLES OF CYTOKINES IN EXPERIMENTAL LIVER INJURY AND REGENERATION

Mechanisms of Hepatocyte Injury

The most common measure of hepatic damage in disease states is the release of hepatocyte intracellular enzymes into the circulation, which reflects increased plasma membrane permeability and cell death. From a clinical standpoint and in experimental *in vivo* models, hepatocyte death is demonstrated by elevated serum levels of such enzymes as the aminotransferases and isocitrate, sorbitol, or lactate dehydrogenases. In tissue cultures of hepatocytes and in suspensions of isolated hepatocytes, release of these same enzymes into the culture medium also reflects cell damage, as does inability to exclude trypan blue or to prevent leakage of K+. However, loss of plasma membrane integrity leading to hepatocyte lysis is usually considered a rather late and irreversible change. In order to better define the mechanisms of hepatotoxicity with various experimental toxins, there have been efforts to describe the cellular alterations that represent earlier manifestations of hepatotoxicity.

Inability to prevent high intracellular free calcium concentration has been postulated to be a common pathologic process mediating hepatotoxicity by a number of agents,[1,2] so that measures of hepatocyte damage have therefore included increased cytosolic calcium. Furthermore, since depletion of intracellular thiols such as reduced glutathione (GSH) is linked to hepatotoxicity[3] and can lead to decreased calcium sequestration intracellularly,[4] GSH depletion is often viewed as another early manifestation of hepatotoxicity. However, apparently disparate results regarding the roles of GSH depletion and high intracellular calcium came from studies in isolated hepatocyte suspensions in which GSH-depleting agents were more toxic in the absence of extracellular calcium.[5] These investigators reconciled their findings with those of others by discovering that content of cellular vitamin E, the most potent lipid-soluble membrane antioxidant, was higher in cells incubated with calcium, and toxicity correlated with low vitamin E content.

GSH, like vitamin E, is important for protection against the damaging effects of free radicals, including toxic oxygen species.[6] GSH depletion is therefore also viewed as a manifestation of "oxidative stress". Oxidative stress, which can cause peroxidation of membrane lipids, plays a role in the toxicity of many agents.[7-13] Measures of lipid peroxidation include generation of ethane or pentane, chemiluminescence, conjugated dienes, malonaldehyde, and fluorescent conjugated Schiff bases.[14,15] However, lipid peroxidation is not always associated with cell toxicity.[10,16] Another phenomenon proposed to cause oxidative stress and which might play a role in hepatotoxicity is the induction of xanthine oxidase, especially by hypoxia, with generation of $\cdot O_2^-$ (the superoxide radical) from xanthine upon reperfusion.[17] Allopurinol, an inhibitor of xanthine oxidase, has been shown to protect against ischemic liver damage[18] and damage caused by the combination of D-galactosamine (D-GLN) and endotoxin[19] (see below). However, the role of xanthine oxidase in generating $\cdot O_2^-$ upon reperfusion has been questioned since liver cells exposed to hypoxia/reoxygenation do not form active xanthine oxidase until after toxicity occurs.[19a]

In addition to reactive oxygen intermediates as potential mediators of toxicity, other investigators[19b] have proposed metabolites of L-arginine, termed "reactive nitrogen intermediates", that might play an important role in the inflammatory process. With regard to cytokines, a link between systemic hypotensive effects[19c] and cytotoxic effects[19d] of TNF and reactive nitrogen intermediates has been made by showing reversal of these phenomena with N^G-monomethyl-L-arginine, an inhibitor of NO production from arginine. Hepatotoxicity after endotoxin or cytokine activation of hepatic Kupffer cells has been linked to products of arginine metabolism[19e] which might cause the release of toxic Fe species from hepatic ferritin stores.[19f]

Although many of the above phenomena have been associated with hepatocyte damage, a continuing controversy remains as to what changes represent the mechanisms of hepatotoxicity. A common problem with almost all the "early" manifestations of damage is in differentiating which alterations represent primary causes of hepatocyte damage and which are secondary phenomena representing the cell's response to hepatocyte injury. For example, decreases in total protein synthesis, RNA synthesis, and changes in numerous other "normal" hepatic functions are often associated with hepatotoxicity, but might actually represent responses that play a protective role rather than contributing to damage. For this reason most of the data presented below will be limited to experimental models in which hepatocellular lysis occurs.

Several reviews[20-22] have described the morphologic changes that occur with a variety of toxic insults to the liver. Early, potentially reversible changes include clumping of nuclear chromatin, mitochondrial inner membrane condensation and loss of matrix granules, dilation of the smooth endoplasmic reticulum, and surface membrane blebbing. Later, irreversible changes include massive mitochondrial swelling with fragmentation of membranes and disintegration of chromatin, presumably from inability of the cell to contain degradative lysosomal enzymes. The liver displays areas of "coagulative necrosis" in which the damaged cells have lost their structure and appear as an amorphous mass of eosinophilic material. The types of injuries that lead to experimental coagulative necrosis, such as hypoxia and CCl_4, seem to involve a loss of cellular energy. Alternatively, areas of cell dropout due to "lytic necrosis" occur, in which all cellular debris is removed presumably by a combination of autolytic enzymes and scavenging phagocytic cells. Areas of cell dropout are often recognized only by a collapse of the reticulin supporting matrix seen on special stains. This form of cell injury may be primarily from processes that rapidly rupture the plasma membrane rather than deplete the cells of energy. A form of individual cell dropout called "apoptosis" has also been described,[23,24] which is thought to be a normal genetically programmed process of cell turnover that begins with chromatin clumping followed by activation of autolytic lysosomal enzymes. Detailed studies of the mechanism by which TNF causes cytotoxicity has been reviewed recently[25a] and seems to be a process either similar or identical to apoptosis.

With regard to this last point, several very different pathobiologic mechanisms of hepatocyte damage have been identified for which there are suggestions that cytokines might be involved. For example, hepatotoxic drugs, xenobiotics, and metals, such as acetaminophen,[25] ethanol,[26,27] D-galactosamine,[28] CCl_4,[29] bromobenzene,[30] and lead-acetate,[31] have distinct metabolic pathways that either generate toxic metabolites and/or inhibit essential cellular processes. Implicating cytokines in several of these chemically induced forms of injury are the observations that agents that down regulate cytokines, such as the prostacyclin-derivative iloprost, can be protective against CCl_4 and bromobenzene hepatotoxicity[32,33] and D-galactosamine hepatotoxicity.[34] Very strong evidence implicating TNF, in particular, in the D-galactosamine (D-GLN) model of hepatic injury[34] will be reviewed below along with our own data implicating TNF in the lead-enhanced endotoxin model. Activation of hepatic mononuclear cells (Kupffer cells) in acetaminophen toxicity[35] also suggests a role for cytokines.

A unifying concept of hepatocellular injury and death has developed,[21] which includes a number of processes such as the generation of free radicals and toxic oxygen species, the related phenomenon of GSH depletion, increased phospholipid breakdown with arachidonate oxidation leading to eicosanoid production, and increased levels of intracellular free calcium. In the complex cascade of events leading to hepatocellular lysis, another potentially "unifying" concept is the importance of gut-derived endotoxins (lipopolysaccharide, LPS). Since LPS is such a potent stimulant of cytokines,[36,37] these data are the most compelling for implicating cytokines in the development of hepatotoxicity.

Hepatic Effects of Endotoxins (LPS)

From a historical perspective, data implicating LPS in liver damage began with observations that gut-derived LPS was important in a number of experimental models of liver injury in rodents, including CCl_4,[38-40] choline-deficiency,[41] lead-acetate,[31] D-galactosamine[42] and frog virus 3.[43] Either elimination of gut bacteria with antibiotics or using germ-free animals, or binding bacterial LPS with the antibiotic polymixin-B, ameliorated the hepatic damage. In a review of this data, Nolan[4a,40] suggested that LPS-stimulated Kupffer cells, the resident hepatic tissue macrophages, might secrete products that mediated the hepatocellular damage. However, Nolan also thought that LPS itself might directly damage hepatocytes if there was blockade of the reticulo-endothelial system allowing LPS to be taken up by hepatocytes. An important model that helped define the importance of the monocyte in mediating endotoxicity was the genetically LPS-resistant C3H/HeJ mouse, whose macrophages were relatively unresponsive to LPS.[44] These mice were also more resistant to hepatotoxic damage by D-GLN. Transplantation of spleen-cells from LPS-sensitive strains to C3H/HeJ mice restored susceptibility to D-GLN hepatotoxicity.[45]

The hepatic effects of LPS administration have been reviewed several times since Nolan's summary in 1975.[40,46-50] It is now well accepted that the liver is responsible for removing gut bacteria-derived LPS, which frequently seeds the portal venous circulation.[40,48,51,52] It appears that the hepatic resident macrophage, the Kupffer cell, is responsible for LPS removal.[53-56] This process prevents the potentially devastating systemic effects of endotoxemia.

Although Kupffer cell uptake of LPS is accepted, there is controversy (reviewed in Reference 49) over how much LPS reaches hepatocytes, how quickly the LPS gets to hepatocytes, and in what form (detoxified?). Fox et al.[56a] showed that *in vitro*, Kupffer cell uptake of LPS differed from uptake by peritoneal macrophages. *In vivo* experiments that showed major parenchymal hepatocyte uptake[57,58] administered the LPS intravenously. Since LPS travels with HDL in the systemic circulation[59] and has complex interactions with serum proteins,[60] there might be a difference in how i.v.-administered LPS is processed compared to LPS presented to the liver from the portal circulation. However, Freudenberg et al.,[54] found predominant Kupffer cell uptake with i.v. administration, so that this controversy has still not been resolved. Although the dose of LPS might also be expected to make a difference in uptake parameters, the studies cited above used widely ranging doses of LPS (100 µg/kg to 20 mg/kg) with conflicting results.

The removal of LPS by Kupffer cells might be regulated by fibronectin. Fibronectin is a plasma glycoprotein synthesized primarily by hepatocytes, but also by Kupffer cells and endothelial cells. Although used by many cell types for matrix support and adhesion, fibronectin has been postulated to be a nonspecific opsonin for macrophages.[61] Fibronectin can bind to LPS[62] and presumably hastens LPS removal by Kupffer cells. It has been known that plasma fibronectin levels decrease in liver failure,[63] which might contribute to the systemic endotoxemia that has been reported in patients with cirrhosis.[64] LPS administration itself can increase plasma fibronectin levels in rats,[65] possibly as an acute phase reactant, and Kang et al.,[66] showed increased hepatocyte synthesis of fibronectin *in vivo* after LPS. Vincent et al.,[67] using isolated hepatocytes and Kupffer cells, showed that although most of the fibronectin comes from hepatocytes, after low-dose LPS administration (100 µg/d × 3d) there was an increase in fibronectin from Kupffer cells, but less than produced by hepatocytes. Because administration of fibronectin was protective in D-GLN-induced liver failure,[68] these authors surmised that maintaining "macrophage function" (i.e., phagocytosis) was responsible for fibronectin's protective effect. However, Richards and Saba[65] showed that the elevated fibronectin levels that occurred after LPS administration quickly returned to normal, even though phagocytic capacity of the Kupffer cells remained significantly elevated. Therefore, the mechanism of fibronectin's protective effect is still not clear. Agents such as glucan, which can maintain phagocytic activity, protect against *E. coli* sepsis,[69,70] but it is not known if this effect involves fibronectin.

It is important to know how LPS is handled by the liver, since it is not yet clear whether LPS can directly damage parenchymal hepatocytes or whether all hepatotoxic effects are mediated through LPS activation of nonparenchymal cells. Experiments with the LPS nonresponsive C3H/HeJ mice do not help elucidate this issue because it is known that LPS binds to C3H/HeJ macrophages and B-cells[71,72] and is therefore probably cleared normally. In a recent review of how the liver handles endotoxins, Fox et al.[73] concluded that gut-derived LPS is initially taken up into Kupffer cells by nonsaturable pinocytosis and then processed mainly by deglycosylation. This modified LPS, which is not yet detoxified, is then somehow transferred to parenchymal hepatocytes where deacylation[73a] might lead to detoxification and excretion. However, the issue of direct hepatocyte injury from LPS was not addressed. We have unpublished evidence that LPS by itself is not toxic from *in vitro* experiments exposing LPS to hepatocytes in culture. In doses from

0.01 up to 100 μg/ml, cultured chick embryo hepatocytes displayed no toxic effects in a variety of serum-free hormonally supplemented media. However, LPS might have direct hepatotoxic effects in the presence of other inflammatory mediators.

Since the majority of cytokines come from mononuclear macrophages, the Kupffer cell, being the resident hepatic sinusoidal macrophage, would be a likely source of cytokines after LPS exposure. Kupffer cells are generally a relatively stable population of cells, but can be recruited from circulating monocytes, especially during pathologic states (reviewed in Reference 74). Kupffer cells have been shown to be a potential source of cytokines[75-78] and are clearly activated in LPS-induced injury.[40,79] Within several hours after LPS administration, swollen activated Kupffer cells and a heavy granulocytic infiltrate are seen in sinusoids along with fibrin and platelet clots.[53,56,80] By 48 h there are six times as many Kupffer cells, suggesting peripheral recruitment, and they are much more active in phagocytosis and more responsive to chemoattractants.[79] Interestingly, Kupffer cells isolated at 48 h after LPS produced less $\cdot O_2^-$ after C5a stimulation than did Kupffer cells from control rats. This last finding might reflect the development of endotoxin "tolerance".

Tolerance to LPS is defined as a blunted response to LPS administration, that is induced by a prior nonlethal exposure to LPS. There are two phases of the tolerant state, which appear to be caused by very different mechanisms. "Late" tolerance to LPS occurs after a week or more of LPS exposure and appears to be mediated by specific antibody to that LPS. "Early" tolerance, however, is much more interesting in that it develops after a single sublethal dose. With "early" tolerance, animals are protected against a normally lethal dose of LPS[81] and display less of a pyrogenic response,[82] less of a decrease in hepatic drug metabolizing enzymes,[83,84] and fewer changes in hepatic glucose metabolism.[85] On a cellular level, the tolerant state is associated with the macrophage's inability *in vitro* to secrete prostaglandins[86] and cytokines[87] in response to LPS. Understanding the mechanisms that lead to "early" LPS tolerance would likely increase our understanding of how LPS causes hepatotoxicity, especially in combination with agents such as D-GLN, lead acetate, and *Propionibacterium acnes*, all of which will be reviewed below.

In a recent review of LPS tolerance, Warren and Chedid[88] present data implicating the acute phase response as mediating this phenomenon. These authors found that acute phase serum neutralized much more LPS than control serum and further suggested that certain acute phase reactants, such as serum amyloid A, might enhance binding of LPS to HDL and promote more rapid clearance of the LPS by an activated reticuloendothelial (RE) system. In the D-GLN-plus-LPS model of hepatotoxicity (see below), it was also recently shown that induction of the acute phase response with a turpentine abscess 24 h prior to D-GLN/LPS decreased lethality.[88a] Further supporting a possible role for acute phase reactants in early LPS tolerance is data from Vogel et al.,[89] which showed that single injections of both recombinant human IL-1 and TNF could induce tolerance in mice, although neither cytokine alone would. Their IL-1-plus-TNF injections induced tolerance to LPS lethality and also reduced serum concentrations of colony-stimulating factor and bone marrow macrophage progenitor cells in response to LPS. Although hepatic responses and acute phase reactants were not evaluated, it can be presumed that an acute phase response was elicited. Wallach et al.[90] showed that either IL-1 or TNF pretreatment could desensitize BALB/c mice "primed" to LPS lethality by BCG and could desensitize mice primed to IL-1 and TNF lethality by actinomycin-D. In similar types of studies, Cross et al.[90a] showed that pretreatment of C3H/HeJ mice with either TNF or IL-1 protected against lethal infection with K4-encapsulated *E. coli*. If early LPS tolerance is indeed mediated by one or more hepatic-derived acute phase glycoproteins, a most important experiment in this regard would be to see whether IL-6, the cytokine that best elicits acute phase reactants, can also induce tolerance. However, it should be noted that monoclonal antibodies against IL-6 were recently shown to protect against LPS and TNF lethality in mice.[90b] This finding might argue against the importance of inducing acute phase reactants.

There are other problems with the explanation that LPS tolerance is mediated by one or more acute phase reactants that detoxify LPS. This is because TNF itself causes lethality and almost certainly is the primary mediator of the lethal effects of LPS.[91-93] However, Lehmann et al.[94] showed in mice that LPS induces tolerance to the lethal effects of TNF, and Fraker et al.[95] and Sheppard et al.[95a] have shown in rats that TNF by itself, given at a dose of 100 µg/kg i.p. twice per d for 3 to 5 d, induced tolerance to both itself and LPS. These data argue that tolerance is not due to more rapid LPS detoxification, but rather due to a TNF-elicited protective mechanism. This protection might still involve an acute phase reactant, possibly one that has potent antioxidant effects or that can block further TNF uptake by susceptible target cells. Fraker et al. also showed that the TNF-elicited tolerance was not due to increased clearance of TNF nor due to development of anti-TNF antibodies, and they showed that the tolerance waned with time, just as did LPS-induced "early" tolerance. It should also be noted, however, that Tracey et al.[96] did not obtain tolerance with repeated injections of higher doses of TNF into rats (250 µg/kg twice per d for 7 d). These investigators did show that this dose of TNF caused marked increase in liver weights, with periportal "pleomorphic mononuclear infiltrations", "small bile duct proliferation", and "focal parenchymal cell necrosis".

Another very plausible mechanism by which tolerance might develop would be the induction of protective cellular proteins by the "tolerizing" exposure to either LPS or TNF. In this regard, Wong et al.[96a] showed that the induction of the mitochondrial enzyme Mn superoxide dismutase (MnSOD) could protect sensitive target cells from TNF cytolysis. Shaffer et al.[96b] using endothelial cells exposed to TNF *in vitro* showed induction of MnSOD, mRNA, protein, and activity. Jaattela et al.[96c] and Kusher et al.[96d] similarly found that induction of heat shock proteins was associated with protection from TNF cytolysis. These cellular protective responses could thus explain resistance to subsequent TNF exposure (i.e., TNF tolerance). However, it is not yet clear whether tolerance to TNF cytotoxicity is identical to tolerance to LPS.

Tolerance to LPS is relevant to the area of hepatotoxicity since tolerant animals are more resistant to the hepatotoxic effects of CCl_4,[97,98] D-GLN,[94] and to cholestasis from taurolithocholate.[99] As mentioned above, there is much data linking the development of LPS tolerance with the macrophage.[86-88] It would seem that tolerance to LPS also could be explained by a temporary refractory state of the macrophage to further production or secretion of toxic cytokines of other toxic mediators. This could occur despite their being in a state of "activation" as measured by increased phagocytosis.[79] Therefore, at the present time, it is not yet known whether the LPS-tolerant state is due more to a refractory state of RE cells which take up and respond to LPS or to the induction of proteins such as MnSOD or heat shock proteins which protect cells from cytokine-mediated damage. Both processes may explain aspects of the tolerant state.

On the other side of LPS tolerance, a number of studies have demonstrated enhanced toxic effects of LPS when macrophages are "primed" with various agents such as *Mycobacterium bovis*,[100] and *Corynebacterium parvum*,[101,102] which is now called *Propionibacterium acnes*[103] (see below). However, these findings seem to be in conflict with other experiments mentioned above,[69,80] suggesting that animals pretreated with glucan, whose macrophages are activated as measured by a number of parameters, seem to be protected against *E. coli* sepsis. In this regard, it is interesting to note the effects of muramyl dipeptide, another macrophage modulator that might be related to some of the above "priming" organisms. This agent can induce Kupffer cell tumoricidal activity (TNF?) when administered in a low dose (1 µg per mouse), but actually inhibits control Kupffer cell tumoricidal activity in higher doses of 10 and 100 µg.[104] Although dose effects of various macrophage modulators might explain conflicting data, there are probably intrinsic differences amongst the many priming agents. For example, Galleli and Chedid[105] noted that tolerance did not develop to muramyl dipeptide's ability to stimulate serum colony-stimulating activity (CSA) as it did with other CSA-inducing agents such as LPS. Furthermore, muramyl dipeptide stimulated CSA activity in LPS-resistant C3H/HeJ mice. Finally, the timing of the macrophage modulating agent makes a difference, as shown by Shiratori et al.,[106] who found that latex particles injected 24 h prior to D-GLN treatment in rats prevented hepatotoxicity, whereas latex particles injected 12 h after augmented hepatotoxicity.

Although hepatic damage could be mediated by cytokines, especially TNF, it is also clear that many other potential mediators are released after LPS administration.[40] Within hepatic sinusoids, conspicuous thrombi with granulocyte and platelet accumulation suggest that there is activation of coagulation factors[107] and release of granulocyte chemoattractants (reviewed in Reference 108). Granulocyte products, such as toxic oxygen species and proteinases, could cause injury, although tissue damage by granulocytic infiltration has never been clearly proven (reviewed in Reference 109). However, Holman and Saba[110] have shown that activated granulocytes, but not Kupffer cells, caused toxicity when these cells were added to cultured rat hepatocytes. Others, in an *in vitro* model of alcoholic hepatitis, have shown that alcohol-exposed Kupffer cells might regulate granulocyte activity by producing a chemotactic factor, whereas normal Kupffer cells inhibited chemotaxis.[111] A role for the platelet thrombi and vascular plugging in mediating endotoxin damage is suggested, since heparin administration partially prevents damage.[112,113] Shibiyama also showed that thrombin alone, infused into the portal vein, could cause hepatotoxicity.

In summarizing the hepatotoxic effects of LPS, it is obvious that there are numerous and complicated events that occur after LPS administration, and many questions persist regarding mechanisms and the role of cytokines. Reviewing other related models of hepatic damage might help clarify some of the above issues.

Viral Models of Hepatocyte Injury

In viral models of hepatic injury, where immunologic responses (both serologic and cellular) are better defined, invoking a role for cytokines in mediating injury might seem unnecessary. Conventional wisdom would hold that hepatocyte damage must be secondary to the virus taking over critical cellular synthetic processes in order to produce viral products. However, mechanisms of viral-induced injury are more complicated than a simple cytopathic effect.[114,115] Host immune-mediated cytotoxicity clearly plays a role in Hepatitis B viral hepatitis[116] and presumably in Hepatitis A infections, since virally infected cultured hepatocyte lines do not show a cytopathic effect, and injury correlates with onset of humoral immunity (reviewed in Reference 117). Although cytotoxic T-lymphocytes are thought to primarily mediate damage to hepatocytes carrying foreign viral antigens, there is evidence that Kupffer cells and endothelial cells, both potential sources of cytokines, can also play a role in parenchymal cell injury.

With murine hepatitis virus infections, Kupffer cells clearly become enlarged and presumably activated early in the infection and are responsible for the initial uptake of virus and possibly early viral replication (reviewed in Reference 118). Hepatotoxicity, however, can be decreased if the Kupffer cells can be kept in a state of "activation" using glucan.[119] Glucan-treated mice develop hepatic granulomas and have enhanced clearance of colloidal particles. Data from these same authors using rats demonstrated that splenic macrophages from glucan-treated animals produce *in vitro* more of a cytotoxic factor, like tumor necrosis factor,[120] and more IL-1 and IL-2.[121] These data suggest that enhanced macrophage cytokine production is associated with protection in the murine hepatitis virus model. In contradistinction, however, Abecassis et al.[122] found that splenic macrophages from mice infected with the murine hepatitis virus had markedly increased production of cytokine-stimulated "procoagulant activity".[123] Suppressing macrophage procoagulant activity with 16,16 dimethyl prostaglandin E_2 (PGE_2) was associated with hepatic protection, even though viral replication was not decreased. These investigators also noted that PGE_2 protected monolayers of hepatocytes infected in culture from the normal cytopathic effect of the virus. This latter finding of a "direct" hepatic cytoprotective effect was also reported for PGE_1 by Mizoguchi et al.[103] in rat hepatocyte cultures exposed to a toxic factor released from Kupffer cells exposed to *P. acnes* and LPS (see below).

In another viral model of hepatic damage, the frog virus model in mice,[43,124] the virus does not proliferate in hepatocytes, but destroys hepatic Kupffer and endothelial cells, which is then followed by hepatocyte lysis. In this model, elimination of endogenous LPS by colectomy or polymixin-B administration was protective, and less damage was seen in LPS-tolerant mice and LPS-resistant C3H/HeJ mice. Protection, however, was incomplete (mortality going from 70–100% down to 13–30%). Surprisingly, indomethacin pretreatment was also protective, even though it is assumed that indomethacin inhibits potentially "protective" prostaglandin formation. Although the above findings in the frog virus model suggest that Kupffer cells normally protect hepatocytes from "direct" LPS toxicity, it is also possible in this model that systemic cytokines and/or other macrophage products are stimulated by LPS that escapes removal by Kupffer cells. Hepatocyte damage might then be mediated by these systemic cytokines (or other mediators) rather than by LPS directly.

D-Galactosamine Model of Hepatic Injury — Role of TNF

The D-GLN model of hepatotoxicity and its association with LPS has been studied in great detail. D-GLN by itself is a classic hepatotoxin. It depletes hepatocytes of uridine nucleotides and uridine 5'-diphosphate hexoses leading to hepatocyte death which is reversible by uridine administration up to 3 h after the D-GLN.[28] Most animals will survive doses up to 1 g/kg, in spite of the liver damage. However, Grun et al.[42] found that colectomized rats had much less liver cell necrosis, even though Liehr et al.[125] found they had just as much uracil nucleotide depletion. Clearly associating the hepatic damage with LPS, Galanos et al.[126] noted that D-GLN markedly augmented the lethal effects of LPS. In rabbits and mice, the lethal dose of LPS ($LD_{50} \approx 100$ µg/kg) could be reduced to < 0.01 µg/kg. This D-GLN-induced sensitivity to endotoxin was shown to be dependent upon the state of the macrophage, in that mice did not die after D-GLN when their macrophages were made LPS tolerant by a prior nonlethal injection of LPS or by *in vitro* exposure of macrophages to LPS before injecting the macrophages into the animal.[127] The tolerance to LPS/D-GLN required that the macrophages be exposed to LPS *in vivo* for at least 1 h before the D-GLN or that *in vitro* LPS-stimulated macrophages similarly be injected at least 1 h before the D-GLN.

As mentioned above with the discussion of LPS tolerance, there is preliminary data that eliciting the acute phase response itself (using a turpentine abscess) will decrease mortality in the D-GLN/LPS model.[89] One might ask then why LPS, which can elicit the acute phase response, augments rather than protects against D-GLN damage. One possible explanation might be demonstrated by the findings of other preliminary studies.[128] These investigators found that the change in hepatic-derived serum lipoproteins, C-reactive protein, and serum amyloid A that occur with the acute phase response (following 1% croton oil, i.m.) were completely opposite of the changes seen with D-GLN hepatitis. D-GLN, therefore, appears to inhibit or alter the development of the acute phase response and thereby block any possible protection.

Determining what macrophage secretory products might be responsible for hepatocyte lysis and lethal outcome in LPS/D-GLN-treated animals has also been studied. Wendel and Tiegs,[129] after noting that leukotrienes were excreted into bile after LPS,[130] examined a number of "anti-inflammatory" agents and found that those compounds that inhibited the 5-lipoxygenase pathway protected against hepatocyte death and mortality, whereas inhibiting prostaglandin formation was not protective. They subsequently showed in NMRI mice[131] that LTD_4 was a mediator of hepatocyte death in the LPS/D-GLN model, since (1) blocking LTC_4 metabolism to LTD_4 prevented hepatocyte death, (2) LTD_4 given with D-GLN instead of LPS caused the same degree of damage, and (3) LTE_4 plus D-GLN was not hepatotoxic. Further studies by this group[19] found that allopurinol, superoxide dismutase, catalase, and a prostacyclin analogue, iloprost, could each ameliorate the hepatic damage. These investigators interpreted their results "as evidence for a LTD_4-induced hepatic ischemia/reperfusion syndrome" as a key pathobiologic part of LPS/D-GLN hepatic damage and lethality. As mentioned above, in the reperfusion model of tissue injury, superoxide anion ($\cdot O_2^-$) is formed, which is itself toxic or leads to formation of other toxic O_2 radicals, such as $\cdot OH$, (reviewed in Reference 11). Shiratori et al.[106] found in rats that Kupffer cell phagocytosis of latex particles after D-GLN worsened hepatotoxicity, which could be reversed by superoxide dismutase, invoking again the possible role of toxic $\cdot O_2^-$.

Possible involvement of TNF as a mediator of hepatotoxicity in the LPS/D-GLN model was first invoked by Lehmann et al.[94] These investigators showed that in non-D-GLN-treated mice, very high doses of recombinant human TNF (500 µg per mouse) could cause 80% mortality. In these studies TNF lethality had a time course until death very similar to mice injected with LPS (24 to 48 h), and lethality occurred in both LPS-sensitive C3H/TifF and LPS-resistant C3H/HeJ mice. They then showed that D-GLN pretreatment markedly sensitized the animals to TNF lethality (with only 0.1 to 1.0 µg TNF required), just as D-GLN sensitized animals to endotoxicity. Tiegs et al.[34] brought their leukotriene-induced ischemia/reperfusion model into agreement with the TNF-induced model by first confirming that TNFα caused hepatic injury in D-GLN-sensitized NMRI mice and then showing that the agents that protected mice against LPS/D-GLN damage (several inhibitors of eicosanoid metabolism, allopurinol, SOD, catalase, and iloprost) did not protect against TNF/D-GLN damage. They suggested that in the sequence of events with LPS/D-GLN damage, TNF is the "terminal mediator" whose production is caused by leukotriene induced ischemia/reperfusion and $\cdot O_2^-$ generation. In other systems, leukotrienes themselves (LTB_4) have been shown to augment monocyte IL-1 and TNF production.[132] Further supporting the role of TNF in this liver injury, Hishinuma et al.[133] reported that treatment with anti-TNFα antibody can also protect against LPS/D-GLN damage.

Lehmann et al.[94] quoted some of their unpublished studies showing that D-GLN does not sensitize animals to increased hepatotoxicity when given simultaneously with a number of macrophage "activators", such as "zymosan, muramyl dipeptide, *P. acnes*. . .", and sensitization by D-GLN is therefore "specific for LPS-induced macrophage products." Because LPS, and in particular the lipid-A component of LPS, must also be considered a macrophage "activator", these findings might point out a major difference between LPS macrophage activation and activation by other types of bacterial components.

In terms of how D-GLN sensitizes hepatocytes to LPS-induced toxicity, it is well documented that agents that inhibit RNA or protein synthesis will markedly enhance LPS toxicity[31] and TNF toxicity.[134] Preliminary data have been reported that D-GLN treatment blocks the induction of hepatic Mn-superoxide dismutase (MnSOD) mRNA by LPS.[135a] Inhibiting a protective cellular response, which the increase in MnSOD might represent, might sensitize hepatocytes to lysis by TNF (see below). Whether D-GLN, by depleting hepatic uridine and thereby inhibiting RNA and protein synthesis, represents a unique mechanism of enhancing LPS and TNF hepatotoxicity is not yet known. Two other agents that also enhance LPS and TNF hepatotoxicity will be reviewed next.

Lead-Enhanced Endotoxin (LPS) Model of Liver Injury

Selye et al.[135] originally reported that a single intravenous injection of lead acetate, in doses that were normally well tolerated, caused markedly enhanced lethality in LPS-treated rats. This observation has been confirmed in a variety of different animal models, including mice,[136] chicks,[137] and baboons.[138] The mechanism for this lead-enhanced LPS toxicity is unclear. The peak synergistic effect of lead with LPS occurs when lead is given directly with the LPS. If lead is given more than 2 to 3 h before or after the LPS, its enhancing effect is dramatically diminished. A variety of anatomic and biochemical effects of lead have been demonstrated, including ultrastructural alterations of the liver and spleen.[139] Lead is taken up mainly by the liver after intravenous injection (approximately 60 to 70% of total injected dose), which may explain, in part, why lead enhances to such a great degree LPS liver toxicity.[140] Lead has been reported by some investigators to impair the detoxification of LPS, thereby enhancing endotoxicity.[141] On the other hand, other investigators have shown that lead did not appreciably alter clearance of chromium-labeled LPS,[31] nor impair hepatic detoxification of LPS either *in vivo* or *in vitro*.[142] Lead has been documented to cause derangement of several hepatic biochemical pathways and energy metabolism by multiple investigators.[143-146] In particular, lead has been shown to cause transient marked hypercalcemia, which may be a mechanism in hepatotoxicity.[147,148] However, lead does not appear to decrease total hepatic protein synthesis as a mechanism of enhanced LPS liver injury.[149] Finally, lead-enhanced endotoxicity can be blocked or ameliorated by a variety of agents, such as methylprednisolone and cysteine.[150]

We utilized the lead-enhanced LPS hepatotoxicity model to further evaluate the role of TNF in liver injury. The objectives of our initial study were (1) to confirm that lead pretreatment enhanced LPS liver injury, (2) to determine whether lead pretreatment increased plasma TNF concentrations, and (3) to determine whether lead could enhance TNF liver injury in a fashion similar to LPS. Rats were administered saline, lead acetate, LPS, TNF, and combinations of LPS plus lead, or LPS plus TNF. Rats were sacrificed and serum enzymes and liver histology evaluated.

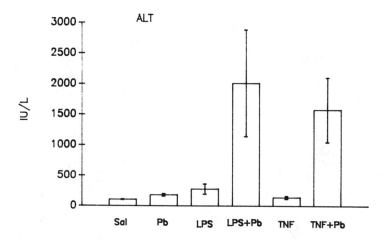

Figure 1. ALT levels in rats administered saline, lead acetate (15 mg/kg IV), LPS (100 μg/kg IV), TNF 2.5 × 106 units IV, or combinations thereof. All animals were sacrificed at 24 h except those given LPS plus lead, which were sacrificed at 14 h to prevent any 24 h mortality. Six rats were in each group. Lead markedly enhanced LPS liver injury and TNF liver injury (see Reference 151).

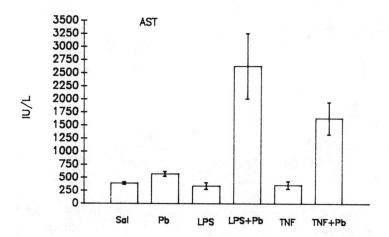

Figure 2. Same design as Figure 1. Note markedly and significantly ($p < 0.01$), increased AST levels in animals receiving either LPS or TNF plus lead.

In a separate experiment, we performed a time course on the effect of LPS plus lead on plasma TNF levels. Low doses of LPS or TNF caused little liver injury, while the same doses of these agents administered with lead caused severe hepatotoxicity (Figures 1 to 3). Lead also caused the peak plasma TNF level to markedly increase

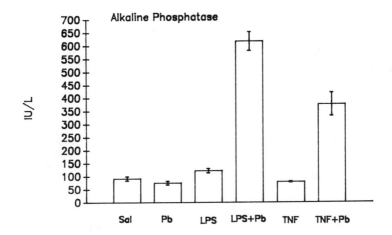

Figure 3. Same design as Figure 1. Note markedly and statistically significantly ($p < 0.01$) elevated alkaline phosphatase levels in animals treated with lead and then given either LPS or TNF.

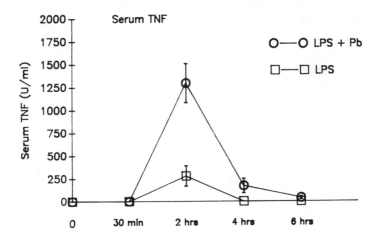

Figure 4. Six rats were administered either LPS (100 μg/kg) or LPS plus lead (15 mg/kg). Note that lead markedly and significantly elevated peak 2 h plasma TNF levels in LPS treated rats.

in response to LPS administration (Figure 4). These data clearly show that lead enhances LPS liver injury when administered simultaneously with LPS, and there are significant increases in plasma TNF levels. The increased plasma TNF levels very likely contribute to the lead-enhanced LPS liver injury. Finally, lead also enhances liver injury with TNF administration (Figures 1 to 3;[151]). Thus, we feel that lead must not only increase plasma TNF levels, but also cause end organ (liver) sensitivity to TNF.

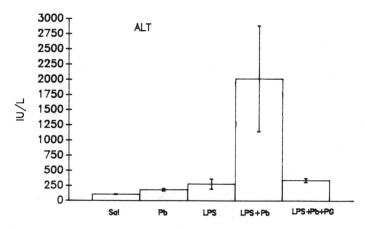

Figure 5. Same design as Figure 1. However, a final group of rats treated with LPS plus lead also were pretreated with prostaglandin E_1 (30 µg/kg IV) 1/2 h before LPS administration. Note the marked blunting of liver injury produced by pretreatment with prostaglandin E_1.

Because prostaglandins are major inhibitors of TNF production, we next asked whether prostaglandin-induced hepatic cytoprotection was really cytokine (TNF) protection. Prostaglandins have been shown to protect against a variety of hepatotoxins, including CCl_4,[152] D-GLN,[34,153] alphanaphthylisothiocyanate,[154] murine hepatitis virus,[122] and *Propionibacterium acnes*.[103] Using the lead-enhanced LPS liver injury model, we demonstrated that rats pretreated with prostaglandin E_1 and given LPS plus lead were protected from liver injury (Figure 5). We then evaluated whether decreased TNF production correlated with liver injury or lethality.[155] Rats injected with LPS (100 µg/kg) plus lead developed markedly elevated peak 2 h plasma TNF levels and were all dead at 24 h (Figure 6). However, when rats were pretreated with PGE_1, there was a marked decrease in TNF production (in all except one rat), and every rat (except that one rat) survived this dose of LPS plus lead. The PGE_1-treated rats also had decreased hepatotoxicity. Because prostaglandins have a variety of metabolic functions besides down regulating TNF production, we used a second drug, dexamethasone, that also decreases TNF production. In animals pretreated with dexamethasone and then injected with LPS plus lead, no mortality was seen. Furthermore, the dexamethasone treated animals showed a major decrease in their peak plasma TNF levels and a marked amelioration of liver injury (data not shown). Thus, we conclude that appropriately timed PGE_1 or dexamethasone administration (prior to LPS injection) ameliorates lead-enhanced LPS liver injury. Furthermore, this protection is associated with decreased peak plasma TNF levels. These studies provide strong evidence that TNF is the major mediator of hepatotoxicity in this model and that hepatotoxicity can be blunted by agents that block or decrease TNF production.

How TNF might mediate hepatocyte damage in both the D-GLN and lead acetate models of hepatic injury is not yet clear. A reasonable presumption is that the

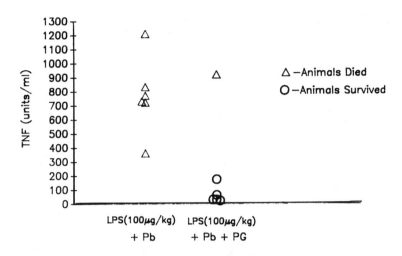

Figure 6. Six rats were treated with LPS (100 μg/kg) plus lead. All rats died at 24 h, and all rats had highly elevated peak 2 h plasma TNF levels. However, when this same dose of LPS plus lead was given to rats pretreated with prostaglandin E$_1$ (30 μg/kg) 1/2 h before LPS administration, only one rat died. That particular rat did not have an attenuation in its peak 2 h plasma TNF level. The five surviving rats had markedly decreased plasma TNF levels, suggesting that the protective effect of prostaglandins may be their property of down regulating TNF production.

mechanism of hepatocyte lysis is similar to the cytotoxicity caused by TNF in sensitive cell lines, such as the L929 fibroblast, whose lysis is used as a bioassay for TNF. Two recent reviews of TNF cytostasis and cytotoxicity have been published, one concerning TNFβ (lymphotoxin) by Paul and Ruddle[156] and the other reviewing TNFα by Larrick and Wright.[25a] Paul and Ruddle proposed that TNF, once taken up by the susceptible cell, might activate a lysosomal endonuclease, which then causes DNA fragmentation eventually leading to cell death. Schmid et al.[157] had shown that target cell DNA fragmentation occurs early after TNF exposure and long before cell lysis occurs. Paul and Ruddle suggested that this was a process akin to "apoptosis" or programmed cell death and presented data that TNF did not lyse cells in the same way as did C5a and other proteins that form pores in the target cell membrane. Ruggiero et al.[158] have shown that cells resistant to TNF lysis might make proteins of an unspecified nature that are protective. Agarwal et al.[159] reported that there was increased ADP-ribosylation in target cells sensitive to TNF, and inhibitors of ADP-ribosylation (e.g., nicotinamide) prevented TNF-induced cytolysis. These data suggest that there might be "protective" proteins that can either prevent or reverse ADP-ribosylation and by so doing prevent activation or effects of endonucleases. In the review of Larrick and Wright[25a] these concepts are developed further with DNA fragmentation being proposed as a possible "final common pathway" by which a number other processes, including several second messenger pathways, oxygen radicals, nitrogen radicals, TNF inducible/suppressible gene products (e.g., NF-κβ), and ADP-ribosylation, might all lead to "apoptosis" of the TNF-sensitive target cell. With regard to oxygen radicals, induction of mitochondrial MnSOD was mentioned above as a possible way cells might protect themselves against TNF damage. Wong et al.[96a] and Shaffer et al.[96b] showed that induction of MnSOD blocked TNF cytolysis and Schmiedeberg et al.[135a] proposed that prevention of the LPS-mediated MnSOD induction in hepatocytes by D-GLN was the cause of D-GLN's enhancement of both LPS and TNF toxicity. However, other "protective" proteins might be involved and a role for induction of heat shock proteins was proposed by Jaattela et al.[96c] and Kusher et al.[96d]

Since inhibitors of transcription (e.g., actinomycin D) sensitize certain resistant cells to TNF,[160,161] it is probably D-GLN's uridine-depleting effect that may block transcription of "protective" genes and thereby sensitize hepatocytes to TNF (and LPS) toxicity. Lead acetate might also sensitize hepatocytes to TNF toxicity by inhibiting synthesis of "protective" proteins. However, measurements of total hepatic protein synthesis after lead exposure have not shown a decrease, and in fact increases are seen.[149] Lead has numerous biologic effects, including inhibition of heme synthesis and inactivation of several sulfhydryl-containing enzymes. Therefore, lead might block other "protective" processes, some possibly involving DNA repair.

Another question regarding hepatotoxicity from TNF is, how long must the TNF be present? Is there a difference between a brief TNF exposure vs. chronic exposure? TNF injected intravenously is rapidly cleared, with most going to the liver and skin (30% of dose to each).[91] TNF is only transiently secreted into the serum after LPS administration with a peak at 1 to 2 h[91,162,163] and further LPS does not stimulate more TNF secretion. The lack of continued TNF secretion could underlie the mechanism of LPS tolerance (see above), but would not explain tolerance to further effects of TNF. However, it would appear that the hepatotoxicity of TNF in the D-GLN and lead models requires only an initial burst of TNF, even though the hepatotoxicity does not occur until hours later. In the elegant murine model developed by Oliff et al.,[164] in which a TNF-secreting cell line was implanted into nude mice, high continuous levels of serum TNF were found, but hepatic alterations or injury were not evaluated.

Although TNF might be considered the "terminal mediator" in the D-GLN/LPS model of hepatotoxicity[34] and the lead/LPS model, we have not been able to show *in vitro* toxicity of either murine or human recombinant TNF in cultured chick embryo hepatocytes (unpublished data). We have, however, recently noted dose-dependent TNF toxicity in the human HepG2 hepatocyte cell line when co-treated with actinomycin D (unpublished data). While the lack of toxicity in the chick hepatocyte cultures could be due to a species difference and inability to internalize mammalian TNF, we have seen toxicity when cultures were treated with a semipurified murine monocytic cell line supernatant stimulated with LPS. Therefore, mediators other than or in addition to TNF may be important for toxicity. In this regard, TNF has been shown to stimulate enhanced granulocyte adherence to rat liver sinusoidal endothelial cells.[165] In view of Holman and Saba's findings[110] that activated granulocytes might mediate hepatotoxicity, rather than activated Kupffer cells, TNF might not be the "terminal" mediator, but certainly appears to be a necessary one.

Propionibacterium Acnes (P. Acnes) Model of Liver Injury

P. acnes is a Gram positive aerobe that has been used as a sensitizing agent for LPS-induced fulminant liver injury in experimental animals. When heat-killed *P. acnes* is injected into mice and a small amount of LPS administered one week later, most mice die of massive liver necrosis.[103] There is a marked infiltration of mononuclear cells in this animal model, and the mononuclear cells are thought to release cytokines that induce the liver injury. Initial studies by Mizoguchi et al.[103] demonstrated that 1 μg of LPS caused massive hepatic necrosis in *P. acnes*-primed mice. However, when these animals were pretreated with prostaglandin E_1, there was dramatic reduction of liver injury and in mortality. When liver adherent cells from *P. acnes*-primed mice were incubated with LPS *in vitro,* they released a toxic factor for cultured rat hepatocytes. Varying doses of PGE_1 reduced hepatocyte injury (as assayed by protein synthesis and trypan blue exclusion). Hepatocytes cultured with PGE_1 also had reduced injury when supernatants of the LPS-stimulated liver adherent cells containing the toxic factor were added to the culture medium. Thus, this study demonstrated that PGE_1 improved survival rate, histology, and liver enzymes in this model of massive liver cell necrosis, and the beneficial effects of PGE_1 were associated with diminished release of a cytotoxic factor from liver adherent cells. Furthermore, cultured hepatocytes also were provided some direct hepatoprotection by PGE_1 when cultured with supernatants from the LPS-stimulated adherent cells.

Nagakawa et al.[165a] used the same animal model and showed that pretreatment of mice with dexamethasone, PGE_2, or antibodies to TNF markedly blunted liver injury. Mice that had attenuated liver injury after pretreatment with dexamethasone or PGE_2 also had lower plasma TNF activity. Immunohistochemical staining demonstrated that the infiltrating mononuclear cells in the liver were Mac2-positive activated macrophages. Lastly, plasma TNF activity was markedly increased 2 to 3 h after injection of LPS in the *P. acnes*-primed mice. These findings again strongly suggest that TNF release from activated macrophages plays a central role in the murine *P. acnes* LPS-induced fulminant liver failure model. The results observed by these investigators are very similar to those observed in our lead enhanced LPS model liver injury.

Liver Regeneration Model

After a variety of types of experimental liver injury, such as that induced by D-GLN, CCl_4, or after partial (67%) hepatectomy, there is systemic endotoxemia.[166] Cornell proposed that gut derived endotoxin is responsible for the endotoxemia that develops after these forms of liver injury and after partial hepatectomy. This systemic endotoxemia then causes release of insulin and glucagon, two hormones that have well-documented hepatotrophic activity.[167] Low dose infusion of LPS into the portal vein does not cause elevation of the insulin/glucagon level; however, infusion of this same low dose of LPS into the systemic circulation does.[168,169] Thus, systemic endotoxemia appears to be important for the elaboration of these two hepatotrophic factors. The cytokines IL-1 and TNF are probable mediators of the insulin/glucagon response following endotoxemia. For example, direct infusion of TNF into experimental animals causes increased production and release of insulin/glucagon.[134] Similarly, administration of low doses of IL-1 *in vitro* stimulates pancreatic insulin secretion.[170] Both IL-1 and TNF stimulate ornithine decarboxylase (ODC) activity, which is necessary for polyamine synthesis and liver regeneration.[171-173] Furthermore, polyamine synthesis increases with the acute phase response initiated by LPS administration.[174,175] Some investigators have shown that pretreatment with modest doses of LPS stimulates liver regeneration.[176]

Studying liver regeneration, Cornell et al.[177] demonstrated that prevention of gut-derived endotoxemia by gut sterilization with oral nonabsorbable antibiotics, polymixin B neutralization of LPS, and induction of LPS tolerance all depressed liver DNA synthesis as well as plasma insulin and glucagon levels following partial hepatectomy. Also, Cornell et al.[177] demonstrated that liver regeneration is depressed after partial hepatectomy in germ free, athymic, and LPS resistant mice. When germ free mice were conventionalized and then underwent 67% hepatectomy, they had a normal liver regeneration response. Thus, there is substantial evidence to suggest that, at least in experimental animals, low-grade systemic endotoxemia may stimulate liver regeneration. This response is likely mediated by cytokines.

How may these findings be reconciled with our previous data that strongly suggest that cytokines, especially TNF, may cause severe liver injury? As mentioned above, it is likely that some of the macrophage and hepatic responses to toxins such as LPS are beneficial, including cytokine release and elicitation of the acute phase response. Most of the time, cytokines like TNF do more to destroy or neutralize toxic agents or invading microorganisms than they do to damage the liver. Furthermore, the liver response protects it against cytokine-mediated damage. Thus, at low doses TNF may be important for protection against pathogens, for tissue repair, etc. But the beneficial/toxic effects of TNF clearly relate to factors such as tissue concentration, duration of exposure, and the presence of modifying agents. When "priming" or enhancing agents such as lead, *P. acnes*, or D-GLN are present, high levels of cytokines may occur and the liver protective responses are overwhelmed and severe liver injury occurs. The concept of TNF being potentially either a beneficial or toxic substance based on factors such as tissue concentration and other mediators is true for many organ systems, and this has been reviewed by Tracey et al.[178]

CYTOKINES IN SELECTED CLINICAL HEPATIC DISEASE STATES

Endotoxin (LPS) in Clinical Hepatic Disease States

We have reviewed a relatively extensive literature suggesting a role for cytokines (especially TNF) in multiple forms of experimental liver injury. There is also a growing body of literature associating cytokines with several forms of liver injury in humans. Indeed, many of the known biologic actions of cytokines are known complications of a variety of types of clinical liver injury.[179,180] These manifestations include fever, neutrophilia, synthesis of acute phase reactants (such as C reactive protein), depression of the serum albumin concentration, anorexia, muscle catabolism, depressed cytochrome P450-mediated hepatic drug metabolism, fibroblast proliferation, bone resorption with loss of bone mass, depression of the serum zinc level, and hypertriglyceridemia.

Heyman et al.[181] stimulated interest in the role of bacterial toxins in human liver injury with their observation that intravenous injection of a typhoid bacterial pyrogen caused a significantly greater febrile response in patients with cirrhosis than in patients with a variety of other disease processes and in normal controls. They suggested that the increased sensitivity to their bacterial pyrogen observed in the patients with chronic liver disease might be caused by impaired reticuloendothelial function, which prevented rapid detoxification of the pyrogen. Initial studies demonstrated both detectable LPS levels in patients with liver disease and considerably higher titers of antibodies to enteric bacteria in patients having liver disease compared to healthy controls.[182,183,183a] Moreover, five patients with Reye's syndrome were all found to have endotoxemia.[184] The mitochondrial dysfunction in this disease resembles that of LPS-induced mitochondrial dysfunction.

Several more recent studies evaluated LPS levels in patients with liver disease, using improved assays for LPS. Bigatello et al.,[185] using the chromogenic limulus assay, observed endotoxemia without sepsis in 36 of 39 cirrhotics, and no endotoxemia in 7 healthy controls. In 11 patients who underwent elective portasystemic shunt, portal vein endotoxemia was higher than that in the inferior vena cava. In 21 patients who developed hepatic encephalopathy after bleeding from esophageal varices, systemic endotoxemia was significantly higher than that observed in well-compensated cirrhotics. LPS levels were positively correlated with serum bilirubin concentrations, prolongation of the prothrombin, and mortality. Lumsden et al.[64] evaluated two groups of patients with liver disease. The first group consisted of 56 patients undergoing angiographic evaluations. The second group of 22 patients underwent surgery for splenorenal or portacaval shunts. In both groups of patients, there was high rate of endotoxemia. In group 1, there was a gradient between hepatic to peripheral vein LPS levels. In group 2, there was a gradient between portal to peripheral vein LPS levels. Portal vein LPS concentrations correlated with prothrombin time, aspartate aminotransferase (AST), Child's grade, and serum creatinine concentration. Bode et al.[186] evaluated endotoxin levels in 88 patients with alcoholic cirrhosis, with 59 having detectable peripheral blood endotoxin levels. They also evaluated 42 patients with nonalcoholic liver disease, and 19 had detectable endotoxin levels. Furthermore, 11 of 24 alcoholics without evidence of chronic liver disease also had endotoxemia. Thus, their study supports the concept that patients with alcoholic liver disease have more frequent and more severe endotoxemia than do patients with nonalcoholic liver disease. This hypothesis is further strengthened by the observation of significantly higher *E. coli* antibody titers in patients with alcoholic cirrhosis compared to nonalcoholic cirrhosis.[187] However, it is important to note that while patients with alcoholic liver disease appear to have more frequent and more severe endotoxemia, patients with nonalcoholic liver disease also frequently have endotoxemia.

It is well documented that alcohol increases intestinal permeability to a variety of agents. For example, Bjarnason et al.[188] used chromium-labeled ethylenediaminetetraacetic acid (EDTA) as a noninvasive marker of intestinal permeability, and they observed that recently intoxicated alcoholics had enhanced urinary excretion of chromium-labeled EDTA. As these alcoholics went through detoxification, their urinary excretion of this marker of intestinal permeability significantly decreased. Similar findings were observed by Robinson et al.[189] using low-molecular-weight polyethylene glycol as a marker of intestinal permeability. Animal studies also document that ethanol increases intestinal permeability.

A variety of other factors probably enhance passage of endotoxin or other toxic materials across the intestine. Examples would include ischemia, disuse, bacterial overgrowth, malnutrition, and even endotoxemia.[190,191] A major postulated mechanism for the liver injury of multisystem organ failure (MSOF) observed in some severe trauma patients is the "gut hypothesis". In this hypothesis, bacteria translocate across the intestine for a variety of different reasons. After making their way to the portal circulation, these bacteria or LPS activate Kupffer cells or circulating macrophages to release cytokines, which then initiate the acute phase response. A suggested explanation of how malnutrition and intestinal disease might cause endotoxemia is the "glutamine/TPN" concept. Glutamine is a major energy source for the small intestine.[192] TPN solutions do not contain glutamine because it is unstable in IV solutions. Thus, patients receiving TPN do not have oral intake to stimulate intestinal function (thus, intestinal disuse), and they do not receive glutamine, the major energy source for the small bowel. Therefore, it is felt by many investigators that TPN administration is a predisposing factor to the liver injury of MSOF. Some investigators have shown that they can decrease the abnormal metabolic response to trauma or injury (e.g., thermal injury in experimental animals) by early enteral feeding.[193-195] It is presumed that this early enteral feeding decreases gut permeability and bacterial translocation, causes less-pronounced release of cytokines and counter regulatory hormones, and therefore causes a less-prominent hypermetabolic/hypercatabolic response.

Studies in rats show that cholestatic liver damage from biliary obstruction may also lead to endotoxemia.[196] Although surgeons used to perform procedures to divert bile flow externally for a variety of types of biliary obstruction, recent controlled trials suggest that external drainage can be associated with increased morbidity.[197,198] Because it appears that bile salts are important in preventing endotoxemia,[199,200] internal stenting and drainage of bile is now preferred.

In summary, there are a host of forms of clinical liver disease in which endotoxemia (both peripheral and portal) occur, with alcoholic liver disease being the most prominent. There are multiple reasons for the development of endotoxemia, such as enhanced intestinal permeability caused by factors such as alcohol and malnutrition, and impaired reticuloendothelial function with impaired clearance of intestinal

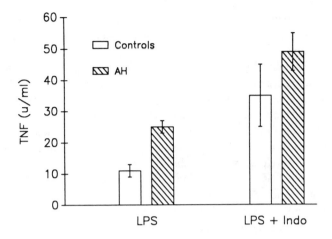

Figure 7. Monocyte TNF activity in 16 alcoholic hepatitis (AH) patients and healthy controls. All values are reported as mean ± S.E. Patients with AH had significant increased LPS-stimulated TNF activity compared to controls (25.3 ± 3.7 vs. 10.9 ± 2.4 units per ml, p <0.005). When LPS-stimulated monocytes were incubated with the cyclooxygenase inhibitor indomethacin (Indo) in order to block prostaglandin production, there was a significant (p <0.05) increase in TNF activity in both AH patients (49.3 ± 6.2 units per ml) and control (35.2 ± 9.8 units per ml) compared to LPS-stimulated monocytes alone (see Reference 180).

toxins. Thus endotoxemia could be a major inducer of cytokine production in patients with liver disease. However, the data on enhanced cytokine activity in patients with clinical liver disease is relatively limited due mainly to the recent availability of reproducible bioassays and radioimmunoassays for the cytokines IL-1, TNF, and IL-6 and our newly developed understanding of the metabolic and immunologic actions of these cytokines. Information concerning cytokine activity/levels in clinical liver disease will now be presented.

Alcoholic Liver Disease

Research by our group demonstrated increased TNF production by monocytes in patients with alcoholic hepatitis.[180] Sixteen patients hospitalized for alcoholic hepatitis who had ingested no alcohol for at least 3 d had peripheral blood monocyte TNF production evaluated in the basal state and after stimulation with endotoxin. Patients had a mean serum bilirubin concentration of 6.9 ± 2.3 mg/dl, albumin of 2.6 ± 0.2 gm/dl, neutrophilia, and elevated C-reactive protein levels. Alcoholic hepatitis patients had increased basal and LPS-stimulated TNF release from monocytes (Figure 7). Furthermore, incubation of monocytes with indomethacin enhanced TNF production from both volunteer and alcoholic hepatitis monocytes. As noted previously, prostaglandins are potent endogenous down regulators of cytokine production, and the indomethacin data implies that the prostaglandin pathway is at least partially intact in alcoholic hepatitis patients. Our results suggested that monocytes from alcoholic hepatitis patients may be stimulated *in vivo* to release TNF and may have a lower threshold for endotoxin-induced TNF release. We speculated that some of the metabolic abnormalities observed during alcoholic hepatitis, such as fever, neutrophilia, hypozincemia, etc., may be TNF mediated, as may be some of the liver injury.

DeViere et al.[201] reported increased LPS stimulated monocyte production of TNF, IL-1, and IL-6 in alcoholic cirrhosis, thus supporting the concept of enhanced monocyte cytokine activity in alcoholic liver disease. Several groups have reported increased serum/plasma cytokine concentrations in patients with alcoholic liver disease. We initially suggested an association between cytokines and both liver injury and metabolic complications of alcoholic liver disease in a 1986 report showing increased serum IL-1 bioactivity in six patients with severe alcoholic hepatitis.[179] Felver et al.[201a] reported increased plasma TNF concentrations in patients with alcoholic hepatitis and showed that high TNF correlated with poor outcome. Bird et al.[201b] reported increased plasma TNF concentrations which correlated with abnormal serum creatinine, bilirubin, blood neutrophil counts, and fever. Khoruts et al.[201c] evaluated serum TNF, IL-1, and IL-6 concentrations in patients with severe alcoholic hepatitis, chronic alcoholism without evident liver disease, and stable micronodular cirrhosis. These authors found elevated plasma TNF, IL-1, and IL-6 only in the patients with hepatitis. When hepatitis patients were followed longitudinally, TNF did not correlate with clinical improvement. However, IL-6 concentrations did. Further unpublished studies by our group have documented in 30 patients with alcoholic hepatitis that plamsa IL-6 concentrations correlate well with both acute phase reactants (C-reactive protein) and the clinical course of liver disease over a 6-month follow-up period.

Thus, it appears that there is increased monocyte production and increased plasma cytokine levels in patients with alcoholic hepatitis. One potential stimulus for this increased cytokine production is endotoxemia observed in these patients. Another possibility we considered is ethanol. However, we observed that incubating monocytes with ethanol or acetaldehyde actually decreased, instead of increased, IL-1 and TNF production.[180] Furthermore, both our group and D'Souza et al.[202] observed that giving large acute doses of ethanol to rats depressed their plasma TNF response to LPS. This is in contrast to our observation that rats chronically fed alcohol have an increased plasma TNF response to endotoxin.[202a] Thus, it appears that acute alcohol ingestion vs. chronic alcohol feeding have different effects on cytokine production.

Currently the only widely accepted mode of therapy for alcoholic liver disease is abstinence. Other forms of therapy that are advocated by some groups include a combination of anabolic steroids and nutritional support, treatment with the antithyroid drug propylthiouracil, colchicine therapy, and corticosteroid therapy. Indeed, a recent multicenter study suggests very positive effects of corticosteroid therapy in alcoholic hepatitis.[203] As we have noted previously, steroids can block the production of cytokines, and the beneficial effect of corticosteroid therapy could theoretically be cytokine mediated. Similarly, colchicine downregulates TNF receptors,[203a] and may be of benefit in alcoholic hepatitis through this mechanism.

Viral Hepatitis

Yoshioka et al.[204] evaluated monocyte TNF production in patients with chronic viral liver disease. Patients with chronic persistent hepatitis (CPH), chronic active hepatitis (CAH), and cirrhosis all had increased monocyte TNF activity after stimulation with recombinant interferon and IL-2, compared to control monocytes. Furthermore, there was a significant correlation between monocyte TNF levels and histologic activity index in patients with CPH and CAH. These findings thus documented both increased monocyte TNF activity in chronic viral liver disease and a relationship between increased monocyte TNF levels and activity of hepatitis. In patients with fulminant hepatic failure, increased monocyte production of both TNF and IL-1 was reported, with higher levels of both cytokines observed in patients who died.[205] Increased monocyte IL-1 activity also was reported by Roger Williams' group in patients with chronic liver disease due to hepatitis B virus.[206] Furthermore, IL-1 activity correlated with severity of fibrosis on liver biopsy.

While there is no accepted form of therapy for acute fulminant viral hepatitis, a study by Sinclair et al.[207] did demonstrate a benefit of using prostaglandin therapy in 17 patients with fulminant hepatitis. Again, prostaglandins have been shown to down regulate TNF production and have been shown to be effective in a variety of forms of experimental injury, presumably through their effects on TNF production.

Primary Biliary Cirrhosis

Little work has been done in the area of primary biliary cirrhosis (PBC). However, since this disease is associated with granuloma formation in the liver, there theoretically could be increased cytokine production in this disease. Supportive of this concept is a recent abstract by Miller et al.[208] They reported elevated plasma TNF levels in patients with PBC, and TNF levels decreased after colchicine therapy. The decrease of TNF with therapy again suggests a possible role for TNF in liver injury.

In summary, increased monocyte or plasma cytokine activity (especially TNF and, more recently, IL-6) has been reported in a variety of types of liver disease, including alcohol, viral, and PBC. This increased cytokine activity has been speculated to play a role in the immunologic, metabolic, and hepatic abnormalities observed in these patients.

SUMMARY

In this review, we have discussed several experimental models in which there is evidence that cytokines can cause hepatic dysfunction, and we have discussed data from a number of clinical studies suggesting that cytokines play a role in the pathobiology of several hepatic diseases. In summarizing the effects of cytokines in experimental hepatic dysfunction, we have limited our review to models in which hepatic dysfunction is assessed by parenchymal cell injury. Furthermore, TNF seems to be the most important cytokine that mediates this effect. However, other cytokines besides TNF, such as IL-1 and IL-6, also have undeniably important effects on hepatic metabolism. Alterations occur in carbohydrate, lipid, and protein metabolism (reviewed in Reference 209), trace metal metabolism (reviewed in Reference 210), hepatic cytochrome P-450 mediated drug metabolism,[211-214] and bile salt metabolism.[215]

In terms of hepatic nutritional and hormonal parameters, Klasing[209] recently reviewed numerous effects of cytokines, distinguishing between effects demonstrated *in vivo* and direct effects on target tissues demonstrable *in vitro*. Increases in plasma insulin and glucagon concentrations, up regulation of ACTH and glucocorticoids, increases in both glucose utilization and gluconeogenesis, increased acute phase protein synthesis, and increased hepatic triglyceride synthesis have all been reported as being cytokine mediated. However, none of these changes would be considered "dysfunctional" and were therefore not reviewed. Decreases in bile salt and organic anion excretion (cholestasis) is a well-known effect of LPS (reviewed in Reference 216) which might be considered dysfunctional, and Ott et al.,[215] using a semipurified monocyte supernatant containing IL-1 in an isolated perfused rat liver, have demonstrated decreased bile flow. However, it is not yet certain if this phenomenon represents hepatic injury. Decreases in hepatic cytochrome P-450-mediated drug (and steroid?) metabolism after LPS (reviewed in Reference 217) and after interferons (reviewed in Reference 218), which could be mediated by IL-1 or TNF[211-214] could also be considered "dysfunctional". However, decreased cytochrome P-450 oxidations might actually represent a mechanism by which the hepatocyte decreases its own generation of $\cdot O_2^-$ and thereby protects itself.

Cytokine administration (IL-1, TNF, or IL-6) induces marked alterations in mineral metabolism, with a depression in the serum zinc and iron concentration and a subsequent elevation in copper levels. The depression in the serum zinc concentration appears to be mediated by cytokine enhancement of metallothionein gene expression, with IL-1 and IL-6 being major cytokine enhancers.[210] Similarly, the increase in serum copper concentration appears to be due to cytokine-enhanced synthesis of ceruloplasmin, the major copper-binding protein.[219] Mechanisms for the decrease in the iron level are less clear. It has been suggested that cytokines cause degranulation of granulocytes, with subsequent release of lactoferrin, which then binds the iron and causes its sequestration in the liver.[220] None of these above changes in mineral metabolism would be considered a manifestation of hepatic dysfunction.

TGFβ is a cytokine that was not reviewed in this chapter and that may have a major role in the regulation of hepatic collagen production and possibly liver regeneration.[221] For example, TGFβ-1 mRNA levels are increased in a murine model of schistosomiasis, and liver cells treated with TGFβ-1 have increased procollagen mRNA. TGFβ-1 also may interact in a regulatory role with IL-1 and TNF. We have not dealt with TGFβ-1 in this chapter because the preponderance of evidence suggests that its major role will be in collagen formation with fibrosis and not on liver cell death. However, this cytokine still warrants mention. Cytokines such as IL-1 and TNF have profound effects on tissues surrounding the hepatocyte, which may precipitate liver injury or dysfunction. Effects of cytokines on Ito cells and Kupffer cells have been noted, but the effects of cytokines on neutrophils[222] and on endothelium[223] with expression of adhesion molecules need to be cited.

In conclusion, it is clear that cytokines (especially TNF) can have profound effects on hepatocyte function in a variety of experimental models of liver cell injury, and cytokines probably play an important role in immunologic and hepatic abnormalities and hepatic dysfunction in several types of clinical hepatitis. Our knowledge concerning the effects of cytokines on liver injury will expand as we better understand how modifying factors, local tissue concentration, receptor number, and duration of exposure to cytokines influence cytokine hepatotoxicity.

ACKNOWLEDGMENTS

The authors wish to thank Dr. Donald Cohen for helpful suggestions in the preparation of this review, Mr. Ron Honchel and Mrs. Betty Miller for technical assistance, and Ms. Sharon Evans and Mr. Steve Mahanes for manuscript preparation. This research was supported by the Department of Veterans Affairs Medical Center, CRC 5MORR02602-4, The Alcoholic Beverage Medical Research Foundation, and NIH 1-R01-NS22712-01A1.

REFERENCES

1. Schanne, F.A.X., A.B. Kane, E.E. Young, and J.L. Farber, Calcium dependence of toxic cell death: a final common pathway. *Science,* 206:700–702, 1979.
2. Orrenius, S. and P. Nicotera, Studies of CA^{2+}-mediated toxicity in hepatocytes. *Klin. Wochenschr.,* 64 (suppl. VII):138, 1986.
3. Mitchell, D.B., D. Acosta, and J.V. Bruckner, Role of glutathione depletion in the cytotoxicity of acetaminophen in a primary culture system of rat hepatocytes. *Toxicology,* 37:127–146, 1985.
4. Thor, H., P. Hartzell, S.-A. Svensson, S. Orrenius, et al., On the role of thiol groups in the inhibition of liver microsomal CA^{2+} sequestration by toxic agents. *Biochem. Pharmacol.,* 34:3717, 1985.
5. Fariss, M.W., G.A. Pascoe, and D.J. Reed, Vitamin E reversal of the effect of extracellular calcium on chemically induced toxicity in hepatocytes. *Science,* 227:751, 1985.
6. Chance B., H. Sies, and A. Boveris, Hydroperoxide metabolism in mammalian organs. *Physiol. Rev.,* 59:527, 1979.
7. Dianzani, M.U., The role of free radicals in liver damage. *Nutri. Soc.,* 46:43, 1987.

8. Poli, G., E. Albano, and M.U. Dianzani, The role of lipid peroxidation in liver damage. *Chem. Phys. Lipids,* 45:117–123, 1987.
9. Kappus, H., A survey of chemicals inducing lipid peroxidation in biological systems. *Chem. Phys. Lipids,* 45:105, 1987.
10. Tribble, D.L., T.Y. Aw, and D.P. Jones, The pathophysiological significance of lipid peroxidation oxidative cell injury. *Hepatology,* 7:377, 1987.
11. Cross, C.E., E. Borish, W. Pryor, B. Ames, R. Saul, J. McCord, and D. Harman, Oxygen radicals and human disease. *Ann. Intern. Med.,* 107:526–545, 1987.
12. Arthur, M.J.P. Reactive oxygen intermediates and liver injury. *J. Hepatol.,* 6:125, 1988.
13. O'Brien, P.J., Radical formation during the peroxidase catalyzed metabolism of carcinogens and xenobiotics. *Free Rad. Biol. Med.,* 4:169, 1988.
14. Tappel, A.L., Measurement of and protection from in vivo lipid peroxidation, in *Free Radicals in Biology,* Academic Press, New York, 1980, 1–51.
15. Smith, M.T., H. Thor, P. Hartzell, and S. Orrenius, The measurement of lipid peroxidation in isolated hepatocytes. *Biochem. Pharmacol.,* 31:19, 1982.
16. Shedlofsky, S.I., H.L. Bonkowsky, P.R. Sinclair, J.F. Sinclair, W.J. Bement, and J.S. Pomeroy, Iron loading of cultured hepatocytes. Effect of iron on 5–aminolaeulinate synthase is independent of lipid peroxidation. *Biochem. J.,* 212:321–330, 1983.
17. McCord, J.M., Oxygen-derived free radicals in postischemic tissue injury. *N. Engl. J. Med.,* 312:159–163, 1985.
18. Adkison, D., M.E. Hollwarth, J.N. Benoit, D.A. Parks, J.M. McCord, and D.N. Granger, Role of free radicals in ischemia-reperfusion injury to the liver. *Acta Physiol. Scand.,* 548:101–107, 1986.
19. Wendel, A., G. Tiegs, and C. Werner, Evidence for the involvement of a reperfusion injury in galactosamine/endotoxin-induced hepatitis in mice. *Biochem. Pharmacol.,* 36:2637, 1987b.
19a. DeGroot, H. and A. Littauer, Hypoxia, reactive oxygen, and cell injury. *Free Rad. Biol. Med.,* 6:541–551, 1989.
19b. Hibbs, J.B., R.R. Taintor, and Z. Vavrin, Macrophage cytotoxicity: role for L-arginine deiminase and imino nitrogen oxidation to nitrate. *Science,* 240:473–475, 1987.
19c. Kilbourn, R.G., S.S. Gross, A. Jubran, J. Adams, O.W. Griffith, R. Levi, and R.F. Lodato, N^6-methyl-L-arginine inhibits tumor necrosis factor-induced hypotension: implications for the involvement of nitric oxide. *Proc. Natl. Acad. Sci. U.S.A.,* 87:3629–3632, 1990.
19d. Keller, R., R. Keist, A. Wechsler, T.P. Leist, and P.H. van der Meide, Mechanisms of macrophage-mediated tumor cell killing: a comparative analysis of the roles of reactive nitrogen intermediates and tumor necrosis factor. *Int. J. Cancer,* 46:682–686, 1990.
19e. Billiar, T.R., R.D. Curran, D.J. Steuhr, F.K. Ferrari, and R.L. Simmons, Evidence that activation of Kupffer cells results in production of L-arginine metabolites that release cell-associated iron and inhibit hepatocyte protein synthesis. *Surgery,* 106:364–372, 1989.
19f. Reif, D.W. and R.D. Simmons, Nitric oxide mediates iron release from ferritin. *Arch. Biochem. Biophys.,* 283:537–541, 1989.
20. Desmet, V.J. and R. De Vos, Structural analysis of acute liver injury, in *Mechanisms of Hepatocyte Injury and Death,* D. Keppler, H. Popper, L. Bianchi, and W. Reutter, MTP Press, London, 1984, 11–30.
21. Keppler, D. and H. Popper, Mechanisms of hepatocellular degeneration and death, in *Recent Advances in Hepatology,* Vol. 2, H.C. Thomas and E.A. Jones, Eds., Churchill Livingstone, Avon, England, 1986, 45–54.

22. Popper, H., Hepatocellular degeneration and death, in *The Liver: Biology and Pathobiology*, 2nd ed., Raven Press, New York, 1988, 1087–1103.
23. Kerr, J.F.R., A.H. Wyllie, and A.R. Currie, Apoptosis: a basic biological phenomenon with wide ranging implications in tissue kinetics. *Br. J. Cancer*, 26:239, 1972.
24. Wyllie, A.H., J.F.R. Kerr, and A.R. Currie, Cell death: the significance of apoptosis. *Int. Rev. Cytol.*, 68:251, 1980.
25. Mitchell, J.R., D.J. Jollow, W.Z. Potter, D.C. Davis, J.R. Gillette, and B.B. Brodie, Acetaminophen-induced hepatic necrosis. I. Role of drug metabolism. *J. Pharmacol. Exp. Ther.*, 187:185, 1973.
25a. Larrick, J.W. and S.C. Wright, Cytotoxic mechanism of tumor necrosis factor-α. *FASEB. J.*, 4:3215–3223, 1990.
26. Lieber, C.S., Alcohol, protein metabolism, and liver injury. *Gastroenterology*, 79:373, 1980.
27. Weiner, F.R., M.J. Czaja, and M.A. Zern, Ethanol and the liver, in *The Liver: Biology and Pathobiology*, 2nd ed., Raven Press, New York, 1988, 1169–1184.
28. Decker, K. and D. Keppler, Galactosamine hepatitis: key role of the nucleotide deficiency period in the pathogenesis of cell injury and cell death. *Rev. Physiol. Biochem. Pharmacol.*, 71:77, 1974.
29. Recknagel, R.O. and E.A. Glende, Carbon tetrachloride hepatotoxicity: an example of lethal cleavage. *Crit. Rev. Toxicol.*, 2:263–274, 1973.
30. Lau, S.S. and V.G. Zannoni, Hepatic microsomal epoxidation of bromobenzene to phenols and its toxicological implication. *Toxicol. Appl. Pharmacol.*, 50:309, 1979.
31. Seyberth, H.W., H. Schmidt-Gayk, and E. Hackenthal, Toxicity, clearance and distribution of endotoxin in mice as influenced by actinomycin D, cycloheximide, a-amanitin and lead acetate. *Toxicon*, 10:491, 1972.
32. Bursch W. and R. Schulte-Hermann, Cytoprotective effect of the prostacyclin derivative iloprost against liver cell death induced by the hepatotoxins carbon tetrachloride and bromobenzene. *Klin. Wochenschr.*, 64 (suppl. VII):47, 1986.
33. Bursch, W., H.S. Taper, M.P. Somer, S. Meyer, B. Putz, and R. Schulte-Hermann, Histochemical and biochemical studies on the effect of the prostacyclin derivative iloprost on CCl_4-induced lipid peroxidation in rat liver and its significance for hepatoprotection. *Hepatology*, 9:830, 1989.
34. Tiegs, G., M. Wolter, and A. Wendel, Tumor necrosis factor is a terminal mediator in galactosamine/endotoxin-induced hepatitis in mice. *Biochem. Pharmacol.*, 38:627, 1989.
35. Laskin, D.L. and A.M. Pilaro, Potential role of activated macrophages in acetaminophen hepatotoxicity. *Toxicol. Appl. Pharmacol.*, 86:204, 1986.
36. Dinarello, C.A., Interleukin-1. *Rev. Infect. Dis.*, 6:51–95, 1984.
37. Le, J. and J. Vilcek, Biology of disease. Tumor necrosis factor and interleukin 1: cytokines with multiple overlapping biological activities. *Lab. Invest.*, 56:234–247, 1987.
38. Leach, B.E. and J.C. Forbes, Sulfonamide drugs as protective agents against carbon tetrachloride poisoning. *Proc. Soc. Exp. Biol. Med.*, 48:361–363, 1941.
39. Formal, S.B., H.E. Noyes, and H. Schneider, Expermental shigella infections. III. Sensitivity of normal, starved and carbon tetrachloride treated guinea pigs to endotoxin. *Proc. Soc. Exp. Biol. Med.*, 103:415–418, 1960.
40. Nolan, J.P., Endotoxin, reticuloendothelial function, and liver injury. *Hepatology*, 1:458, 1981.
40a. Nolan, J.P., The role of endotoxin in liver injury. *Gastroenterology*, 69:1346–1356, 1975.
41. Rutenburg, A.M., E. Sonnenblick, I. Koven, et al., The role of intestinal bacteria in the development of dietary cirrhosis in rats. *J. Exp. Med.*, 106:1–13, 1957.

42. Grun, M., H. Liehr, and U. Rasenack, Significance of endotoxemia in experimental "galactosamine-hepatitis" in rats. *Acta Hepato. Gastroenterol.,* 24:64, 1977.
43. Gut, J.-P., S. Schmitt, A. Bingen, M. Anton, and A. Kirn, Probable role of endogenous endotoxins in hepatocytolysis during murine hepatitis caused by frog virus 3. *J. Infect. Dis.,* 149:621, 1984.
44. Glode, L.M., S.E. Mergenhagen, and D.L. Rosenstreich, Significant contribution of spleen cells in mediating the lethal effects of endotoxin in vivo. *Infect. Immun.,* 14:626, 1976.
45. Chojkier, M. and J. Fierer, D-galactosamine hepatotoxicity is associated with endotoxin sensitivity and mediated by lymphoreticular cells in mice. *Gastroenterology,* 88:115, 1985.
46. Utili, R., C.O. Abernathy, and H.J. Zimmerman, Minireview: endotoxin effects on the liver. *Life Sci.,* 20:553, 1977.
47. Nolan, J.P. and D.S. Camara, Endotoxin, sinusoidal cells, and liver injury, in *Progress in Liver Disease,* Grune & Stratton, New York, 1982, 361–376.
48. Nolan, J.P., The role of intestinal endotoxins in gastrointestinal and liver diseases, in *Bacterial Endotoxins: Pathophysiological Effects, Clinical Significance, and Pharmacological Control,* Alan R. Liss, New York, 1988, 147–159.
48a. Nolan, J.P., Intestinal endotoxins as mediators of hepatic injury — an idea whose time has come again. *Hepatology,* 10:887–889, 1989.
49. Kaplowitz, N., S. Maitra, D. Rachmilewitz, and D. Eberle, Hepatocellular uptake of endotoxin in the rat, in *Mechanisms of Hepatocyte Injury and Death,* D. Keppler, H. Popper, L. Bianchi, and W. Reuttler, Eds., MTP Press, London, 1984, 215–223.
50. Dunn, M. and T. Brewer, Non-viral liver infections, in *The Liver Annual,* Elsevier, New York, 1986, 225–244.
51. Ravin, H.A., D. Rowley, C. Jenkins, and J. Fine, On the absorption of bacterial endotoxin from the gastro-intestinal tract of the normal and shocked animal. *J. Exp. Med.,* 112:783–792, 1960.
52. Nolan, J.P., The role of endotoxin in liver injury. *Gastroenterology,* 69:1346, 1975.
53. DePalma, R., J. Coil., J.H. Davis, and W.D. Holden, Cellular and ultrastructural changes in endotoxemia: a light and electron microsopic study. *Surgery,* 62:505, 1967.
54. Freudenberg, M.A., N. Freudenberg, and C. Galanos, Time course of cellular distribution of endotoxin in liver, lungs and kidneys of rats. *Br. J. Exp. Pathol.,* 63:56, 1982.
55. Mathison, J.C. and R.J. Ulevitch, The clearance, tissue distribution, and cellular localization in intravenously injected lipopolysaccharide in rabbits. *J. Immunol.,* 123:2133, 1979.
56. Ruiter, D.J., J. Van Der Meulen, A. Brouwer, M.J.R. Hummel, B.J. Mauw, J.C.M. Van Der Ploeg, and E. Wisse., Uptake by liver cells of endotoxin following its intravenous injection. *Lab. Invest.,* 45:38, 1981.
56a. Fox, E.S., P. Thomas, and S.A. Broitman, Comparative studies of endotoxin uptake by isolated rat Kupffer and peritoneal cells. *Infect. Immun.,* 55:2962, 1987.
57. Zlydaszyk, J.C. and R.J. Moon, Fate of ^{51}CR-labeled lipopolysaccharide in tissue culture cells and livers of normal mice. *Infect. Immun.,* 14:100, 1976.
58. Maitra, S.K., D. Rachmilewitz, D. Eberle, and N. Kaplowitz, The hepatocellular uptake and biliary excretion of endotoxin in the rat. *Hepatology,* 1:401, 1981.
59. Freudenberg, M.A., T.C. Bog-Hansen, U. Back, and C. Galanos, Interaction of lipopolysaccharides with plasmna high-density lipoprotein in rats. *Infect. Immun.,* 28:373–380, 1980.

60. Tesh, V.L., S. Vukajlovich, and D. Morrison, Endotoxin interactions with serum proteins: relationship to biological activity, in *Bacterial Endotoxins: Pathophysiological Effects, Clinical Significance, and Pharmacological Control*, Alan R. Liss, New York, 1988, 47–62.
61. Mosher, D.F., Fibronectin and liver disease. *Hepatology*, 6:1419–1421, 1986.
62. Porvaznik, M., M.E. Cohen, S.W. Bockowski, E.F. Mueller, II, and M.R. Wirthlin, Jr., Enhancement of cell attachment to a substrate coated with oral bacterial endotoxin by plasma fibronectin. *J. Periodont. Res.*, 17:154–159, 1982.
63. Almasio, P.L., R.D. Hughes, and R. Williams, Characterization of the molecular forms of fibronectin in fulminant hepatic failure. *Hepatology*, 6:1340, 1986.
64. Lumsden, A.B., J.M. Henderson, and M.H. Kutner, Endotoxin levels measured by a chromogenic assay in portal hepatic and peripheral venous blood in patients with cirrhosis. *Hepatology*, 8:232, 1988.
65. Richards, P.S. and T.M. Saba, Effect of endotoxin on fibronectin and Kupffer cell activity. *Hepatology*, 5:32–37, 1985.
66. Kang, Y.-H., T. McKenna, L.P. Watson, R. Williams, and M. Holt, Cytochemical changes in hepatocytes of rats with endotoxemia or sepsis: localization of fibronectin, calcium and enzymes. *J. Histochem. Cytochem.*, 36:665–678, 1988.
67. Vincent, P.A., E. Cho, and T.M. Saba, Effect of repetitive low-dose endotoxin on liver parenchymal and Kupffer cell fibronectin release. *Hepatology*, 9:562, 1989.
68. Moriyama, T., H. Aoyama, S. Ohnishi, and M. Imawari, Protective effects of fibronectin in galactosamine-induced liver failure in rats. *Hepatology*, 6:1334–1339, 1986.
69. Williams, D.L., I.W. Browder, and N.R. Di-Luzio, Immuno therapeutic modification of Escherichia-coli induced experimental peritonitis and bacteremia by glucan. *Surgery*, 93(3):448, 1983.
70. Williams, D.L., E.R. Sherwood, I.W. Browder, R.B. McNamee, E.L. Jones, and N.R. Di-Luzio, The role of complement in glucan-induced protection against septic shock. *Circ. Shock.*, 25:53–60, 1988.
71. Kabir, S. and D.L. Rosenstreich, Binding of bacterial endotoxin to murine spleen lymphocytes. *Infect. Immun.*, 15:156–164, 1977.
72. Fogler, W.E., J.E. Talmadge and I.J. Fidler, The activation of tumoricidal properties in macrophages of endotoxin responder and non-responder mice by liposome-encapsulated immunomodulators. *J. Reticuloendothel. Soc.*, 33:165–174, 1983.
73. Fox, E.S., S.A. Broitman, and P. Thomas, Bacterial endotoxins and the liver. *Lab. Invest.*, 63:733–741, 1990.
73a. Fukuda, I., K. Tanamoto, S. Kanegasaki, Y. Yajima, and Y. Goto, Deacylation of bacterial lipopolysaccharide by rat hepatocytes *in vitro*. *Br. J. Exp. Pathol.*, 70:267–273, 1989.
74. Jones, E.A. and J.A. Summerfield, Kupffer cells, in *The Liver: Biology and Pathobiology*, 2nd ed., I.M. Arias, W.B. Jakoby, H. Popper, D. Schachter, and D.A. Shafritz, Eds., Raven Press, New York, 1980, 683–704.
75. Leser, H.G., K.M. Dobatin, H. Northoff, and D. Gemsa, Kupffer cells, enriched nonparenchymal liver cells and peritoneal macrophages differ in PGE release, production of IL1 and tumor cytotoxicity, in *Sinusoidal Liver Cells*, Elsevier/North Holland, Amsterdam, 1982, 369–382.
76. Decker T., M. Lohmann-Matthes, and G.E. Gifford, Cell-associated tumor necrosis factor (TNF) as a killing mechanism of activated cytotoxic macrophages. *J. Immunol.*, 138:957–962, 1987.

77. Van Bossuyt, H., C. Desmaretz, B. Rombaut, and E. Wisse, Response of cultured rat Kupffer cells to lipopolysaccharide. *Arch. Toxicol.*, 62:316, 1988a.
77a. Van Bossuyt, H. and E. Wisse, Structural changes produced in Kupffer cells in the rat liver by injection of lipopolysaccharide. *Cell Tissue Res.*, 251:205, 1988b.
78. Decker, K., U. Karck, and T. Peters, Regulatory circuit in endotoxin-exposed Kupffer cells involving tumor necrosis factor and prostaglandin E_2. *Hepatology*, 8(Abstr.):1230, 1988.
79. Pilaro, A.M. and D.L. Laskin, Accumulation of activated mononuclear phagocytes in the liver following lipopolysaccharide treatment of rats. *J. Leukoc. Biol.*, 40:29–41, 1986.
80. Hirata, K., A. Kaneko., K. Ogawa., H. Hayasaka, and T. Onoe, Effect of endotoxin on rat liver: analysis of acid phosphatase isozymes in the liver of normal and endotoxin-treated rats. *Lab. Invest.*, 43:165, 1980.
81. Beeson, P.B., Tolerance to bacterial pyrogens. II. Role of reticulo-endothelial system. *J. Exp. Med.*, 86:39, 1947.
82. Greisman, S.E., F.A. Carozza, and J.D. Hills, Mechanisms of endotoxin tolerance. I. Relationship between tolerance and reticuloendothelial system phagocytic activity in the rabbit. *J. Exp. Med.*, 117:663–674, 1963.
83. Williams, J.F., Induction of tolerance in mice and rats to the effect of endotoxin to decrease the hepatic microsomal mixed-function oxidase system. Evidence for a possible macrophage-derived factor in the endotoxin effect. *Int. J. Immunopharmacol.*, 7:501, 1985.
84. Sasaki, K., M. Ishikawa-Saitoh, and G. Takayanagi, Effect of lipopolysaccharide (from Escherichia coli) on the hepatic drug-metabolizing activities in successively LPS-treated mice. *Jpn. J. Pharmacol.*, 34:241–248, 1984.
85. Lang, C.H. and J.A. Spitzer, Glucose kinetics and development of endotoxin tolerance during long-term continuous endotoxin infusion. *Metabolism*, 35:469–474, 1987.
86. Rogers, T.S., P.V. Haluska, W.C. Wise, and J.A. Cook, Differential alteration of lipoxygenase and cyclooxygenase metabolism by rat peritoneal macrophages induced by endotoxin tolerance. *Prostaglandins*, 31:639, 1986.
87. Bagby, G.J., C.B. Corll, J.J. Thompson, and L.A. Wilson, Lipoprotein lipase-suppressing mediator in serum of endotoxin treated rats. *Am. J. Physiol.*, 251:E470, 1986.
88. Warren, H.S. and L.A. Chedid, Strategies for the treatment of endotoxemia: significance of the acute-phase response. *Rev. Infect. Dis.*, 9(5):S630, 1987.
88a. Alcorn, J., J. Fierer, and M. Chojkier, Endotoxic death prevented by acute phase response in D-galactosamine-treated mice. *Hepatology*, 8(Abstr.):1250, 1988.
89. Vogel, S.N., E.N. Kaufman, M.D. Tate, and R. Neta, Recombinant interleukin-1a and recombinant tumor necrosis factor a synergize in vivo to induce early endotoxin tolerance and associated hematopoietic changes. *Infect. Immun.*, 56:2650, 1988.
90. Wallach, D., H. Holtman, H. Engelmann, and Y. Nophar, Sensitization and desensitization to lethal effects of tumor necrosis factor and IL-1. *J. Immunol.*, 140:2994–2999, 1988.
90a. Cross, A.S., J.C. Sadoff, N. Kelly, E. Bernton, and P. Gemski, Pretreatment with recombinant murine tumor necrosis factor-α/cachectin and murine interleukin-1α protects mice from lethal bacterial infection. *J. Exp. Med.*, 169:2021–2027, 1989.
90b. Starnes, H.F. Jr., M.K. Pearce, A. Tewari, J.H. Yim, J.C. Zou, and J.S. Abrams, Anti-IL-6 monoclonal antibodies protect against lethal *E. coli* infection and lethal tumor necrosis factor-α challenge in mice. *J. Immunol.*, 145:4185–4191, 1990.
91. Beutler, B., I.W. Milsark, and A.C. Cerami, Passive immunization against cachectin/tumor necrosis factor protects mice from lethal effect of endotoxin. *Science*, 229:869, 1985.

92. Beutler, B. and A.C. Cerami, Cachectin: more than a tumor necrosis factor. *N. Engl. J. Med.*, 316:379, 1987.
93. Tracy, K.J., B. Beutler, S.F. Lowry, J. Merryweather, S. Wolpe, I.W. Milsark, et al., Shock and tissue injury induced by recombinant human cachectin. *Science*, 234:470–474, 1986.
94. Lehmann, V., M.A. Freudenberg, and C. Galanos, Lethal toxicity of lipopolysaccharide and tumor necrosis factor in normal and D-galactosamine-treated mice. *J. Exp. Med.*, 165:657, 1987.
95. Fraker, D.L., M.C. Strovroff, M.J. Merino, and J.A. Norton, Tolerance to tumor necrosis factor in rats and the relationship to endotoxin tolerance and toxicity. *J. Exp. Med.*, 168:95, 1988.
95a. Sheppard, B.C., D.L. Fraker, and J.A. Norton, Prevention and treatment of endotoxin and sepsis lethality with recombinant human tumor necrosis factor. *Surgery*, 106:156–162, 1989.
96. Tracy, K.J., H. Wei, K.R. Manogue, Y. Fong, et al., Cachectin/tumor necrosis factor induces cachexia, anemia and inflammation. *J. Exp. Med.*, 167:1211, 1988.
96a. Wong, G.H.W., and D.V. Goeddel, Induction of manganous superoxide dismutase by tumor necrosis factor: possible protective mechanism. *Science*, 242:941–944, 1989.
96b. Shaffer, J.B., P.T. Treanor, and P.J. Del Vecchio, Expression of bovine and mouse endothelial cell antioxidant enzymes following TNFα exposure. *Free Rad. Biol. Med.*, 8:497–502, 1990.
96c. Jaattela, M., K. Saksela, and E. Saksela, Heat shock protects WEHI-164 target cells from cytolysis by tumor necrosis factors alpha and beta. *Eur. J. Immunol.*, 19:1413–1417, 1989.
96d. Kusher, D.I., C.F. Ware, and L.R. Gooding, Induction of the heat shock response protects cells from lysis by tumor necrosis factor. *J. Immunol.*, 145:2925–2931, 1990.
97. Nolan, J.P. and M.V. Ali, Endotoxin and the liver. II. Effect of tolerance on carbon tetrachloride induced injury. *J. Med.*, 4:28–38, 1973.
98. Williams, J.F., Carbon tetrachloride hepatotoxicity in endotoxin tolerant and polymyxin B-treated rats. *Int. J. Immunopharmacol.*, 10:975, 1988.
99. Utili, R., L.E. Adinolfi, G.B. Gaeta, M.F. Tripodi, and D. Alvaro, Effect of endotoxin tolerance on drug hepatotoxicity: amelioration of taurolithocholate cholestasis in the perfused rat liver. *Digestion*, 36:74, 1987.
100. Peavy, D.L., R.E. Baughn, and D.M. Musher, Effects of BCG infection on the susceptibility of mouse macrophages to endotoxin. *Infect. Immunol.*, 24:59–64, 1979.
101. Tanner, A., A. Keyhani, R. Reiner, G. Holdstock, and R. Wright, Proteolytic enzymes released by liver macrophages may promote hepatic injury in a rat model of hepatic damage. *Gastroenterology*, 80:647–654, 1981.
102. Arthur, M.J.P., I.S. Bentley, A.R. Tanner, P.K. Saunders, G.H. Millward-Sadler, and R. Wright, Oxygen-derived free radicals promote hepatic injury in the rat. *Gastroenterology*, 89:1114–22, 1985.
103. Mizoguchi, Y., H. Tsutsui, K. Miyajima, Y. Sakagami, S. Seki, K. Kobayashi, S. Yamamoto, and S. Morisawa, The protective effects of prostaglandin E_1 in an experimental massive hepatic cell necrosis model. *Hepatology*, 7:1184–1188, 1987.
104. Phillips, N.C., J. Rioux, and M.-S. Tsao, Activation of murine Kupffer cell tumoricidal activity by liposomes containing lipophilic muramyl dipeptide. *Hepatology*, 8:1046–1050, 1988.

105. Galelli, A. and L. Chedid, Induction of colony-stimulating activity (CSA) by a synthetic muramyl peptide (MDP): synergism with LPS and activity in C3H/HeJ mice and in endotoxin-tolerized mice. *J. Immun.*, 137:3211–3215, 1986.
106. Shiratori, Y., T. Kawase, S. Shiina, K. Okano, T. Sugimoto, et al., Modulation of hepatotoxicity by macrophages in the liver. *Hepatology,* 8:815, 1988.
107. Balis, J.U., E.S. Rappaport, L. Gerber, J. Fareed, F. Buddingh, and H.L. Messmore, A primate model for prolonged endotoxin shock. Blood-vascular reactions and effects of glucocorticoid treatment. *Lab Invest.,* 38:511–523, 1978.
108. Cybulsky M.I., M.K.W. Chan, and H.Z. Movat, Neutrophil emigration. Quantitation, kinetics, and role of mediators, in *Cellular and Molecular Aspects of Inflammation,* Plenum, New York, 1988, 41–56.
109. Weiss S.J., Tissue destruction by neutrophils. *N. Engl. J. Med.,* 320(6):365, 1989.
110. Holman J.M. and T.M. Saba, Hepatocyte injury during post-operative sepsis: activated neutrophils as potential mediators. *J. Leukoc. Biol.,* 43:193–203, 1988.
111. Fainsilber, Z., L. Feinman, S. Shaw, and C.S. Lieber, Biphasic control of polymorphonuclear cell migration by Kupffer cells: effect of exposure to metabolic products of ethanol. *Life Sci.,* 43:603–608, 1988.
112. Margaretten W., D.G. McKay, and L.L. Phillips, The effect of heparin on endotoxin shock in the rat. *Am. J. Pathol.,* 57:61, 1967.
113. Shibayama, Y., Sinusoidal circulatory disturbance by microthrombosis as a cause of endotoxin-induced hepatic injury. *J. Pathol.,* 151:315, 1987.
114. Dienstag, J.L., Studies of cell-mediated immunity in chronic hepatitis B virus infection: the elusive goal of virus and host antigen specificity, in *Advances in Hepatitis Research,* Masson, Paris, 1984, 163–167.
115. Chisari, F.V., Hepatic immunoregulator molecules and the pathogenesis of hepatocellular injury in viral hepatitis, in *Advances in Hepatitis Research,* Masson, Paris, 1984, 168–182.
116. Mondelli, M., N. Naumov, and A.L.W.F. Eddleston, The immuno-pathogenesis of liver cell damage in chronic hepatitis B virus infection, in *Advances in Hepatitis Research,* Masson, Paris, 1984, 144–151.
117. Lemon, S.M., Type A viral hepatitis. New developments in an old disease. *N. Engl. J. Med.,* 313:1059, 1985.
118. Sabesin, S.M. and R.S. Koff, Pathogenesis of experimental viral hepatitis. *N. Engl. J. Med.,* 290:944, 1974.
119. Williams, D.L. and N.R. Di Luzio, Glucan-induced modification of murine viral hepatitis. *Science,* 208:67, 1980.
120. Sherwood, E.R., D.L. Williams, and N.R. Di-Luzio, Glucan stimulates production of antitumor cytolytic/cytostatic factor(s) by macrophages. *J. Biol. Response Mod.,* 5:504–526, 1986.
121. Sherwood, E.R., D.L. Williams, R.B. McNamee, E.L. Jones, and I.W. Browder, Enhancement of interleukin-1 and interleukin-2 production by soluble glucan. *Int. J. Immunopharmacol.,* 9:261–267, 1987.
122. Abecassis, M., J.A. Falk, L. Makowka, V.J. Dindzens, R.E. Falk, and G.A. Levy, 16,16 Dimethyl prostaglandin E_2 prevents the development of fulminant hepatitis and blocks the induction of monocyte/macrophage procoagulant activity after murine hepatitis virus strain 3 infection. *J. Clin. Invest.,* 80:881–889, 1987.
123. Levy, G.A., J.L. Leibowitz, and T.S. Edgington, Induction of monocyte procoagulant activity by murine hepatitis virus type 3 parallels disease susceptibility in mice. *J. Exp. Med.,* 154:1150–1163, 1981.

124. Pereira, C.A., A.-M. Steffan, and A. Kirn, Kupffer and endothelial liver cell damage renders A/J mice susceptible to mouse hepatitis virus type 3. *Virus Res.*, 1:557, 1984.
125. Liehr, H., M. Grun, H.-P. Seelig, R. Seelig, W. Reutter, and W.-D. Heine, On the pathogenesis of galactosamine hepatitis: indications of extrahepatocellular mechanisms responsible for liver cell death. *Arch. B. Cell Pathol.*, 26:331, 1978.
126. Galanos, C., M.A. Freudenberg, and W. Reutter, Galactosamine-induced sensitization to the lethal effects of endotoxin. *Proc. Natl. Acad. Sci. U.S.A.*, 76:5939, 1979.
127. Freudenberg, M.A. and C. Galanos, Induction of tolerance to lipopolysaccharide (LPS)-D-galactosamine lethality by pretreatment with LPS is mediated by macrophages. *Infect. Immun.*, 56:1352, 1988.
128. Cabana, V.G., M.A. Linderman, J.S. Siegel, S.W. Weidman, and S.M. Sabesin, Galactosamine hepatitis and the acute phase response: models of altered lipoprotein metabolism. *Hepatology*, 8(Abstr.):1340, 1988.
129. Wendel, A. and G. Tiegs, Protection by ebselen (PZ 51) against galactosamine/endotoxin-induced hepatitis in mice. *Biochem. Pharmacol.*, 34:2115–2118, 1986.
130. Hagmann, W., C. Denzlinger, and D. Keppler, Role of peptide leukotrienes and their hepatobiliary elimination in endotoxin action. *Circ. Shock*, 14:223–235, 1984.
131. Wendel, A. and G. Tiegs, Leukotriene D4 mediates galactosamine/endotoxin-induced hepatitis in mice. *Biochem. Pharmacol.*, 36:1867, 1987a.
132. Gagnon, L., L.G. Filion, C. Dubois, and M. Rola-Pleszczynski, Leukotrienes and macrophage activation: augmented cytotoxic activity and enhanced interleukin 1, tumor necrosis factor and hydrogen peroxide production. *Agents Actions*, 26(1/2):141–147, 1989.
133. Hishinuma, I., J.I. Nagakawa, K. Kirota, K. Miyamoto, K. Tsukidate, T. Yamanaka, K.I. Katayama, and I. Yamatsu, Involvement of tumor necrosis factor-α in development of hepatic injury in galactosamine-sensitized mice. *Hepatology*, 12:1187–1191, 1990.
134. Warren, R.S., H. Fletcher, H.F. Starnes, Jr., et al., Hormonal and metabolic response to recombinant human tumor necrosis factor in rat: in vitro and in vivo. *Am. J. Physiol.*, 255:E206–E212, 1988.
135. Selye, H., B. Tuchweber, and L. Bertok, Effect of lead acetate on the susceptibility of rats to bacterial endotoxins. *J. Bacteriol.*, 91:884–890, 1966.
135a. Schmiedeberg, P., C. Klein, and M.J. Czaja, Galactosamine liver injury blocks a cellular protective response to tumor necrosis factor-α toxicity. *Hepatology*, 12(abstr.):903, 1990.
136. Rippe, D.F. and L.J. Berry, Metabolic manifestations of lead acetate sensitization to endotoxin in mice. *J. Reticuloendothel. Soc.*, 13:527–735, 1973.
137. Truscott, R.B., Endotoxin studies in chicks: effect of lead acetate. *Can. J. Comp. Med.*, 34:134–137, 1970.
138. Holper, K., R.A. Trejo, L. Brettschneider, and N.R. Diluzio, Enhancement of endotoxic shock in lead sensitized sub-human primates. *Surg. Gynecol. Obstet.*, 136:593, 1973.
139. Hoffman, E.O., R.A. Trejo, N.R. Di Luzio, and J. Lamberty, Ultrastructural alterations of liver and spleen following acute lead administration in rats. *Exp. Mol. Pathol.*, 17:159–170, 1972.
140. Cook, J.A., W.J. Dougherty, and T. Holt, Distribution of ^{203}Pb during lead-potentiated endotoxic shock. *Exp. Mol. Pathol.*, 34:253–263, 1981.
141. Trejo, R.A. and N.R. Di Luzio, Impaired detoxification as a mechanism of lead acetate-induced hypersensitivity to endotoxin. *Proc. Soc. Exp. Biol. Med.*, 136:889–893, 1971.
142. Filkins, J.P. and B.J. Buchanan, Effects of lead acetate on sensitivity to shock, intravascular carbon and endotoxin clearances, and hepatic endotoxin detoxification. *Proc. Soc. Exp. Biol. Med.*, 142:471–475, 1973.

143. Kuttner, R.E. and W. Schumer, Altered carbohydrate metabolism in endotoxin-tolerant rats after lead sensitization. *Circ. Shock*, 13:233–240, 1984a.
144. Kuttner, R.E. and T. Ebata, The mediated effect of endotoxin and lead upon hepatic metabolism. *Surg. Gyn. Obstet.*, 159:319–324, 1984b.
145. Cornell, R.P. and J.P. Filkins, Depression of hepatic gluconeogenesis by acute lead administration. *Proc. Soc. Exp. Biol. Med.*, 147:371–376, 1974.
146. Taki, Y., Y. Shimahara, and W. Isselhard, Derangement of hepatic energy metabolism in lead-sensitized endotoxicosis. *Eur. Surg. Res.*, 17:140–149, 1985.
147. Kato, Y., S. Takimoto, and H. Ogura, Mechanisms of induction of hypercalcemia and hyperphosphatemia by lead acetate in the rat. *Calcif. Tissue Res.*, 24:41–46, 1977.
148. Talmage, R.V., C.J. Vander Wiel, and H. Norimatsu, A reevaluation of the cause of acute hypercalcemia following intravenous administration of lead acetate. *Calcif. Tissue Res.*, 26:149–153, 1978.
149. Nicholls, D.M., M.L. Wassenaar, G.R. Girgis, and M.J. Kuliszewski, Does lead exposure influence liver protein synthesis in rats? *Comp. Biochem. Physiol.*, 78C:403–408, 1984.
150. Cook, J.A. and N.R. Di Luzio, Protective effect of cysteine and methylprednisolone in lead acetate-endotoxin induced shock. *Exp. Mol. Pathol.*, 19:127–138, 1973.
151. Honchel, R., L. Marsano, D. Cohen, S. Shedlofsky, and C.J. McClain, Lead enhances lipopolysaccharide and tumor necrosis factor liver injury. *J. Lab. Clin. Med.*, 117:202–208, 1991.
152. Stachura, J., A. Tarnawski, K.J. Ivey, T. Mach, J. Bogdal, J. Szczudrawa, and B. Klimczyk, Prostaglandin protection of carbon tetrachloride-induced liver cell necrosis in the rat. *Gastroenterology*, 81:211–217, 1981.
153. Stachura, J., A. Tarnawski, J. Szczudrawa, J. Bogdal, T. Mach, B. Klimczyk, and S. Kirchmayer, Cytoprotective effect of 16, 16′Dimethyl prostaglandin E_2 and some drugs on an acute galactosamine induced liver damage in rat. *Folia Histochem. Cytochem.*, 18:311–318, 1980.
154. Ruwart, M.J., B.D. Rush, N.M. Friedle, J. Stachura, and A. Tarnawski, 16,16–Dimethyl-PGE_2 protection against a-napthylisothiocyanate-induced experimental cholangitis in the rat. *Hepatology*, 4:658–660, 1984.
155. Honchel, R., S. Shedlofsky, D. Cohen, L. Marsano, E. Lee, and C. McClain, Is hepatic cytoprotection really cytokine protection? in *Physiological and Pathological Effects of Cytokines*, Dinarello, C., M. Kluger, M. Powanda, and J. Oppenheim, Eds., Wiley-Liss, New York, 1990, 275–280.
156. Paul, N.L. and N.H. Ruddle, Lymphotoxin. *Ann. Rev. Immunol.*, 6:407–438, 1988.
157. Schmid, D.S., J.P. Tite, and N.H. Ruddle, DNA fragmentation: manifestation of target cell destruction mediated by cytotoxic T-cell lines, lymphotoxin-secreting helper T-cell clones, and cell-free lymphotoxin-containing supernatant. *Proc. Natl. Acad. Sci. U.S.A.*, 83:1881–1885, 1986.
158. Ruggiero, V., K. Latham, and C. Baglioni, Cytostatic and cytotoxic activity of tumor necrosis factor on human cancer cells. *J. Immunol.*, 138:2711–2717, 1987.
159. Agarwal, S., B.-E. Drysdale, and H.S. Shin, Tumor necrosis factor-mediated cytotoxicity involves ADP-ribosylation. *J. Immunol.*, 140:4187–4192, 1988.
160. Chen, A.R., K.P. McKinnon, and H.S. Koren, Lipopolysaccharide (LPS) stimulates fresh human monocytes to lyse actinomycin D-treated WEHI-164 target cells via increased secretion of a monokine similar to tumor necrosis factor. *J. Immunol.*, 135:3978, 1985.

161. Kornbluth, R.S. and T.S. Edgington, Tumor necrosis factor production by human monocytes is a regulated event: induction of TNF-a-mediated cellular cytotoxicity by endotoxin. *J. Immunol.*, 137:2585–2591, 1986.
162. Waage, A., Production and clearance of tumor necrosis factor in rats exposed to endotoxin and dexamethasone. *Clin. Immunol. Immunopathol.*, 45:348–355, 1987.
163. Ferraiolo, B.L., J.A. Moore, D. Crase, P. Gribling, H. Wilking, R.A. Baughman, Pharmacokinetics and tissue distribution of recombinant human tumor necrosis factor-α in mice. *Drug Metab. Dispos.*, 16:270–275, 1988.
164. Oliff, A., D. Defeo-Jones, M. Boyer, D. Martinez, D. Kiefer, G. Vuocolo, A. Wolfe, and S.H. Socher, Tumors secreting human TNF/cachectin induce cachexia in mice. *Cell*, 50:555–563, 1987.
165. Schlayer, H.J., U. Karck, U. Ganter, R. Hermann, and K. Decker, Enhancement of neutrophil adherence to isolated rat liver sinusoidal endothelial cells by supernatants of lipopolysaccharide-activated monocytes. *J. Hepatol.*, 5:311, 1987.
165a. Nagakawa, J., I. Hishinuma, K. Hirota, K. Miyamoto, T. Yamanaka, K. Tsukdata, K. Katayama, and I. Yamatsu, Involvement of tumor necrosis factor-α in pathogenesis of activated macrophage-mediated hepatitis in mice. *Gastroenterology*, 99:758–765, 1990.
166. Cornell, R.P., Endotoxin-induced hyperinsulinemia and hyperglucagonemia after experimental liver injury. *Am. J. Physiol.*, 241 (Endocrinol. Metab. 4):E428–E435, 1981.
167. Cornell, R.P., Gut-derived endotoxin elicits hepatotrophic factor secretion for liver regeneration. *Am. J. Physiol.*, 249 (Regulatory Integrative Comp. Physiol. 18): R551–R562, 1985.
168. Wolter, J., H. Liehr, and M. Grun, Hepatic clearance of endotoxins: differences in arterial and portal venous infusion. *J. Reticuloendothel. Soc.*, 23:145–152, 1978.
169. Cornell, R.P., Role of the liver in endotoxin-induced hyperinsulinemia and hyperglucagonemia in rats. *Hepatology*, 3:188–192, 1983.
170. Spinas, G.A., T. Mandrup-Poulsen, J. Molvig, L. Baek, K. Bendtzen, C.A. Dinarello, and J. Nerup, Low concentrations of interleukin-1 stimulate and high concentrations inhibit insulin release from isolated rat islets of Langerhans. *Acta Endocrin. (Copenh.)*, 113:551–558, 1986.
171. Endo, Y., R. Suzuki, and K. Kumagai, Induction of ornithine decarboxylase in the liver and spleen of mice by interleukin 1-like factors produced from a macrophage cell line. *Biochim. Biophys. Acta*, 838:343–350, 1985.
172. Endo, Y., K. Matsushima, K. Onozaki, et al., Role of ornithine decarboxylase in the regulation of cell growth by IL-1 and tumor necrosis factor. *J. Immunol.*, 141:2342–2348, 1988.
173. Luk, G.D., Essential role of polyamine metabolism in hepatic regeneration. *Gastroenterology*, 90:1261–1267, 1986.
174. Taffet, S.M. and M.K. Haddox, Bacterial lipopolysaccharide induction of ornithine decarboxylase in the macrophage-like cell line RAW264: requirement of an inducible soluble factor. *J. Cell. Physiol.*, 122:215–220, 1985.
175. Scalabrino, G., M.E. Ferioli, R. Piccoletti, et al., Activation of polyamine biosynthetic decarboxylases during the acute phase response of rat liver. *Biochem. Biophys. Res. Commun.*, 143:856–862, 1987.
176. Simek, J., L. Sobotka, Z. Cervinkova, and J. Smejkalova, The stimulatory effect of E. coli endotoxin on DNA synthesis in regenerating rat liver. *Sb. Ved. Pr. Lek. Fak. Karlovy Univerzity Hradci Kralove*, 21:589–593, 1978.

177. Cornell, R.P., B.L. Liljequist, and K.F. Bartizal, Depressed liver regeneration after partial hepatectomy of germfree, athymic, and LPS-resistant mice: possible roles for gut-derived endotoxin and monokines. *Hepatology,* 11:916–922, 1990.
178. Tracy, K.J., H. Vlassara, and A. Cerami, Cachectin/tumour necrosis factor. *Lancet,* 1:1122–1125, 1989.
179. McClain, C.J., D.A. Cohen, C.A. Dinarello, J.G. Cannon, S.I. Shedlofsky, and A.M. Kaplan, Serum interleukin-1 (IL-1) activity in alcoholic hepatitis. *Life Sci.,* 39:1479–1485, 1986.
180. McClain, C.J. and D.A. Cohen, Increased tumor necrosis factor production by monocytes in alcoholic hepatitis. *Hepatology,* 9:349–351, 1989a.
181. Heyman, A. and P.B. Beeson, Influence of various disease states upon the febrile response to intravenous injection of typhoid bacterial pyrogen. *J. Lab. Clin. Med.,* 34:1400–1403, 1949.
182. Triger, D.R., M.H. Alp, and R. Wright, Bacterial and dietary antibodies in liver disease. *Lancet,* 1:60–63, 1972.
183. Bjorneboe, M., H. Prytz, and F. Orskov, Antibodies to intestinal microbes in serum of patients with cirrhosis of the liver. *Lancet,* 1:58–60, 1972.
183a. Liehr, B.K., H. Prytz, J. Hoest-Christensen, and B. Korner, Portal venous and systemic endotoxemia in patients without liver disease, and systemic endotoxemia in patients with cirrhosis. *Scand. J. Gastroenterol.,* 2:857–863, 1976.
184. Cooperstock, M.S., R.P. Tucker, and J.V. Baublis, Possible pathogenic role of endotoxin in Reyes syndrome. *Lancet,* 1:1271–1274, 1975.
185. Bigatello, L.M., S.A. Broitman, L. Fattori, M. Di Paoli, M. Pontello, G. Bevilacqua, and A. Nespoli, Endotoxemia, encephalopathy, and mortality in cirrhotic patients. *Am. J. Gastroenterol.,* 82:11–15, 1987.
186. Bode, C., V. Kugler, and J.C. Bode, Endotoxemia in patients with alcoholic and non-alcoholic cirrhosis and in subjects with no evidence of chronic liver disease following acute alcohol excess. *J. Hepatol.,* 4:8–14, 1987.
187. Prytz, H., H. Bjorneboe, F. Orskov, and M. Hilden, Antibodies to *Escherichia coli* in alcoholic and non-alcoholic patients with cirrhosis of the liver or fatty liver. *Scand. J. Gastroenterol.,* 8:433–438, 1973.
188. Bjarnason, I., K. Ward, and T.J. Peters, The leady gut of alcoholism: possible route of entry for toxic compounds. *Lancet,* p. 179–182, 1984.
189. Robinson, G.M., F.H. Orrego, Y. Israel, P. Devenyi, and B.M. Kapur, Low-molecular-weight polyethylene glycol as a probe of gastrintestinal permeability after alcohol ingestion. *Dig. Dis. Sci.,* 26:971–977, 1981.
190. Carrico, C.J., J.L. Meakins, J.C. Marshall, D. Fry, and R.V. Maier, Multiple-organ-failure syndrome. *Arch. Surg.,* 121:196–208, 1986.
191. Border, J.R., Hypothesis: sepsis, multiple systems organ failure, and the macrophage. *Arch. Surg.,* 123:285–286, 1988.
192. Souba, W.W., R.J. Smith, and D.W. Wilmore, Glutamine metabolism by the intestinal tract. *J. Parenter. Enter. Nutr.,* 9:608–617, 1985.
193. Saito, H., O. Trocki, J.W. Alexander, R. Kopcha, T. Heyd, and S.N. Joffe, The effect of route of nutrient administration on the nutritional state, catabolic hormone secretion, and gut mucosal integrity after burn injury. *J. Parenter. Enter. Nutr.,* 11:1–7, 1987.
194. Andrassy, R.J., Preserving the gut mucosal barrier and enhancing immune reponse. *Contemp. Surg.,* 32(2–A):21–25, 1987.

195. Moore, E.E., Early postinjury enteral feeding: attenuated stress response and reduced sepsis. *Contemp. Surg.*, 32 (2–A):28–32, 1987.
196. Gouma, D.J., J.C.U. Coelho, J.D. Fisher, J.F. Schlegel, Y.F. Li, and F.G. Moody, Endotoxemia after relief of biliary obstruction by internal and external drainage in rats. *Am. J. Surg.*, 151:479, 1986.
197. McPherson, G.A.D., I.S. Benjamin, H.I.F. Hodgson, N.B. Bowley, D.I. Allison, and L.H. Blumgart, Preoperative percutaneous transhepatic biliary drainage. *Br. J. Surg.*, 71:371–5, 1984.
198. Hatfield, A.R.W., J. Terblanche, S. Fataar, L. Kernoff, R. Tobias, A.H. Girdwood, R. Harries-Jones, and I.N. Marks, Preoperative external biliary drainage in obstructive jaundice. *Lancet,* ii:896–899, 1982.
199. Cahill, C.J., Prevention of postoperative renal failure in patients with obstructive jaundice — the role of bile salts. *Br. J. Surg.*, 70:590–595, 1983.
200. Bailey, M.E., Endotoxin, bile salt and renal function in obstructive jaundice. *Br. J. Surg.*, 63:774–778, 1976.
201. Deviere, J., J. Content, C. Denys, P. Vandenbussche, L. Schandene, J. Wybran, and E. Dupont, Excessive *in vitro* bacterial lipopolysaccharide-induced production of monokines in cirrhosis. *Hepatology,* 11:628–634, 1990.
201a. Felver, M.E., E. Mezey, M. McGuire, M.C. Mitchell, H.F. Herlong, G.A. Veech, and R.L. Veech, Plasma tumor necrosis factor-α predicts decreased long-term survival in severe alcoholic hepatitis. *Alcohol. Clin. Exp. Res.,* 14:255–259, 1990.
201b. Bird, G.L.A., N. Sheron, A.K.J. Goka, G.J. Alexander, and R.S. Williams, Increased plasma tumor necrosis factor in severe alcoholic hepatitis. *Ann. Int. Med.,* 112:917–920, 1990.
201c. Khoruts, A., L. Stahnke, C.J. McClain, G. Logan, and J.I. Allen, Circulating tumor necrosis factor, interleukin-1, and interleulin-6 concentrations in chronic alcoholics. *Hepatology,* 13:267–276, 1991.
202. D'Souza, N.B., G.J. Bagby, S. Nelson, C.H. Lang, and J.J. Spitzer, Acute alcohol infusion suppresses endotoxin-induced serum tumor necrosis factor alcoholism. *Clin. Exper. Res.*, 13:295–298, 1989.
202a. Honchel, R., L. Marsano, D. Cohen, S. Shedlofsky, and C. McClain, A role for tumor necrosis factor in alcohol enhanced endotoxin liver injury, in *Physiological and Pathological Effects of Cytokines,* Dinarello, C., M. Kluger, M. Powanda, and J. Oppenheim, Eds., Wiley-Liss, New York, 1990, 171–176.
203. Carithers, R.L., H.F. Herlong, A.M. Diehl, E.W. Shaw, B. Combes, H.J. Fallon, and W.C. Maddrey, Methylprednisone therapy in patients with severe alcoholic hepatitis. *Ann. Intern. Med.,* 110:685–690, 1989.
203a. Ding, A.H., F. Porteu, E. Sanchez, and C.F. Nathan, Downregulation of tumor necrosis factor receptors on macrophages and endothelial cells by microtubule depolymerizing agents. *J. Exp. Med.,* 171:715–721, 1990.
204. Yoshioka, K., S. Kakumu, M. Arao, Y. Tsutsumi, and M. Inoue, Tumor necrosis factor alpha production by peripheral blood mononuclear cells of patients with chronic liver disease. *Hepatology,* 10:769–773, 1989.
205. Muto, Y., K.T. Nouri-Aria, A. Meager, G.J.M. Alexander, A.L.W.F. Eddleston, and R. Williams, Enhanced tumor necrosis facter and interleukin-1 in fulminant hepatic failure. *Lancet,* July 9, 1988.

206. Anastassakos, C., G.J.M. Alexander, R.A. Wolstencroft, J.A. Avery, B.C. Portmann, G.S. Panayi, D.C. Dumonde, A.L.W.F. Eddleston, and R. Williams, Interleukin-1 and interleukin-2 activity in chronic hepatitis B virus infection. *Gastroenterology*, 94:999–1005, 1988.
207. Sinclair, S.B., P.D. Greig, L.M. Blendis, M. Abecassis, E.A. Roberts, M.J. Phillips, R. Cameron, and G.A. Levy, Biochemical and clinical response of fulminant viral hepatitis to administration of prostaglandin E. *J. Clin. Invest.*, 84:1063–1069, 1989.
208. Miller, L.C., C.A. Dinarello, R.A. Dempsey, and M.M. Kaplan, Circulating cytokines in primary biliary cirrhosis (PBC): Colchicine lowers interleukin-2 (IL-2) and tumor necrosis facter/cachectin (TNF) levels. *Gastroenterology*, 96:A630, 1989.
209. Klasing, K.C., Nutritional aspects of leukocytic cytokines. *J. Nutr.*, 118:1436–1446, 1988.
210. Huber, K.L. and R.J. Cousins, Maternal zinc deprivation and interleukin-1 influence metallothionein gene expression and zinc metabolism of rats. *J. Nutr.*, 118:1570–1576, 1988.
211. Ghezzi, P., B. Saccardo, and M. Bianchi, Recombinant tumor necrosis factor depresses cytochrome P450-dependent microsomal drug metabolism in mice. *Biochem. Biophys. Res. Commun.*, 136:316–321, 1986b.
212. Ghezzi, P., B. Saccardo, P. Villa, V. Rossi, M. Bianchi, and C.A. Dinarello, Role of interleukin-1 in the depression of liver drug metabolism by endotoxin. *Infect. Immun.*, 54:837–840, 1986a.
213. Shedlofsky, S.I., A.T. Swim, J.M. Robinson, V.S. Gallicchio, D.A. Cohen, and C.J. McClain, Interleukin-1 (IL-1) depresses cytochrome P450 levels and activities in mice. *Life Sci.*, 40:2331–2336, 1987.
214. Bertini, R., M. Bianchi, P. Villa, and P. Ghezzi, Depression of liver drug metabolism and increase in plasma fibrinogen by interleukin 1 and tumor necrosis factor: a comparison with lymphotoxin and interferon. *Int. J. Immunopharmacol.*, 10(5):525–530, 1988.
215. Ott, M.T., M. Vore, D.E. Barker, W.E. Strodel, and C.J. McClain, Monokine depression of bile flow in the isolated perfused rat liver. *J. Surg. Res.*, 47:248–250, 1989.
216. Zimmerman, H.J., C.O. Abernathy, R. Utili, G.B. Gaeta, and L. Adinolfi, Endotoxin action on bile formation, in *Mechanism of Hepatocy Injury and Death*, D. Keppler, H. Popper, L. Biancha, and W. Reutler, Eds., MTP Press, Lancaster, England, 1984.
217. Kasai, N. and K. Egawa, Effect of endotoxin on cytochrome P-450 activity, in *Handbook of Eendotoxin*, Vol. 3, *Cellular Biology of Endotoxin*, L.J. Berry, Ed., Elsevier, New York, 1985.
218. Mannering, G.J., K.W. Renton, R. Azhary, and L.B. Deloria, Effects of interferon-inducing agents on hepatic cytochrome P-450 drug metabolizing systems. *Ann. N.Y. Acad. Sci.*, 350:314–331, 1980.
219. Barber, E.F. and R.J. Cousins, Interleukin-1–stimulated induction of ceruloplasmin synthesis in normal and copper-deficient rats. *J. Nutr.*, 118:375–381, 1988.
220. Goldblum, S.E., D.A. Cohen, M. Jay, C.J. McClain, Interleukin-1–induced depression of iron and zinc: role of granulocytes and lactoferrin. *J. Physiol.*, 252:E27–E32, 1987.
221. Sporn, M.B., A.B. Roberts, L.M. Wakefield, and B. De Crombrugghe, Some recent advances in the chemistry and biology of transforming growth factor-beta. *J. Cell Biol.*, 105:1039–1045, 1987.
222. Phillips, M.R., S.B. Abramson, and G. Weissman, Neutrophil adhesion and autoimmune vascular injury. *Clin. Asp. Autoimmun.*, 3:6–15, 1989.
223. Mantovani, A. and E. Dejana, Modulation of endothelial function by interleukin-1. A novel target for pharmacological intervention. *Biochem. Pharmacol.*, 36:301–305, 1987.

10. Cytokines and Acute Phase Protein Expression

JACK GAULDIE AND HEINZ BAUMANN

INTRODUCTION

The homeostatic response of the body to trauma or infection consists of an orderly and orchestrated series of events that result in the halting of the process of injury, the protection of the rest of the organism against further injury, and the initiation of repair processes aimed at returning the body to normal function. The general term used to describe this reaction is "inflammation" and encompasses the clinical presentation of heat, swelling, pain, and redness as manifestations of the host response. While the general picture of inflammation as being a destructive process has been broadly accepted, it has recently been recognized that the systemic inflammatory response is a normal healing and repair mechanism and becomes pathologic only when control systems become disordered or chronicity is apparent. One of the more striking changes that occurs during inflammation is a dramatic alteration in the plasma concentration of a series of proteins, commonly called acute phase proteins, which are primarily made by the liver. The changes appear independent of the cause of inflammation whether it be trauma or burns, bacterial, viral, or parasitic infection, or tissue necrosis associated with neoplastic growth.[1-3] There are two early reports that highlight these plasma protein changes. The first was described by Fahraeus in 1921, in which he demonstrated that the sedimentation rate of erythrocytes was increased in blood taken from individuals undergoing an inflammatory challenge.[4] This change in erythrocyte sedimentation rate, or ESR, is now recognized as being primarily caused by increased plasma levels of several acute phase proteins, in particular fibrinogen. The second is a report by Tillet and Francis in 1930, who described the ability of inflammation-derived plasma to aggregate pneumococcal C-polysaccharide due to the presence of C-reactive protein.[5] Subsequently, the group headed by Avery described a comprehensive response, that they termed the "acute phase response", during which they recognized the rapidly changing nature of the plasma constituents.[6]

Systemic aspects of the acute phase response include fever, leukocytosis, alteration of plasma heavy metal concentrations (particularly iron and zinc), alteration in circulating corticosteroid levels, and depending on the initiating event, the release of potent low-molecular-weight mediators, including prostaglandins, leukotrienes, and vasoactive amines. However, the most well-studied and understood aspect of the acute phase response is the hepatic acute phase protein reaction.[2] The fact that this reaction has been strongly conserved through evolution in vertebrate species indicates that this response plays a major and primary role in systemic homeostasis.[1]

When one considers that the initiation of the hepatic response may be as a result of inflammation due to trauma or infection at a distal tissue site, it is clear that there must be a plasma-borne mediator(s) that regulates the response. Homburger in 1945 first showed that a soluble component extracted from purile exudate was active in increasing the plasma fibrinogen level.[7] Kampschmidt, in the late 60s and early 70s, described an entity called leukocyte endogenous mediator, which was able to raise both the level of fibrinogen in the plasma and the temperature of rats when administered systemically.[8-10] Koj, in a 1974 review, suggested that this mediator was secreted at the site of inflammation and specifically interacted with the liver to cause the induction of the acute phase proteins.[3] This activity was suggestive of a hormone. However, it is only recently that we have recognized the presence of a family of polypeptide molecules that are released from a variety of cells and that interact with specific receptors on target cells and tissues, resulting in metabolic changes. These molecules, including interleukins, lymphokines, interferons, and growth factors, may more generally be termed cytokines, referring to a class of intermediate-molecular-weight polypeptide hormone-like molecules that transmit signals from one cell to another and form the basic language of communication between cells in both normal and abnormal processes.

In considering those cytokines that could be classed as hepatocyte-stimulating cytokines, we must restrict ourselves to considering those molecules that interact specifically with hepatocytes and induce acute phase protein gene changes similar to those seen *in vivo* during inflammation. In the last few years, a series of developments have led to significant progress in the understanding of the hepatic acute phase reaction: (1) *in vitro* hepatocyte and hepatoma cell culture systems have been developed that reproduce the *in vivo* liver cell reaction; (2) acute phase protein genes have been cloned, and the cellular and molecular mechanisms by which they are regulated have been described; (3) a series of mediators or cytokines that cause the regulation of acute phase protein genes in the liver have been identified and cloned; (4) these hepatocyte-stimulating factors have been shown to have pleiotropic actions and mediate both differentiation and proliferation of hemopoietic cells as well as other parenchymal cells. These advances indicate that studies on the regulation of the hepatic acute phase reaction could contribute to developments in a variety of other fields, including immunology, endocrinology, and developmental biology. In addition, the recognition of the pleiotropic nature of many of these mediators has brought a wide variety of disciplines together to focus on understanding acute phase protein gene regulation and the importance that this plays in the physiology of homeostasis.

Table 1. Acute Phase Proteins

Protein		Species	Function
Major changes			
C-reactive protein	CRP	Man, rabbit	Opsonin
Serum amyloid A protein	SAA	Man, rat, mouse	(?)
Cysteine proteinase inhibitor (thiostatin)	CPI, TST	Rat	Antiproteinase
α_2 Macroglobulin	α_2M	Rat	Antiproteinase cytokine transport
α_1 Acid glycoprotein	α_1AGP	Most species	Transport
Medium to minor changes			
α_1 Proteinase inhibitor (α_1 Antitrypsin)	α_1Pi	Man, rat	Antiproteinase
α_1 Antichymotrypsin (Contraspin)	α_1 ACHY	Most species	Antiproteinase
Fibrinogen	FBG	Most species	Coagulation
Haptoglobin	HP	Most species	Binds and removes hemoglobin
Hemopexin	HPX	Most species	Binds heme
Ceruloplasmin	CER	Most species	0^2 scavenger, transport
Complement C3	C3	Mouse	opsonin
Serum amyloid P	SAP	Man, mouse	(?)
Negative changes			
Albumin	Alb	Most species	
Transferrin	Tf	Most species	Transport
α_1 Inhibitor III		Rat	Antiproteinase

ROLE OF ACUTE PHASE PROTEINS

The majority of acute phase plasma proteins are derived from the hepatocyte, but some may be produced at extrahepatic sites. Although collectively they share the acute phase nature of their induction, these proteins differ in their physicochemical properties and known functional activities. For each positive and negative acute phase protein, the kinetics and magnitude of change and the absolute level of expression are characteristic.[11,12] Most of them are glycoproteins, and an examination of their function implicates them in the control of inflammation and participation in the healing process (Table 1). While there are substantial differences in the species specificity of various acute phase proteins, common functions, maintained through evolution, are performed by the various proteins,[13] which suggests that these functions are important for the maintenance of the species. From Table 1 it is obvious that the proteins can both mediate and be consumed by the inflammatory process, including fibrinogen, haptoglobin, hemopexin, α_1-acid glycoprotein, complement component C3, members of the serpin family (including α_1-antichymotrypsin and α_1-antitrypsin and the pentaxins), C-reactive protein, and serum amyloid P component. Some proteins, such as C-reactive protein in man and rabbit (CRP), α_2-macroglobulin in rat, and serum amyloid A protein in man and mouse (SAA), rise dramatically, up to several hundredfold, while other proteins, including fibrinogen, haptoglobin, and α_1-acid glycoprotein in most species, or the major acute phase globulin of the rat, a cysteine proteinase inhibitor (thiostatin) or t-kininogen,[14-16] rise four to five times the level of normal. Other proteins, including albumin, transferrin, and α and β lipoproteins, decrease in circulation during this response.[2,3,17,18]

Despite extensive examination, the true physiologic role of a number of the major acute phase proteins remains undiscovered. Some proteins perform an obvious protective role. Complement C3 and C-reactive protein opsonize bacteria, immune complexes, and foreign particles and act through the reticuloendothelial system. Fibrinogen has a well-recognized role in blood coagulation and is consumed in the homeostatic response to trauma and therefore must be rapidly replaced in the circulation. A major function displayed by a number of the acute phase proteins is that of the inhibition of various serine and cysteine proteinases. Alpha$_2$-macroglobulin (α_2M), which is most noted as an acute phase protein in rat and which is expressed constitutively at high levels in the human, is a broad-spectrum antiproteinase. α_2M is apparently essential, as there are no reports of humans with homozygous deficiencies,[19,20] and thus would appear to be an indispensable human protein. In man, α_1-proteinase inhibitor (α_1-antitrypsin) has specificity towards neutrophil elastase, while α_1-antichymotrypsin has specificity towards leukocyte cathepsin G.[21] In the rat, the major α_1-acute phase protein, also called thiostatin, is an inhibitor of cysteine proteinases.[15,16] Thus, the broad nature of inhibition against a wide variety of proteinases is maintained across the species by different inhibitors, and the general nature of the acute phase protein response is that of controlling proteolytic activity and inhibiting the trauma and destruction associated with inflammation.

In addition to proteinase inhibition, these proteins are also known to be involved in reducing the effects of endotoxin shock and modulating the activation of macrophages,[22,23] as well as reducing the chemotaxis of monocytes and activation of lymphocytes.[24] Moreover, most of the proteinase inhibitors have been shown to be capable of modulating immune reactions *in vitro*. Both α_1-proteinase inhibitor and α_2-macroglobulin, as well as α_1-chymotrypsin, have been shown to modify the activity of natural killer cells and antibody-dependent cell-mediated cytotoxicity.[25-30] α_1-Proteinase inhibitor can suppress the mitogen-induced proliferation of lymphocytes, while complexes between the inhibitors and specific proteinases can interact with receptors on macrophages and suppress macrophage activation, Ia antigen expression, and possibly antigen presentation by these cells.[31,32]

Such immunosuppressive activities have also been suggested for α_1-acid glycoprotein, since this molecule, previously called orosomucoid, can modify the *in vitro* proliferative effect of lymphocytes stimulated by lectins or lipopolysaccharides,[33-35] while C-reactive protein may modulate immune responses during inflammation.[36] It is interesting to note that CRP has been proposed as a scavenger and opsonizing factor for chromatin fragments released from damaged cells, since it can bind to chromatin and mediate its solubilization.[37,38] However, as is the case for many other acute phase proteins, the main physiologic function of CRP remains undescribed.

A more recent and relevant finding indicates that while α_2-macroglobulin may be a potent proteinase inhibitor, it may also be a plasma carrier for many of the cytokine hormones. Thus, platelet-derived growth factor, transforming growth factor β, interleukin-1, and interleukin-6 have all been shown to have some binding affinity for α_2-macroglobulin.[39-43] While these effects have to date only been shown *in vitro*, it is interesting to speculate here that α_2-macroglobulin may provide both a carrier and protective function for these important cytokines, since they are generated in proteinase-rich environments, and binding to α_2-macroglobulin may protect them from proteolytic damage, allowing some of them to be transported to target tissues and to initiate the healing response. This suggests that one of the roles for α_2-macroglobulin may be as a hormone carrier, in addition to its role as a proteinase inhibitor, although the general nature of this has recently been questioned.[161]

Thus, acute phase proteins appear to play important roles in controlling the random destruction associated with the inflammatory response, mediating the clearance and tissue repair associated with activation of the reticuloendothelial system, and modulating the immune response initiated subsequent to the inflammatory response against the invading organisms. These behaviors are seen as anti-inflammatory and homeostatic, albeit nonspecific. Since the proteins are induced rapidly, they can be present and play a broad protective role during the early stages of the host response to invasion, prior to the initiation of the specific immune response that will subsequently protect the host from further attack.

While the majority of plasma-derived acute phase proteins are synthesized by the hepatocyte, recent evidence demonstrates that a number of the acute phase proteins are synthesized by mononuclear cells.[44,45] This synthesis is at a remarkably low level compared to that of the hepatocyte, but appears to be under the same regulated induction by cytokines as is the liver.[46] The liver may contribute in an overwhelming way to the altered levels seen in plasma during inflammation; however, the fact that other cells can make even limited amounts of proteinase inhibitors may be very important in modifying the microenvironment around cells involved in activation or immune regulation. Consideration of the major regulation of synthesis brings us to examine the hepatocyte and the various signals that mediate alterations in gene transcription associated with the onset of the acute phase response.

CYTOKINE REGULATION OF ACUTE PHASE PROTEIN SYNTHESIS

Early work showing that endogenous mediators could initiate the acute phase protein response *in vivo* was followed by similar investigations using isolated hepatocytes and hepatoma cells *in vitro*. By these approaches and with the recent availability of purified and cloned material, identification of the major cytokine regulators of acute phase gene expression has been achieved. Thus, a combination of corticosteroid and a few specific cytokines can account for the hepatic regulation seen *in vivo,* as well as most of the other aspects of the acute inflammatory response.

Corticosteroid

Corticosteroids do not appear to exhibit major hepatocyte-stimulating activity on their own. However, they do participate in the regulation of many genes expressed in the liver. If one initiates an inflammatory response in hypophysectomized rats having low levels of endogenous corticosteroids, typical acute phase proteins such as α_2-macroglobulin are not induced. Administration of exogenous glucocorticoids prior to inflammatory challenge restores a full acute phase protein response.[47] Moreover, the ability of cytokines such as interleukin-6 to induce the full acute phase response *in vivo* is dependent on an appropriate glucocorticoid level being present. This adequate level of corticosteroid appears to be found naturally in some strains of rats,[48] while it must be augmented exogenously in others.[49] The situation *in vitro* is somewhat similar. Primary hepatocytes, particularly those of the rat, require corticosteroids to elicit acute phase gene regulation by any of the cytokines.[50,51]

Some hepatoma cells appear to be able to respond to corticosteroids alone.[52,54,55] However, these hepatoma cells have recently been shown to be capable of producing endogenous IL-6 as well as leukemia inhibitory factor[53] (see section entitled Leukemia Inhibitory Factor, below), and the addition of corticosteroid would allow these endogenously produced cytokines to cause autocrine regulation. Thus, corticosteroids have a prominent and permissive synergistic effect in combination with the other cytokines in regulating acute phase protein expression.[50,51,54-57] This aspect of synergistic action of corticosteroids is all the more likely given that acute phase protein genes, such as α_1-acid glycoprotein and haptoglobin, possess glucocorticoid regulatory elements in the 5'-untranslated region of the genes.[58,59] These regions are distinct from the regions coding for cytokine induction or for liver cell expression.

Interleukin-1/Tumor Necrosis Factor

The realization that interleukin-1 can regulate the expression of acute phase protein genes came with the identification of leukocyte endogenous mediator as being the same molecule that had been cloned through its ability to stimulate thymocytes[60-62] and called interleukin-1 (IL-1). Two molecular forms exist, α and β, both with a molecular weight of 17 kD.[62,63] Recombinant-derived material or highly purified natural IL-1 causes increased expression of acute phase proteins, such as serum amyloid A and fibrinogen, when injected into mouse or rat.[64] The time course of change in fibrinogen levels suggested direct action on the liver, and until a series of other hormones were identified, interleukin-1 was thought to be the major regulator of the hepatic acute phase response.[61] Similar to results seen with IL-1, when recombinant tumor necrosis factor (TNF) was administered *in vivo*, acute phase protein changes were seen with approximately similar kinetics. Since TNF and IL-1 have many common pleiotropic activities,[63] it was believed that these two potent proinflammatory cytokines, released from activated macrophages, mediated the overall acute phase response and were directly responsible for the hepatic acute phase protein response.

However, with the development of *in vitro* assay systems using mouse and rat primary hepatocytes and rat and human hepatoma cells lines (reviewed in Reference 56), it became obvious that neither IL-1 nor TNF could regulate the full set of acute phase proteins *in vitro*, even in the presence of corticosteroid. IL-1, both purified natural and recombinant α and β forms, stimulates a limited number of acute phase protein genes. There are difficulties in interpreting data from cell culture. Differences between species as well as differences between primary hepatocyte and hepatoma cell cultures exist in the response to various mediators. Moreover, there is variability even between subclones of the same hepatoma cell.[56] Thus, several cell systems must be used to fully describe the regulatory control or modulation exhibited by any of the individual mediators.

Given these restrictions, IL-1 causes stimulation of complement component factors B and C3, serum amyloid P protein in mouse and human cells,[66-70] as well as α_1-acid glycoprotein in mouse, rat, and human hepatocytes and hepatoma cells.[51,71-75] IL-1-mediated stimulation of serum amyloid A protein was seen in mouse hepatoma cells[65] and was also evident on SAA and factor B genes incorporated into mouse fibroblasts by transfection.[76] In addition to positive regulation, IL-1 suppresses the synthesis of albumin and therefore contributes to both positive and negative regulation of acute phase protein genes. There appears to be no differences in how IL-1α or IL-1β regulate these genes, albeit some minor differences have been reported for the regulation of α_1-acid glycoprotein.[48] The ability of IL-1 to stimulate hepatocyte gene regulation is not restricted to soluble forms of the interleukin, as membrane-bound IL-1α on mouse macrophages appears able to increase the synthesis of complement C3 and reduce the synthesis of albumin in hepatoma cells.[77]

In all instances, tumour necrosis factor α (TNF) behaves similarly to IL-1, in that there are only a limited number of acute phase protein genes (the same ones regulated by IL-1) that can be directly stimulated by TNF.[67,73,74] Combinations of the two cytokines do not elicit any different response than they do individually. Despite the fact that IL-1 and TNF administered *in vivo* cause increased expression of all acute phase proteins, including fibrinogen, neither of these hormones increased the expression of fibrinogen *in vitro*. In fact, IL-1 has a negative effect on the expression of fibrinogen *in vitro* in human and rat cells, as well as similar negative effects on the expression of α_2-macroglobulin and cysteine proteinase inhibitor in the rat.[49,51,72,73,78-84]

Interleukin-6

In the early 1980s, Fuller and colleagues demonstrated that monocyte-conditioned medium contained an activity, called hepatocyte stimulating factor (HSF), which regulated fibrinogen synthesis in isolated rat hepatocytes.[78,84] Others, including ourselves, confirmed the presence of this additional mediator and showed that monocyte-conditioned medium could regulate most of the acute phase protein genes in isolated rat hepatocytes and hepatoma cells in a coordinated manner similar to that seen *in vivo*. Moreover, corticosteroids were found to augment this effect *in vitro*.[50,51,74,78-80,84-87] By using a combination of antibody neutralization, and purified monocyte-derived and recombinant-derived material, we showed that the human monocyte-derived mediator that regulated all of the acute phase protein genes was identical to a previously cloned molecule, variously called interferon-β2, B-cell stimulating factor 2, hybridoma growth factor, and 26K protein, and now collectively referred to as interleukin-6.[72] The polypeptide has a molecular weight of 23 to 30 kD and is the product of a single gene present on human chromosome 7 and mouse chromosome 5.[88-92] The human IL-6 is glycosylated,[93] while the murine and rat are not.[94-97]

The designation of an interleukin recognizes that the molecule is released from a variety of cells and has pleiotropic activity both in inflammation and immunity, although there are undoubtedly other activities, such as in cell differentiation,[98-100] that are not yet fully explored. The identification of IL-6 as an HSF was rapidly confirmed by others showing that the recombinant human molecule, expressed with or without appropriate glycosylation, caused stimulation of many of the known positive acute phase proteins and reduction of albumin and transferrin in several species (Figure 1).[58,79-81,100-107] The recombinant material, or natural purified material derived from monocytes or fibroblasts, acts directly on the hepatocyte through a specific receptor; this signal is communicated to the cis-acting regulatory elements of the acute phase protein genes, resulting in altered gene transcription and concentration of the appropriate mRNA species.[53,59,101,105,110,111,114] Recombinant materials from human, mouse, and rat are active across the species, although there may be differences in the level of sensitivity with fully homologous interactions.[94,105,112] While interleukin-6 affects most of the acute phase proteins, when combined with other cytokines (see below), its effect on some of the genes is modulated. There is no doubt, however, that IL-6 is the major direct regulator of the hepatic acute phase response both *in vitro* and *in vivo* (Figure 2). Since we have recently shown that hepatocytes undergoing an acute phase response do not express IL-6 themselves[113] and since exogenous IL-6 at the picomolar level causes gene regulation through a specific high-affinity receptor, it should be viewed as a metabolic hormone, along with the other cytokines.

Leukemia Inhibitory Factor

In addition to IL-1, TNF, and IL-6 (all of which are released by activated monocytes, in addition to other cells, such as fibroblasts and endothelial cells), there are other molecules that possess hepatocyte stimulating activity. We have previously shown that a molecule released from human keratinocytes, and in particular, squamous carcinoma cell line (COLO-16) elicited a similar response to that seen with IL-6.[74,114] However, we were unable to demonstrate the presence of the mRNA for

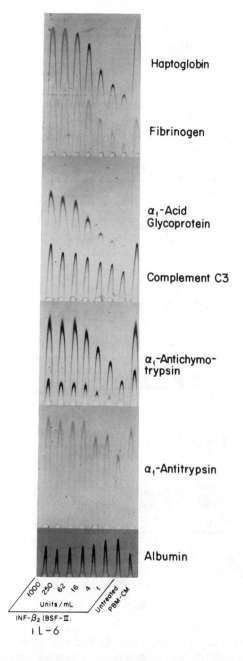

Figure 1. *In vitro* action of rhuman IL-6. Human HepG2 cells were exposed to increasing amounts of IL-6. Protein synthesis was detected by immunoelectrophoresis with specific antibodies against human acute phase proteins. (Modified from Gauldie, J. et al., in *Monokines and Other Non-Lymphocyte Cytokines*, M.C. Powanda et al., Eds., Alan R. Liss, New York, 1988, 15–20. With permission.)

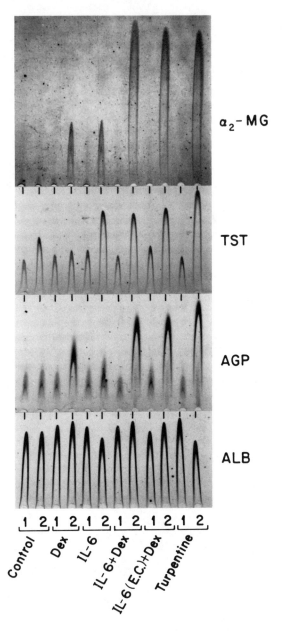

Figure 2. *In vivo* action of rhuman IL-6. Buffalo rats were given two injections of IL-6, i.p., and the serum levels of acute phase proteins were measured before (1) and 24 h after (2) the injections. Dexamethasone (μg/kg) was simultaneously injected in some animals. α2-MG — α2-macroglobulin; TST — thiostatin; AGP — α1-acid glycoprotein; ALB — albumin. (From Marinkovic, S. et al., *J. Immunol.*, 142:808–812, 1989. With permission.)

IL-6 in COLO-16 or to abrogate the biologic activity of COLO-HSF with a neutralizing antibody to human IL-6.[114] Recently, using a similar approach to that taken for the identification of IL-6 as an HSF, we have shown[116] that COLO-HSF is identical to a previously cloned molecule described as T-cell-derived leukemia inhibitory factor (LIF), also known as differentiation inhibitory activity.[117-120] This molecule has an apparent molecular weight ranging from 25,000 to 45,000 and has been shown to be released by ascites cells, stimulated T-cells, as well as Buffalo rat liver cells and, more recently, by some rat hepatoma and macrophage cell lines.[121] Like IL-6, we are not able to show the presence of the LIF message in hepatocytes undergoing acute phase response stimulation, and thus LIF is also unlikely to be an autocrine stimulator for hepatic acute phase protein expression. Recombinant-derived LIF causes the stimulation of the same sets of genes as does IL-6 (Figure 3), but these two molecules work through separate receptors on the hepatocyte membrane.[114] Initial results indicate that IL-6 is a more effective regulator than is LIF; however, the relative potency of these major regulators must await examination of appropriately expressed and purified homologous material. LIF and IL-6 can act in an additive way on tissue culture cells, but unlike IL-6, LIF has not yet been detected in circulation. Therefore, the role of LIF in physiologic inflammatory reactions must remain speculative.

Other Modulating Cytokines

An analysis of other factors that may modify the response of the hepatocyte in culture has shown that several previously described cytokines can modulate the stimulatory effect of the other cytokines. Thus, cyclic adenosine monophosphate (cAMP)-dependent factors[123] and protein kinase C-activating factors[80,109] can modulate the expression of some of the stimulated acute phase protein genes. In addition, interferon-γ (IFNγ), epidermal growth factor (EGF) as well as transforming growth factor β (TGFβ) can all modify the regulation caused by IL-6.[124] Interferon-γ appears to directly cause some stimulation of α_2-macroglobulin in human hepatoma cells (not normally an acute phase protein in the human), but decreases the IL-6-mediated stimulation of α_1-chymotrypsin, haptoglobin, and fibrinogen in the same cells.[122] Human IFN was not active on primary rat hepatocytes and thus is different from other direct mediators in showing species specificity. Human TGFβ directly stimulates the synthesis of α_1-antichymotrypsin in human hepatoma cell lines, but only to a limited extent, and causes not nearly as dramatic an increase as is seen with IL-6. α_1-Proteinase inhibitor was also stimulated to a limited degree, and the output of fibrinogen and albumin were modulated by TGFβ. EGF had little direct effect on the secretion of acute phase proteins, but at relatively low levels was able to reverse the IL-6-mediated inhibition of albumin synthesis.[124] These additional cytokines, which modulate the major regulatory control *in vitro*, have not yet been demonstrated to be relevant *in vivo,* but may dictate the final level of gene expression and protein synthesis of the individual acute phase proteins elicited under different inflammatory conditions. Situations involving the rapid release of IL-6 and IL-1 are

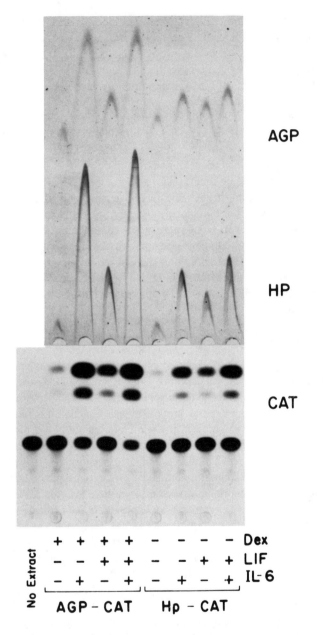

Figure 3. *In vitro* stimulation of human HepG2 cells with rhuman IL-6 and rhuman LIF preparations. Upper panel shows the stimulation of endogenous human α1-acid glycoprotein (AGP) and haptoglobin (HP) genes detected by immunoelectrophoresis. Lower panel shows stimulation of transfected rat gene constructs pAGP (3 × DRE)-140CAT (109,130) and pHP(400)-CAT, (160) detected by CAT activity. (From Baumann, H. and G.G. Wong, *J. Immunol.,* 143:1163–1167, 1989. With permission.)

likely to show a different pattern of plasma changes than situations involving the release of not only IL-1 and IL-6, but raised levels of EGF and TGFβ or interferon-γ. This differential activity in modulation of the major stimulation could account for the variation that one sees in acute phase protein levels in different disease states.[17,18]

HEPATOCYTE RESPONSE

Gene Regulation

The acute phase protein response *in vivo* or in cultured cells occurs in a kinetically coordinated fashion indicating that common regulatory mechanisms exist at the level of the hepatocyte.[56,110,111] The time course of stimulation (kinetics of mRNA accumulation and increased protein synthesis) is characterized by an initial lag period and starts within 4 h with detectable protein synthesis and secretion within 12 h from the onset of inflammation. There are concomitant changes in plasma concentrations, which are comparable in magnitude to those of protein production, mRNA accumulation, and gene transcription.[2,12,57,87,101,125] The overall regulation of acute phase protein expression appears to be primarily at the gene transcriptional level,[57,101,125] but post-transcriptional control processes may play additional important roles. Post-transcriptional events in acute phase protein regulation[126] have not been studied to any great extent, unlike the post-transcriptional regulation of other mRNA types, such as changes in mRNA stability as found for the various cytokine mRNAs.[127] Moreover, post-translational events, such as altered glycosylation, may further influence the final accumulation in the plasma of the various acute phase proteins.[107,128,129]

A number of the acute phase genes have been cloned and sequenced, and the 5' flanking gene region, which generally contains the regulatory elements responsible for both tissue specificity and inducibility, has been studied in both cell transfection and transgenic systems. As a result, several genes, including α_1-glycoprotein, rat α_1-I3 inhibitor, haptoglobin, and fibrinogen, have been shown to contain a glucocorticoid-responsive element (GRE).[53,58,78] This GRE accounts for the synergistic activity of corticosteroids on the expression of the acute phase protein genes. In addition, IL-1- and IL-6-responsive elements have been mapped to the 5'-region of the rat α_1-AGP, α_2-macro, and cysteine proteinase inhibitor genes as well as the rat and human haptoglobin genes.[59,78,101,114,130] Potential IL-6-responsive transcription control elements have recently been described by several groups and have been shown to consist of a consensus sequence (T/A)T(C/G)TGGGA(A/T). This sequence is present in the acute phase protein genes rat α_2-macroglobulin, α_1-acid glycoprotein, fibrinogen, and cysteine proteinase inhibitor, as well as in human haptoglobin and C-reactive protein.[53,105,131,132] Thus, it is likely that this cis-acting element interacts with a nuclear factor(s) leading to gene regulation. It has also been suggested that the transcriptional factor NFκB might be involved in control of acute phase protein genes.[133] No details are yet available as to a consensus sequence for an IL-1-responsive element. None of the other cytokines have been studied in great detail with regard to their respective cis-located gene regulatory elements.

Attempts to define acute phase protein gene regulation have also been carried out using transgenic mice. These experiments, while more limited in scope, determine whether the foreign gene is appropriately regulated *in vivo*. Studies of acute phase induction in mice in which the human genes for α1AGP, SAA, α_1-antitrypsin, and CRP have been introduced, highlight an important finding relating to tissue and developmental specificity as well as species differences in acute phase protein responses.[134-136]

For instance, the human CRP-transfected gene maintained its response to inflammatory mediators, while the endogenous murine CRP gene was not induced, indicating that it is likely the cis-acting control elements on the murine CRP gene that render it nonresponsive to inflammatory signals, rather than some intrinsic difference in receptor/signal transduction mechanisms of the murine hepatocyte.

Metabolic Changes

Kinetics of Response

There is an initial lag period that occurs after tissue injury before the plasma acute phase protein response is evident. This is taken to reflect the time required to synthesize not only suitable amounts of hepatocyte stimulating mediators as well as corticosteroids, but also the acute phase proteins.[56] Kinetics are established *in vivo* by measuring changes in plasma levels of the acute phase proteins or quantifying the levels of acute phase protein mRNAs in the liver of experimental animals at various times after initiation of inflammation.[12,125] Maximal stimulation of gene transcription precedes mRNA accumulation, and peak mRNA expression precedes the increases in plasma protein levels. The plasma response for most acute phase proteins is maximum between 18 to 24 h, although some proteins show earlier changes (8 to 12 h), such as murine SAA and rat fibrinogen. The kinetics *in vitro* closely mimic the changes seen *in vivo*. Hepatocytes and hepatoma cells respond to IL-1 or IL-6 by an immediate and coordinated rise in most of the acute phase protein mRNAs. Maximal expression is reached within 12 to 24 h, and then the levels decline at varying rates dependent on transcriptional activity of the particular gene and on the stability of the specific mRNA. Genes that require the presence of glucocorticoid for full acute phase regulation, such as rat α_2 macroglobulin and α_1 acid glycoprotein, show a lag period of several hours before seeing any rise. However, they still reach maximum levels by 12 h.[56,107,114]

Carbohydrate Metabolism

In addition to increased synthesis of acute phase proteins, the hepatocyte responds to inflammatory cytokines by altering the levels of a series of enzymes involved in carbohydrate metabolism. During the acute phase, there are changes seen in the glycosylation pattern of many acute phase proteins.[129] Differences in a glycoprotein heteroglycan antennary structure occur at the biosynthetic level and are mediated by different mechanisms than those controlling the synthesis of the protein.[130] Concomitant with changes in the glycosylation pattern, there is a characteristic rise in glycosyl transferases, in particular sialyltransferase,[137] which may account for the modulation in carbohydrate content of the acute phase proteins. These changes can be induced *in vitro* by the same mediators (IL-1, IL-6, and glucocorticoid) that induce protein synthesis.

A second effect of acute inflammation is the perturbation of the normal hormonal control of hepatic glucose homeostasis, contributing to the hypoglycemia seen after severe trauma, septicemia, or burns. This interruption of gluconeogenesis is associated with an inhibition of normal hormonal induction of phosphoenolpyruvate carboxykinase (PEPCK), the rate-limiting enzyme in glyconeogenesis. This inhibition may be directly mediated *in vitro* by IL-6.[138] Thus, the same mediator that induces acute phase protein synthesis and altered glycosylation causes at the same time alterations in the carbohydrate metabolism of the hepatocyte. Other liver metabolic enzymes are decreased in acute phase, including catalase and fatty acid synthetase, but it is not known whether these changes are also mediated by IL-6.[139]

Amino Acid Metabolism

Since the protein synthetic capacity of the hepatocyte is dramatically altered during the acute phase, it is to be expected that substrates for protein synthesis must also be regulated. Amino acid uptake by the hepatocyte is directly stimulated by IL-6 as well as by mediators such as glucagon, while other mediators, including TGFβ and EGF, modulate this stimulation.[102,124] The stimulation in amino acid uptake is consistent with the increased protein being synthesized and suggests that the entire protein synthesizing machinery of the acute phase hepatocyte becomes more efficient.

Signal Transduction

The mechanisms whereby the mediators initiate these metabolic and synthetic changes in the hepatocyte have only recently been studied. Some groups have claimed that the action of IL-6 could be replaced by phorbol esters and other known inducers of the protein kinase pathway.[80] However, in our hands, phorbol ester pretreatment of hepatoma cells did not interfere with the action of IL-6, and the action of these inflammatory mediators is independent of the protein kinase C signal transduction system. Experiments on HepG2 human hepatomas and on rat H-35 and Fao hepatomas indicate that although the protein kinase C pathway is not particularly relevant to acute phase protein gene induction, the pathway serves as a modulating influence controlling some fine tuning of the level of expression of the genes.[109,114]

CYTOKINE RECEPTORS

Recently, cell receptors for two of the main hepatocyte-stimulating cytokines, IL-1 and IL-6, have been isolated and cloned.[140,141] Recombinant molecules have been used to determine the affinity and cellular distribution of these receptors. A single class of receptor with high affinity for IL-6 (kd = 3.4×10^{-10} M) has been identified on B-cells, and the same or similar receptor appears to be on hepatocytes and hepatoma cells.[114] In addition, high affinity receptors for IL-1 have been found on hepatocytes.[142] Both receptors belong to the immunoglobulin supergene family, with the IL-6 receptor requiring a second transmembrane and cytoplasmic protein for signal transduction.[143] Using a cDNA probe to the human IL-6 receptor, it was shown that the amount of IL-6 receptor mRNA increases during mediator stimulation.[144] This up regulation is specific to the liver, since the same receptor is down regulated or untouched during inflammatory activation of monocytes. The number of receptors on the hepatocyte or hepatoma cells for either IL-1 or IL-6 is very low (400 to 600 IL-6 receptors on primary rat hepatocytes) (Figure 4), whereas the same hormones have much higher receptor density on other cells, in particular transformed tumor cell lines (more than 10,000 receptors for IL-6 per cell on a human myeloma cell, U266).[114,145]

CYTOKINE INTERACTION

Target Tissue

Since the hepatocyte possesses receptors for multiple hormones and responds directly to exposure to these same hormones, it is likely that the final status of protein regulation and modulation of the metabolic state seen in acute phase responses will depend not only on whether specific cytokines are present, but also on whether the same cytokines have an additive or inhibitory action on each other. When we examined the interaction of IL-1, TNF, and IL-6, as well as glucocorticoid, on acute phase protein gene regulation, several general patterns emerge. First, each of the cytokines has a specific receptor on the hepatocyte, and the presence of corresponding cis-acting regulatory gene regions is assumed. Second, mixtures of cytokines result in a different pattern of gene regulation from that seen by adding the cytokines independently. Third, the sequence and kinetics of cytokine interaction at the hepatocyte "syntax" are important in determining the final status of acute phase protein gene regulation. Finally, there are major regulatory cytokine hormones and minor modulatory hormones that dictate the level of gene activation in the hepatocyte.

The major regulatory cytokine is IL-6, released by activated monocyte/macrophages and stromal cells, including endothelial and epithelial cells and fibroblasts. LIF can qualitatively substitute for IL-6 in stimulating acute phase protein genes. When we examine how the other main inflammatory cytokines, IL-1 and TNF, interact

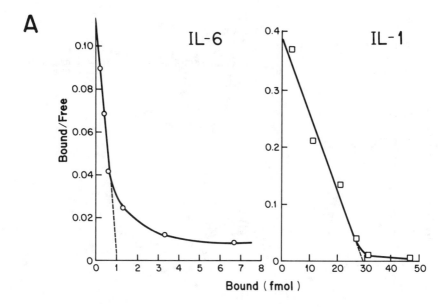
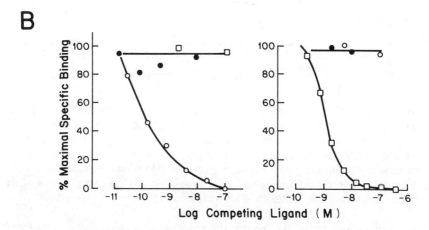

Figure 4. Binding and competition of HSF-III/LIF, IL-6, and IL-1 to HepG2 cells. (A) HepG2 cell monolayers were incubated for 4 h at 11°C with increasing concentrations of ^{125}I-IL-6 (○) or ^{125}I-IL-1α (□). The binding of label to the cells was expressed according to Scatchard. (B) HepG2 cells were incubated with a fixed concentration of ^{125}I-IL-6 (9×10^{-11} M) or ^{125}I-IL-1 (1.5×10^{-9} M) and various concentrations of unlabeled IL-6 (○), HSF-III/LIF (●) and IL-1 (□). The conditions were the same as in (A). The nonspecific binding was measured in the presence of 1×10^{-7} M unlabeled IL-6 and IL-1. No evidence for IL-1 or HSF-III/LIF competition for IL-6 binding or vice versa.

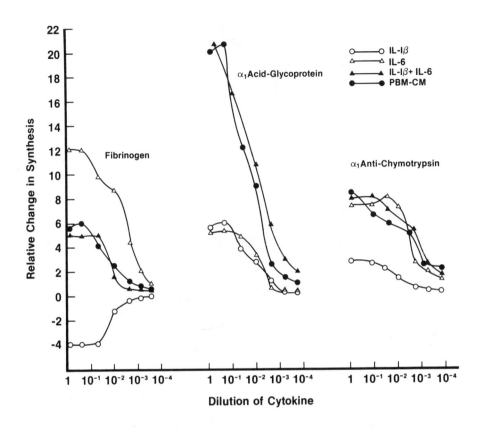

Figure 5. Effect of combination of IL-1 and IL-6 *in vitro* on acute phase protein expression. Human HepG2 hepatoma cells were exposed to IL-1 alone, IL-6 alone, or combinations in various doses and compared to the stimulation seen with peripheral blood monocyte-conditioned medium (PBM-CM). The synthesis of acute phase proteins was monitored by immunoelectrophoresis. (Modified from Gauldie, J. et al., in *Monokines and Other Non-Lymphocyte Cytokines*, M.C. Powanda et al., Eds., Alan R. Liss, New York, 1988, 15–20. With permission.)

with the major regulators, we see that the acute phase proteins fall into two main groups. One set, including the opsonins and transport proteins, are stimulated by both IL-6 and IL-1. The second set, which includes all of the antiproteinases and fibrinogen, are stimulated only by IL-6 or LIF (Figure 5, Table 2). The stimulation seen with combinations of cytokines on group 1 genes is both additive and synergistic and best seen in the response of α_1 acid glycoprotein (Figure 3). This synergistic action is evident in all species studied and indicates that to achieve maximum stimulation *in vivo*, both cytokines need to be delivered to the hepatocyte during the inflammatory response.[58,79,80] Similar synergy between IL-1 and IL-6 is seen in the stimulation of thymocytes by these cytokines.[146,147]

Table 2. Acute Phase Protein Genes Regulated by Inflammatory Cytokines

Group I Induced by IL-1 and IL-6	Group II Induced by IL-6 only
α_1 acid glycoprotein	α_2-Macroglobulin
C3	Fibrinogen
Factor B	α_1-pPoteinase inhibitor
SAA	Haptoglobin (human)
SAP	α_1-Antichymotrypsin
Haptoglobin (rat)	Ceruloplasmin
Hemopexin	Cysteine proteinase inhibitor
C-reactive protein (human)	C1 esterase inhibitor
Cytokine interaction	
Synergy IL-1 and IL-6	Inhibition IL-1 on IL-6
α_1 Acid glycoprotein	Fibrinogen
C3	α_2-Macroglobulin
Haptoglobin (rat)	Cysteine proteinase inhibitor

The action of IL-1 or TNF on the expression of group 2 genes is modulatory, in that IL-1 inhibits the IL-6-mediated stimulation of such proteins as fibrinogen in all species and α_2 macroglobulin and cysteine proteinase inhibitor in the rat.[51,72,73,79,84] Other hormones, including stimulatory corticosteroids and modulatory hormones and growth factors, such as IFNα, TGFβ and EGF, as well as factors that modulate the protein kinase C pathway of the hepatocyte, modify the action of the major stimulators of gene regulation such that the secretion of acute phase proteins and their accumulation in plasma is the result of the liver receiving a complex mixture of signals and being able to decipher the language or syntax of the message.

Inflamed Tissue

Considering that inflammation at a distant tissue results in the response of the liver, it is important to understand how the mediators of the hepatic response are generated in the tissue to fully appreciate the mechanism of control of the acute phase response. Activated macrophage/monocytes are a potent source of both IL-1 and the IL-6. However, evidence has accumulated that *in vitro*, IL-1 and TNF are potent stimulators of the release of IL-6 and other cytokines from stromal cells, such as fibroblasts and endothelial cells.[90,93,148-150] Thus, it is likely that the activation of the surveillance and phagocytic cell, the macrophage/monocyte, by trauma or infection results not only in the release of liver-modulating hormones from that cell, but also causes activation and recruitment of the adjacent stromal cell population, the most abundant cell in the body, by locally released IL-1/TNF, resulting in an augmentation of the IL-6 signal and a rapid rise in the circulating level of this important mediator. Since IL-1 and TNF elicit proinflammatory actions[61-63] and cause release of IL-6 and since IL-6 is the major regulator of the synthesis of antiproteinases

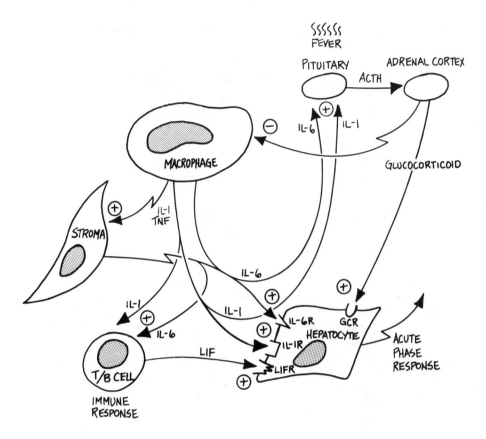

Figure 6. Scheme showing the interaction of inflammatory and stromal cells in the cytokine cascade and initiation of the hepatic response. The coordinated involvement of many cells is required to deliver the correct sequence of signals to the hepatocyte and result in the acute phase response.

and fibrinogen, IL-6 may be seen to mediate the anti-inflammatory or homeostatic response of the body.

Not only does the locally released IL-1 cause increased IL-6 release, but evidence has been presented that these two mediators are both important in the febrile response[151] and in stimulation of the adrenal/pituitary axis, resulting in an increase in the circulating level of glucocorticoids,[152,153,159] thus providing a mechanism for the generation of the third required hormone signal for hepatic acute phase responses. Figure 6 attempts to summarize these interactions occurring during an acute phase response.

Since most of the evidence for cytokine interaction in the tissue has been documented from tissue culture studies, one must be cautious in extrapolating these results to *in vivo*, where actual concentrations of particular cytokines may not reach physiologic relevance. For example, IFNα and IL-1 levels in circulation have been particularly difficult to detect and must be assumed to be very low, even after inflammatory stimulation by such mechanisms as endotoxin administration. On the other hand, IL-6 and TNF are easily detected in the plasma during inflammation and reach physiologic stimulatory levels within a short time after challenge.[154] IL-6 has been shown to reach considerable plasma levels after burns and during ongoing inflammatory responses, as in inflammatory joint disease and in transplant rejection.[155-158] Moreover, we have shown in a rat model of inflammation that hepatocytes undergoing an acute phase response do not make their own IL-6 and therefore must have the mediator delivered exogenously.

In vivo evidence for glucocorticoid stimulation by IL-6 has recently been provided,[159] but whether this effect is a direct action of IL-6 has not been proven. This dilemma is all the more notable when we examine the data for stimulation of fibrinogen *in vivo*. There is no doubt that administration of IL-1 to animals results in increased plasma levels of fibrinogen. However, since IL-1 does not directly stimulate the expression of fibrinogen in hepatocytes and in fact acts as a negative modulator of stimulation, the increase can only come from the participation of IL-6, the major inducer of fibrinogen synthesis. This participation must arise from the release of IL-6 from stromal cells after activation by the exogenously administered IL-1. Still, the direct effect of inflammatory cells entering the liver during the acute phase response and releasing mediators locally cannot be ruled out. Thorough examination by means of immunohistochemistry and *in situ* hybridization is needed before the mechanisms of tissue interaction are understood in the generation of the hepatic response.

CONCLUSIONS

The cascade of events that follow trauma or invasion of the body by infectious organisms results in the production of acute phase proteins by the liver. Mediation of these events involve both inflammatory and stromal cells as well as a family of acute phase cytokine hormones with multipotent and overlapping biological activities. These cytokines interact both in the tissue undergoing the inflammatory response as well as at the target tissue, the hepatocyte. Each acute phase protein gene is regulated by a specific combination of cytokines and the sequence and concentration of these hormones arriving at the liver is important in defining the final expression of the proteins.

We have established a role for a number of these regulatory cytokines in the stimulation of acute phase protein genes, but there are undoubtedly other hormones with liver-modulating activities that are present and that refine the stimulation. Determining the relative importance and physiologic role of these various mediators will be necessary before considering how to modulate their function and thereby mediate the acute phase response for therapeutic means. In addition, the continued study of acute phase protein gene regulation by cytokines will benefit our understanding of basic cell and molecular biology as we continue to examine the molecular events and the signals involved in regulating the increased and decreased transcriptional activities.

REFERENCES

1. Kushner, I., The phenomenon of the acute phase response. *Ann. N.Y. Acad. Sci.*, 389:39–48, 1982.
2. Koj, A., Biological functions of acute phase proteins, in *The Acute Phase Response to Injury and Infection*, A.H. Gordon and A. Koj, Eds., Elsevier, Amsterdam, 1985, 145–160.
3. Koj, A., Acute phase reactants — their synthesis, turnover and biological significance, in *Structure and Function of Plasma Proteins*, Vol. 1, A.C. Allison, Plenum, New York, 1974, 73–131.
4. Fahraeus, R., The suspension stability of the blood. *Acta Med. Scand.*, 55:1–228, 1921.
5. Tillett, W.S. and T. Francis, Serological reactions in pneumonia with a non-protein somatic fraction of pneumococcus. *J. Exp. Med.*, 52:561–571, 1930.
6. Abernathy, T.J. and O.T. Avery, The occurrence during acute infection of a protein not normally present in blood. I. Distribution of the reactive protein in patients sera and the effect of calcium on the flocculation reaction with the C polysaccharide of pneumococcus. *J. Exp. Med.*, 73:173–182, 1941.
7. Homburger, F., A plasma fibrinogen — increasing factor obtained from sterile abscesses in dogs. *J. Clin. Invest.*, 24:43–45, 1945.
8. Kampschmidt, R.F. and H.F. Upchurch, Effects of bacterial endotoxins on plasma iron. *Proc. Soc. Exp. Biol. Med.*, 110:191–193, 1962.
9. Kampschmidt, R.F. and H.F. Upchurch, Effect of leukocytic endogenous mediator on plasma fibrinogen and haptoglobin. *Proc. Soc. Exp. Biol. Med.*, 146:904–907, 1974.
10. Kampschmidt, R.F., H.F. Upchurch, and L.A. Pulliam, Characterization of a leukocyte-derived endogenous mediator responsible for increased plasma fibrinogen. *Ann. N.Y. Acad. Sci.*, 389:338–353, 1982.
11. Schreiber, G., Synthesis, processing and secretion of plasma proteins by the liver (and other organs) and their regulation, in *The Plasma Proteins*, Vol. 5, F.W. Putnam, Ed., Academic Press, New York, 1987, 293–363.
12. Schreiber, G., A. Tsykin, A.R. Aldred, T. Thomas, W.-P. Fung, P.W. Dickson, T. Cole, H. Birch, F.A. De Jong, and J. Milland, The acute phase response in the rodent. *Ann. N.Y. Acad. Sci.*, 557:61–85, 1989.
13. Pepys, M.B. and M.L. Baltz, Acute phase proteins with special reference to C-reactive protein and related proteins (pentaxins) and serum amyloid A protein. *Adv. Immunol.*, 34:141–212, 1983.
14. Cole, T., A.S. Inglis, C.M. Roxburgh, G.J. Howlett, and G. Schreiber, Major acute phase α_1-protein of the rat is homologous to bovine kininogen and contains the sequence for bradykinin: its synthesis is regulated at the mRNA level. *FEBS Lett.*, 182:57–61, 1985.
15. Esnard, F. and F. Gauther, Rat α_1-cysteine proteinase inhibitor. *J. Biol. Chem.*, 258:12443–12447, 1983.
16. Fung, W.-P. and G. Schreiber, Structure and expression of the genes for major acute phase α-protein (thistatin) and kininogen in the rat. *J. Biol. Chem.*, 262:9298–9308, 1987.
17. Gauldie, J., L. Lamontagne, and A. Stadnyk, Acute phase response in infectious disease. *Surv. Synth. Path. Res.*, 4:126–151, 1985.

18. Kushner, I. and A. Mackiewicz, Acute phase proteins as disease markers. *Dis. Markers*, 5:1–11, 1987.
19. Bergquist, D. and I.M. Nilsson, Hereditary α_2-macroglobulin deficiency. *Scand. J. Haematol.*, 23:433–436, 1979.
20. Harpel, P.C. and M.S. Brower, α_1-Macroglobulin. An introduction. *Ann. N.Y. Acad. Sci.*, 421:1–9, 1983.
21. Travis, J. and G.S. Salvesen, Human plasma proteinase inhibitors. *Ann. Rev. Biochem.*, 52:655–709, 1983.
22. Van Vugt, H., J. van Gool, and L. de Ridder, α_2-Macroglobulin of the rat, an acute phase protein mitigates the early course of endotoxin shock. *Br. J. Exp. Pathol.*, 67:313–319, 1986.
23. Hoffman, M., S.R. Feldman and S.V. Pizzo, α_2-Macroglobulin "fast" forms inhibit superoxide production by activated macrophages. *Biochim. Biophys. Acta.*, 760:421–423, 1983.
24. James, K., Alpha$_2$-macroglobulin and its possible importance in immune system. *Trans. Biochem. Sci.*, 5:43–47, 1980.
25. Hudig, D., T. Haverty, C. Fulcher, D. Redelman, and J. Mendelsohn, Inhibition of human natural cytotoxicity by macromolecular antiproteases. *J. Immunol.*, 126:1569–1574, 1981.
26. Ades, E.W., A. Hinson, C. Chapuis-Cellier, and P. Arnaud, Modulation of the immune response by plasma protease inhibitors. I. Alpha$_2$-macroglobulin and alpha$_1$–antitrypsin inhibit natural killing and antibody-dependent cell-mediated cytotoxicity. *Scand. J. Immunol.*, 15:109–113, 1982.
27. Gravagna, P., E. Gianazza, P. Arnaud, M. Neels, and E.W. Ades, Modulation of the immune response by plasma protease inhibitors. III. Alpha$_1$–antichymotrypsin inhibits human natural killing and antibody-dependent cell-mediated cytotoxicity. *J. Reticuloendothel. Soc.*, 32:125–130, 1982.
28. Breit, S.N., D. Wakefield, J.P. Robinson, E. Luckhurst, P. Clark, and R. Penny, The role of α_1-antitrypsin deficiency in the pathogenesis of immune disorders. *Clin. Immunol. Immunopathol.*, 35:363–380, 1985.
29. Simon, P.L., J.B. Willoughby and W.F. Willoughby, Inhibition of T-cell activation by alpha-1-antiprotease is reversed by purified rabbit interleukin-1, in *Interleukins, Lymphokines and Cytokines*, J.J. Oppenheim, S. Cohen, and M. Landy, Eds., Academic Press, New York, 1983, 487–493.
30. Hooper, D.C., C.J. Steer, C.A. Dinarello, and A.C. Peacock, Inhibition of human natural cytotoxicity by macromolecular antiproteases. *J. Immunol.*, 126:1569–1574, 1981.
31. Hoffman, M., S.R. Feldman, and S.V. Pizzo, α_2-Macroglobulin 'fast' forms inhibit superoxide production by activated macrophages. *Biochim. Biophys. Acta*, 760:421–423, 1983.
32. Hoffman, M.R., S.V. Pizzo, and J.B. Weinberg, Modulation of mouse peritoneal macrophage Ia and human peritoneal macrophage HLA-DR expression by α_2-macroglobulin 'fast' forms. *J. Immunol.*, 139:1885–1890, 1987.
33. Chiu, K.M., R.F. Mortensen, A.P. Osmand, and H. Gewurz, Interactions of alpha$_1$–acid glycoprotein with the immune system. I. Purification and effects upon lymphocyte responsiveness. *Immunology*, 32:997–1001, 1977.
34. Bennett, M. and K. Schmid, Immunosuppression by human α_1-acid glycoprotein: importance of the carbohydrate moiety. *Proc. Natl. Acad. Sci. U.S.A.*, 77:6109–6113, 1980.

35. Cheresh, D.A., D.H. Haynes, and J.A. Distasio, Interaction of an acute phase reactant, α_1-acid glycoprotein (orosomucoid) with the lymphoid cell surface: a model for nonspecific immune suppression. *Immunology*, 51:541–548, 1984.
36. Robey, F.A., K. Ohura, S. Futaki, et al., Proteolysis of human C-reactive protein produces peptides with potent immunomodulating activity. *J. Biol. Chem.*, 262:7053–7057, 1987.
37. Robey F.A., K.D. Jones, T. Tanaka, et al., Binding of C-reactive protein to chromatin and nucleosome core particles. *J. Biol. Chem.*, 259:7311–7316, 1984.
38. Robey, F.A., K.D. Jones, and A.D. Steinberg, C-reactive protein mediates the solubilization of nuclear DNA by complement *in vitro*. *J. Exp. Med.*, 161:1344–1356, 1985.
39. Huang, J.S., S.S. Huang, and T.F. Deuel, Specific covalent binding of platelet-derived growth factor to human plasma α_2-macroglobulin. *Proc. Natl. Acad. Sci. U.S.A.*, 81:342–346, 1984.
40. Teodorescu, M., J.L. Skosey, C. Schlesinger, et al., Covalent disulfide binding of IL1 to α_2-macroglobulin, in *Monokines and Other Non-Lymphocytic Cytokines*, M.C. Powanda, Ed., Alan R. Liss, New York, 1988, 209–212.
41. O'Connor-McCourt, M.D. and L.M. Wakefield, Latent transforming growth factor β in serum: a specific complex with α_2-macroglobulin. *J. Biol. Chem.*, 262:14090–14099.
42. Huang, S.S., P. O'Grady, and J.S. Huang, Human transforming growth factor β. α_2-Macroglobulin is a latent form of transforming growth factor β. *J. Biol. Chem.*, 263:1535–1541, 1988.
43. Matsuda, T., T. Hirano, S. Nagasawa, and T. Kishimoto, Identification of α_2-macroglobulin as a carrier protein for IL-6. *J. Immunol.*, 142:148–152, 1989.
44. Perlmutter, D.H., F.S. Cole, P. Kilbridge, T.H. Rossing, and H.R. Colten,. Expression of the alpha-1–proteinase inhibitor gene in human monocytes and macrophages. *Proc. Natl. Acad. Sci. U.S.A.*, 82:795–799, 1985.
45. Barbey, C., J.A. Pierce, E.J. Campbell, and D.H. Perlmutter. Lipopolysaccharide modulates the expression of α_1 proteinase inhibitor and other serine proteinase inhibitors in human monocytes and macrophages. *J. Exp. Med.*, 166:1041–1054, 1987.
46. Perlmutter, D.H., IFNβ2/IL-6 is one of several cytokines that modulate acute phase gene expression in human hepatocytes and human macrophages. *Ann. N.Y. Acad. Sci.*, 557:332–342, 1989.
47. Northemann, W., H. Ueberberg, and G.H. Fey, Coordinated regulation of the rat alpha-2-macroglobulin gene by inflammatory signals and glucocorticoids. *Life Sci. Adv.*, 7:7–10, 1988.
48. Geiger, T., T. Andus, J. Klapproth, et al., Induction of rat acute phase proteins by interleukin 6 in vivo. *Eur. J. Immunol.*, 18:717–721, 1988.
49. Marinkovic, S., G.P. Jahreis, G.G. Wong, and H. Baumann, Interleukin-6 modulates the synthesis of a specific set of acute phase plasma proteins *in vivo*. *J. Immunol.*, 142:808–812, 1989.
50. Koj, A., J. Gauldie, E. Regoeczi, D.N. Sauder, and G.D. Sweeney, The acute-phase response of cultured rat hepatocytes. System characterization and the effect of human cytokines. *Biochem. J.*, 224:505–514, 1984.
51. Baumann, H., C. Richards, and J. Gauldie, Interaction among hepatocyte-stimulating factors, interleukin 1, and glucocorticoids for regulation of acute phase plasma proteins in human hepatoma (HepG2) cells. *J. Immunol.*, 139:4122–4128, 1987.
52. Gross, V., T. Andus, T.A. Tran-Thi, et al., Induction of acute phase proteins by dexamethasone in primary rat hepatocyte cultures. *Exp. Cell. Res.*, 151:46–54, 1984.

53. Fey, G.H., M. Hattori, W. Northemann, L.J. Abraham, M. Baumann, T.A. Braciak, R.G. Fletcher, J. Gauldie, F. Lee, and M.F. Reymond, Regulation of rat liver acute phase genes by interleukin-6 and production of hepatocyte stimulating factors by rat hepatoma cells. *Ann. N.Y. Acad. Sci.*, 557:317–331, 1989.
54. Baumann, H. and W.A. Held, Biosynthesis and hormone-regulated expression of secretory glycoproteins in rat liver and hepatoma cells. Effect of glucocorticoids and inflammation. *J. Biol. Chem.*, 256:10145–10155, 1981.
55. Baumann, H., G.L. Firestone, T.L. Burgess, K.W. Gross, K.R. Yamamoto, and W.A. Held, Dexamethasone regulation of alpha-1-acid glycoprotein and other acute phase reactants in rat liver and hepatoma cells. *J. Biol. Chem.*, 258:563–570, 1983.
56. Baumann, H., Hepatic acute phase reaction in vivo and in vitro. *In Vitro Cell. Dev. Biol.*, 25:115–126, 1989.
57. Kulkarni, A.B., R. Reinke, and P. Feigelson, Acute-phase mediators and glucocorticoids elevate alpha 1-acid glycoprotein gene transcription. *J. Biol. Chem.*, 260:15386–15389, 1985.
58. Baumann, H. and L.E. Maquat, Localization of DNA sequences involved in dexamethasone-dependent expression of the rat α_1-acid glycoprotein gene. *Mol. Cell. Biol.*, 6:2551–2561, 1986.
59. Baumann, H., K.R. Prowse, S. Marinkovic, K.-A. Won, and G.P. Jahreis, Stimulation of hepatic acute phase response by cytokines and glucocorticoids. *Ann. N.Y. Acad. Sci.*, 557:280–296, 1989.
60. Bornstein, D.L., Leukocytic pyrogen: a major mediator of the acute phase reaction. *Ann. N.Y. Acad. Sci.*, 389:323–337, 1982.
61. Dinarello, C.A., The biology of interleukin 1 and comparison to tumor necrosis factor. *Immunol. Lett.*, 16:227–232, 1987.
62. Dinarello, C.A., Biology of interleukin 1. *FASEB J.*, 2:108–115, 1988.
63. Le, J. and J. Vilcek, Biology of disease. Tumor necrosis factor and interleukin 1: cytokines with multiple overlapping biological activities. *Lab. Invest.*, 56:234–248, 1987.
64. Kampschmidt, R.F. and M. Mesecher, Interleukin-1 from P388D: effects upon neutrophils plasma iron, and fibrinogen in rats, mice, and rabbits. *Proc. Soc. Exp. Biol. Med.*, 179:197–200, 1985.
65. Ganapathi, M.K., D. Schultz, A. Mackiewicz, D. Samols, S.-I. Hu, A. Brabenec, S.S. MacIntyre, and I. Kushner, Heterogeneous nature of the acute phase response. Differential regulation of human serum amyloid A, C-reactive protein, and other acute phase proteins by cytokines in Hep 3B cells. *J. Immunol.*, 141:564–569, 1988.
66. Ramadori, G., J.D. Sipe, C.A. Dinarello, S.B. Mizel, and H.R. Colten, Pretranslational modulation of acute phase hepatic protein synthesis by murine recombinant interleukin 1 (IL-1) and purified human IL-1. *J. Exp. Med.*, 162:930–942, 1985.
67. Perlmutter, D.H., C.A. Dinarello, P.I. Punsal, and H.R. Coltent, Cachectin/tumor necrosis factor regulates hepatic acute phase gene expression. *J. Clin. Invest.*, 78:1349–1354, 1986.
68. Perlmutter, D.H., R.C. Strunk, G. Goldberger, and F.S. Cole, Regulation of complement proteins C2 and factor B by interleukin-1 and interferon-gamma acting on transfected L cells. *Mol. Immunol.*, 23:1263–1266, 1986.
69. Perlmutter, D.H., G. Goldberger, C.A. Dinarello, S.B. Mizel, and H.R. Colten, Regulation of class III major histocompatibility complex gene products by interleukin 1. *Science*, 232:850–852, 1986.
70. Le, P.T. and R.F. Mortensen, In vitro induction of hepatocyte synthesis of the acute phase reaction mouse serum amyloid P-component by macrophages and IL-1. *J. Leukoc. Biol.*, 35:587–603, 1984.

71. Gauldie, J., D.N. Sauder, K.P.W.J. McAdam, and C.A. Dinarello, Purified interleukin-1 (IL-1) from human monocytes stimulates acute-phase protein synthesis by rodent hepatocytes in vitro. *Immunology,* 60:201–207, 1987.
72. Gauldie, J., C. Richards, D. Harnish, P. Lansdorp, and H. Baumann, Interferon beta 2/B-cell stimulatory factor type 2 shares identity with monocyte-derived hepatocyte-stimulating factor and regulates the major acute phase protein response in liver cells. *Proc. Natl. Acad. Sci. U.S.A.,* 84:7251–7255, 1987.
73. Koj, A., A. Kurdowska, D. Magielska-Zero, H. Rokita, J.D. Sipe, J.M. Dayer, S. Demczuk, and J. Gauldie, Limited effects of recombinant human and murine interleukin 1 and tumour necrosis factor on production of acute phase proteins by cultured rat hepatocytes. *Biochem. Int.,* 14:553–560, 1987.
74. Baumann, H., V. Onorato, J. Gauldie, and G.P. Jahreis, Distinct sets of acute phase plasma proteins are stimulated by separate human hepatocyte-stimulating factors and monokines in rat hepatoma cells. *J. Biol. Chem.,* 262:9756–9768, 1987.
75. Geiger, T., T. Andus, J. Klapproth, H. Northoff, and P.C. Heinrich, Induction of alpha 1-acid glycoprotein by recombinant human interleukin-1 rat hepatoma cells. *J. Biol. Chem.,* 263:7141–7146, 1987.
76. Woo, P., J. Sipe, C.R. Dinarello, and H.R. Colten, Structure of a human serum amyloid A gene and modulation of its expression in transfected L cells. *J. Biol. Chem.,* 262:15790–15795, 1985.
77. Beuscher, H.U., R.J. Fallon, and H.R. Colten, Macrophage membrane interleukin 1 regulates the expression of acute phase proteins in human hepatoma Hep 3B cells. *J. Immunol.,* 139:1896–1901, 1987.
78. Woloski, B.M.R.N.J. and G.M. Fuller, Identification and partial characterization of hepatocyte-stimulating factor from leukemia cell lines: comparison with interleukin-1. *Proc. Natl. Acad. Sci.,* 82:1443–1447, 1985.
79. Darlington, G.J., D.R. Wilson, and L.B. Lachman, Monocyte-conditioned medium, interleukin-1, and tumour necrosis factor stimulate the acute phase response in human hepatoma cells in vitro. *J. Cell Biol.,* 103:787–793, 1986.
80. Evans, E., G.M. Courtois, P.L. Kilian, G.M. Fuller, and G.R. Crabtree, Induction of fibrinogen and a subset of acute phase response genes involves a novel monokine which is mimicked by phorbol esters. *J. Biol. Chem.,* 262:10850–10854, 1987.
81. Gauldie, J., C. Richards, D. Harnish, and H. Baumann, Interferon beta-2 is identical to monocytic HSF and regulates the full acute phase protein response in liver cells, in *Monokines and Other Non-Lymphocyte Cytokines,* M.C. Powanda, J.J. Oppenheim, M.J. Kluger, C.A. Dinarello, Eds., Alan R. Liss, New York, 1988, 15–20.
82. Gauldie, J., C. Richards, W. Northemann, G. Fey, and H. Baumann, IFNβ2/BSF2/IL-6 is the monocyte-derived HSF that regulates receptor-specific acute phase gene regulation in hepatocytes. *Ann. N.Y. Acad. Sci.,* 557:46–59, 1989.
83. Darlington, G.J., D.R. Wilson, M. Revel, and J.H. Kelly, Response of liver genes to acute phase mediators. *Ann. N.Y. Acad. Sci.,* 557:310–316, 1989.
84. Ritchie, D.G. and G.M. Fuller, Hepatocyte-stimulating factor: a monocyte-derived acute-phase regulatory protein. *Ann. N.Y. Acad. Sci.,* 408:490–502, 1983.
85. Bauer, J., M. Birmelin, G.H. Northoff, W. Northemann, T.A. Tran-Thi, H. Ueberberg, K. Decker, and P.C. Heinrich, Induction of rat alpha-2-macroglobulin in vivo and in hepatocyte primary cultures: synergistic action of glucocorticoids and a Kupffer cell derived factor. *FEBS Lett.,* 177:89–94, 1984.
86. Goldman, N.D. and T.Y. Liu, Biosynthesis of human C-reactive protein in cultured hepatoma cells is induced by a monocyte factor(s) other than interleukin-1. *J. Biol. Chem.,* 262:2363–2368, 1987.

87. Otto, J.M., H.E. Grenett, and G.M. Fuller, The coordinated regulation of fibrinogen gene transcription by hepatocyte-stimulating factor and dexamethasone. *J. Cell. Biol.*, 105:1067–1072, 1987.
88. Sehgal, P.B., L.T. May, I. Tamm, et al., Human β2 interferon and B-cell differentiation factor BSF-2 are identical. *Science*, 235:731–732, 1987.
89. Billiau, A., Interferon β2 as a promoter of growth and differentiation of B cells. *Immunol. Today*, 8:84–87, 1988.
90. Van Damme J., G. Opdenakker, R.J. Simpson, et al., Identification of the human 25 kD protein, interferon β2, as a B cell hybridoma/plasmacytoma growth factor induced by interleukin 1 and tumor necrosis factor. *J. Exp. Med.*, 165:914–919, 1987.
91. Wong, G.G. and S.C. Clark, Multiple actions of interleukin 6 within a cytokine network. *Immunol. Today*, 9(5):137–139, 1988.
92. Nordan, R.P., B.A. Mock, L.M. Neckers, and S. Rudikoff, The role of plasmacytoma growth factor in the *in vitro* responses of murine plasmacytoma cells. *Ann. N.Y. Acad. Sci.*, 557:200–205, 1989.
93. May, L.T., U. Santhanam, S.B. Tatter, J. Ghrayeb, and P.B. Sehgal, Multiple forms of human interleukin-6: phosphoglycoproteins secreted by many different tissues. *Ann. N.Y. Acad. Sci.*, 557:114–121, 1989.
94. Northemann, W., T.A. Braciak, M. Hattori, F. Lee, and G.H. Fey, Structure of the rat interleukin 6 gene and its expression in macrophage-derived cells. *J. Biol. Chem.*, 264:16072–16082, 1989.
95. Van Snick, J., S. Cayphas, J.P. Szikora, J.C. Renauld, E. Van Roost, T. Boon, and R.J. Simpson, cDNA cloning of murine interleukin HP1: homology with human interleukin 6. *Eur. J. Immunol.*, 18:193–197, 1988.
96. Chiu, C.-P., C. Moulds, R.L. Coffmann, D. Rennick, and F. Lee, Multiple biological activities are expressed by a mouse interleukin 6 cDNA clone isolated from bone marrow stromal cells. *Proc. Natl. Acad. Sci. U.S.A.*, 85:7099–7103, 1988.
97. Fuller, G.M. and H.E. Grenett, The structure and function of the mouse hepatocyte stimulating factor. *Ann. N.Y. Acad. Sci.*, 557:31–45, 1989.
98. Shabo, Y., J. Lotem, M. Rubinstein, M. Revel, S.C. Clark, S.F. Wolf, R. Kamen, and L. Sachs, The myeloid blood cell differentiation-inducing protein MGI-2A is interleukin-6. *Blood*, 72:2070–2073, 1988.
99. Chiu, C.-P. and F. Lee, IL-6 is a differentiation factor for M1 and WEHI-3B myeloid leukemic cells. *J. Immunol.*, 142:1909–1915, 1989.
100. Sachs, L., J. Lotem, and Y. Shabo, The molecular regulators of macrophage and granulocyte development: role of MGI-2/IL-6. *Ann. N.Y. Acad. Sci.*, 557:417–437, 1989.
101. Morrone, G., G. Ciliberto, S. Oliviero, R. Arcones, L. Dente, J. Content, and R. Cortese, Recombinant interleukin 6 regulates the transcriptional activation of a set of human acute phase genes. *J. Biol. Chem.*, 263:12554–12558, 1988.
102. Bereta, J., A. Kurdowska, A. Koj, T. Hirano, T. Kishimoto, J. Content, W. Fiers, J. Van Damme, and J. Gauldie, Different preparations of natural and recombinant human interleukin-6 (IFN-β2, BSF-2) similarly stimulate acute phase protein synthesis and uptake of α-aminoisobutyric acid by cultured rat hepatocytes. *Int. J. Biochem.*, 21:361–366, 1989.
103. Moshage, H.J., H.M.J. Roelofs, J.F. Van Pelt, B.P.C. Hazenberg, M.A. Van Leeuwen, P.C. Limburg, L.A. Aarden, and S.H. Yap, The effect of interleukin-1, interleukin-6 and its interrelationship on the synthesis of serum amyloid A and C-reactive protein in primary cultures of adult human hepatocytes. *Biochem. Biophys. Res. Commun.*, 155:112–117, 1988.

104. Ganapathi, M.K., L.T. May, D. Schultz, A. Brabenec, J. Weinstein, P.B. Sehgal, and I. Kushner, Role of interleukin-6 in regulating synthesis of C-reactive protein and serum amyloid A in human hepatoma cell lines. *Biochem. Biophys. Res. Commun.*, 157:271–277, 1988.
105. Castell, J.V., T. Andus, D. Kunz, and P.C. Heinrich, Interleukin-6: the major regulator of acute-phase protein synthesis in man and rat. *Ann. N.Y. Acad. Sci.*, 557:86–101, 1989.
106. Andus, T., T. Geiger, T. Hirano, H. Northoff, U. Ganter, J. Bauer, T. Kishimoto, and P.C. Heinrich, Recombinant human B cell stimulatory factor 2 (BSF-2/IFN-beta2) regulates beta-fibrinogen and albumin mRNA levels in Fao-9 cells. *FEBS Lett.*, 221:18–22, 1987.
107. Andus, T., T. Geiger, T. Hirano, T. Kishimoto, T.A. Tran-Thi, K. Kecker, and P.C. Heinrich, Regulation of synthesis and secretion of major rat acute-phase proteins by recombinant human interleukin-6 (BSF-2/IL-6) in hepatocyte primary cultures. *Eur. J. Biochem.*, 173:287–293, 1988.
108. Baumann, H. and U. Mueller-Eberhard, Synthesis of hemopexin and cysteine protease inhibitor is coordinately regulated by HSF-II and interferon-beta$_2$ in rat hepatoma cells. *Biochem. Biophys. Res. Commun.*, 146:1218–1226, 1987.
109. Baumann, H., H. Isseroff, J.J. Latimer, and G.P. Jahreis, Phorbol ester modulates interleukin 6– and interleukin 1–regulated expression of acute phase plasma proteins in hepatoma cells. *J. Biol. Chem.*, 263:17,390–17,396, 1988.
110. Fey, G.H. and G.M. Fuller, Regulation of acute phase gene expression by inflammatory mediators. *Mol. Biol. Med.*, 4:323–338, 1987.
111. Fey, G. and J. Gauldie, The acute phase response of the liver in inflammation, in *Progress in Liver Diseases*, Vol. 9, H. Popper and F. Schaffner, Eds., W.B. Saunders, Philadelphia, 1990, 89–116.
112. Prowse, K.R. and H. Baumann, Interleukin-1 and interleukin-6 stimulate acute-phase protein production in primary mouse hepatocytes. *J. Leukoc. Biol.*, 45:55–61, 1989.
113. Gauldie, J., W. Northemann, and G.H. Fey, IL-6 functions as an exocrine hormone in inflammation: hepatocytes undergoing acute phase responses require exogenous IL-6. *J. Immunol.*, 144:3804–3808, 1990.
114. Baumann, H. and J. Gauldie, Regulation of hepatic acute phase plasma protein genes by hepatocyte stimulating factors and other mediators of inflammation. *Mol. Biol. Med.*, 7:147–160, 1990.
115. Baumann, K., K.-A. Won, and G.P. Jahreis, Human hepatocyte-stimulating factor III and interleukin-6 are structurally and immunologically distinct but regulate the production of the same acute phase plasma proteins. *J. Biol. Chem.*, 264:8046–8051, 1989.
116. Baumann, H. and G.G. Wong, Hepatocyte-stimulating factor III shares structural and function identity with leukemia inhibitory factor. *J. Immunol.*, 143:1163–1167, 1989.
117. Smith, A.G., J.K. Heath, D.D. Donaldson, G.G. Wong, J. Moreau, M. Stahl, and D. Rogers, Inhibition of pluripotential embryonic stem cell differentiation by purified polypeptides. *Nature*, 336:688–690, 1989.
118. Moreau, J.-F., D.D. Donaldson, F. Bennett, J. Witek-Giannotti, S.C. Clark, and G.G. Wong, Leukaemia inhibitory factor is identical to the myeloid growth factor human interleukin for DA cells. *Nature*, 336:690–692, 1988.
119. Gearing, D.P., N.M. Gough, J.A. King, D.J. Hilton, N.A. Nicola, R.J. Simpson, E.C. Nice, A. Kelson, and D. Metcalf, Molecular cloning and expression of cDNA encoding a murine myeloid leukemia inhibitory factor. *EMBO J.*, 6:3995–4002, 1987.

120. Gough, N.M., D.P. Gearing, J.A. King, T.A. Willson, D.J. Hilton, N.A. Nicola, and D. Metcalf, Molecular cloning and expression of the human homologue of the murine gene encoding myeloid leukemia-inhibitory factor. *Proc. Natl. Acad. Sci. U.S.A.*, 85:2623–2627, 1988.
121. Northemann, W., M. Hattori, G. Baffet, T.A. Braciak, R.G. Fletcher, L.J. Abraham, J. Gauldie, M. Baumann, and G.H. Fey, Production of interleukin 6 by hepatoma cells. *Mol. Biol. Med.*, 7:273–285, 1990.
122. Magielska-Zero, D., J. Bereta, P.V. Czuba-Pelech, J. Gauldie, and A. Koj, Inhibitory effect of recombinant human interferon gamma on synthesis of some acute phase proteins in human hepatoma Hep G2 cells. *Biochem. Int.*, 17:17–23, 1988.
123. Baumann, H., G.P. Jahreis, and K.C. Gaines, Synthesis and regulation of acute phase plasma proteins in primary cultures of mouse hepatocytes. *J. Cell Biol.*, 97:866–876, 1983.
124. Rokita, H., J. Bereta, A. Koj, A.H. Gordon, and J. Gauldie, Epidermal growth factor and transforming growth factor-β differently modulate the acute phase response elicited by interleukin-6 in cultured liver cells from man, rat and mouse. *Comp. Biochem. Physiol.*, 95A:41–45, 1990.
125. Birch, H. and G. Schreiber, Transcriptional regulation of plasma protein synthesis during inflammation. *J. Biol. Chem.*, 261:8077–8080, 1986.
126. Shiels, B.R., W. Northeman, M.R. Gehring, and G.H. Fey, Modified nuclear processing of α_1-acid glycoprotein RNA during inflammation. *J. Biol. Chem.*, 262:12826–12831, 1987.
127. Cosman, D., Control of messenger RNA stability. *Immunol. Today*, 8:16–17, 1987.
128. Mackiewicz, A., M.K. Ganapathi, D. Schultz, et al., Monokines regulate glycosylation of acute phase proteins. *J. Exp. Med.*, 166:253–258, 1987.
129. Mackiewicz, A. and I. Kushner, Role of IL-6 in acute phase protein glycosylation. *Ann. N.Y. Acad. Sci.*, 557:515–517, 1989.
130. Prowse, K.R. and H. Baumann, Hepatocyte-stimulating factor, beta-2 interferon, and interleukin-1 enhance expression of the rat alpha-1-acid glycoprotein gene via a distal upstream regulatory region. *Mol. Cell. Biol.*, 8:42–51, 1988.
131. Fowlkes, D.M., N.T. Mullis, C.M. Comeau, and G.R. Crabtree, Potential basis for regulation of the coordinately expressed fibrinogen genes: homology in the 5' flanking regions. *Proc. Natl. Acad. Sci. U.S.A.*, 81:2313–2316, 1984.
132. Tsuchiya, Y., M. Hattori, K. Hayashida, H. Ishibashi, H. Okubo, and Y. Sakaki, Sequence analysis of the putative regulatory region of rat α_2-macroglobulin gene. *Gene*, 57:73–80, 1987.
133. Burt, M.R., J.R. Cheshire, and P. Woo, Identification of cis-acting sequences responsible for phorbol ester induction of human serum amyloid A gene expression. *Mol. Cell. Biol.*, in press, 1989.
134. Sifers, R.N., J.A. Carlson, S.M. Clift, F.J. DeMayo, D.W. Bullock, and S.L.C. Woo, Tissue specific expression of the human alpha-1–antitrypsin gene in transgenic mice. *Nucleic. Acids Res.*, 15:1459–1475, 1987.
135. Dente, L., U. Ruther, M. Tripodi, E.F. Wagner, and R. Cortese, Expression of human alpha-1-acid glycoprotein genes in cultured cells and in transgenic mice. *Genes Dev.*, 2:259–266, 1988.
136. Ciliberto, G., R. Arcone, E.F. Wagner, and U. Ruther, Inducible and tissue-specific expression of human C-reactive protein in transgenic mice. *EMBO J.*, 6:4017–4022, 1987.

137. Jamieson, J.C., G. Lammers, R. Janzen, and B.M. Woloski, The acute phase response to inflammation: the role of monokines in changes in liver glycoproteins and enzymes of glycoprotein metabolism. *Comp. Biochem. Physiol.(B)*, 87:5–11, 1987.
138. Hill, M.R., R.D. Stith, and R.E. McCallum, Mechanism of action of interferon-β2/interleukin-6 on induction of hepatic liver enzymes. *Ann. N.Y. Acad. Sci.*, 557:502–505, 1989.
139. Koj, A., The role of interleukin-6 as the hepatocyte stimulating factor in the network of inflammatory cytokines. *Ann. N.Y. Acad. Sci.*, 557:1–8, 1989.
140. Sims, J.E., C.J. March, D. Cosman, M.B. Widmer, H.R. Macdonald, C.J. McMaham, C.E. Grubin, J.M. Wignall, J.L. Jackson, S.M. Call, D. Friend, A.R. Alpert, S. Gillis, D.L. Urdal, and S.K. Dower, cDNA expression of cloning of the IL-1 receptor, a member of the immunoglobulin superfamily. *Science*, 241:585–589, 1989.
141. Yamasaki, K., T. Taga, Y. Hirata, H. Yawata, Y. Kawanishi, B. Seed, T. Taniguchi, T. Hirano, and T. Kishimoto, Cloning and expression of the human interleukin-6 (BSF-2/IFN beta-2) receptor. *Science*, 241:825–828, 1988.
142. Urdal, D.L., S.M. Call, J.L. Jackson, et al., Affinity purification and chemical analysis of the interleukin 1 receptor. *J. Biol. Chem.*, 263:2870–2877, 1988.
143. Hirano, T., T. Taga, F. Yamasaki, T. Matsuda, K. Yasukawa, Y. Kirata, H. Yawata, O. Tanable, S. Akira, and T. Kishimoto, Molecular cloning of the cDNAs for IL-6 and its receptor. *Ann. N.Y. Acad. Sci.*, 557:167–180, 1989.
144. Bauer, J., G. Lengyel, T.M. Bauer, G. Acs, and W. Gerok, Regulation of interleukin-6 receptor expression in human monocytes and hepatocytes. *FEBS Lett.*, 249:27–30, 1989.
145. Taga, T., Y. Kawanishi Y., R.R. Hardy, et al., Receptors for B cell stimulatory factor 2: Quantitation, specificity, distribution and regulation of their expression. *J. Exp. Med.*, 166:967–981, 1987.
146. Elias, J.A., G. Trinchieri, J.M. Beck, P.L. Simon, P.B. Sehgal, L.T. May, and J.A. Kern, A synergistic interaction of IL-6 and IL-1 mediates the thymocyte-stimulating activity produced by recombinant IL-1–stimulated fibroblasts. *J. Immunol.*, 142:509–514, 1989.
147. Houssiau, F.A., P.G. Coulie, D. Olive, and J. Van Snick, Synergistic activation of human T cells by interleukin 1 and interleukin 6. *Eur. J. Immunol.*, 18:653–655, 1988.
148. Jirik, F.R., T.J. Podor, T. Hirano, T. Kishimoto, D.J. Loskutoff, D.A. Carson, and M. Lotz, Bacterial lipopolysaccharide and inflammatory mediators augment IL-6 secretion by human endothelial cells. *J. Immunol.*, 142:144–147, 1989.
149. Sironi, M., F. Breviario, P. Proserpio, A. Biondi, A. Vecchi, J. Van Damme, E. Dejana, and A. Mantovani, IL-1 stimulates IL-6 production in endothelial cells. *J. Immunol.*, 142:549–553, 1989.
150. Kohase, M., L.T. May, I. Tamm, J. Vilcek, and P.B. Sehgal, A cytokine network in human diploid fibroblasts: interactions of beta interferons, tumor necrosis factor, platelet-derived growth factor and interleukin-1. *Mol. Cell. Biol.*, 7:273–280, 1987.
151. Helle, M., J.P.J. Brakenhoff, E.R. De Groot, and L.A. Aarden, Interleukin 6 is involved in interleukin 1-induced activities. *Eur. J. Immunol.*, 18:957–959, 1988.
152. Woloski, B.M.R.N.J., E.M. Smith, W.J. Meyer, III, G.M. Fuller, and J.E. Blalock, Corticotropin-releasing activity of monokines. *Science*, 230:1035–1037, 1985.
153. Besedovsky, H., A. Del Rey, E. Sorkin, and C.A. Dinarello, Immunoregulatory feedback between interleukin 1 and glucocorticoid hormones. *Science*, 238:652–654, 1986.
154. Jablons, D.M., J.J. Mule, J.K. McIntosh, P.B. Sehgal, L.T. May, C.M. Huang, S.A. Rosenberg, and M.T. Lotze, IL-6/IFN-β-2 as a circulating hormone. Induction by cytokine administration in humans. *J. Immunol.*, 142:1542–1547, 1989.

155. Houssiau, F.A., K. Bukasa, C.J.M. Sindic, J. Van Damme, and J. Van Snick, Elevated levels of the 26K human hybridoma growth factor (interleukin 6) in cerebrospinal fluid of patients with acute infection of the central nervous system. *Clin. Exp. Immunol.*, 71:320–323, 1988.
156. Van Oers, M.H.J., A.A.P.A.M. Van Der Heyden, and L.A. Aarden, Interleukin 6 (IL-6) in serum and urine of renal transplant recipients. *Clin. Exp. Immunol.*, 71:314–319, 1988.
157. Houssiau, F.A., J.-P. Devogelaer, J. Van Damme, C. Nagant de Deuxchaisnes, and J. Van Snick, Interleukin-6 in synovial fluid and serum of patients with rheumatoid arthritis and other inflammatory arthritides. *Arthr. Rheum.*, 31:784–788, 1988.
158. Nijsten, M.W.N., E.R. DeGroot, H.J. TenDuis, J.H. Klaren, C.E. Hack, and L.A. Aarden, Serum levels of interleukin 6 and acute phase responses. *Lancet*, ii:921, 1987.
159. Naitoh, Y., J. Fukata, T. Tominaga, Y. Nakai, S. Tamai, K. Mori, and H. Imura, Interleukin-6 stimulates the secretion of adrenocorticotropic hormone in conscious, freely-moving rats. *Biochem. Biophys. Res. Commun.*, 155:1459–1463, 1988.
160. Marinkovic, S. and H. Baumann, Cytokine regulation of the rat haptoglobin gene. *Ann. N.Y. Acad. Sci.*, 557:492–494, 1989.
161. James, K. Interactions between cytokines and alpha$_2$-macroglobulin. *Immunol. Today*, 11:163–166, 1990.

11. Cytokines and Shock

JOSEPH G. CANNON

SEPTIC SHOCK: A CONSEQUENCE OF MODERN MEDICINE

Septic shock is an often fatal syndrome characterized by a sudden onset of hypotension, fever, respiratory dysfunction (tachypnea, hypocapnea, hypoxia), tachycardia, abnormal white cell counts, thrombocytopenia, and renal or hepatic failure.[1] The incidence of septic shock has grown over the past several decades in spite of improved antibiotic therapies.[2,3] As an illustration, less than 1 case of Gram negative bacteremia per 1000 admissions was reported at Boston City Hospital in 1935, but by 1972 the incidence rate had grown to over 11 cases per 1000 admissions.[4] This trend is actually a consequence of improved critical care for potentially fatal conditions, such as severe burns, multiple trauma, cardiorespiratory diseases, and cancers. Although patients now stand a better chance of surviving their primary illness, they are at great risk of infection because of the size of their wounds and because of pharmacologically, trauma-, or disease-induced immunosuppression.[5]

In shock resulting from hypovolemia or failing myocardial function, poor tissue perfusion leads to cell damage in peripheral tissues. In contrast, septic shock is initiated by peripheral cellular alterations of function that lead to systemic dysfunction. Some of this functional alteration may be caused directly by infectious organisms or their products, but a significant portion of the pathogenesis is mediated by endogenous host factors,[6-8] particularly cytokines. It seems paradoxical for the host to produce factors that would be directly responsible for its demise. One rationale is that host tissues are "innocent bystanders" that become damaged as the host mobilizes its defenses against an overwhelming infection. An alternative viewpoint is that mammalian defense systems did not evolve to cope with an injury of such magnitude as an 80% body surface area, full-thickness burn. The fact that a patient is still alive in such a situation is because of technological intervention: there has been no opportunity for appropriate adaptations to develop through natural selection.[9]

This review consists of a brief outline of the clinical features associated with septic shock and an exploration of the postulated roles of cytokines and other host factors in mediating these responses. An emphasis will be placed on the interdependencies of these mediators and on the concept that the cytokines represent one of several parallel pathways that probably contribute to septic shock. It will be shown that the contribution of an individual cytokine *in vivo* is difficult to assess because each induces the production of other cytokines or modulates their biological activity.

Interleukin-1 (IL-1) and tumor necrosis factor (TNF) have been studied most intensively as mediators of septic shock. Although there are differences in the production and secretion characteristics between the two forms of IL-1 (α and β) and between the two forms of TNF (α-cachectin and β-lymphotoxin), the α and β forms of each cytokine bind to common receptors (see review in References 10 and 11) Therefore, for the sake of simplicity, there will be no differentiation between α and β forms in this review. Furthermore, although T-cell cytotoxic activity and immune complexes may contribute to septic shock[12] and are modulated by cytokines,[13] this review will focus on the nonspecific inflammatory properties of the cytokines.

CLINICAL FEATURES OF SEPTIC SHOCK

Cardiovascular

A fall in systemic vascular resistance (SVR) is the first cardiovascular event in septic shock. Autonomic baroreceptor reflexes maintain blood pressure temporarily by increasing cardiac output. But in ultimately fatal cases, SVR continues to fall despite increased sympathetic efferent drive, and cardiac output becomes insufficient to maintain blood pressure, which exacerbates hypotension and reduces tissue perfusion. For detailed reviews of the clinical aspects of septic shock, please refer to the article by Harris et al.[1] and the book edited by Root and Sande.[14]

Respiratory

Often before a decrease in blood pressure is evident, hyperventilation and respiratory alkalosis develop. A sterotyped sequence of events known as the Adult Respiratory Distress Syndrome (ARDS), or "Shock Lung", often accompanies septic shock as well.[15] In ARDS the pulmonary vascular endothelium becomes more permeable and pulmonary vascular resistance increases. This leads to interstitial and alveolar edema, which in turn causes reduced lung compliance and impedes gas exchange. The ensuing hypoxia and increased work of breathing contribute to respiratory muscle fatigue and, ultimately, respiratory failure.

Leukocyte Redistribution

Total white blood cell counts can be abnormally high *or* low depending upon the type of infection, the time point in the course of infection, and the age and chronic health of the patient. Usually a leukocytosis is observed. The increase in cell number is initially due to demargination of neutrophils, followed by release of immature band cells from bone marrow reserves. Prolonged elevations can occur, which indicate that bone marrow hematopoiesis has accelerated. Extreme cases, leukemoid reactions, can exhibit total white cell counts on the order of 100,000 cells per mm^3.

Bacteremia in immunocompromised individuals can sometimes result in neutropenia. This may be due to poor bone marrow reserves, inadequate neutrophil production in the bone morrow, neutrophil aggregation and adherence to endothelium, or damage to blood cells by microbial factors.

Irrespective of the number of freely circulating cells, increased neutrophil and platelet adhesion to vascular endothelium and clot formation impede perfusion in organs such as the lungs, spleen, liver, and kidney; a condition termed disseminated intravascular coagulation (DIC). The adherent cells then release cytotoxic oxygen radicals, arachidonic acid metabolites, and proteases, which damage the endothelium.

Metabolic/Endocrine Responses

Sepsis is often accompanied by severe wasting of lean body mass.[16] Skeletal muscle catabolism and hepatic acute phase plasma protein synthesis dramatically increase. Skeletal muscle apparently serves as a reservoir of amino acids that can be mobilized for incorporation in acute phase plasma proteins and deaminated for gluconeogenesis. Carbohydrate metabolism is characterized by hyperglycemia, glycogenolysis, increased Cori cycle activity, reduced oxidative metabolism, and reduced insulin sensitivity.[17] Alterations in lipid metabolism include elevations in serum triacylglycerol and free fatty acids.[18]

Septic patients exhibit a generalized neuroendocrine stress response, including elevated plasma catecholamine, ACTH, and cortisol concentrations. In addition, growth hormone insulin and glucagon levels are elevated in the bloodstream.[19]

Organ Damage or Dysfunction

Kidney and/or liver failure are frequent contributors to a fatal outcome in septic shock. The microvascular injury observed in these organs is similar to that described in the lungs; in addition, hepatic sinusoids and renal tubules become obstructed with cellular casts. Functional alterations in hepatic detoxification rates (cytochrome P-450 activity) and renal sodium excretion rates are observed during sepsis.

MEDIATORS OF PATHOGENESIS

Certain pathogens or toxins produced by these pathogens cause direct damage to host cells through specific disruption of plasma membranes, transport systems, or metabolic enzymes.[20] In general, however, the pattern of clinical events in septic shock is relatively constant regardless of underlying bacterial, viral, or parasitic infection.[2,21] This implies that infectious organisms or their products activate host mechanisms that are the final common pathways of pathogenesis. Host factors implicated in cellular damage and microvascular abnormalities include oxygen radicals, proteases, vasoactive amines and peptides,[22,23] and lipid metabolites, including prostaglandins,[24] leukotrienes, and platelet activating factor.[8,25,26] Table 1

Table 1. Endogenous Mediators That Contribute to Septic Shock

	1	2	3	4	5	6
Thromboxane A_2	+			+		
Leukotriene B_4		+	+	+	+	+
PAF	+		+	+		+
C3a, C5a		+	+	+	+	+
O_2 radicals					+	
PGE_2/PGI_2	+					
Bradykinin	+					+
Substance P	+					+

Column 1 = vasodilation; 2 = vasoconstriction; 3 = chemotaxis; 4 = platelet aggregation; 5 = cytotoxic; 6 = permeability.

shows that each factor has several biological activities that contribute to septic shock, and many have overlapping (redundant) actions. Mononuclear phagocytic cells (including blood monocytes, Kupffer cells, and astrocytes) readily produce these mediators upon direct stimulation by pathogenic stimuli such as endotoxin.[27] Other host cells (endothelial neutrophils, platelets, and mast cells) are relatively less sensitive to the direct action of exogenous stimuli[28] and instead release these mediators through the influence of pathogen-activated complement cascades,[29] coagulation cascades, or through the action of cytokines.[30,31] Studies have sought to protect laboratory animals from lethal sepsis by blocking specific mediators or effectors. Blocking coagulation,[32] complement,[33] or cytokines[34] has provided protection in models of sepsis or ARDS, indicating that diverse pathways lead to common pathological results. Moreover, these pathways are interrelated: C5a induces IL-1,[35] TNF up-regulates complement receptors,[36] and both IL-1 and TNF have procoagulant activity.[37] In such a system with many redundant and interdependent factors, there may be no single mediator that is the essential determinant of pathogenesis.

INVOLVEMENT OF CYTOKINES IN SEPTIC SHOCK

The potential involvement of cytokines in the pathogenesis of septic shock will be reviewed at three levels. First, cytokines may have direct cytotoxic action on cell targets or release of the factors listed in Table 1. TNF and IL-1 are examples of this case. Second, cytokines that have relatively few direct inflammatory actions themselves may induce other cytokines that are inflammatory. The induction of TNF and IL-1 during IL-2 therapy for cancer patients is an example. Third, some cytokines can modulate the action of other cytokines. Interferon γ (IFN-γ) and transforming growth factor β (TGFβ) are important examples of this last case. In addition, cytokines such as IL-1 and TNF act at all three levels to promote septic shock.

TNF as an Archetypical Shock-Inducing Cytokine

C3H/HeJ mice, which have a genetic inability to produce TNF (and IL-1) in response to endotoxin, are resistant to doses of endotoxin that are lethal to most other strains of mice. If monocytes from endotoxin-susceptible strains are transferred to the C3H/HeJ mice by bone marrow grafts, the C3H/HeJ mice then become susceptible to endotoxin.[38] Adminstration of sufficient quantities of recombinant TNF will induce shock in several species.[39,40] Mortality in septic shock has been reduced in mice,[41] rabbits,[42] and baboons[34] by passive immunization of the animals with antibodies directed against TNFα. These data clearly indicate an important role for TNF in septic shock. As shown in Table 2, TNF has a direct effect on virtually every biological response associated with septic shock.

Nevertheless, other experiments show that TNF is not the sole determinant of shock. A non-shock-inducing synthetic derivative of lipid A (the toxic moiety of endotoxin) caused an increase in serum TNF levels in mice similar to the levels induced by the natural form of lipid A. However, the mortality induced by the synthetic derivative was less than natural lipid A (46% vs. 63%) and only half as much as for endotoxin itself (96%),[43] indicating that other factors must contribute to the lethal effect.

Shock-Like Responses in Cytokine Clinical Trials

Hypotension, fever, and the other clinical signs associated with the septic shock syndrome develop in patients receiving recombinant IL-2,[44,45] TNF,[46,47] and IFNγ.[48] Although IL-2 has been implicated in pulmonary edema and vasoconstriction[49] and stimulates PGI_2 from the vascular endothelium,[50] current evidence supports the concept that much of the pathology may be mediated by IL-2-induced TNF and IL-1. Cancer patients enrolled in Phase II clinical trials responded to injections of 10^5 U/kg rIL-2 with pronounced increases in plasma TNF and modest increases in plasma IL-1β.[51] In another clinical trial, recombinant TNF administration (300 µg/m^2) induced increases in IL-6 of up to 82 ng/ml.[52] In this same report, patients receiving IL-2 therapy responded first with a rise in TNF and then a rise in IL-6 after the TNF had peaked, which could be interpreted as secondary and tertiary manifestations of a cytokine cascade.

The modulatory interactions between cytokines have an important impact on cytokine-induced toxicity. The shock, DIC, and organ damage caused by combinations of TNF and IL-1, or TNF and IFNγ were much greater than the additive toxicity induced by each agent separately when tested in laboratory animals.[53,54] Of course, during infection all three cytokines, and additional factors, would be secreted.

Regulation of cell surface receptors for one cytokine by another cytokine may account for some of the modulatory interactions: the synergy between TNF and IFNγ can be attributed to up regulation of TNF receptors by IFNγ.[55] But contrary to the observed synergism of IL-1 and TNF in shock *in vivo*, IL-1 has been shown to *down* regulate TNF receptors *in vitro*.[56] The synergy in shock may be due to IL-1 and TNF mediating independent, but complementary, actions. For example, IL-1

Table 2. Biological Activities of Cytokines that may Contribute to Septic Shock

	IL-1	IL-2	GM-CSF	IL-4	IL-6	IL-8	TNF	IFNγ	MIP-1	TGFβ
In vivo										
Hypotension	+	+					+	+		
Fever	+	+	+		+		+		+	
Edema	+	+				+	+			
Leukocytosis	+		+							
Wasting	+				+		+			
In vitro										
Direct cytotoxicity	+		+				+			
Cell-mediated	+		+	+			+	+		+
Proteases, O₂ radicals				+		+				
Chemotaxis			+	+		+			+	
Vascular adhesion	+						+			+

Some cytokines have priming effects (especially GM-CSF); some are inhibitory (see individual references).
Key: IL-1,[10] IL-2,[143] GM-CSF,[92] IL-4,[144] IL-6,[88,145] IL-8,[60] TNF,[11] IFNγ,[48] MIP-1,[146] TGFβ.[147,148] Immunoglobulins are also instrumental in cell activation and cytotoxic mechanisms; therefore, cytokines which influence Ig production, such as IL-5[149] and IL-7,[150] may play indirect roles in the pathogenesis of septic shock.

Table 3. Intermodulation of Cytokines

Modulating Cytokine	IL-1	IL-2	CSFs	IL-6	IL-8	TNF	INFγ	PDGF
IL-1	1	2,A	3	4	5	6,b	7	8,21
IL-2	9		10			11	12	
GM-CSF	13					13	14	
IL-4	15							
TNF	16			4	5			8
IFNγ	17		10			C		8
PDGF	d				18			
TGFβ	19					20		

Plain number = induces or enhances cytokine production. Underlined number = inhibits cytokine production. Capitol letter = upregulates receptors for cytokine. Lower case letter = down regulates receptors for cytokine.
1. References 151,152. 2. Reference 153. 3. Reference 154. 4. Reference 155. 5. Reference 60. 6. Reference 156. 7. Reference 157. 8. Reference 158. 9. Reference 51. 10. Reference 159. 11. Reference 160. 12. Reference 161. 13. Reference 162. 14. Reference 163. 15. Reference 164. 16. Reference 165. 17. Reference 166. 18. Reference 167. 19. Reference 148. 20. References 168. 21. Reference 129. A. Reference 169. b. Reference 56. C. Reference 55. d. Reference 170.

might by expected to increase numbers of neutrophils, while TNF activates them. The influence of cytokines on the production or receptor expression for other cytokines is summarized in Table 3.

TISSUE-SPECIFIC ACTIONS OF CYTOKINES

Vascular Endothelium

The most pervasive actions of the cytokines in the context of septic shock are directed at the vascular endothelium. The endothelial cells themselves produce cytokines in response to bacterial stimulation,[57,58] as do smooth muscle cells.[59] Furthermore, cytokine-endothelial cell interactions account for most of the cellular mechanisms leading to DIC. IL-1 and TNF induce chemotactic proteins, including IL-8,[60,61] which draw neutrophils and monocytes to an inflammatory focus. IL-1 and TNF also induce adhesion molecules (ELAM-1, ICAM-1)[62] on the endothelial surface and promote transendothelial passage of inflammatory cells.[63] Furthermore, IL-1 and TNF induce plasminogen activator inhibitor-1[64-66] and procoagulant activity,[37] promoting fibrin deposition, platelet aggregation, and clotting. In addition, cytokines induce endothelial cells and phagocytic cells to release platelet-activating factor (PAF),[31] eicosanoids, oxygen radicals,[67] and amines with chemotactic and cytotoxic activities.

Vascular Contractility

Both IL-1 and TNF cause a fall in systemic vascular resistance within 30 min of infusion into laboratory animals. The doses needed are generally on the order of 5 µ/kg, which is about three orders of magnitude higher than the doses required to cause fever and immunopotentiation.[68] This underscores the concept that these cytokines become pathological only when very high levels are attained.

The exact mechanism responsible for the fall in SVR is difficult to pinpoint. Isolated vascular strips lose their constrictive response to catecholamines when incubated with IL-1,[69] but this phenomenon takes 2 to 3 h to develop, is prostaglandin-independent, and is protein-synthesis dependent.[70] This action does not explain *in vivo* studies in which an inhibitor of prostaglandin synthetase given 15 min before infusion with IL-1 or TNF blocks the fall in SVR.[53] Other *in vitro* actions of IL-1 and TNF that are hard to reconcile with the *in vivo* actions include up regulation of β adrenoreceptors[71] and B_1 kinin receptors[72] by IL-1, and inhibition of endothelium-derived relaxing factor (EDRF) production by TNF.[73] All of these actions would be expected to promote vasoconstriction rather than vasorelaxation. It may be that cytokine- (and bacterial-) induced prostaglandins, PAF, and thromboxanes are the dominant regulators of vascular tone in the early stages of shock.[74]

Cardiac Function

If septic shock persists, left ventricular ejection fraction decreases, although cardiac output remains elevated due to increased end-diastolic volume and increased chronotropic sensitivity to β-adrenergic stimulation.[75] Studies with isolated hearts and with cultured fetal cardiac cells have shown that circulating factors are responsible for this reversible cardiac function. Purification of one suppressor factor indicates that it is a protein of about 2 kD that can be induced from blood monocytes.[76] Recombinant TNF also causes a depression of fetal cell contractility.[77] It is not known whether the 2 kD factor is a fragment of TNF or if it is a completely independent factor. Inhibition of left ventricular ejection fraction has been observed following IL-2 administration to humans for cancer therapy.[44] However, this may be another manifestation of IL-2 inducing other cytokines, because the depressant effect of IL-2 is not observed in an isolated, perfused heart preparation.[78]

Other changes in organ function that appear to be regulated by cytokines, rather than being consequences of cellular damage, include reduction of hepatic cytochrome P-450 activity[79] and increased renal sodium excretion.[80]

Metabolism

The abnormalities observed in lipid metabolism during sepsis may be attributed, in part, to inhibition of the enzymes lipoprotein lipase, acetyl-CoA carboxylase, and fatty acid synthetase by TNF and IL-1.[11] Reduced oxidative metabolism brought on by sepsis or by infusion of IL-1 and TNF may be related to a decrease in hepatic mitochondrial pyruvate dehydrogenase.[81]

The hormonal regulators of energy metabolism are dramatically elevated during sepsis; this includes epinephrine, cortisol, glucagon, growth hormone, and insulin. Infusing epinephrine, cortisol, and glucagon into healthy subjects brings on a metabolic condition mimicking sepsis,[82] indicating that changes in the hormonal milieu are causal in the development of the metabolic abnormalities. Cytokines apparently influence these hormonal changes: IL-1 induces secretion of insulin and glucagon from pancreatic islets[83] and promotes ACTH and growth hormone release from the pituitary.[84] It is probably the action of these hormones on hepatocytes that stimulates gluconeogenesis, rather than a direct action of IL-1, because IL-1 does not have an effect *in vitro* on hepatocyte cultures.[85] Infusion of TNF in laboratory animals preferentially increases glucose utilization by macrophage-rich tissues, thus channeling energy substrates to immune cells, a beneficial adaption in sepsis.[86]

Recombinant IL-1 and TNF trigger breakdown of skeletal muscle when infused in intact animals.[87] Fong et al.[88] showed that although the weight loss by animals receiving chronic IL-1 or TNF injections could be (outwardly) explained by cytokine-induced anorexia, these animals exhibited accelerated skeletal muscle protein loss and a preservation of liver protein. In contrast, pair-fed or -starved controls exhibited a loss of liver proteins and preservation of skeletal protein. The fact that IL-1- or TNF-induced proteolysis is difficult to observe *in vitro* suggests that counterregulatory hormones, such as insulin or cortisol, or other cytokines may be necessary cofactors.[89] Both IL-1 and TNF activate specific muscle enzymes involved in amino acid metabolism.[90]

CIRCULATING CYTOKINES IN SEPTIC SHOCK

In animal models, live bacteria or sufficient doses of endotoxin to cause severe hypotension or lethality have been administered to study the time course of cytokine release into the circulation. Dramatic increases in TNF activity (up to 20 ng/ml have been recorded[91] with peak TNF activity attained within 90 to 120 min. Up to 1000-fold increases in colony-stimulating factor (CSF) activity have been observed 3 to 6 h following endotoxin infusion.[92] Changes in circulating IL-1 and IFNγ activity have generally been more modest.[91]

In alternative experiments, human volunteers have been infused with endotoxin. Although these human studies by necessity do not approach the magnitude of a clinical septic shock situation (often no fall in blood pressure is observed), they offer the distinct advantage of gathering data on human responses.[93] In these investigations, TNF was shown to increase 10- to 100-fold (ca. 300 pg/ml) within 90 min of injection,[94] and IL-1 levels doubled (ca. 70 pg/ml) within 3 h.[95] IL-6 reached 11 ng/ml at 2 h following infusion,[88] and interferon activity increased eightfold by 3 h postinfusion.[96]

Both bioassays and immunoassays have been employed to determine cytokine levels in clinical episodes of septic shock.[95,97,98] Several reports indicate that elevated circulating TNF levels often correlate with severity of disease or fatal outcome. Although immunoassays are theoretically more specific, increased lipid, immunoglobulin, and acute phase protein concentrations in serum from septic individuals can introduce nonspecific interferences in immunoassays as well as bioassays.[99] Also, immunoassays may detect biologically inert breakdown products of cytokines. The fact that elevated TNF levels in sepsis have been measured by both bioassay and immunoassay reduces concerns about these problems. Measurements of whole plasma or serum by radioimmunoassays sometimes yield results that would suggest that other cytokines (IL-1, IFN-γ) also show a positive correlation with severity of disease or mortality. But when interfering plasma factors have been removed from plasma by chromatography or extraction, the IL-1 levels measured either by bioassay[100] or immunoassay[95,98] correlated with survival. Likewise IFN-γ measured by ELISA was not elevated in meningococcal septic shock.[101]

Measuring blood concentrations alone is probably not sufficient to adequately assess the status of a particular cytokine *in vivo*. Michie et al.[102] have shown that during a continuous infusion of recombinant TNF into humans, the TNF can only be detected in the circulation for the first few hours. Apparently clearance rates for this cytokine increase to the point that it is eliminated from the circulation as fast as it is infused. It may, therefore, be necessary to evaluate TNF flux (by measuring TNF appearance in urine or breakdown products) to get a more accurate indication of its role in septic shock and inflammation.

In a sterile model of tissue trauma (eccentric exercise-induced muscle damage), IL-1β was elevated in the circulation for less than 24 h, although the metabolic responses to the muscle damage, increased plasma CK and urinary 3 methylhistidine, peaked at 5 and 12 d postdamage, respectively. Immunohistochemical analysis of muscle biopsies indicated that IL-1β was localized within the muscle tissue for at least 5 d,[103] another indication that circulating cytokine levels do not fully define cytokine status.

REGULATION OF CYTOKINES: POSITIVE AND NEGATIVE FEEDBACK

Based on data from *in vitro* experimentation, a number of regulatory feedback mechanisms can be postulated for cytokine production and action. It is important to emphasize, however, that some of these mechanisms are purely theoretical. It is not known if the concentrations used in some cell culture experiments are actually attained *in vivo* or if clearance mechanisms and other factors not present *in vitro* are important influences. Furthermore, transformed cell lines may have different receptor characteristics. The types of regulation that seem feasible include feedback of (a) cytokine-induced end-stage mediators such as prostaglandin E_2 (PGE_2), (b) cytokine-induced acute phase proteins, (c) cytokine-neuropeptide interactions, and (d) cytokine-induced cytokines.

Cytokine-Induced Mediators

Addition of cyclooxygenase inhibitors to mononuclear cell cultures augments production of cytokines in response to endotoxin and other stimuli. PGE_2 has been identified as an inhibitor of a post-transcriptional step in IL-1 production;[10] therefore, when endotoxin-induced PGE_2 production is blocked, IL-1 production is enhanced. PGE_2, therefore, mediates a negative feedback loop; leukotriene B_4, on the other hand, enhances IL-1 and TNF production[104] and therefore may mediate a positive feedback loop.

Cytokine Regulation by Acute Phase Proteins

The mechanisms by which acute phase proteins modulate cytokine activity may be as numerous as the acute phase proteins themselves, but this discussion will be limited to three examples. IL-1 stimulates hepatic synthesis of C reactive protein (CRP), which in turn binds to damaged tissue or fragments. The CRP/tissue complex activates macrophages, and IL-1 secretion is increased; thus, a positive feedback loop drives cytokine production as long as damaged tissue is present.[105]

IL-1, TGFβ, and IL-6 are bound in the circulation by α_2 macroglobulin.[106-108] IL-1 retains biological activity in this bound state.[109] Therefore, the binding of 17 kD IL-1 to 450 kD α_2 macroglobulin would be expected to reduce its clearance by renal filtration and thus prolong its biological effectiveness in the circulation. Soluble receptors for TNF have been reported that inhibit TNF bioactivity[110] and may thus serve as counterregulatory factors.

As IL-1 and TNF stimulate release of proteases and oxygen radicals by phagocytic cells, they also, along with IL-6, stimulate hepatic production of antiproteases[111] and intracellular oxygen radical scavengers.[112] These acute phase proteins are presumed to help protect host tissue from damage and limit the action of proteases to the site of infection.

Cytokine-Neuropeptide Interactions

IL-1 induces secretion of ACTH from the pituitary, which in turn stimulates adrenal secretion of corticosteroids.[113] Whether IL-1 acts directly on the pituitary or indirectly via stimulation of hypothalamic CRF secretion is controversial.[114] ACTH directly inhibits macrophage cytotoxicity, but more importantly, ACTH-induced glucocorticoids have potent inhibitory actions on many aspects of the inflammatory response, including production of IL-1, IL-2, TNF, and IFNγ.[115] In addition, glucocorticoids down regulate TNF receptors[116] and up regulate IL-1 receptors.[117] TNF also causes increases in the levels of these hormones, when infused in animals, but this may be an indirect effect; TNF does not act directly on pituitary cells in culture.[118]

Neuropeptides released within an inflammatory site by nerve terminals modulate cytokines. Substance P and substance K induce IL-1, TNF, and IL-6 production.[119] Neurotensin enhances endotoxin-stimulated IL-1 production[120] and augments

M-CSF-induced colony formation.[121] In addition, neurotensin directly increases vasodilation, vascular permeability, degranulation, chemotaxis, and phagocytosis.[122]

There is the possibility that some neuropeptides may be produced within an inflammatory site by immune cells, based on evidence that mRNA for pro-opiomelanocorticotrophic (POMC) peptide is present in macrophages.[123] The POMC peptide is the precursor for ACTH, β endorphin, and α melanocyte-stimulating hormone (αMSH). αMSH inhibits fever,[124] a wide range of other actions of IL-1,[125,126] C5a-mediated neutrophil migration,[127] and TNF-induced fever and cytotoxicity.[128]

Cytokine-Induced Cytokines

Certain biological activities originally thought to be caused directly by IL-1 and TNF have, upon closer investigation, been found to be mediated by "secondary" cytokines induced by IL-1 and TNF (which might be termed primary cytokines for the purposes of this discussion). Examples of IL-1 activity that now appear to be caused by IL-1-induced secondary cytokines include chemotaxis, which is mediated by IL-8,[60] and smooth muscle cell proliferation mediated by platelet-derived growth factor (PDGF).[129] The induction of secondary cytokines by primary cytokines would provide a mechanism for amplification. On the other hand, some secondary cytokines (e.g., TGFβ) inhibit the action of primary cytokines such as IL-1 (vascular adhesion), thus providing a mechanism of negative feedback.

CLINICAL INTERVENTIONS IN SEPTIC SHOCK

Inhibition of Cytokines

Patients lapsing into septic shock receive pressors, ventilation, and other interventions to temporarily support arterial pressure, respiration, and other individual consequences of septic shock, but these measures do not treat the underlying cause(s). It has often been stated that the only truly effective treatment is to somehow avoid the onset of septic shock in the first place.[1] Short of that, attempts have been made to neutralize bacterial toxins before serious injury has resulted. Clinical trials using antiserum to J5, a nontoxic core glycolipid form of endotoxin with epitopes common to most other forms of endotoxin, have shown some increases in survival rates.[130] However, septic shock in general and cytokine production in particular can be induced by a large number of other infectious organisms, apart from Gram negative bacteria.

The comprehensive inhibition of cytokine production and action by corticosteroids *in vitro* would seemingly make them strong candidates for septic shock therapy. Indeed, pretreatment of mice with dexamethasone does reduce both serum IL-1[131] and serum TNF[132] responses to infectious challenge. Over the years, various therapeutic regimens involving corticosteroids, alone or in combination with vasopressors or anticoagulants, have been reported to increase short-term survival when tested in animal models of shock. However, when a group of researchers were brought together to perform their procedures in dogs under standardized conditions, and survival was followed for 7 d, none of the 15 therapies tested resulted in significantly increased survival rates compared to the untreated control group.[133] Likewise, corticosteroid treatment of humans provides no benefit when compared to placebo in double-blind, placebo-controlled trials.[134,135]

Treatment with ibuprofen blocked or reversed much of the cardiovascular and pulmonary effects of IL-1 and TNF in a rabbit model.[53] Ibuprofen administration to healthy human subjects significantly reduced cardiovascular and metabolic responses to endotoxin.[136] Other nonsteroidal anti-inflammatory drugs with more specific inhibitory actions on individual eicosinoid mediators have reduced lethality in animal models of septic shock (reviewed in Reference 8). Their usefulness in clinical septic shock has yet to be established.

Treatment with Cytokines

Most of this review has been devoted to enumerating the pathological actions of cytokines in septic shock. However, as pointed out in the section describing circulating cytokines, there is evidence that the ability to produce IL-1 is associated with an increased chance of survival.[95,98,100]

Van der Meer et al.[137] have shown that preteatment of neutropenic mice with a single, low dose (8 ng per mouse) of IL-1 provided significant protection against lethal *Pseudonomas* infection. IL-1 appears to be 10 to 100-times more potent in providing protection than IL-6 or TNF.[138] IL-1 and TNF have been shown to provide protection against other challenges, such as lethal radiation, as well.[139]

IL-2 administration to immunosuppressed mice (due to burn injury) reduced the mortality of the animals after cecal ligation and puncture.[140] In another form of trauma-induced immunosuppression, hemorrhagic shock, IFNγ restored the efficacy of antibiotic treatment of staph infection,[141] and GM-CSF therapy restored neutrophil fuction in AIDS patients.[142] These observations indicate a potential usefulness for cytokine therapy when a patient is immunodeficient and at greater risk of infection and subsequent development of septic shock. In other words, cytokines may offer a potential treatment to avoid the onset of sepsis and septic shock.

ACKNOWLEDGMENTS

I am grateful to Dr. Jeffrey Gelfand for critically reviewing this paper and Susan Coelho for typing the manuscript.

REFERENCES

1. Harris, R.L., D.M. Musher, K. Bloom, J. Gathe, L. Rice, B. Sugerman, T.W. Williams, Jr., and E.J. Young, Manifestations of sepsis. *Arch. Intern. Med.*, 147:1895, 1987.

2. Sanford, J.P., Epidemiology and overview of the problem, in *Septic Shock*, R.K. Root and M.A. Sande, Eds., Churchill Livingstone, New York, 1985, 1–11.
3. Zimmerman, J.J. and K.A. Dietrich, Current perspectives on septic shock. *Pediatr. Clin. North Amer.*, 34:131, 1987.
4. McGowan, J.E., Jr., M.W. Barnes, and M. Finland, Bacteremia at Boston City Hospital: occurrence and mortality during 12 selected years (1935–1972) with special reference to hospital-acquired cases. *J. Infect. Dis.*, 132:316, 1975.
5. Hansbrough, J.F., R.L. Zapata-Sirvent, and V.M. Peterson, Immunomodulation following burn injury. *Surg. Clin. North Amer.*, 67:69, 1987.
6. Morrison, D.C. and J.L. Ryan, Endotoxins and disease mechanisms. *Ann. Rev. Med.*, 38:417, 1987.
7. Halliwell, B. A radical approach to human disease. Oxygen radicals and tissue injury. 139: 1988.
8. Feuerstein, G. and J.M. Hallenbeck, Prostaglandins, leukotrienes and platelet activating factor in shock. *Ann. Rev. Pharmacol. Toxicol.*, 27:301, 1987.
9. Burke, J.F., personal communication.
10. Dinarello, C.A., Interleukin-1 and its biologically related cytokines. *Adv. Immunol.*, 44:153, 1989.
11. Beutler, B. and A. Cerami, The endogenous mediator of endotoxic shock. *Clin. Res.*, 35:192, 1987.
12. Steinmuller, D., J.D. Tyler, M.E. Snider, R.L. Noble, B.L. Riser, H.F. Maassat, and S.J. Gallie, Tissue destruction resulting from the interaction of cytotoxic T cells and their targets. *Ann. N.Y. Acad. Sci.*, 532:106, 1988.
13. Plate, J.M.D., T.L. Lukaszewska, G. Bustamente, and R.L. Hayes, Cytokines involved in the generation of cytotoxic effector T lympohcytes. *Ann. N.Y. Acad. Sci.*, 532:149, 1988.
14. Root, R.K. and M.A. Sande, Eds., *Septic Shock*, Churchill Livingstone, New York, 1985.
15. Seeger, W. and H.G. Lasch, Septic lung, in *Perspectives on Bacterial Pathogenesis and Host Defense*, B. Urbaschek, Ed., University of Chicago Press, Chicago, 1988, 140–149.
16. Beisel, W.R. and R.W. Wannemacher, Gluconeogenesis, ureagenesis and ketogenesis during sepsis. *J. Parenter. Enter. Nutr.*, 4:277, 1980.
17. Filkins, J.P., Monokines and the metabolic pathophysiology of septic shock. *Fed. Proc.*, 44:300, 1985.
18. Gallin, J.I., D. Kaye, and W.M. O'Leary, Serum lipids in infection. *N. Engl. J. Med.*, 281:1081, 1969.
19. Frayn, K.N., Hormonal control of metabolism in trauma and sepsis. *Clin. Endocrinol.*, 24:577, 1986.
20. Lubran, M.M., Bacterial toxins. *Ann. Clin. Lab. Sci.*, 18:58, 1988.
21. Natanson, C., R.L. Danner, R.J. Elin, J.M. Hosseini, K.W. Peart, S.M. Banks, T.J. MacVittie, R.I. Walker, and J.E. Parrillo, Role of endotoxemia in cardiovascular dysfunction and mortality. Escherichia coli and staphylococcus aureus challenges in a canine model of human septic shock. *J. Clin. Invest.*, 83:243, 1989.
22. Kumakura, S., I. Kano, and S. Tsurufuji, Role of bradykinin in the vascular permeability response induced by carrageenin in rats. *Br. J. Pharmacol.*, 93:739, 1988.
23. Proud, A.P. and A.P. Kaplan, Kinin formation: mechanisms and role in inflammatory disorders. *Ann. Rev. Immunol.*, 6:49, 1988.

24. Rossi, V., F. Breviario, P. Ghezzi, E. Dejana, and A. Mantovani, Prostacyclin synthesis induced in vascular cells by interleukin-1. *Science,* 229:174, 1985.
25. Chang, S.W., C.O. Feddersen, P.M. Henson, and N.F. Voelkel, Platelet activating factor mediates hemodynamic changes and lung injury in endotoxin-treated rats. *J. Clin. Invest.,* 79:1498, 1987.
26. Harlan, J.M. and R.K. Winn, The role of phospholipase products in the pathogenesis of vascular injury in sepsis, in *Septic Shock,* R.K. Root and M.A. Sande, Eds., Churchill Livingstone, New York, 1985, 83–104.
27. Nathan, C.F., Secretory products of macrophages. *J. Clin. Invest.,* 79:319, 1987.
28. Ryan, J.L., Microbial factors in pathogenesis: lipopolysaccharides, in *Septic Shock,* R.K. Root and M.A. Sande, Eds., Churchill Livingstone, New York, 1985, 13–25.
29. Goldstein, I.M., Host factors in pathogenesis: the complement system — potential pathogenic role in sepsis, in *Septic Shock,* R.K. Root and M.A. Sande, Eds., Churchill Livingstone, New York, 1985, 41–60.
30. Ferrante, A., M. Nandoskar, E.J. Bates, D.H.B. Goh, and L.J. Beard, Tumour necrosis factor beta (lymphotoxin) inhibits locomotion and stimulates the respiratory burst and degranulation of neutrophils. *Immunology,* 63:507, 1988.
31. Bussolino, F., F. Brevario, C. Tretta, A. Aglietta, A. Mantovani, and E. Dejana, Interleukin 1 stimulates platelet activating factor production in cultured human endothelial cells. *J. Clin. Invest.,* 77:2027, 1986.
32. Coleman, R.W., D.N. Flores, R.A. DeLaCadena, C.F. Scott, and L. Cousens, Recombinant alpha-1-antitrypsin Pittsburgh attenuates experimental Gram negative septicemia. *Am. J. Pathol.,* 130:418, 1988.
33. Gelfand, J.A., M. Donelan, A. Hawiger, and J.F. Burke, Alternative complement pathway activation increases mortality in a model of burn injury in mice. *J. Clin. Invest.,* 70:1170, 1982.
34. Tracey, K.J., Y. Fong, D.G. Hesse, K.R. Manogue, A.T. Lee, G.C. Kuo, S.F. Lowry, and A. Cerami, Anti-cachectin/TNF monoclonal antibodies prevent septic shock during lethal bacteremia. *Nature,* 330:662, 1987.
35. Okusawa, S., C.A. Dinarello, K.B. Yancey, S. Endres, T.J. Lawley, M.M. Frank, J.F. Burke, and J.A. Gelfand, C5a induction of human interleukin-1. Synergistic effect with endotoxin or interferon-γ. *J. Immunol.,* 139:2635, 1987.
36. Berger, M., E.M. Wetzler, and R.S. Wallis, Tumor necrosis factor is the major monocyte product that increases complement receptor expression on mature human neutrophils. *Blood,* 71:151, 1988.
37. Bevilacqua, M.P., J.S. Pober, G.R. Majeau, W.Fiers, R.S. Cotran, and M.A. Gimbrone, Jr., Recombinant tumor necrosis factor induces procoagulant activity in cultured human vascular endothelium: characterization and comparison with interleukin-1. *Proc. Natl. Acad. Sci. U.S.A.,* 83:4533, 1986.
38. Michalek, S.M. and R.N. Moore, The primary role of lymphoreticular cells in the mediation of host responses to bacterial endotoxin. *J. Infect. Dis.,* 141:55, 1980.
39. Tracey, K.J., B. Beutler, S.F. Lowry, and A. Cerami, Shock and tissue injury induced by recombinant human cachectin. *Science,* 234:470, 1986.
40. Weinberg, J.R., D.J.M. Wright, and A. Guz, Interleukin-1 and tumor necrosis factor cause hypotension in the conscious rabbit. *Clin. Sci.,* 75:251, 1988.
41. Beutler, B., I.W. Milsark, and A.C. Cerami, Passive immunization against cachectin/tumor necrosis factor protects mice from lethal effect of endotoxin. *Science,* 229:869, 1985.

42. Mathison, J.C., E. Wolfson, and R.J. Ulevitch, Participation of tumor necrosis factor in the mediation of Gram negative bacterial lipopolysaccharide-induced injury in rabbits. *J. Clin. Invest.*, 81:1925, 1988.
43. Kiener, P.A. F. Marek, G. Rodgers, P.F. Lin, G. Warr, and I. Desiderio, Induction of tumor necrosis factor, IFN-γ and acute lethality in mice by toxic and non-toxic forms of lipid A. *J. Immunol.*, 141:870, 1988.
44. Ognibene, F.P., S.A. Rosenberg, M. Lotze, J. Skibber, M.M. Parker, J.H. Shelhammer, and J.E. Parrillo, Interleukin-2 administration causes reversible hemodynamic changes and left ventricular dynfunction similar to those seen in septic shock. *Chest*, 94:750, 1988.
45. Gaynor, E.R., L. Vitek, L. Sticklin, S.P. Creekmore, M.E. Ferrara, J.X. Thomas, Jr., S.G. Fisher, and R.I. Fisher, The hemodynamic effects of treatment with interleukin-2 and lymphokine-activated killer cells. *Ann. Intern. Med.*, 109:953, 1988.
46. Salby, P., S. Hobbs, C. Viner, E. Jackson, A. Jones, D. Newell, A.H. Calvert, T.McElwain, K. Fearon, J. Humphreys, and T. Shiga, Tumour necrosis factor in man: clinical and biological observations. *Br. J. Cancer*, 56:803, 1987.
47. Creagan, E.T., J.S. Kovach, C.G. Moertel, S. Frytak, and L.K. Kvols, A phase I clinical trial of recombinant human tumor necrosis factor. *Cancer*, 62:2467, 1988.
48. Perez, R., A. Lipton, H.A. Harvey, M.A. Simmonds, P.J. Romano, S.L. Imboden, and G. Giudice, A phase I trial of recombinant human gamma interferon in patients with advanced malignancy. *J. Biol. Response Mod.*, 7:309, 1988.
49. Ferro, T.J., A. Johnson, J. Everitt, and A.S. Malik, IL-2 induces pulmonary edema and vasoconstriction independent of circulating lymphocytes. *J. Immunol.*, 142:1969, 1989.
50. Frasier-Scott, K., H. Hatzakis, D. Seong, C.M. Jones, and K.K. Wu, Influence of natural and recombinant interleukin-2 on endothelial cell arachidonate metabolism. *J. Clin. Invest.*, 82:1877, 1988.
51. Mier, J.W., G. Vachino, J.W.M. van der Meer, R.P. Numerof, S. Adams, J.G. Cannon, H.A. Bernheim, M.B. Atkins, D.R. Parkinson, and C.A. Dinarello, Induction of circulating tumor necrosis factor as the mechanism for the febrile response to interleukin-2 in cancer patients. *J. Clin. Immunol.*, 8:426, 1988.
52. Jablons, D.M., J.J. Mule, J.K. McIntosh, P.B. Sehgal, L.T. May, C.M. Huang, S.A. Rosenberg, and M.T. Lotze, IL-6/IFN-β-2 as a circulating hormone. Induction by cytokine administration in humans. *J. Immunol.*, 142:1542, 1989.
53. Okusawa, S., J.A. Gelfand, T. Ikejima, R.J. Connolly, and C.A. Dinarello, Interleukin-1 induces a shock-like state in rabbits. Synergism with tumor necrosis factor and the effect of cyclooxygenase inhibition. *J. Clin. Invest.*, 81:1162, 1988.
54. Talmadge, J.E., O. Bowersox, H. Tribble, S.H. Lee, H.M. Shepard, and D. Liggitt, Toxicity of tumor necrosis factor is synergistic with γ interferon and can be reduced with cycloxygenase inhibitors. *Am. J. Pathol.*, 128:410, 1987.
55. Aggarwal, B.B., T.E. Eessalu, and P.E. Hass, Characterization of receptors for human tumour necrosis factor and their regulation by gamma-interferon. *Nature*, 318:665, 1985.
56. Holtman, H. and D. Wallach, Down regulation of the receptors for tumor necrosis factor by interleukin-1 and 4 beta-phorbol-12-myristate-13-acetate. *J. Immunol.*, 139:1161, 1987.
57. Libby, P., J.M. Ordovas, K.R. Auger, A.H. Robbins, L.K. Kiring, and C.A. Dinarello, Endotoxin and TNF induce IL-1 gene expression in adult human vascular endothelial cells. *Am. J. Pathol.*, 124:179, 1986.

58. Jirik, F.R., T.J. Podor, T. Hirano, T. Kishimoto, D.J. Loskutoff, D.A. Carson, and M. Lotz, Bacterial lipopolysaccharide and inflammatory mediators augment IL-6 secretion by human endothelial cells. *J. Immunol.*, 142:144, 1989.
59. Warner, S.J.C. and P. Libby, Human vascular smooth muscle cells. Target for and source of tumor necrosis factor. *J. Immunol.*, 142:100, 1989.
60. Larsen, C.G., A.J. Anderson, E. Appella, J.J. Oppenheim, and K. Matsushima, The neutrophil-activating protein (NAP-1) is also chemotactic for T lymphocytes. *Science*, 243:1464, 1989.
61. Strieter, R.M., S.L. Kunkel, H.J. Showell, D.G. Remick, S.H. Phan, P.A. Ward, and R.M. Marks, Endothelial cell gene expression of a neutrophil chemotactic factor by TNF-α, LPS and IL-1β. *Science*, 243:1467, 1989.
62. Pober, J.S., Activation of vascular endothelium by tumor necrosis factor, in *Tumor Necrosis Factor/Cachectin and Related Cytokines*, Karger, Basel, 1988, 74–81.
63. Moser, R., B. Schleiffenbaum, P. Groscurth, and J. Fehr, Interleukin-1 and tumor necrosis factor stimulate human vascular endothelial cells to promote transendothelial neutrophil passage. *J. Clin Invest.*, 83:444, 1989.
64. Nachman R.L., K.A. Hajjar, R.L. Solverman, and C.A. Dinarello, Interleukin-1 induces endothelial cell synthesis of plasminogen activator inhibitor. *J. Exp. Med.*, 163:1595, 1986.
65. van Hinsbergh, V.W.M., T. Kooistra, E.A. van den Berg, H.M.G. Pricen, W. Fiers, and J.J. Emeis, Tumor necrosis factor increases the production of plasminogen activator inhibitor in human endothelial cells in vitro and in rats in vivo. *Blood*, 72:1467, 1988.
66. Suffredini, A.F., P.C. Harpel, and J.E. Parrillo, Promotion and subsequent inhibition of plasminogen activation after administration of intravenous endotoxin to normal subjects. *N. Engl. J. Med.*, 320:1165, 1989.
67. Lewis, M.S., R.E. Whatley, P. Cain, T.M. McIntyre, S.M. Prescott, and G.A. Zimmerman, Hydrogen peroxide stimulates the synthesis of platelet activating factor by endothelium and induces endothelial cell-dependent neutrophil adhesion. *J. Clin. Invest.*, 82:2045, 1988.
68. Cannon, J.G., B.D. Clark, P. Wingfield, V. Schmeissner, C. Losberger, C.A. Dinarello, and A.R. Shaw, Rabbit IL-1: cloning, expression, biologic properties and transcription during endotoxemia. *J. Immunol.*, 142:2299, 1989.
69. McKenna, T.M., D.W. Reusch, and C.O. Simpkins, Macrophage conditioned medium and interleukin-1 suppress vascular contractility. *Circ. Shock*, 25:187, 1988.
70. Beasley, D., R.A. Cohen, and N.G. Levinsky, Interleukin-1 inhibits contraction of vascular smooth muscle. *J. Clin. Invest.*, 83:331, 1989.
71. Stern, L. and G. Kunos, Synergistic regulation of pulmonary β-adrenergic receptors by glucocorticoids and interleukin-1. *J. Biol. Chem.*, 263:15876, 1988.
72. Deblois, D., J. Bouthillier, and F. Marceau, Effect of glucocorticoids, monokines and growth factors on the spontaneously developing responses of the rabbit isolated aorta to dis-Arg-bradykinin. *Br. J. Pharmacol.*, 93:969, 1988.
73. Aoki, N., M. Siegfried, and A.M. Lefer, Anti-EDRF effect of tumor necrosis factor in isolated, perfused cat carotid arteries. *Am. J. Physiol.*, 256:H1509, 1989.
74. Raajmakers, J.A.M., C. Beneker, E.C.G. van Geffen, T.M.H. Meisters, and P. Power, Inflammatory mediators and β-adrenoceptor. *Agents Actions*, 26:45, 1989.
75. Smith, L.W., S.L. Winbery, L.A. Barker, and K.H. McDonough, Cardiac function and chronotropic sensitivity to β-adrenergic stimulation in sepsis. *Am. J. Physiol.*, 251:H405, 1986.

76. Parrillo, J.E., C. Burch, J.H. Shelhamer, M.M. Parker, C. Natanson, and W. Schuette, A circulating myocardial depressant substance in humans with septic shock. *J. Clin. Invest.,* 76:1539, 1985.
77. Hollenberg, S.M., R.E. Cunnion, M. Lawrence, J.L. Kelly, and J.E. Parrillo, Tumor necrosis factor depresses mycardial cell function: results using an *in vitro* assay of myocyte performance. *Clin. Res.,* 37:528A, 1989.
78. Stein, D.G., P.A. Sobta, and P.A. Scanlon, The effects of interleukin-2 on the isolated, perfused, working rat heart. *Chest,* 92:915, 1987.
79. Ghezzi, P., R. Bertini, M. Bianchi, A. Erroi, P. Villa, and A. Mantovani, Interleukin-1 and tumor necrosis factor depress cytochrome P450-dependent lever drug metabolism in mice, in *Monokines and Other Non-lymphocyte Cytokines,* M.C. Powanda, J.J. Oppenheim, M.J. Kluger, and C.A. Dinarello, Eds., Alan R. Liss, New York, 1988, 337–342.
80. Beasley, D., C.A. Dinarello, and J.G. Cannon, Interleukin-1 induces natriuresis in concious rats: role of renal prostaglandins. *Kidney Int.,* 33:1059, 1988.
81. Tredget, E.E., Y.M. Yu, S. Shong, R. Burini, S. Okusawa, J.A. Gelfand, C.A. Dinarello, V.R. Young, and J.F. Burke, Role of interleukin-1 and tumor necrosis factor on energy metabolism in rabbits. *Am. J. Physiol.,* 255:E760, 1988.
82. Bessey, P.Q., J.M. Watters, T.T. Aoki, and D.W. Wilmore, Combined hormonal infusion simulates the metabolic response to injury. *Ann. Surg.,* 200:264, 1984.
83. Spinas, G., T. Mandrup-Poulsen, J. Molvig, L. Back, K. Bendtzen, C.A. Dinarello, and J. Nerup, Low concentrations of interleukin-1 stimulate and high concentrations inhibit insulin release from isolated islets of Langerhans. *Acta Endocrinol.,* 113:551, 1986.
84. Bernton, E.W., J.E. Beach, J.W. Holaday, R.C. Smallridge, and H.G. Fein, Release of multiple hormones by a direct action of interleukin-1 on pituitary cells. *Science,* 238:519, 1987.
85. Roh, M.S., L.L. Moldawer, L.G. Ekman, C.A. Dinarello, B.R. Bistrian, M. Jeevanandam, and M.F. Brennan, Stimulatory effect of interleukin-1 upon hepatic metabolism. *Metabolism,* 35:419, 1986.
86. Meszaros, K., C.H. Lang, G.J. Bagby, and J.J. Spitzer, Tumor necrosis factor increases in vivo glucose utilization of macrophage-rich tissues. *Biochem. Biophys. Res. Commun.,* 149:1, 1987.
87. Flores, E.A., B.R. Bistrian, J.J. Pomposalli, C.A. Dinarello, G.L. Blackburn and N.W. Istfan, Infusion of tumor necrosis factor/cachectin promotes muscle catabolism in the rat. A synergistic effect with interleukin-1. *J. Clin. Invest.,* 83:1614, 1989.
88. Fong, Y., L.L. Moldawer, M. Marano, H. Wei, S.B. Tatter, R.H. Clarick, U. Santhanam, D. Sherris, L.T. May, P.B. Sehgal, and S.F. Lowry, Endotoxemia elicits increased circulating β_2-IFN/IL-6 in man. *J. Immunol.,* 142:2321, 1989.
89. Moldawer, L.L., G. Svaninger, J. Gelin, and K.G. Lundholm, Interleukin-1 and tumor necrosis factor do not regulate protein balance in skeletal muscle. *Am. J. Physiol.,* 253:C766, 1987.
90. Nawabi, M.D., K.P. Block, M.C. Chakrabarti, et al., Administration of endotoxin, tumor necrosis factor or interleukin-1 activates skeletal muscle branched-chain α-keto acid dehydrogenase. *J. Clin. Invest.,* 85:256, 1990.
91. Hesse, D.G., K.J. Tracey, Y. Fong, K.R. Mansgue, M.A. Palladino, Jr., A. Cerami, G.T. Shires, and S.F. Lowry, Cytokine appearance in human endotoxemia and primate bacteremia. *Surg. Gynecol. Obstet.,* 166:147, 1988.
92. Metcalf, D., Granulocyte-macrophage colony stimulating factors. *Science,* 229:16, 1985.

93. Wolff, S.M., Biological effects of endotoxin in man. *J. Infect. Dis.*, 128:S259, 1973.
94. Michie, H.R., K.R. Manogue, D.R. Spriggs, A. Revhaug, S. O'Dwyer, C.A. Dinarello, A. Cerami, S.M. Wolff, and D.W. Willmore, Detection of circulating tumor necrosis factor after endotoxin administration. *N. Engl. J. Med.*, 318:1481, 1988.
95. Cannon, J.G., R.G. Tompkins, J.A. Gelfand, H.R. Michie, G.G. Stanford, J.W.M. van der Meer, S. Endres, G. Lonnemann, J. Corsetti, B. Chernow, D.W. Wilmore, S.M. Wolfe, J.F. Burke, and C.A. Dinarello, Circulating interleukin-1 and tumor necrosis factor in septic shock and experimental endotoxin fever. *J. Infect. Dis.*, 161:79, 1990.
96. Sauter, C. and C. Wolfensberger, Interferon in human serum after injection of endotoxin. *Lancet*, 2:852, 1980.
97. Waage, A., A. Halstensen, and T. Espevik, Association between tumor necrosis factor in serum and fatal outcome in patients with meningococcal disease. *Lancet*, 1:355, 1987.
98. Casey, L., R., Balk, S. Simpson, et al., Plasma tumor necrosis factor, interleukin-1 beta, and endotoxin in patients with sepsis, in *The Physiological and Pathological Effects of Cytokines*, Dinarello, C.A., M.J. Kluger, M.C. Powanda, and J.J. Oppenhiem, Eds., Wiley-Liss, New York, 1990.
99. Cannon, J.G., J.W.M. van der Meer, D. Kwiatkowski, S. Endres, G. Lonnemann, J.F. Burke, and C.A. Dinarello, Interleukin-1β in human plasma: optimization of blood collection, plasma extraction and radioimmunoassay methods. *Lymphokine Res.*, 7:457, 1988.
100. Luger, A., H. Graf, H.P. Schwarz, H.K. Stummvoll, and T.A. Luger, Decreased serum interleukin-1 activity and monocyte interleukin-1 production in patients with fatal sepsis. *Crit. Care Med.*, 1:458, 1986.
101. Waage, A., P. Brandtzaeg, A. Halstensen, P. Kierulf, and T. Espevik, The complex pattern of cytokines in serum from patients with meningoccal septic shock. *J. Exp. Med.*, 169:333, 1989.
102. Michie, H.R., D.R. Spriggs, K.R. Manogue, M.L. Sherman, A. Revhaug, S.T. O'Dwyer, K. Arthur, C.A. Dinarello, A. Cerami, S.M. Wolff, D.W. Kufe, and D.W. Wilmore, Tumor necrosis factor and endotoxin induce similar metabolic responses in human beings. *Surgery*, 104:280, 1988a.
103. Cannon, J.G., R.A. Fielding, M.A. Fiatarone, S.F. Orencole, C.A. Dinarello, and W.J. Evans, Increased interleukin-1β in human skeletal muscle following exercise. *Am. J. Physiol.*, 257:R451, 1989.
104. Gagnon, L., L.G. Filion, C. Dubois, and M. Rola-Pleszczynski, Leukotrienes and macrophage activation: augmented cytotoxic activity and enhanced interleukin-1, tumor necrosis factor and hydrogen peroxide production. *Agents Actions*, 26:141, 1989.
105. Yamada, Y., K. Kimball, S. Okusawa, et al., Cytokines, acute phase proteins and tissue injury: C-reactive protein opsinizes dead cells for debridement and stimulates cytokine production, *Ann. N.Y. Acad. Sci.*, 587:351, 1990.
106. Teodorescu, M., J.L. Skosey, C. Schlesinger, and J. Wallman, Covalent disulfide binding of IL-1 to α_2 macroglobulin, in *Monokines and Other Non-Lymphocytic Cytokines*, M.C. Powanda, J.J. Oppenhiem, M.J. Kluger, and C.A. Dinarello, Eds., A.R. Liss, New York, 1988, 209–212.
107. O'Connor-McCourt, M.D. and L.M. Wakefield, Latent transforming growth factor β in serum. A specific complex with α_2 macroglobulin. *J. Biol. Chem.*, 262:14090, 1987.
108. Matsuda, T., T. Hirano, S. Nagasawa, and T. Kishimoto, Identification of alpha 2-macroglobulin as carrier protein for IL-6. *J. Immunol.*, 142:148, 1989.

109. Barth, W. and T.A. Luger, Indentification of α_2 macroglobulin as a cytokine binding plasma protein. *J. Biol. Chem.*, 264:5818, 1989.
110. Englemann, H., D. Novick, and D. Wallach, Two tumor necrosis factor-binding proteins purified from human urine. Evidence for immunological cross-reactivity with cell surface tumor necrosis factor receptors. *J. Biol. Chem.*, 265:1351, 1990.
111. Dickinson, A.M., B.K. Shenton, A.H. Alomran, P.K. Donnelly, and S.J. Proctor, Inhibition of natural killing and antibody-dependent cell-mediated cytotoxicity by the plasma protease inhibitor α_2 macroglobulin ($\alpha 2M$) and $\alpha 2M$ protease complexes. *Clin. Immunol. Immunopath.*, 36:259, 1985.
112. Wong, G.H.W. and K.J. Goeddel, Induction of manganous superoxide dismutase by tumor necrosis factor: possible protective mechanism. *Science*, 242:941, 1988.
113. Besedovsky, H., A. del Rey, E. Sorkin, and C.A. Dinarello, Immunoregulatory feedback between interleukin-1 and glococorticoid hormones. *Science*, 233:652, 1986.
114. Lumpkin, M.D., The regulation of ACTH secretion by IL-1. *Science*, 238:452, 1987.
115. Munck, A., P.M. Guyre, and N.J. Holbrook, Physiological functions of glucocorticoids in stress and their relation to pharmacological actions. *Endocr. Rev.*, 5:25, 1984.
116. Kull, F.C., Jr., Reduction in tumor necrosis factor affinity and cytotoxicity by glucocorticoids. *Biochem. Biophys. Res. Commun.*, 153:402, 1988.
117. Akahoshi, T., J.J. Oppenheim, and K. Matsushima, Induction of high affinity interleukin 1 receptors on human peripheral blood lymphocytes by glucocorticoid hormones. *J. Exp. Med.*, 167:924, 1988.
118. Kehrer, P., D. Turnill, I.M. Dayer, A.F. Muller, and R.C. Gaillard, Human recombinant interleukin-1β and α, but not recombinant tumor necrosis factor alpha stimulate ACTH release from rat anterior pituitary cells in vitro in a phostaglandin E_2 and cAMP independent manner. *Neuroendocrinology*, 48:160, 1988.
119. Lotz, M., J. H Vaughn, and D.A. Carson, Effect of neuropeptides on production of inflammatory cytokines by human monocytes. *Science*, 241:1218, 1988.
120. Lemaire, I., Neutrotensin enhances IL-1 production by activated alveolar macrophages. *J. Immunol.*, 140:2983, 1988.
121. Moore, R.N., A.P. Osmand, V.A. Dunn, J.G. Joshi, J.W. Koontz, and B.T. Rouse, Neutrotensin regulation of macrophage colony-stimulating factor-stimulated in vitro myelopoiesis. *J. Immunol.*, 142:2689, 1989.
122. Goldman, R., Z. Bar Shavit, and D. Romeo, Neurotensin modulates human neutrophil locomotion and phagocytic capability. *FEBS Lett.*, 159:63, 1983.
123. Lolait, S.J., J.A. Clements, A.J. Markwick, C. Cheng, M. McNally, A.L. Smith, and J.W. Funder, Pro-opiomelanocortin mRNA and posttranslational processing of beta endorphin in spleen macrophages. *J. Clin. Invest.*, 77:1776, 1986.
124. Clark, W.G., M. Holdeman, and J.M. Lipton, Analysis of the antipyretic action of alpha-melanocyte stimulating hormone in rabbits. *J Physiol. (London)*, 359:459, 1985.
125. Cannon, J.G., J.B. Tatro, S. Reichlin, and C.A. Dinarello, α Melanocyte stimulating hormone inhibits immunostimulatory and inflammatory actions of interleukin-1. *J. Immunol.*, 137:2232, 1986.
126. Daynes, R.A., B.A. Robertson, B. Cho, D.K. Burnham, and R. Newton, Alpha melanocyte stimulating hormone exhibits target cell selectivity in its capacity to affect interleukin-1-inducible responses in vivo and in vitro. *J. Immunol.*, 139:103, 1987.
127. Mason, M.J. and D. Van Epps, Modulation of IL-1, tumor necrosis factor and C5a-mediated murine neutrophil migration by α melanocyte stimulating hormone. *J. Immunol.*, 142:1646, 1989.

128. di Giovine, F.S., Y. Townsend, W.I. Cranston, and G.W. Duff, Alpha melanocyte stimulating hormone inhibits tumor necrosis factor. *Eur. J. Clin. Invest.*, 18:A53, 1988.
129. Raines, E.W., S.K. Dower, and R. Ross, Interleukin-1 mitogenic activity for fibroblasts and smooth muscle cells is due to PDGF-AA. *Science*, 243:393, 1989.
130. Ziegler, E.J., J.A. McCutchan, J. Fierer, et al., Treatment of gram negative bacteremia and shock with human antiserum to a mutant Escherichia coli. *N. Engl. J. Med.*, 307:1225, 1982.
131. Staruch, M.J. and D.D. Wood, Reduction of serum interleukin-1-like activity after treatment with dexamethasone. *J. Leukoc. Biol.*, 37:193, 1985.
132. Zuckerman, S.H., J. Shellhaas, and L.D. Butler, Differential regulation of lipopolysaccharide-induced interleukin-1 and tumor necrosis factor synthesis: effects of endogenous and exogenous glucocorticoids and the role of the pituitary-adrenal axis. *Eur. J. Immunol.*, 19:301, 1989.
133. Hinshaw, L.B., Application of animal shock models to the human. *Circ. Shock,* 17:205, 1985.
134. Bone, R.C., C.J. Fisher, T.P. Clemmer, G.J. Slotman, C.A. Metz, and R.A. Balk, A controlled clinical trial of high dose methylprednisoline in the treatment of severe sepsis and septic shock. *N. Engl. J. Med.*, 317:653, 1987.
135. Hinshaw, L.B. and the Septic Shock Study Cooperative, Effect of high dose glucocorticoid therapy on mortality in patients with clinical signs of systemic sepsis. *N. Engl. J. Med.*, 317:659, 1987.
136. Revhaug, A., H.R. Michie, J. Mek. Mason, J.M. Watters, C.A. Dinarello, S.M. Wolff, and D.W. Wilmore, Inhibition of cyclooxygenase attenuates the metabolic response to endotoxin in humans. *Arch. Surg.*, 123:162, 1988.
137. van der Meer, J.W.M., M. Barza, S.M. Wolff, and C.A. Dinarello, A low dose of recombinant interleukin-1 protects granulocytopenic mice from lethal gram-negative infection. *Proc. Natl. Acad. Sci. U.S.A.*, 85:1620, 1988.
138. van der Meer, J.W.M., M. Helle, and L. Aarden, Comparison of the effects of recombinant interleukin 6 and recombinant interleukin-1 on nonspecific resistance to infection. *Eur. J. Immunol.,* 19:413, 1989.
139. Neta, R., S.D. Douches, and J.J. Oppenheim, Interleukin-1 is a radioprotector. *J. Immunol.*, 136:2483, 1986.
140. Gough, D.B., N.M. Moss, A. Jordan, J.T. Grbic, M.L. Rodrick, and J.A. Mannick, Recombinant interleukin-2 (rIL-2) improved immune response and host resistance to septic challenge in thermally injured mice. *Surgery,* 104:292, 1988.
141. Livingston, D.H. and M.A. Malangoni, Interferon-γ restores immune competence after hemorrhagic shock. *J. Surg. Res.*, 45:37, 1988.
142. Baldwin, G.C., J.C. Gasson, and S.G. Quan, Granulocyte-macrophage colony stimulating factor enhances neutrophil function in acquired immunodeficiency syndrome patients. *Proc. Natl. Acad. Sci. U.S.A.,* 85:2763, 1988.
143. Isner, J.M. and W.A. Dietz, Cardiovascular consequences of recombinant DNA-technology: interleukin-2. *Ann. Intern. Med.*, 109:933, 1988.
144. Boey, H., R. Rosenbaum, J. Castracane, and L. Borish, Interleukin-4 is a neutrophil activator. *J. Allergy Clin. Immunol.*, 83:978, 1989.
145. Wick, T.R., J. delCastillo, and K. Guo, In vivo hematologic effects of recombinant interleukin-6 on hematopoiesis and circulating numbers of RBCs and WBCs. *Blood,* 73:108, 1989.

146. Davatelis, G., S.D. Wolpe, D. Sherry, J.-M. Dayer, R. Chicheportiche, and A. Cerami, Macrophage inflammatory-protein-1: a prostaglandin-independent endogenous pyrogen. *Science,* 243:1066, 1989.
147. Gamble, J.R. and M.A. Vadas, Endothelial adhesiveness for blood neutrophils is inhibited by transforming growth factor β. *Science,* 242:97, 1988.
148. Wahl, S.M., D.A. Hunt, L.M. Wakefield, N. McCartney-Francis, L.M. Wahl, A.B. Roberts, and M.B. Sporn, Transforming growth factor type β induces monocyte chemotaxis and growth factor production. *Proc. Natl. Acad. Sci. U.S.A.,* 84:5788, 1987.
149. Miyajima, A., S. Miyatake, J. Schreure, J. DeVries, N.Arai, T. Yokota, and K. Arai, Coordinate regulation of immune and inflammatory responses by T cell-derived lymphokines. *FASEB J.,* 2:2462, 1988.
150. Goodwin, R.G., S. Lupton, A. Schmierer, K.J. Hjerrild, R. Jerzy, W. Clevenger, S. Gillis, D. Cosman, and A.E. Namen, Human interleukin 7: molecular cloning and growth factor activity on human and murine B-lineage cells. *Proc. Natl. Acad. Sci. U.S.A.,* 86:302, 1989.
151. Dinarello, C.A., T. Ikejima, S.J.C. Warner, S.F. Orencole, G. Lonnemann, J.G. Cannon, and P. Libby, Interleukin-1 induces interleukin-1. I. Induction of circulating interleukin-1 in rabbits in vivo and in human mononuclear cells in vitro. *J. Immunol.,* 139:1902, 1987.
152. Manson, J.C., J.A. Symons, F.S. DiGiovine, S. Poole, and G.W. Duff, Autoregulation of interleukin-1 production. *Eur. J. Immunol.,* 19:261, 1989.
153. Mizel, S.B., Interleukin 1 and T cell activation, *Immunol. Rev.,* 63:51, 1982.
154. Zucali, J.R., C.A Dinarello, M.A. Gross et al., Interleukin 1 stimulates fibroblasts to produce granulocyte-macrophage colony stimulating activity and prostaglandin E_2. *J. Clin. Invest.,* 77:1857, 1986.
155. van Damme, J., G. Opdenakker, R.J. Simpson, M.R. Rubira, S. Cayphas, A. Vink, B. Billiau, and J. van Snick, Identification of the human 26-kD protein interferon beta-2 as a B cell hybridoma/plasma cytoma growth factor conduced by IL-1 and TNF. *J. Exp. Med.,* 165:914, 1987.
156. Ikejima, T., S. Okusawa, P. Ghezzi, et al. IL-1 induces TNF in human PBMC *in vitro* and a circulating TNF-like activity in rabbits. *J. Infect. Dis.* 162:215, 1990.
157. van Damme, J., M. De Ley, G. Opdenakker, et al., Homogeneous interferon-inducing factor is related to endogenous pyrogen and interleukin-1. *Nature,* 314:266, 1985.
158. Suzuki, H., K. Shibano, M. Okane, I. Kono, Y. Matsui, K. Yamana, and H. Hashiwagi, Interferon γ modulates messenger RNA levels of C-Sis PDGF-B chain, PDGF-A chain and IL-1β genes in human vascular endothelial cells. *Am. J. Pathol.,* 134:35, 1989.
159. Oster, W., A. Lindemann, R. Mertelsmann, and F. Herrmann, Production of macrophage-, granulocyte-, granulocyte-macrophage-, and multi-colony-stimulating factor by peripheral blood cells. *Eur. J. Immunol.,* 19:543, 1989.
160. Nedwin, G.E., L.P. Svedersky, T.S. Bringman, et al., Effect of interleukin 2, interferon-gamma and mitogens on the production of tumor necrosis factors alpha and beta. *J. Immunol.,* 135:24, 1985.
161. Kawase, I., C.G. Brooks, and K. Kuribayashi, Interleukin 2 induces gamma-interferon production: participation of macrophages and NK-line cells. *J. Immunol.,* 131:288, 1983.
162. Sisson, S.D. and C.A. Dinarello, Production of interleukin-1α, interleukin-1β and tumor necrosis factor by human mononuclear cells stimulated with granulocyte-macrophage colony stimulating factor. *Blood,* 72:1368, 1988.

163. Moore, R.N., H.S. Larsen, D.W. Horohov, et al., Endogenous regulation of macrophage proliferative expansion by colony-stimulating factor-induced interferon, *Science*, 223:178, 1984.
164. te Velde, A.A., J.P.G. Klomp, B.A. Yard, J.E. deVries, and C.G. Figdor, Modulation of phenotypic and functional properties of human peripheral blood monocytes by IL-4. *J. Immunol.*, 10:1548, 1988.
165. Dinarello, C.A., J.G. Cannon, S.M. Wolff, H.A. Bernheim, B. Beutler, A. Cerami, I.S. Figari, M.A. Palladino, Jr., and J.V. O'Connor, Tumor necrosis factor (cachectin) is an endogenous pyrogen and induces production of interleukin 1. *J. Exp. Med.*, 163:1433, 1986.
166. Boraschi, D., S. Censhini, and A. Tagliabue, Interferon gamma reduces macrophage suppressive activity by inhibiting prostaglandin E_2 release and inducing interleukin-1 production, *J. Immunol.*, 133:764, 1984.
167. Ray, A., S.B. Tatter, L.T. May, and P.B. Sehgal, Activation of the human β_2-interferon/hepatocyte stimulating factor/interleukin-6 promoter by cytokines, viruses and second messenger agonists. *Proc. Natl. Acad. Sci. U.S.A.*, 85:6701, 1988.
168. Wiseman, D.M., P.J. Polverini, D.W. Kamp, and S.J. Leibovich, Transforming growth factor-beta is chemotactic for human monocytes and induces their expression of angiogenic activity. *Biochem. Biophys. Res. Commun.*, 157:793, 1988.
169. Kaye, J., S. Gillis, and S. Mizel, Growth of a cloned helper T cell line induced by a monoclonal antibody specific for antigen receptor: Interleukin 1 is required for the expression of receptors for interleukin 2. *J. Immunol.*, 133:1339, 1984.
170. Bonin, P.D. and J.P. Singh, Modulation of interleukin-1 receptor expression and interleukin-1 response in fibroblasts by platelet-derived growth factor. *J. Biol. Chem.*, 263:11052, 1988.

Index

A

A23187, 65, 100
A375 tumor cells, 75
A431 cell line, 37
AA861, 69
Abscess formation, 9
Acetaminophen, 237
Acetyl-CoA carboxylase, 314
ACTH, see Adrenocorticotropin hormone
Actin filaments, 124–125, 199
Actin nuclear RNAs, 64
Actinomycin, 76
Actinomycin D, 217, 241, 251
Activated T-cells, 4, 15
Active osteoblasts, 146
Acute fulminant viral hepatitis, 258
Acute inflammation, 2, 4, 6, 12
Acute inflammatory properties, 9
Acute inflammatory responses, 9
Acute lung injury, see Pulmonary vascular endothelial injury
Acute pathologic responses, 23
Acute phase proteins, 67, 74, 91, 275–305
  Amino acid metabolism, 289
  carbohydrate metabolism, 288–289
  corticosteroids, 280
  cytokine interaction, 290–295
  cytokine receptors, 290
  cytokine regulation of synthesis of, 279–287
  gene regulation, 287–288
  glycosylation pattern, 289
  hepatocyte response, 287–289
  inflamed tissue, 293–295
  interleukin-1, 280–281
  interleukin-6, 281-295
  kinetics of response, 288
  leukemia inhibitory factor, 280, 282–285
  metabolic changes, 288–289
  other modulating cytokines, 285–287
  role of, 277–279
  shock, 317
  signal transduction pathways, 289
  target tissue, 290–293
  tumor necrosis factor, 280–281
Acute phase reactants, 2–3, 73, 240–241, 254, 257
Acute phase response, 38, 191, 244, 255, 275–276, 295
Acute pulmonary vascular endothelial injury, see Pulmonary vascular endothelial injury
ADCC, see Antibody-dependent cell-mediated cytotoxicity
Adenosine monophosphate (AMP), 60
Adenosine triphosphate (ATP), 100
Adenylate cyclase, 100, 151
Adenylate cyclase production, 178
Adipocytic cells, 146
Adjuvant arthritis, 21, 62, 64, 68–69, 71–73, 75, 169
ADP-ribosylation, 250
Adrenal glands, 64
Adrenal-pituitary axis, 18, 294
Adrenocorticotropin hormone (ACTH), 18, 38, 64, 74, 309
Adult respiratory distress syndrome (ARDS), 191, 204–205, 216–217, 308
Agonists of second messenger pathways, 94–101
AIDS neutropenia, 21
AIDS patients, 40, 319
Alcoholic cirrhosis, 254–255, 257
Alcoholic hepatitis, 242, 256–258
Alcoholic liver disease, 256–258
Alkaline phosphatase, 155, 248
Allergy, 21
Allopurinol, 245
All-*trans*-retinoic acid, 115–116, 120, 123, 125, 131, 133–134
α-fetoprotein gene, 63
Alphanaphthylisothiocyanate, 249
Alveolar edema, 308

Alveolar macrophages, 197, 205
Alzet minipumps, 11–12
Amines, 314
Amino acid metabolism, 289
AML-M5, 43
AML-193, 44
Ammonium sulfate precipitability, 47
AMP, see Adenosine monophosphate
Analgesic compounds, 71–72
Angiogenesis, 11–13, 199
  new blood vessels, 11
  tumor and lymphocyte-induced, 21
Angiogenic agents, 21
Angiogenic reaction, 11
Anorexia, 254, 315
Antagonists, 5, 22, 63, 94–101
Antiarthritic compounds, 71–72
Antiarthritic properties, 67
Antibodies, 22
  against cytokines, 48–49
Antibody-dependent cell-mediated cytotoxicity (ADCC), 199
Antifibrotic agents, 20
Antigen-activated B-cells, 45
Antigen challenge, 8
Antigen-sensitized lymphocytes, 8
Antigen-stimulated T-cell proliferation, 37
Antigen stimulation, 3
Antihuman rTNFα murine monoclonal antibody, 203
Antihuman rTNFα polyclonal antibody, 203
Anti-inflammatory agents, 5, 20
Anti-inflammatory disease, 21
Anti-inflammatory properties, 64
Antimalarial compounds, 71–72
Antipodal responses to corticosteroids, 63
Antiproliferative activity, 16
Antiproliferative agents, 5
Antiproliferative effects, 15, 38
Antiproteases, 317
Antipyretic, 38
Antisense oligonucleotides, 77–78
Antisense polynucleotides, 77
Antisense technology, 78
Anti-TNFα antibodies, 203
Anti-TNFα monoclonal antibody, 18
Antitumor activity, 5
Antiviral effects, 5
Aortic endothelial cells, 174
AP-1, 117
AP-1 complex, 117
Apoptosis, 237, 250–251

Arachidonate metabolism, 60, 101, 199, 202, 213
  activation in endothelium, 174
  cytokines, effect of, 201–202
  inhibitors of cyclooxygenase and lipoxygenase pathways of, 65–69
Arachidonate metabolites, see Arachidonic acid
Arachidonic acid, 45, 92–93, 95, 191–192, 200
Arachidonic acid cascade, 60, 66
Arachidonic acid metabolism, see Arachidonate metabolism
ARDS, see Adult respiratory distress syndrome
Arend's inhibitor, 44
Arthralgia, 22
Arthritic diseases, 4, 181
  cytokines, 109–143
  growth factors, 109–143
  pathogenesis of, 18
Arthritides, 17
Arthritis, 59, 62, 66, 68–69, 72, 124, 133
Arthritogenic potential, 16
Arthus hypersensitivity, 8
Articular cartilage, 17–18
Articular destruction, 18
Articular diseases, 47
Aspirin, 42, 60, 73, 114
Astrocytes, 171, 182
Astroglial cells, 174
ATP, see Adenosine triphosphate
Auranofin, 70–71
Aurothioglucose, 70
Aurothiomalate, 70
Autoamplification loops, 61–62
Autocrine mechanisms, 145
Autocrines, 110, 119
Autodamping feedback system, 42
Autoimmune arthritis, 18
Autoimmune diseases, 35, 45, 183
Autoimmune lupus nephritis, 21
Autoimmune neuropathies, 182
Autoimmune response, 109
Autoimmunity, 1
Autolytic enzymes, 237

B

B-cell differentiation factor, 35
B-cell growth, 9
B-cell growth factor, 35

INDEX   333

B-cells, 7, 91, 109
B-lymphocytes, 4, 170, 174, 183
Bacteremia, 18, 309
Bacterial cell components, 1
Bacterial infection, 21
Bacterial overgrowth, 255
Bacterial pyrogen, 254
BALB/3T3 fibroblast proliferation, 177
BALB/3T3 fibroblasts, 174
BALB/3T3 mouse embryonic fibroblast cell line, 176
BALB/3T3 proliferation, 178
BALF, see Bronchoalveolar lavage fluid
Basement membrane, 7
Basic calcium phosphate (BCP) crystals, 113, 121, 136
Basic FGF, 21
BCP, see Basic calcium phosphate crystals
Biliary obstruction, 255
Binucleate cells, 17
Bioassays, 15, 316
Biological responses, 60
Bisbenzylisoquinoline, 72
BK, see Bradykinin
Blood vessels, 1
Body temperature, 60
Bone, 17–18
  calcitonin, 149
  cell types, 145
  colony-stimulating factors, 154–155
  cortisol, 149
  cytokines, 150–157
  differentiation inducing factors, 157
  functions, 145
  glucocorticoids, 149
  growth factors, 155–157
  growth hormone, 149
  interactions among factors, 157
  interferon gamma, 153–154
  interleukin-1, 150–152
  local factors, 150–157
  parathyroid hormone, 147–148
  prostaglandins, 150–152
  sex steroids, 149
  systemic hormones which influence, 147–149
  thyroid hormone, 149
  tumor necrosis factor, 152–153
  1,25(OH)2 vitamin D, 148–149
Bone cells
  factors regulating function of, 147
  origin, 146–147

Bone destruction, 4, 145, 174
Bone formation, see Bone
Bone loss, 145
Bone marrow, 14
Bone marrow cells, 147
Bone marrow macrophage progenitor cells, 240
Bone marrow progenitor cells, 2
Bone marrow release, 198
Bone marrow transplantation, 21
Bone metabolism, 145–168, see also Bone
Bone mineralization, 151
Bone remodeling, 145–146
Bone resorption, 17–18, 118, 122, 146, 149, 174, 254, see also Bone
Bone turnover, 153
Bordetella pertussis toxin, 213
Bovine serum albumin, 39
Bowel necrosis, 203–204
Bradykinin (BK), 169, 173–174, 182
  interactions between interleukin-1 and, 176–177, 181
  mediator of inflammation, 174
Bradykinin receptors, 173–174
Bromobenzene, 237–238
Bronchoalveolar lavage fluid (BALF), 197–198, 205
Burns, 46, 289, 295, 307
BW-755C, 65, 69, 200, 213

C

C5 cleavage products, 191–192, 204
Cachectin, 4
Cachexia, 4
CAH, see Chronic active hepatitis
Calcified matrix, 146, 156
Calcitonin, 149, 153
Calcitonin gene-related peptide (CGRP), 171
Calcium, 94–100
Calcium ionophore, 65, 100
Calmidazolum, 100
Calmodulin kinase, 95–100
cAMP, 60, 62, 94–95, 97, 100–101, 151
cAMP-dependent pathway, 101
Cancer, 69, 73, 111, 214, 307, 310–311
Cancer metastasis, 137
Capillary leak syndrome, 19
Carbohydrate metabolism, 288–289
Carbohydrate side chains, 38
Cardiac function, 314

Cardiorespiratory diseases, 307
Carrageenan, 75
Cartilage degradation, 17, 60, 71
Cartilage destruction, 4, 16, 72, 174
Cartilage matrix integrity, 17
Cartilage repair, 21
CAT, see Chloramphenicol acetyl transferase
Catabolin, 17, 60, 71
Catalase, 245, 289
CCl_4, 237–238, 241, 249
cDNA, 38, 41, 44, 89–90
cDNA clone, 114–116
cDNA sequence, 37
CDw18 complex, 196
Cell death, 236, 250
Cell differentiation, 282
Cell dropout, 237
Cell fusion, 121
Cell growth, 123
Cell-mediated immune responses, 8
Cell-mediated reactivity, 15
Cell morphology, 124
Cell proliferation, 40, 71, 111, 113, 118–121, 124, 137
   rheumatoid arthritis, 109–143
Cell turnover, 237
Cellular activation processes, 177
Cellular infiltrates, 7
Central nervous system, 35
Central nervous system dysfunction, 23
Central neuroendocrine pathways, 64
Cerebral malaria, 21
Cerebral scarring, 182
c-fos, 64, 94–95, 117, 119–122, 136
cGMP, 97, 101
CGRP, see Calcitonin gene-related peptide
Checks and balances, 35
Chemokinesis, 198
Chemotactic factor production, 181
Chemotactic factors, 3, 181
Chemotactic lymphokines, 3
Chemotaxis, 11, 13, 122, 172, 198, 242, 278, 318
Chick embryo hepatocytes, 240, 251
Chloramphenicol acetyl transferase (CAT), 134–135
Chloroquine, 71–72
Choline deficiency, 238
Chondroblast, 146
Chondrocyte activators, 181

Chondrocyte collagenase synthesis, 16
Chondrocytes, 60, 169, 178
Chromatin clumping, 237
Chromatofocusing, 39
Chromogenic limulus assay, 254
Chronic active hepatitis (CAH), 258
Chronic asthmatic reactions, 14
Chronic degenerative joint disease, 64
Chronic diseases, 4, 10–11, 23
Chronic fatigue immunodeficiency syndrome, 22
Chronic granulomatous disease, 20–21
Chronic granulomatous inflammatory reactions, 23
Chronic granulomatous response, 6, 11
Chronic inflammation, 6–7, 9, 12, 174
Chronic inflammatory disease, 7, 9–10, 23, 35, 121
   cytokines as mediators of, 1–33
Chronic inflammatory reaction, 2
Chronic inflammatory response, 170
Chronic inflammatory tissue, 11
Chronic lung inflammation, 175
Chronic malignant B-cell leukemias, 45
Chronic persistent hepatitis (CPH), 258
Chronic phase, 2, 8
Chronicity, 6
Ciprofloxacin, 75–76
Circulating cytokines, 22, 315–316
Circulating leukocytes, 22
Circulating levels of cytokines, 22
Circulating macrophages, 255
Circulatory shock, 201, 204
Circulatory system, 1
Cirrhosis, 239, 254, 258
c-jun, 117, 120, 136
CKS 17, 39, 46, 74–75
CKS 17-BSA, 39
Class I antigen, 5
Class II antigens, 5
Clinical diseases, 5–6
Clinical hepatic disease states, 253–258
Clinical immunopathological conditions, 22
Clinical interventions in septic shock, 318–319
Clinical signs, 22
Clinical syndromes, 22
Cloning, 44, 47, 89–90, 115
Clotting, 313
Clozic, 60
CMV induced IL-1 inhibitor, 37

c-myc, 64, 94–95, 119, 121–122, 131, 133, 136
c-myc mRNA, 131, 133
Coagulation cascades, 310
Coagulative necrosis, 237
"Cocktails", 77
Colchicine, 257–258
Collagen synthesis, 15, 148, 154
Collagenase, 43, 60, 63, 110, 114, 116–118, 120, 130–131, 136, 169
  production of, 113, 133
  transforming growth factor β, effects of, 122
Collagenase gene, 116, 134–135
Collagenase mRNA, 114–115, 120–121, 134
Collagenase mRNA synthesis, 122
Collagenase synthesis, 113–115
CO/LO, see Cyclooxygenase and lipoxygenase pathways
Colony-stimulating factor 1 (CSF-1), 8, 21, 154–155, 177
Colony-stimulating factors (CSF), 21, 154–155, 204, 240, 315
Colony-stimulating growth factors, 20
Comitogenic response, 66
Comitogenic signal, 61
Complement, 181, 191
Complex nuclear transcription factor, 117
Connective tissue, 11, 16–17, 122, 169
Connective tissue activation, 169
Connective tissue cells, 2–3
Connective tissue matrix, 137
Continuous feedback loop, 170
Contra IL-1, 37
Cori cycle activity, 309
Corticosteroids, 62–64, 257–258, 280, 317–319
Corticosterone, 18
Corticotropin releasing factor (CRF), 18, 64
Cortisol, 149–150, 309, 315
Corynebacterium parvum, 242
Cozzolino's inhibitor, 41
CP-66, 66–67, 248
CP-96345, 183
CPH, see Chronic persistent hepatitis
CR3, 199
CRF, see Corticotropin releasing factor
CRI, 199
C-reactive protein (CRP), 67, 73, 244, 257, 275, 317

Cross-reactivity, 170
Croton oil, 8
CRP, see C-reactive protein
Crystals, 113, 121
CS-2, 75
CSF, see Colony-stimulating factors
CTLL-2 assay, 47
Cyclic nucleotides, 100–101
Cycloheximide, 217
Cyclooxygenase, 45, 151, 201
Cyclooxygenase and lipoxygenase (CO/LO) pathways of arachidonic acid metabolism inhibitors of, 65–69
Cyclooxygenase blockers, 42
Cyclooxygenase inhibitors, 60–62, 64, 67, 213
Cyclooxygenase pathway, 101
Cyclooxygenation, 60, 65–69
Cyclophilin, 22
Cyclosporin A, 22, 72–73
Cytochalasin B, 121
Cytochrome P450-mediated hepatic drug metabolism, 254
Cytokine assays, 5
Cytokine biology, 2–5
Cytokine combination therapy, 20
Cytokine-disease relationship, 6
Cytokine-driven pathological responses, 9
Cytokine-induced cytokines, 318
Cytokine-induced systemic disease, 23
Cytokine interactions, 23
Cytokine levels, 22
Cytokine-neuropeptide interactions, 317–318
Cytokine pathology, 21
Cytokine patterns, 22
Cytokine production, 5
Cytokine receptors, 290
Cytokine regulators, 279–280
Cytokine replacement therapy, 20
Cytokine research, 21–22
Cytokine therapy, 20
Cytokines, see also specific topics
  acute phase protein expression, 275–305
  arthritic diseases, 109–143
  bone metabolism (resorption and formation), 145–168
  chronic inflammatory disease, 1–33
  definition, 59, 276
  disease therapy, 19–21
  factors influencing, 59
  hepatic dysfunction, 235–273

induction of production of other cytokines, 35
low molecular weight inhibitors, 59–87
naturally occurring inhibitors, 35–57
neurogenic inflammation, involvement in, 169–189
pathologic effects, 23
production, 22
pulmonary vascular endothelial injury, role in, 191–234
rheumatoid arthritis, 109–143
second messenger pathways, 89–107
shock, 307–329
synergism, 9, 12, 59, 177, 292, 311
systemic disease, 18–19
systemic effects, 22
Cytomegalovirus, 37
Cytosolic glucocorticoid receptors, 64
Cytotoxic drug treatment, 20
Cytotoxic injury, 21
Cytotoxicity, 91

D

D10.G4.1 cell line, 66–67
DAG, see Diacylglycerol
Dayer's inhibitor, 43–44
Deacylation, 239
Deendothelialization, 10
Deglycosylation, 43, 239
Degradative enzymes, 146
Degranulation, 318
Delayed-type hypersensitivity, 8, 14–15
Demineralization, 18
Dendritic cell, 112
Dermal fibroblasts, 61–62, 120
desArg-BK, 173
desArg-(Leu)-BK, 173
Dexamethasone, 63–64, 115–116, 131, 133–134, 249, 252, 318
D-GLN, see D-Galactosamine
Diacylglycerol (DAG), 94, 96, 100, 112
Dialysis, 63
DIC, see Disseminated intravascular coagulation
DIFs, see Differentiation inducing factors
Differentiation inducing factors (DIFs), 157
Differentiation inhibitory activity, 285
Dihomogamma linoleic acid, 45
Dinitrofluorobenzene, 74
Disease, 4–8, 18, 22–23, 316

Disease-inducing capacity of cytokines, 21
Disease intensity, 22
Disease mechanisms, 21
Disease syndromes, 21
Disease therapy, 19–21
Disseminated intravascular coagulation (DIC), 9, 20, 309
Disulfide-binding interaction, 41
DNA, 136
DNA sequence cloning, 47
DNA synthesis, 60, 121, 154
DNA turnover, 176
DNase I, 38
Doxorubicin, 76
D-Phe-Thi-BK, 173
D-Pro-D-Trp-SP(DPDT), 179
DTH granuloma, 21
Duff's inhibitor, 41

E

E-5110, 69–70
EBV, see Epstein-Barr virus
Echinomycin, 76
Ectopic hormones, 145
Edema, 2, 4, 13, 22, 204, 207–209, 212–213, 308
EDRF, see Endothelium-derived relaxing factor
EGF, see Epidermal growth factor
Eicosanoid metabolism, 60
Eicosanoid synthesis, 69
Eicosanoids, 314
E-LAM 1, see Endothelial-leukocyte adhesion molecule 1
ELISA, 5, 46–47, 69
Ellner factor, 37
Endocrine elements, 59
Endocrine systems, 35
Endocrinological changes, 2
Endocytosis, 121
Endogenous IFN response cycle, 20
Endogenous pyrogen-EP, 3, 60, 62
Endoglycosidase F, 43
Endothelial cell adhesiveness, 11
Endothelial cell growth factor, 156–157
Endothelial cell-leukocyte adhesion, 9–10
Endothelial cells, 4–5, 7–8, 48, 60–61, 110, 174, 239, 243, 293, 313
Endothelial-leukocyte adhesion molecule 1 (E-LAM 1), 12, 195, 200–201

Endothelial procoagulant "tissue factor" synthesis, 4
Endothelium, 2, 196
 cytokines, effects of, 199–201
 hyperadhesiveness of, 3
Endothelium-derived relaxing factor (EDRF), 314
Endotoxemia, 235, 252–256
Endotoxic shock, 18
Endotoxin, 9–11, 191, 236–237, 295, 310–311, 315
 clinical hepatic disease states, 253–256
 hepatic effects of, 238–243
 tolerance, 240
Endotoxin shock, 278
Enzymatic destruction of tissues, 72
Eosinophil chemoattractants, 14
Eosinophil leukocytes, 13
Eosinophilia, 14, 19
Eosinophilic cytoplasmic inclusions, 13
Eosinophils, 14
Epidermal cells, 16, 41
Epidermal growth factor (EGF), 113, 116–117, 119, 123–125, 131, 156, 176, 285
Epinephrine, 315
Epithelioid cell/giant cell formation, 7
Epithelioid cells, 2, 7
Epithelioid granulomata, 7
Epstein-Barr virus (EBV), 37
ERE, see Estrogen response element
Erythrocyte sedimentation rate (ESR), 275
Erythropoiesis, 154
Escherichia coli, 10, 203, 206, 239, 241–242, 254
ESR, see Erythrocyte sedimentation rate
Estrogen, 149, 152
Estrogen receptors, 149
Estrogen response element (ERE), 136
Ethanol, 237, 257
Ethylene vinyl acetate (EVA) copolymer, 10–16
Etiocholanalone, 62
ETYA, 65
EVA, see Ethylene vinyl acetate copolymer
EVA disks, 10, 12
EVA-IFNδ, 14
EVA-IL-1, 11–12
Experimental allergic encephalomyelitis, 21
Extracellular matrix, 110–111, 122

Extrapulmonary tissues, 205–206
Extravasation of inflammatory serum proteins, 169
Extravasation of serum proteins, 181

F

Fatigue, 22
Fatty acid synthetase, 289, 315
FDCP-1 cells, 75
Febrile response, 294
Fever, 3, 5, 22, 38, 48, 60, 62, 74, 254, 257, 275, 307, 311, 314
FGF, see Fibroblast growth factors
Fibrin deposition, 313
Fibrinogen, 280, 294–295
Fibroblast activation, 180
Fibroblast growth factors (FGF), 156, 176
Fibroblast growth factor-induced DNA synthesis, 178
Fibroblast-like cells, 109
Fibroblast proliferation, 70–71, 113, 119, 136, 254
Fibroblastic cells, 146
Fibroblasts, 38, 43, 60, 110, 112, 115–116, 118, 131, 169, 293
 bradykinin receptors, 174
 chemotaxis of, 122
 proliferative and destructive potential, 131
Fibronectin, 73, 110, 239
Fibrosarcoma cell line, 174
Fibrosis, 11, 21, 258
Fibrotic reaction, 12
Fibrotic response, 11, 14
Fluid leakage, 19
Focal parenchymal cell necrosis, 241
Foreign body giant cells, 15
FPL 55712, 65
Frog virus, 238
Frog virus model, 243–244
Fulminant hepatic failure, 258
Fungal metabolite, 72–73

G

G protein-coupled receptor reactions, 177
G proteins, 94, 96, 213
D-Galactosamine (D-GLN), 236–238, 240–241, 244–246, 249
Ganglioside, 45
Gastric tissues, 170

G-CSF, see Granulocyte colony-stimulating factor
Gel filtration, 39–40, 47–48
Gene expression, 93
Gene expression induction, 93
Gene expression regulation, see Second messenger pathways
Gene regulation, 287–288
Gene transcription, 63, 92, 117, 287
General syndrome of just being sick, 22
Genomic sequence cloning, 44
Glioblastoma factor, 36–37
Glomerulonephritis, 7
Glucagon, 252–253, 289, 309, 315
Glucan, 239, 242–243
Glucocorticoid hormones, 113
Glucocorticoid-induced protein, 45
Glucocorticoid receptors, 63
Glucocorticoid regulatory DNA sequences, see Glucocorticoid regulatory element (GRE)
Glucocorticoid regulatory element (GRE), 134
Glucocorticoid-responsive element (GRE), 287
Glucocorticoids, 45, 63–64, 294, 317
  antagonism by, 112–119
  bone, influence on, 149
  mechanisms controlling effects of, 133–136
Glucocorticosteroids, 62
Gluconeogenesis, 289
Glutamine, 255
Glycogenolysis, 309
Glycosaminoglycan release, 60
Glycosylation pattern, 289
Glycosylation sites, 38
GM-CSF, see Granulocyte-macrophage colony-stimulating factor
Gold compounds, 69–71
Gouty arthritis, 121
Granular lymphocytes, 61, 91, 214
Granulation reaction, 15
Granulation tissue, 6
Granulocyte adherence, 192
Granulocyte colony-stimulating factor (G-CSF), 21, 154–155, 198
Granulocyte depletion, 210
Granulocyte effector cells, 213, 217
Granulocyte infiltration, 171
Granulocyte-macrophage colony-stimulating factor (GM-CSF), 11, 21, 44, 154–155, 198, 200
Granulocyte-monocyte colony-stimulating factor, 35
Granulocyte products, 242
Granulocyte surface adhesion molecules, 196
Granulocytes, 35, 195, 197–198, 212
  cytokines, effect of, 198–199
  pulmonary vascular endothelial injury, 191–192
Granulocytopenia, 207
Granuloma formation, 7, 258
Granulomas, 6, 45
Granulomatous cells, 7
Granulomatous data, 7
Granulomatous inflammation, 2
Granulomatous lesion, 14
Granulomatous macrophage response, 13
Granulomatous macrophages, 6
Granulomatous reactions, 11
Granulomatous response, 6, 13
Granulomatous tissue, 11
GRE, see Glucocorticoid regulatory element, Glucocorticoid-responsive element
Growth factors, 4, 14, 116–117, 124, 130
  arthritic diseases, 109–143
  bone, influence on, 155–157
Growth hormone, 149, 315
GSH, see Reduced glutathione
GTP, see Guanosine triphosphate
Guanosine triphosphate (GTP) binding proteins, 94, 96
Gut hypothesis, 255

H

H161.29, 44
^{3}H-dexamethasone, 63
Head trauma, 204
Heat lability, 47
Heat shock, 113, 116, 120–122
Heat shock proteins, 241–242, 251
Hematologic malignancies, 153, 157
Hematopoietic progenitor cells, 146–147
Hematopoietic tissues, 8
Hemodynamic alterations, 215
Hemodynamic parameters, 203–204
Hemopoietic growth factors, 20
Hemorrhage, 213
Hemosiderin, 13
Heparin, 242
Hepatectomy, 252–253
Hepatic dysfunction, 235–273

alcoholic liver disease, 256–258
clinical hepatic disease states, cytokines in, 253–258
endotoxin
  clinical hepatic disease states, 253–256
  hepatic effects of, 238–243
D-galactosamine model of hepatic injury, 244–246
hepatocyte injury
  mechanisms of, 235–238
  viral models of, 243–244
lead-enhanced endotoxin model of liver injury, 246–252
lipopolysaccharide
  clinical hepatic disease states, 253–256
  hepatic effects of, 238–243
liver injury and regeneration, role of cytokines in, 235–253
liver regeneration model, 252–253
morphologic changes, 237
primary biliary cirrhosis, 258
Propionibacterium acnes model of liver injury, 252
role of liver, 235
viral hepatitis, 258
Hepatic failure, 307, 309
Hepatic glucose homeostasis, 289
Hepatic injury, see Hepatic dysfunction
Hepatitis A infections, 243
Hepatitis B viral hepatitis, 243
Hepatitis B virus, 258
Hepatocyte death, 236, 245
Hepatocyte injury
  mechanisms of, 235–238
  viral models of, 243–244
Hepatocyte intracellular enzymes, 235
Hepatocyte response, 187–189
Hepatocyte stimulating factor (HSF), 281–282, 285
Hepatocytes, 2
Hepatomas, 174
Hepatotoxic drugs, 237–238
Hepatotoxicity, 235–237, see also Hepatic dysfunction
Hiruden, 196
Histamine, 3, 169, 171, 173, 177–178
Histamine receptors, 178
Histiocytes, 10
HL-60, 44
Homeostasis, 2, 235
Homeostatic neuroimmunoendocrine feedback mechanism, 64
Homeostatic response, 275, 294

Homodimers, 117
Hormonal elements, 59
Hormone carrier, 279
Hormone-like actions, 35
Host antitumor immune response, 75
Host defense, 2, 5
Host defense mechanisms, 235
Host factors, 307, 309
Host immune-mediated cytotoxicity, 243
Host response, 59
h-ras, 116
HSF, see Hepatocyte stimulating factor
^{3}H-thymidine, 36–37
HT1004, 97
HTLV-I, see Human T-lymphotropic virus-I (HTLV-I)
Human T-lymphotropic virus-I (HTLV-I), 39
Hybridoma growth factor, 35
Hydrocortisone, 62–63
Hydrolysis, 64
Hydroxyapatite, 156
Hydroxychloroquine, 71–72
Hydroxyurea, 153
Hyperalgesia, 75
Hypercalcemia, 18, 145, 151–153, 157
Hyperglycemia, 309
Hypermetabolic/hypercatabolic response, 255
Hyperoxia, 213
Hypersensitivity, 8, 13
Hyperthermic response, 60, see also Fever
Hypertriglyceridemia, 254
Hyperventilation, 308
Hypocapnea, 307
Hypoferremia, 206
Hypoglycemia, 289
Hypotension, 171, 173, 203–204, 214–215, 307–308, 311, 315
Hypothalamus, 60, 64
Hypovolemia, 307
Hypoxemia, 212
Hypoxia, 236–237, 307–308
Hypozincemia, 206, 257

I

Ia antigen expression, 73
Ibuprofen, 42, 61, 65, 68, 76, 101, 213, 319
ICAM-1, see Intercellular adhesion molecule
IgE, 169

IGF, see Insulin-like growth factor entries
IL-1, see Interleukin-1
IL-1 binding proteins, 40
IL-1 inhibitors, 6
IL-2, see Interleukin-2
IL-2 receptors, 4, 7, 46
  pathogenic role, 8
  soluble, 8
IL-2R antibody, 8
IL-3, see Interleukin-3
IL-3 Ab, 21
IL-4 Ab, 21
Iloprost, 237, 245
IM-9B-lymphoblastoid cell line, 170
Immune cells
  activation by substance P, 172
  inflammatory response, 169
Immune complexes, 1, 7, 44
Immune lymph node cells, 45
Immune-mediated diseases, 23
Immune reaction, 3
Immune response, 2, 22
Immune suppression syndrome, 37
Immune system 4, 8
Immunoassays, 316
Immunoblotting, 67
Immunocytochemical localization, 7
Immunocytochemical staining, 7
Immunocytochemical technique, 10
Immunocytochemistry, 5
Immunogenic substances, 1
Immunoglobulin, 182–183
Immunoglobulin production, 9
Immunohistochemical analysis, 316
Immunohistochemistry, 295
Immunoinflammatory disease, 3, 14–15, 23
Immunoinflammatory response, 7
Immunologic mechanisms, 182–183
Immunological responses, 74
Immunology, 1, 3
Immunopathology, 59
Immunopotentiation, 314
Immunoregulation, 21
Immunoregulatory molecules, 73
Immunoregulatory properties, 69
Immunosuppression, 8, 39–40, 46, 307
Immunosuppressive properties, 64, 74
Immunosuppressive retroviruses, 46
In situ hybridization, 5–6, 295
Increased vascular permeability, 9
Indomethacin, 42, 60, 64, 66, 69, 71, 74, 92–93, 101, 157, 174, 256–257

Infiltrating lymphocytes, 8
Infiltrating macrophages, 8
Inflamed tissue, 2, 293–295
Inflammation, 59, 73, 122, 183, 295, 316
  acute, 2, 4, 6, 12
  alleviation, 64, 73
  chronic, 6–7, 9, 12, 174
  cytokines as mediators of, 1–33
  defined, 275
  granulomatous, 2
  local, 18
  suppression of, 133
Inflammatory cytokines, 2
Inflammatory disease, 1, 6, 9, 68
Inflammatory disorders, 45
Inflammatory environment, 18
Inflammatory exudate, 2
Inflammatory fluids, 6, 22
Inflammatory joint disease, 47, 295
Inflammatory lesions, 6
Inflammatory leukocytes, 10
Inflammatory phagocytes, 11
Inflammatory potential, 15
Inflammatory process, 23, 145, 236
Inflammatory properties, 17, 112
Inflammatory proteins, 115
Inflammatory reaction, 3, 109
Inflammatory response, 1, 4, 8, 11–12, 64, 74, 169, 295
Inflammatory site, 8, 16
Influenza virus induced IL-1 inhibitor, 40
Inhibition of cytokines, 318–319
Inhibitors of cytokines, see Low molecular weight inhibitors; Naturally occurring inhibitors
Injurious stimuli, 7
Inositol triphosphate (IP_3), 94, 96, 100, 112
Insulin, 149, 252–253, 315
Insulin-like growth factor 1 (IGF-1), 147–149
Insulin-like growth factor 2 (IGF-2), 149
Insulin-like growth factors, 156
Insulin sensitivity, 309
Intercellular adhesion molecule (ICAM-1), 16, 201
Interferon gamma, 7–9, 15, 91, 157, 204, see also Interferons
  acute phase proteins, 285
  bone, influence on, 153–154
  cell growth, 123
  lesions, 15
  MIF family, 8

Interferon inducers, 20–21
Interferons, 3, 5, 13, see also Interferon gamma
  enhanced production, 6
  mediators of chronic inflammatory disease, 1–33
  pathological responses, 16
  therapeutic approaches to disease, 21
  toxicity, 20
Interleukin-1(IL-1), 3–5, 113, 116, 136, 154–157, see also specific topics
  acute inflammatory responses, 9
  acute phase proteins, 280–281
  acute phase response, 191
  bone, influence on, 150–152
  bradykinin stimulation of production, 181
  cell growth, 123
  control of production at level of second messenger pathways, 89–107, see also Second messenger pathways
  detection, 6
  enhanced production, 6
  gene expression regulation, see Second messenger pathways
  interactions with
    bradykinin, 174, 176–177
    substance P, 175, 176
  low molecular weight inhibitors, 59–87, see also Low molecular weight inhibitors
  naturally occurring inhibitors, see Naturally occurring inhibitors
  neurogenic inflammation, 169, 174–175
  production by granulomatous cells, 7
  production by neurokinins, 180
  pulmonary vascular endothelial injury, 206–209
  responsive elements, 287
  rheumatoid arthritis, role in, 178
  shock, 308
  substance P stimulation of production of, 179
  synovial cells, effects on, 118–120
  systemic effects, 35
  therapeutic approaches to disease, 21
Interleukin-2 (IL-2), 3–5, 7, 12–14
  disease therapy, 20
  enhanced production, 6
  pathological levels at inflammatory site, 8
  production, 8
  rheumatoid arthritis lesions, 8

  therapeutic approaches to disease, 21
  toxicity, 19–20
Interleukin-2 lymphokine-activated killer cell-mediated pulmonary vascular endothelial injury, 214–215
Interleukin-3 (IL-3), 14, 154–155, 204
Interleukin-4, 45
Interleukin-5, 14
Interleukin-6, 59, 61, 67
  acute phase proteins, 280–295
  systemic effects, 35
Interleukin-6 receptor, 290
Interleukin-6 responsive elements, 287
Interleukin-6 responsive transcription elements, 287
Interleukins, 3, 6, 9, see also specific types and other topics
  mediators of chronic inflammatory disease, 1–33
Interstitial collagen, 110
Interstitial edema, 209, 308
Intestinal permeability, 255
Intracellular calcium, 122, 136
Intraluminal hydrostatic pressures, 202
Intravascular coagulation/thrombosis, 4
IP_3, see Inositol triphosphate
Ischemia, 255
Ischemic liver damage, 236
Ito cells, 260
IX 207-887, 78

J

Joint erosion, 181
Joint inflammation, 71, 73

K

Kallikrein, 110
Kallikrein formation, 171
Kawasaki disease, 22
Kawasaki disease sera, 201
Keratinocyte inhibitors, 41–42
Keratinocytes, 174
Ketoconazole, 92
Killer cell activation, 45
Kinetics of response, 288
Kinins, 3
Kupffer cells, 238–240, 242–244, 252, 255, 260

L

L929 cells, 75
Lactate dehydrogenase release, 213

LAF, see Lymphocyte activating factor
LAK, see Lymphocyte-activated killer cells, Lymphokine-activated killer cells
Laminin, 110
Large granular lymphocytes, see Granular lymphocytes
Lead-acetate, 237, 240
Lead-enhanced endotoxin model of liver injury, 246–252
Lean body mass, wasting of, 309
Lectin-stimulated thymocyte proliferation, 37
LEM, see Leukocyte endogenous mediator
Lepromatous leprosy, 8–9, 20–21
Leprosy, 15
Lethality, 240–241, 245, 315
Leukemia inhibitory factor (LIF), 157, 280, 282–285
Leukemias, 35, 73
Leukemoid reactions, 308
Leukocyte adhesion, 10
Leukocyte adhesion molecules, 9, 11
Leukocyte chemoattractants, 11
Leukocyte endogenous mediator (LEM), 3
Leukocyte-endothelial cell interactions., 4
Leukocyte migration, 10
Leukocyte redistribution septic shock, 308–309
Leukocytes, 2–3, 15, 62
  adherence, 12
  adherence to microvasculature, 1
  adhesion to vascular endothelium, 2
  infiltration, 2
  phagocytic, 2
  polymorphonuclear, 3
Leukocytic pyrogen, 60
Leukocytosis, 3, 275, 308
Leukopenia, 20
Leukotriene B_4, 65, 93, 171, 201, 204
Leukotriene C_4, 65
Leukotriene D_4, 204
Leukotrienes, 3, 309
Levamisole, 68–69
LIF, see Leukemia inhibitory factor
Lining cells, 146
Lipid metabolism, 309
Lipid metabolites, 309
Lipid peroxidation, 236
Lipopolysaccharide (LPS), 4, 11, 40, 42, 61, 66, 89, 91, 96, 155
  clinical hepatic disease states, 253–256
  hepatic effects of, 238–243
  mRNA expression, 90
  tolerance, 240
Lipoprotein lipase, 314
Liposomes, 78
Lipoxygenase, 201
Lipoxygenase metabolites, 215
Lipoxygenation, 65–69
Liver, 235–273, see also Hepatic dysfunction
Liver injury, see Hepatic dysfunction
Liver regeneration, 253
Liver regeneration model, 252–253
Local administration, 5
Local inflammation, 18
Local inflammatory disease 23
Local pathological response, 15
Local slow-release, 13
Localization of cytokines, 5–9
Low molecular weight IL-1 inhibitor, 40
Low molecular weight inhibitors, 59–87
  antimalarial compounds, 71–72
  cyclooxygenase and lipoxygenase pathways of arachidonic acid metabolism, 65–69
  Cyclosporin A, 72–73
  future approaches, 77–78
  gold compounds, 69–71
  methotrexate, 73–74
  miscellaneous, 75–77
  nonsteroid anti-inflammatory drugs, 60–62
  peptides, 74–75
  steroids, 62–64
LPS, see Lipopolysaccharide
LPS 0.55.B5, 10
Lung injury, 204–206
  inducers of, 204
  local levels of cytokines, 204
Lung lymph flow, 212
Lupus nephritis, 20
Lymphatics, 1
Lymphocyte accumulation, 15–16
Lymphocyte-activated killer (LAK) cells, 20–21
Lymphocyte activating factor (LAF), 3, 35
Lymphocyte activating factor (LAF) assay, 36–40, 42–44, 61
Lymphocyte activation, 278
Lymphocyte adhesion molecule, 16

Lymphocyte cell lines, 149
Lymphocyte-induced angiogenesis, 21
Lymphocyte infiltration, 13, 15–16, 19
Lymphocyte proliferation, 3–4, 37, 39–40
Lymphocyte recruitment, 15
Lymphocyte subsets, 22
Lymphocytes, 10–11, 13, 17, 112, 157
Lymphoid cell activation, 169
Lymphoid cells, 7
Lymphoid-macrophage-eosinophil granulomatous response, 14
Lymphokine-activated killer (LAK) cells, 214–215
Lymphokines, 3, 5, 8–9, 13, 15, 22, 136
Lymphotoxin, 90, 214
Lys-D-Pro-Thr, 75
Lysosomes, 121
Lys-Pro-Thr, 75
Lytic necrosis, 237

M

M20, 42–43
α 2-Macroglobulin, 40–41, 279, 317
Macrophage activating factor, 5
Macrophage activating factor (MAF) lymphokine, 9
Macrophage activation, 6, 278
Macrophage activation factor (MAF), 3
Macrophage activators, 245
Macrophage-derived IL-1 inhibitors, 42
Macrophage granuloma, 11
Macrophage granulomatous lesions, 11
Macrophage infiltrates, 8
Macrophage infiltration, 7
Macrophage membrane, 6
Macrophages, 2–7, 10–13, 17, 40, 43–44, 47, 73, 91, 109–110, 171
  fusion, 9
  neurokinin receptors, 171
  substance P receptors, 171
MAF, see Macrophage activating factor, Macrophage activation factor
Malaise, 22
Malignancy, 137
Malignancy syndrome, 145
Malignant cells, 124
Malignant tissue, 113
Malnutrition, 255
Mammary carcinoma, 63
Marrow cell malignancies, 145
Mast cell degranulation, 169

Mast cell stimulating factor, 8
Mast cells, 171, 181
Matrix degradation, 137
  rheumatoid arthritis, 109–143
Matrix destruction, 114
MC3T3-E1 cell line, 151–152, 155–157
MCF, see Mononuclear cell factor
MDNCF, see Monocyte-derived neutrophil chemotactic factor
Mediators, 1–33, 111, 114, 118, 120, 137, 150, 172, 178, 251, 295, 309–310, see also specific types
  release of, 169
  shock, 308, 317
Medullary carcinoma, 149
Mefloquine, 71–72
α-Melanocyte stimulating hormone, 38, 74, 318
Memory T-cells, 22
Meningococcal septic shock, 316
Menopause, 145, 149, 152
Mercurial compounds, 110
Mesenchymal cells, 122
Metabolic changes, 288–289
Metabolic disturbances, 4
Metabolism shock, 314–315
Metabolism products, 59
Metalloproteinase expression, 114, 116, 131
Metalloproteinase gene cloning, 115
Metalloproteinase gene expression, 117
Metalloproteinase levels, 121
Metalloproteinase mRNA, 131
Metalloproteinase production, 111, 113, 133, 136
Metalloproteinase synthesis, 118–120, 122
Metalloproteinases, 112, 130
Metals, 237
Metastatic tumor cells, 145
Methotrexate, 73–74
MHC II antigen expression, 15
Microenvironment, 6
Microglial cells, 182
Microtubule depolymerization, 121
Microvascular blood flow, 202
Microvasculature, 1
MIF, see Migration inhibition factor
MIF antibody 7D 10, 8
MIF family, 8
Migration inhibition factor (MIF), 8
  antigen-dependent, 8

antigen-independent, 8
infiltrating macrophages, 8
therapeutic potential, 9
Migration inhibitory lymphokines, 3
Mitochondrial dysfunction, 254
Mitochondrial pyruvate dehydrogenase, 315
Mn superoxide dismutase (MnSOD), 241–242, 246
Molecular biology of disease, 22
Monoblastic leukemia cells, 43
Monoclonal antibodies, 44
Monoclonal antibody staining, 7
Monocyte-derived neutrophil chemotactic factor (MDNCF), 197
Monocyte generation in hematopoietic tissues, 8
Monocyte infiltration, 6
Monocyte-mediated antitumor cytotoxic responses, 75
Monocyte stimulating factor, 154–155
Monocyte toxicity, 75
Monocytes, 2, 4, 11, 13, 35, 60, 72, 109
 chemotaxis of, 122, 278
 control of IL-1 and TNFα gene expression in, 90–94
 resting, 91, 93
Monocytic cell lines, 43, 47
Mononuclear cell factor (MCF), 118
Mononuclear cell factor (MCF) assay, 43–44
Mononuclear cell infiltration, 16
Mononuclear cells, 6, 40, 46, 172
Mononuclear infiltrates, 13
Mononuclear leukocyte, 12
Mononuclear phagocytes, 2, 6, 11
Monosodium urate monohydrate crystals, 113, 120–122
Mortality, 316
mRNA, 64, 130, 136
mRNA expression
 induction of IL-1 by IL-2, 91
 regulation of, 90
 regulation of TNFα, 91–92
 role of adherence in induction of, 90–91
mRNA stability, 94
mRNA synthesis, 6
Multi-CSF, 21
Multinucleated giant cells, 2, 7, 9, 154
Multiple colony stimulating factor, 21, 35
Multiple myeloma, 146, 153, 157
Multiple sclerosis, 22, 182
Multiple trauma, 307

Multisystem organ failure, 255
Muramyl dipeptide, 242, 245
Murine hepatitis virus, 249
Murine hepatitis virus infections, 243
Muscle catabolism 254
Myalgia, 22
Mycobacterium bovis, 242
Mycoplasma contamination, 37
Myelodysplastic syndrome, 21
Myeloid cell system 154
Myeloid cells, 147, 198
Myeloid leukemia cell line, 157
Myelomonocytic cell line, 43
Myeloperoxidase, 197
Myocardial function, 307

N

Naive T-cells, 22
Nasal passages, 178
Naproxen, 66, 174
Natriuresis, 62
Natural killer cells, 5, 45
Naturally occurring inhibitors, 35–57
 antibodies against cytokines, 48–49
 Arend's inhibitor, 44
 CKS 17, 39
 CMV induced IL-1 inhibitor, 37
 Contra IL-1, 37
 Cozzolino's inhibitor, 41
 Dayer's inhibitor, 43–44
 Duff's inhibitor, 41
 Ellner factor, 37
 glioblastoma factor, 36–37
 IL-1, 36
 IL-1 binding proteins, 40
 IL-2 inhibitors, 44–45
  nonreceptor, 47–48
  production, 45–46
  responsiveness, 45–46
  serum and synovial, 46–47
 influenza virus induced IL-1 inhibitor, 40
 keratinocyte inhibitors, 41–42
 low molecular weight IL-1 inhibitor, 40
 M20, 42–43
 α-macroglobin, 40–41
 macrophage-derived IL-1 inhibitors, 42
 α-melanocyte stimulating hormone, 38
 neutrophil IL-1 inhibitor, 40
 P388D1 factor, 42
 respiratory synovial virus induced IL-1 inhibitor, 40

Rosenstreich's IL-1 inhibitor, 37–38
soluble receptors for cytokines, 49
submandibular gland IL-1 inhibitor, 39
TNF inhibitors, 48
uromodulin, 38–39
NDGA, see Nordihydroguiaretic acid
Necrosis, 4, 9, 11
Negative-feedback control, 6
Negative feedback loop, 316–318
Negative retinoic acid response elements, 136
Neonatal fibroblasts, 122
Neovascularization, 199
Neuramidase, 45
Neuroactive inflammatory substances, 169
Neuroendocrine influences, 74
Neuroendocrine stress response, 309
Neurogenic inflammation
  axon-reflex model, 171–172
  bradykinin, 169, 173–174
  interactions between neurotransmitters and cytokines, 175–178
  interleukin-1, 174–175
  involvement of cytokines in, 169–189
  neurokinin A and B, 169–170
  neurokinins, 170
  neuropeptides responsible for, 169–174
  promotion of cytokine release by neuropeptides, 178–181
  proposed model of cytokine involvement, 181–182
  substance P, 169–175
  tachykinins, 170
  tumor necrosis factor, 174–175
Neuroimmune feedback loops 74
Neurokinin A (NK-A), 169–170
Neurokinin B (NK-B), 169–170
Neurokinin receptors, 171
Neurokinins
  neurogenic inflammation, 170
  production of interleukin-1, 180
Neuromodulator-immunomodulator interactions, 182
Neuronal tissue, 170
Neurons, 64
Neuropathies, 183
Neuropeptides, 317–318
  neurogenic inflammation, 169–174
  promotion of cytokine release by, 178–181
Neurotensin, 75, 317–318
Neurotransmitters, 169, 175–178
Neutral metalloproteinases, 110

Neutropenia, 212–213, 219
Neutrophil collagenase, 110
Neutrophil IL-1 inhibitor, 40
Neutrophil infiltration, 4
Neutrophil receptor for IL-1, 198
Neutrophilia, 38, 198, 206–207, 213, 254, 256–257
Neutrophils, 11, 198–199, 313
NK-A, see Neurokinin A
NK-B, see Neurokinin B
Nociception, 173
Nociceptive nerve fibers, 169
Nociceptive responses, 75
Nociceptors, 172
Nonalcoholic liver disease, 254–255
Nonimmune mechanisms, 2
Nonimmune-mediated disease, 23
Nonmalignant proliferative diseases, 111
Nonreceptor inhibitors of IL-2, 47–48
Nonspecific inflammatory response, 8
Nonspecific tissue injury, 9
Nonsteroid anti-inflammatory drugs (NSAIDs), 60–62, 66–67, 69, 71, 73, 114, 133
Nordihydroguaiaretic acid (NDGA), 65, 68, 76
Northern blot analysis, 98–99, 130–132
NSAIDs, see Nonsteriod anti-inflammatory drugs
Nude mice, 130, 152
Nyctohemeral rhythm, 20

O

1,25(OH)2 vitamin D, 148–149, 154–157
ODC, see Ornithine decarboxylase
Oligonucleotide, 78
Oligonucleotide probes, 44
Oncogene products, 115
Oncogenes, 116, 148, 149
Organ dysfunction, 4
Ornithine decarboxylase (ODC), 253
Osteoarthritis, 67
Osteoblasts, 145–146, see also Bone origin, 146–147
Osteocalcin, 155
Osteoclasts, 145–146, see also Bone origin, 146–147
Osteocytes, 146
Osteogenic cells, 146
Osteolytic bone lesions, 153
Osteomyelitis, 145
Osteopenia, 145

Osteopetrosis, 146
Osteoporosis, 145–146, 149, 152
Osteoprogenitor, 146
Osteosarcoma cell line, see MC3T3-E1 cell line
Overproduction of cytokines, 35
Oxidative metabolism, 309, 315
Oxidative stress, 236
Oxygen intermediates, 199
Oxygen radicals, 309, 314

P

P15E, 74
P388D1 factor, 42, 178–180, 197
PAF, see Platelet activating factor
Pain
 peripheral transmission, 173
 production of, 174
Pannus, 71
Paracrine mechanisms, 145
Parasitic infection, 21
Parasitic reactions, 14
Parathyroid hormone (PTH), 147–148, 153–157
Parenchymal cell injury, 243
Pathogen-activated complement cascades, 310
Pathogenesis of disease, 21
Pathogenic microorganisms, 59
Pathogenic potential, 16, 23
Pathogenic role, 8
Pathogenicity, 5
Pathological effects of cytokines, 9–18, 23
Pathological events related to cytokines, 5–19
Pathological lesions, 10
Pathological mediators, 3
Pathological processes, 5, 21
Pathological production of IL-1, 4
Pathological responses, 5–7, 9, 16, 20, 22
Pathological roles of cytokines, 5
Pathological tissues and fluids, 5–9, 18, 23
Pathophysiology of disease, 59
Pathophysiology of inflammatory disease, 9
PBL, see Peripheral blood leukocytes
PDGF, see Platelet-derived growth factor
Pefloxacin, 75–76
Penicillamine, 74

PEPCK, see Phosphoenol pyruvate carboxykinase
Peptide-BSA complex, 39
Peptides, 74–75, 309
Peptidoleukotriene receptor antagonist, 65
Peripheral blood leukocytes (PBL), 91
Peripheral blood monocytes, 66, 68, 92, 97, 180
Peripheral mononuclear cells, 47
Peritoneal exudate lymphocytes, 16
Peritoneal exudates, 72
Peritoneal macrophages, 60, 66, 73, 91, 97
Perivascular edema, 207, 213
Perivascular lymphocyte accumulation, 15
Perivascular lymphocytes, 12
Perivascular mononuclear cells, 17
Persistent diseases, 8
Peyer's patches, 172
PG, see Prostaglandins
PGE_1, 252
PGE_2, 38, 42–43, 45, 60–61, 100, 130, 171, 201, 203, 243
 excessive production of, 110
 production, 114, 118
 role, 92
 suppression of production, 133
PGE_2 synthesis, 68, 113
$PGF_{2\alpha}$, 201
6-Keto-$PGF_{1\alpha}$, 201
PGH, see Prostaglandin H
PGI_2, 201
PHA, see Phytohemagglutinin-stimulated T-cell proliferation
Phagocyte activation, 172
Phagocytes, 2, 4
Phagocytic cells, 313
Phagocytic leukocytes, 2, 6
Phagocytosis, 7, 121, 136, 179, 198, 239–240, 242, 318
Phallicidin, 217
Pharmacologic agents, 113
Phenidone, 76
Phenylbutazone, 73
Phorbol ester, 44, 46, 91, 93, 117, 136, 289
Phorbol ester response element, 117
Phorbol myristate acetate (PMA), 95–96, 112–118, 120, 134–135, 191
Phorbol responsive sequence, 120
Phosphatidylinositol-4,5-biphosphate (PIP_2), 100

Phosphoenolpyruvate carboxykinase (PEPCK), 289
Phospholipase A_2, 60, 200–202, 213
Phospholipase C activation, 178
Phospholipid metabolism, 121
Phospholipids, 60
Phosphorylation, 95
Physiological status, 59
Phytohemagglutinin (PHA)-stimulated T-cell proliferation, 45
Pinocytosis, 239
PIP_2, see Phosphatidylinositol-4,5-biphosphate
Piroxicam, 61, 66, 69
Pituitary, 18
Pituitary-adrenal axis, 35
Plasma
  IL-1 binding proteins, 40
  rheumatoid arthritis patients, 8
Plasma catecholamine, 309
Plasma membrane permeability, 235–236
Plasma protein extravasation, 171
Plasmin, 110
Plasminogen activator, 60, 63, 113
Plasminogen activator inhibitor, 200
Plasminogen activator inhibitor-1, 313
Platelet activating factor (PAF), 169, 171, 191–192, 199–200, 202–204, 309, 313–314
Platelet aggregation, 313
Platelet-derived growth factor (PDGF), 4, 117, 119, 121, 131, 156, 176, 200, 318
Platelet microthrombi, 10
Platelets, 122
Pleomorphic mononuclear infiltrations, 241
Pleural effusions, 19
PMA, see Phorbol myristate acetate
PMN, see Polymorphonuclear leukocytes
Poly I:C, 21
Polymixin-B, 238, 244, 253
Polymorphonuclear cells, 2, 172
Polymorphonuclear infiltration, 11
Polymorphonuclear leukocyte infiltration, 9
Polymorphonuclear leukocytes (PMN), 3, 6, 9, 11, 40, 109
Polymorphonuclear phagocyte, 6, 11
Polyoma virus, 116
Polypeptide hormones, 111
Polypeptides, 149

POMC, see Proopiomelanocorticotrophic peptide
Positive feedback loop, 182, 316–318
PPIase, 22
Prednisolone, 64, 74
Preprocollagenase, 110
Pressor therapy, 215
Primary biliary cirrhosis, 258
Primary cytokines, 318
Primed environment, 23
Priming signals, 93
Procoagulant activity, 9, 196, 200, 243, 313
Procollagenase, 110, 122
Progenitor cells, 20
Proinflammatory activities, 59, 69
Proinflammatory agents, 9
Prolactin, 73
Proliferating macrophages, 7
Proopiomelanocortin gene, 63, 74
Proopiomelanocorticotrophic (POMC) peptide, 318
Propionibacterium acnes, 240, 242, 243, 245, 249, 252
Propylthiouracil, 257
Prostacyclin, 174
Prostaglandin synthesis, 60–62, 71
Prostaglandin synthesis inhibitors, 61, 68
Prostaglandin H (PGH) synthetase, 151
Prostaglandin synthetase inhibitors, 66
Prostaglandin therapy, 258
Prostaglandins (PG), 3, 60, 92, 100–101, 169, 249, 309, 314, see also PGE entries
  alcoholic hepatitis, 257
  bone, influence on, 150–152
  bradykinin action, 173
  bradykinin release of, 169, 174
Prostanoids, 204
Proteases, 192, 309
Protective proteins, 251
Protein C, 200
Protein S, 200
Protein kinase A, 97
Protein kinase C, 39, 46, 95–100, 112, 289
Protein-protein interactions, 117
Protein synthesis, 22, 45, 66
Proteinase inhibition, 278
Proteinase inhibitors, 279
Proteinase K, 38
Proteinases, 111, 137, 242

Proteoglycan degradation, 71
Proteoglycan synthesis, 16–17
Proteoglycans, 110
Proteolytic enzyme secretion, 169
Proteolytic enzymes, 63
Protooncogene mRNA, 119
Protooncogene productions, 117
Protooncogenes, 94, 112, 121, 130–131, 136
Pseudomonas infection, 319
Psoriasis, 111
PTH, see Parathyroid hormone
Pulmonary artery hypertension, 210, 212
Pulmonary edema, 192, 207–208, 212–213
Pulmonary hypersensitivity granulomas, 45
Pulmonary hypertension, 204
Pulmonary leukostasis, 192–198, 207, 212–213
Pulmonary tissues, 170
Pulmonary vascular endothelial injury
  arachidonate metabolism, effect of cytokines on, 201–202
  cytokine-induced pulmonary leukostasis, 192–198
  endothelium, effects of cytokines on, 199–201
  extrapulmonary tissues, 205–206
  granulocytes, 191–192, 198–199
  hemodynamic parameters, effects of cytokines on, 203–204
  IL-1-induced, 206–209
  IL-2 lymphokine-activated killer cell-mediated, 214–215
  lung injury, 204–206
  PAF synthesis, cytokine-induced, 202–203
  role of cytokines in, 193–234
  TNFα-induced, 209–213
Pulse-chase experiments, 114–115
Pyrazolo-quinolines, 72
Pyrexia, 74

Q

Quinolone antibiotics, 75

R

Rabbit collagenase gene, 116, 134–135
Rabbit fibroblasts, 124
Radiation injury, 21
Radioimmunoassay, 5, 61, 316

Radioligand binding studies, 63
RAR, see Retinoic acid receptor
RE, see Reticuloendothelial system
Reactive nitrogen intermediates, 236
Reactive oxygen intermediates, 236
Receptor-ligand interaction, 118
Receptors, 170–174
Recombinant cytokines, 10, 36
Reduced glutathione (GSH), 236
Regulatory feedback mechanisms shock, 316–318
Renal cytokine trap hypothesis, 39
Renal failure, 307, 309
Reperfusion model of tissue injury, 245
Respiratory alkalosis, 308
Respiratory dysfunction, 307
Respiratory failure, 308
Respiratory synovial virus induced IL-1 inhibitor, 40
Resting osteoblasts, 146
Reticular cells, 146
Reticuloendothelial (RE) system, 240
Retinoic acid, 135–136
Retinoic acid receptor (RAR), 136
Retinoids, 113–114
  antagonism by, 112–118
  mechanisms controlling effects of, 133–136
Retroviral infections, 39
Reye's syndrome, 254
Rheumatoid arthritis, 7, 9, 21–22, 46, 67, 71, 73, 174–175
  bone destruction, 145
  cell proliferation, 109–143
  Duff's inhibitor, 41
  IL-1 role, 178
  IL-2 receptor levels, 8
  inflammatory aspect, 133
  kinin formation, 174
  matrix degradation, 109–143
  nonreceptor inhibitors of IL-2, 47–48
  pathophysiology, 118, 122, 137
  proliferative/destructive component, 133
  serum and synovial inhibitors of IL-2, 46–47
  synovial tissue, 109
Rheumatoid arthritis lesions, 8
Rheumatoid arthritis tissues, 8
Rheumatoid factor-producing B-cells, 181–182
Rheumatoid factors, 169
Rheumatoid lesion, 124
Rheumatoid synovial cells, see Synovial cells

Rheumatoid synovitis, 6
Rheumatoid synovium, see Synovial cells;
  Synovial tissue
rhIL-1, 10, 14
rhIL-2, 10, 15, 16
rIFN gamma, 10
rIL-1, 17
RNA slot-blot analysis, 131, 133
ROHA-9, 37
ROS 17/2.8 cell line, 153–154
Rosenstreich's IL-1 inhibitor, 37–38

S

SAP, see Serum amyloid-P
Sarcoid granuloma, 7
Sarcoid lesions, 7
Sarcoidosis, 7–8, 22
Schistosome hypersensitivity, 6
Schwartzman-like response, 4
Schwartzman reaction, 9, 196
Scleroderma, 111
SDS gel electrophoresis, 44
SDS-PAGE, 43
Second messenger pathways, 182, 250–251
  agonists, 94–101
  antagonists, 94–101
  arachidonic acid metabolites, 92–93
  calcium, 94–100
  calmodulin kinase, 95–100
  control of gene expression in monocytes, 90–94
  control of IL-1 and TNFα production at level of, 89–107
  cyclic nucleotides, 100–101
  interleukin genes and proteins, 80–90
  mRNA expression, regulations of, 90–94
  overview, 94–95
  potential therapeutic intervention with inhibitors of, 101
  protein kinase C, 95–100
  tumor necrosis factor genes and proteins, 89–90
Second messengers, 76, 112
Secondary cytokines, 318
Secretion, 7
Sensitivity, 36
Sensitized lymphocytes, 3
Sepsis, 3, 9, 191, 235
Septic shock, see Shock
Septicemia, 289
Septum, 170

Serine proteases, 110
Serotonin (5-HT), 3, 173, 177–178
Serum
  IL-1 binding proteins, 40
  IL-2 inhibitors, 46–47
Serum albumin concentration, 254
Serum amyloid-A, 240, 244, 280
Serum amyloid-P (SAP), 74
Serum and synovial inhibitors of IL-2, 46–47
Serum zinc level, 254
Severe burn injury, 22
Severe trauma, 289
Sex steroids, 149
Shock, 4, 18, 49, 62, 191
  acute phase proteins, 317
  cardiac function, 314
  cardiovascular events, 308
  circulating cytokines in, 315–316
  clinical features, 308–309
  clinical interventions, 318–319
  consequence of modern medicine, 307–308
  cytokine-induced cytokines, 318
  cytokine-neuropeptide interactions, 317–318
  cytokines and, 307–329
  feedback mechanisms, 316–318
  incidence, 307
  inhibition of cytokines, 318–319
  initiation, 307
  involvement of cytokines in, 310–313
  leukocyte redistribution, 308–309
  mediators of pathogenesis, 309–310, 317
  metabolic/endocrine responses, 309
  metabolism, 314–315
  organ damage or dysfunction, 309
  regulation of cytokines, 316–318
  respiratory events, 308
  symptoms, 307
  tissue-specific actions of cytokines, 313–315
  treatment with cytokines, 319
  tumor necrosis factor, 311
  vascular contractility, 314
  vascular endothelium, 313–314
Shock lung, 203, 308
Sialated glucopeptide, 45
Signal transduction pathways, 93–96, 118, 289
Silica, 47
SK&F 86002, 68–69, 101
Skeleton, remodeling of, 145

Skin, 41
Skin diseases, 16
Sleep-pattern disturbances, 22
Slow release, 23
Slow-release effects, 12
Slow-release IFNδ, 15
Slow-release IL-1, 11
Slow-release system, 11, 18, 20
Slow-release technology, 10
Small bile duct proliferation, 241
Smooth muscle cell proliferation, 318
Smooth muscle contractions, 171, 174
SOD, see Superoxide dismutase
Solid tumor hypercalcemia syndrome, 153
Solid tumors, 46
Soluble receptors for cytokines, 49
SP, see Substance P
Spirogermanium, 75–76
Splenic macrophages, 62, 64, 72
Squamous carcinoma cell line, 282
Squamous cell carcinoma, 155
src, 116
Staph infection, 319
Staphylococcal organisms, 62
Steroid receptor-ligand complex, 63
Steroid/thyroid hormone receptor superfamily, 63
Steroids, 62–64, 115
Stimulating factor, 8
Stress, 18, 121, 191, 204, 235
Stressors, 18
Striatum, 170
Stromal cells, 293–295
Stromelysin, 63, 110, 115–117, 120, 122, 131, 136
Stromelysin mRNA, 116, 120–121
Submandibular gland IL-1 inhibitor, 39
Substance P (SP), 169–172, 175, 317
  activation of immune cells by, 172
  inflammatory mediator release by, 171–172
  interactions between interleukin-1 and, 175–176
  interleukin-1 production, stimulation of, 179
  intraarticular infusion, 169
  promotion of cytokine release, 178–180
Substance P receptors, 170–171
Superoxide anion, 198–199, 245
Superoxide dismutase (SOD), 245
SVR, see Systemic vascular resistance
Syndromes, 4
Synergism, see Cytokines

Synovial cell function, 133–136
Synovial cells, 62, 110–111
  interleukin-1, effects of, 118–120
  phorbol myristate acetate, effects of, 112–118
  transforming growth factor β, effects of, 122–133
  tumor necrosis factor, effects of, 118–120
Synovial fibroblasts, 63, 71, 109, 113, 121–122, 174
Synovial fluids, 67, 182
  arthritis patients, 59
  IL-1 binding proteins, 40
  IL-2 inhibitors 46–47
  rheumatoid arthritis patients, 8
Synovial hyperplasia, 176, 178
Synovial lymphoid nodules, 46
Synovial macrophage, 112
Synovial tissue, 60, 113
  nondiseased individuals, 109
  rheumatoid arthritis, 109
Synoviocyte activators, 181
Synoviocytes, 60, 63, 171, 178
Synovitis, 17, 122
Synovium, 7
Systemic administration, 5
Systemic disease, 18–19
Systemic effects, 8, 13, 22
Systemic hormones, 147–149
Systemic infection, 109
Systemic lupus erythematosis, 22
Systemic treatment, 14
Systemic vascular resistance (SVR), 308, 314

T

T-cell clones, 45
T-cell dysfunction, 45
T-cell factors, 14
T-cell granulomatous reactions, 14
T-cell growth factor, 35, 44–45
T-cell-lymphokine synthesis inhibition, 22
T-cell mediation, 73
T-cell proliferation assay, 39
T-cell subsets, 22
T-cells, 7–8, 13, 22, 40, 63, 72, 91, 109, 170, 180
T-cells, rabbit, 172
T helper cells, 7, 172, 183
T helper/suppressor/cytotoxic lymphocytes, 7

INDEX

T lymphocyte proliferation, 4
T lymphocytes, 7
Tac antigen, 44, 46–47
Tachycardia, 307
Tachykinins, 170, 182
Tachypnea, 307
Tamm-Horsfall protein, 38
Tamm-Horsfall/uromodulin, 39
Target sites, 59
Target tissue, 290–293
Targeting macromolecule, 78
Tetrandine, 72
TFP, see Trifluoperazine dichloride
TGFβ, see Transforming growth factor β
TH 1, 22
TH 2, 22
Therapeutic agents, 133
Therapeutic intervention, 59, 67
Therapeutic regimens, 318–319
Therapeutic value of cytokines, 23
Thermal injury, 204
THP-1-derived molecule, 47–48
Thrombin, 191, 196, 204
α Thrombin, 200
Thrombocytopenia, 19, 307
Thrombogenic vascular endothelium, 4
Thrombohemorrhagic lesion, 4, 18
Thrombohemorrhagic response, 9
Thymic hypoplasia, 18
Thymidine kinase, 37
Thymidine turnover, 177
Thymocyte proliferation, 42–43, 61, 65
Thymocyte proliferation assay, 37–38
Thymocytes, 40
Thymoma cells, 46
Thyroid hormone, 136, 149
Thyroid hormone receptor, 136
Tilorone, 21
TIL, see Tumor infiltrating lymphocytes
TIMP, see Tissue inhibitor of metalloproteinases
Tissue destruction, 109
Tissue inhibitor of metalloproteinases (TIMP), 122
Tissue injury, 1, 6, 11
Tissue interaction, 295
Tissue invasion, 109
Tissue perfusion, 307–308
Tissue plasma activator inhibitor, 4
Tissue trauma, 4, 316
Toxic oxygen intermediates, 192
Toxic oxygen radicals, 199
Toxic oxygen species, 242

Toxicity, 19–20
Transcription, 117, 134, 136
Transforming growth factor β (TGF β), 37, 113, 131, 137
  acute phase proteins, 285
  bone cell function, 148
  synovial cells, effects on, 122–133
Transforming growth factors α and β, 156
Transin, 115–116
Transplant rejection, 295
Transplantation, 21
Trauma, 191
Trifluoperazine dichloride (TFP), 100–101
Triggering signals, 93
Trypsin, 110
Tuberculin hypersensitivity, 8
Tuftsin, 172
Tumor cell invasion, 137
Tumor cell lines, 60
Tumor destruction, 21
Tumor-induced angiogenesis, 21
Tumor infiltrating lymphocytes (TIL), 20–21
Tumor invasion, 118
Tumor necrosis factors (α and β), 3–5, 59, 67, 113, 136, 154–157, see also specific topics
  acute inflammatory responses, 9
  acute phase proteins, 280–281
  acute phase response, 191
  bone, influence on, 152–153
  cell growth, 123
  control of production at level of second messenger pathways, 89–107, see also Second messenger pathways
  detection, 6
  gene expression regulation, see Second messenger pathways
  inhibitors, 48
  mediators of chronic inflammatory disease, 1–33
  neurogenic inflammation, 169, 174–175
  pulmonary vascular endothelial injury, 209–213
  shock, 308
  shock inducing, 311
  slow-release, 12
  synovial cells, effects on, 118–120
  systemic effects, 35
  therapeutic approaches to disease, 21
Tumor regression, 20
Tumor therapy, 4, 19–20

Tumors, 4, 9, 75, 145
Type 1 procollagen message, 63
Type II collagen, 18
Type IV collagen, 110

U

U937 cells, 37
UMR 106 cell line, 152
Urinary inhibitor of IL-1, 38
Uromodulin, 38–39

V

Vascular contractility shock, 314
Vascular endothelial cells, 9, 169
Vascular endothelium, 9
  adhesion of leukocytes to, 2
  shock, 313–314
  thrombogenic potential of, 4
Vascular leak syndrome, 214–215
Vascular permeability, 171, 318
Vascular smooth muscle, 61, 173, 203–204
Vascular tissues, 170
Vascularization, 59
Vasculitis, 4, 15
Vasoactive amines, 169, 181, 309

Vasoconstriction, 2, 314
Vasodilation, 2, 318
Vasorelaxation, 314
Viral hepatitis, 258
Vitamin A analogues, see Retinoids
Vitamin E, 236
v-sis, 131

W

Weight change, 22
Weight loss, 315
Western blotting studies, 41, 61, 114
Wheal and flare response, 171
Wheal formation, 171
White cell count abnormality, 307–308
Wound healing, 4, 122

X

Xanthine oxidase, 236
Xenobiotics, 237

Z

Zymosan, 66, 199, 245